Approved: The Life Cycle of Drug Development

Narendra Chirmule • Vihang Vivek Ghalsasi

Editors

Approved: The Life Cycle of Drug Development

 Springer

Editors
Narendra Chirmule
SymphonyTech Biologics
University of Pennsylvania
Philadelphia, PA, USA

Vihang Vivek Ghalsasi
Faculty of Applied Sciences
and Biotechnology
Shoolini University of Biotechnology
and Management Sciences
Bajhol, Himachal Pradesh, India

ISBN 978-3-031-81786-1   ISBN 978-3-031-81787-8   (eBook)
https://doi.org/10.1007/978-3-031-81787-8

This Springer imprint is published by the registered company Springer Nature Switzerland AG
The registered company address is: Gewerbestrasse 11, 6330 Cham, Switzerland

If disposing of this product, please recycle the paper.

Foreword

Drug Development from the Patient Perspective

Drug development is a process laden with hope, uncertainty, and the promise of improved quality of life. Patients, often grappling with debilitating illnesses, view drug development as a beacon of hope, offering potential cures or better management of their conditions. The journey from laboratory research to an accessible medication is seen through the lens of personal and collective narratives of suffering and resilience. For many, each phase of clinical trials represents not just scientific progress but a tangible step toward relief from chronic pain, debilitating symptoms, or life-threatening diseases. Patients often participate in trials, contributing their time and bodies to the advancement of medical science, driven by the desire to find solutions not only for themselves but also for future sufferers. However, this perspective is also fraught with challenges. The waiting periods, potential side effects, and the uncertainty of trial outcomes can be daunting. There is a deep reliance on transparency and communication from pharmaceutical companies and healthcare providers, as patients need to be informed and reassured about the safety, efficacy, and potential risks associated with new treatments. Ultimately, drug development is more than a scientific endeavor for patients; it is a lifeline intertwined with their daily lives and futures. The inspirational narratives below are three very personal journeys, which describe the challenges in management of their disease and access to treatments.

Rohan Lepps
(Caregiver + Patient Advocate), Goa, India

"We don't meet people by accident; they cross our paths for a reason."—Unknown

Meeting Prof. Narendra Chirmule affirms these words. I vividly remember standing in a hospital corridor as my wife's, Meghna, condition rapidly deteriorated. The outlook from doctors and advice seemed bleak; yet, there I was, unwilling to accept that we were out of options. Despite my extensive research into oncology and the human body, I felt a desperate need for a miracle. It was in this moment of urgency, just after reading *Good Genes Gone Bad*, that a friend from business school connected me to Prof. Chirmule, a fortuitous encounter that would soon impact us profoundly.

Caregiving, at some point, becomes an emotionally challenging endeavor, driven by adrenaline and determination to understand and counteract the biology at play. I recognized a pattern: people entering our lives, each one contributing to the journey. Yet, as time passed, I needed someone with both the insight and expertise to help us consider all remaining options for Meghna. With the clarity that follows grief, I've come to realize that my deep dive into oncology was not only an attempt to save Meghna but also a path to share knowledge, develop purpose, and help others navigating similar challenges. Cancer, with all its brutality, shapes not only the patients but also those who care for them, testing resilience and character at every turn.

In the complex and ever-evolving battle against cancer, the volume of information available can be overwhelming. The challenge lies in determining which data and options to pursue. The wealth of information from Big Pharma can provide hope but often requires careful scrutiny. Having a guide to navigate the maze of drug development, clinical trials, and treatment paths can be invaluable for both medical professionals and caregivers. Books like this offer critical support, providing insight that busy physicians or unreliable online sources often cannot—particularly for those with limited medical training who need a reliable compass in a field where misinformation can be costly.

Before introducing this book and its benefits, I'd like to share a personal, bittersweet experience. Bitter because by the time I learned about a promising new drug, Tarlatamab, Meghna's performance scores had already disqualified her from participating in the clinical trial. Sweet because the FDA has recently granted fast-track approval to Tarlatamab, an innovative BiTE DLL3 drug by Amgen Pharma, which targets small cell lung carcinoma (SCLC). This approval represents a significant step forward for high-grade neuroendocrine carcinomas (HGNEC), a class that includes other rare and aggressive cancers. Given the lack of awareness around HGNEC, this milestone offers hope for future advancements, building on treatments originally designed for SCLC.

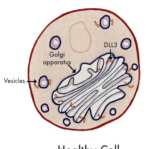

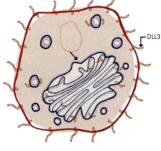

Healthy Cell   SCLC Cell

| ~85% of patients with SCLC have cell surface expression of DLL3. |
| Scientist discover DLL3 expression in most small cell carcinomas... potentially opening the doors to better management of this rare and aggressive form of cancer. |
| Drug designed to attach itself to this DLL3 expressing tumor cell and engaging the body's own T cells to identify and redirect its troops to fight the cancer. |

Over time, I've deepened my understanding of the human body, driven by the need to confront the challenge that cancer presents—a challenge that one simply cannot ignore. This journey has highlighted just one dimension of our health. In a post-COVID world, we have come to recognize not only the value of existence but also the fragility of our bodies, reminding us of the urgency and importance of survival. Skeptical or not, Big Pharma plays a significant role in this survival. Behind its walls work some of the world's most brilliant scientists, dedicated to understanding biological threats and developing effective, practical defenses.

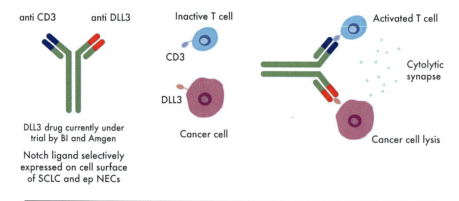

anti CD3   anti DLL3   Inactive T cell   Activated T cell

CD3

Cytolytic
synapse

DLL3

DLL3 drug currently under
trial by BI and Amgen   Cancer cell   Cancer cell lysis

Notch ligand selectively
expressed on cell surface
of SCLC and ep NECs

> One of the most promising results seen of late in the fight against small cell
> carcinoma – paving the way for newer developments, engaging this drug and
> radiotherapeutic allowing for further precision targeted therapies... and even better
> efficacy – End Result –
>
> **FDA fast tracks approval of Amgen's Drug "tarlatamab"**

The drug development process, however, is complex and often unseen by the public. From identifying the nature of a pathogen to testing a potential solution in a petri dish, then through the rigors of clinical trials and, ultimately, to the point where a life-saving drug reaches the market—this journey is both extensive and intricate. This book provides a structured, scientifically grounded overview of the "battle stations" of drug development, encompassing every phase, from preliminary research to clinical trials and final regulatory approval, before reaching patients with the support of medical professionals.

In his latest book, Prof. Chirmule meticulously explains each stage of the drug development lifecycle: drug design, process development, manufacturing, pharmacology, biomarkers, toxicology, clinical trials, statistical analyses, quality assurance, and regulatory requirements, all aligned with FDA guidelines. This comprehensive guide serves as a valuable resource for students, educators, and industry professionals, while also shedding light on the growing influence of AI and machine learning in accelerating drug discovery. By enabling innovative solutions, these technologies not only extend lifespan but also improve our quality of life.

On a personal level, the past four years have reshaped my perspective on health. Today, I view health as central to overall well-being, necessitating a holistic approach that nurtures the mind, body, and spirit. Understanding what strengthens each of these aspects—and, conversely, what undermines them—has allowed me to transition from caregiver to someone focused on overall wellness. The experiences of those fighting for survival remind me of the privilege of awareness and the ability to guide others through complex health challenges. Drugs, ultimately, have a crucial place in this journey, and this book details their profound path from idea to indispensable medicine.

A Patient Journey

Somesh Sambhani, Patient Advocate, Mumbai, India

"It took plain courage to be a chemotherapist in the 1960s and certainly the courage of conviction that cancer would eventually succumb to drugs." Vincent DeVita, an investigator at the National Cancer Institute (NCI) in the United States who would later become its director, penned these words. Yet, what DeVita did not capture was the "forced courage" it takes to be a cancer patient, a firsthand victim of what the biologist Paul R. Ehrlich once called life—a "chemical accident."

That is how I view my own life: as a casualty of a chemical accident. Of course, too much travel, too many stressors, adverse situations, unfortunate genetics, or even just bad luck may have played their part. I was 49 when I was diagnosed with multiple myeloma, a diagnosis that arrived just as I started a new role as India Head for a German multinational in January 2020. My first day in the role also marked the day I was handed an addition to my physical resume: multiple myeloma.

Initially, I could have given in to the thought that life was simply cruel. But soon, other challenges—unexpected and uninvited, much like the cancer itself—demanded my attention, putting that theory on hold.

Looking back, I see that perhaps I had missed the warning signs. About three years before my diagnosis, in 2017, I noticed I was losing muscle mass but not weight. My hemoglobin levels were also dropping. I began to look frail, my rib cage rattling with bone pain that I mistook for muscle pain. My general practitioner prescribed medications to address the hemoglobin issue, but myeloma, adept at disguising itself, evaded detection even in annual health checks.

Later, I began experiencing sharp spinal pain. Twice in a year, I felt a cracking sensation in my spine while traveling, despite not lifting anything heavy. Although my GP suggested an MRI, I avoided it due to claustrophobia. By 2019, I began physiotherapy, which offered little relief. It reached a point where I could barely hold my head upright without supporting my chin or forehead, and my arms vibrated uncontrollably. Finally, my wife insisted I overcome my claustrophobia and get an MRI. That test marked the beginning of a series that would confirm multiple myeloma and rule out tuberculosis.

The diagnosis left both of us in a state of shock, mingled with disbelief and a hint of denial. Yet, the morning after my diagnosis—the second day in my new role—I realized I could not continue with the demands of the job. Cancer, combined with the onset of the COVID-19 pandemic, changed everything. I found myself confronting not just labels but severe illness.

In the days that followed, I began exploring my treatment options, going through a litany of tests—bloodwork, PET scans, CT scans, biopsies, and more. The process felt both surreal and ironic. The oncologists assured me I'd be fine, even humorously noting that if one had to have cancer, this was a "good" one to have. Humor has a way of reversing course, though, and my health deteriorated rapidly. The standard VRD (bortezomib) protocol proved intolerable; each dose brought on a high-grade fever of 105 degrees, landing me in the ICU repeatedly. Subsequent hospitalizations due to pneumonia and other complications followed.

Through these experiences, I gained a close-up view of the overloaded hospital system, witnessing both commercial pressures and instances of medical negligence. My survival was far from assured, and ventilation was not an option for me. Seeking a different perspective, I pursued second opinions and found an oncologist who approached my condition with a broader, more adaptable outlook—a decision that would prove to be life-changing. By March and April of 2020, as COVID surged, I transitioned to a new medical ecosystem with one of India's most experienced oncologists. My treatment shifted from VRD to the KTD protocol, and since then, I've made consistent progress.

My myeloma remained under control, and my medications were scaled back based on biannual SEPs (serum electrophoresis). In early 2023, however, severe back pain returned, and an MRI confirmed a collapse in my L4-L5 vertebrae, signaling that the myeloma was active once more in that area. I underwent spine surgery in March, followed by radiation treatments. In June, I resumed a full regimen with a new combination, KPD, which I will complete shortly. What comes next remains uncertain, as multiple myeloma has a tendency to reappear in various forms and intensities.

In 2020, I had delayed the option of a bone marrow transplant (BMT) or autologous stem cell transplant (ASCT), hoping that my condition—categorized as lower risk—would remain stable. But life's pressures inevitably crept back in. Although I was fortunate to reclaim my former job, family responsibilities weighed heavily, especially with aging parents. Health, despite being a priority, often took a back seat to other demands—a pattern I now recognize.

Today, the treatment landscape for multiple myeloma is evolving with innovative drugs and technologies. Four years is a long time in medical advancements; CAR-T therapy, approved in the United States, has now reached India at more affordable rates. Indian patients also eagerly await the arrival of generic Darzalex, as the current imported option remains prohibitively expensive for many.

Through these years, I have gained both specific and broad insights. Cancer has indeed been a wake-up call, prompting me to identify stressors across physical, mental, emotional, spiritual, and relational domains. Gradually, I've adapted to the dual roles of caregiver and self-advocate, taking on the responsibility of my own health management and knowledge-seeking. I now see how the breakdown of my support network pushed me to become my own anchor, equipped with the resilience and information needed to face this journey.

A Patient Perspective on Drug Development and Treatment for Hemophilia

Murali Pazhayannur, Ph.D., from my home, in Aurora, IL, USA

Question: Could you share a bit about yourself and your experience with hemophilia?

Answer: I was diagnosed with hemophilia at the age of 4 years after a tonsillectomy. My blood lacks certain clotting factors, factor VIII for hemophilia A, which can lead to prolonged bleeding even from minor injuries. Managing hemophilia involves regular intravenous infusions of clotting factor concentrates to prevent or treat bleeding episodes. More recently I am being treated with a new bi-specific monoclonal antibody injected subcutaneously. I do not have FVIII inhibitors which are neutralizing antibodies to the infused FVIII. Many more details of my life with hemophilia are detailed in the first chapter of the book *Good Genes Gone Bad* [1], and several videos on my YouTube channel (UCbBQav0oAVw5ugYculd9tfw).

Hemophilia has greatly influenced various aspects of my life, including my healthcare journey and interactions with pharmaceutical companies.

Drug Development: Patient Perspectives

Question: What should drug developers consider from a patient perspective during the drug development process?

Answer: Patients like me require drugs that not only effectively manage our condition but also integrate seamlessly into our daily lives. For instance, considering the portability of the medication is crucial for those of us who work full time and like to travel. Additionally, smaller quantities of medication make it easier to transport and administer. Stability at room temperature eliminates the need for complex storage solutions, ensuring the medication remains effective and accessible. Furthermore, the route of administration should be convenient, whether it's through injections, infusions, or newer delivery methods like patches or oral formulations. As of this writing, I am a patient with hemophilia approaching the mid-60s with significant venous access issues. I certainly am seeking an adjustable pen-injector in terms of both portability and ease of use. Moreover, providing comprehensive education and training materials to patients and caregivers, along with accessible support systems such as helplines and online communities, can greatly enhance the patient's experience. Along these lines, getting the patient to a level of comfort with the product where he/she understands the value the patient should expect from using/switching to this new product is important. Access to such a liaison person is key to the patient.

Social and Cultural Considerations

Question: What social and cultural factors should be considered when developing hemophilia treatments?

Answer: In addition to the medical aspects, drug developers need to consider the social and cultural context in which patients live. This includes ensuring that educational materials are available in languages and formats that are accessible to diverse populations. Training programs should be culturally sensitive and tailored to the specific needs of different communities.

Providing access to support systems, including helplines staffed by individuals who understand the cultural nuances of the patient population, can improve patient engagement and adherence to treatment regimens. Financial considerations are also paramount, with information about insurance coverage, a manufacturer-supported trial program, copay assistance programs, and access to specialty pharmacies being essential components of patient support.

Interactions with Pharmaceutical Industry

Question: What topics and agendas do you typically discuss when interacting with representatives of the pharmaceutical industry?

Answer: When interacting with individuals in pharmaceutical companies, I often share my firsthand experiences with different treatments and therapies. This includes detailing how the medication has affected my daily life. I discuss my current treatment regimen, route of administration, dosing, volume infused per dose, storage rules (room temperature, refrigeration) supplies needed, pros and cons while traveling, how well it works with joint bleeds, and other conveniences or inconveniences. It's crucial for pharmaceutical companies to understand the real-world impact of their products on patients' lives, as well as to solicit feedback on areas for improvement or innovation.

Reasons for Switching Medications

Question: What prompted you to switch medications after using the same one for 15-18 years?

Answer: My decision to switch medications after using the same one for such a long time was influenced by several factors. Firstly, advancements in hemophilia treatment options led to the availability of new therapies that offered improved convenience and efficacy. In my case, transitioning from a recombinant protein to a bi-specific monoclonal antibody, Hemlibra® presented advantages such as subcutaneous administration, which was particularly beneficial due to my chronically problematic venous access. This, plus the convenient and flexible dosing schedule of weekly, every other week or even monthly, made it so perfect for me. Additionally, gaining a deeper understanding of the pharmacokinetic profiles of newer medications played a role in my decision-making process, as did considerations regarding potential lifestyle changes and treatment preferences.

Challenges in Clinical Trials

Question: What challenges do patients face in participating in clinical trials from their perspective?

Answer: Clinical trial participation can present several challenges from a patient's perspective. The initial screening process can be extensive, requiring individuals to meet specific criteria related to their medical history, comorbidities, and concomitant medications (unknown drug-drug interaction). Recent advancements have led to clinical trials of gene therapy products for hemophilia. In addition to the above factors, trial aspirants could be excluded based on the presence of antibody to the vector protein used (commonly AAV strains). Some applicants express concerns about the long-term side effects of steroid use required in the study. This can lead to potential barriers to enrollment for some patients. Additionally, the follow-up requirements for clinical trials often involve frequent clinic visits, extensive data collection, and adherence to strict protocols, which can be disruptive to daily life. Keeping detailed logs and maintaining compliance with study procedures can require significant burden (time and effort) from patients, highlighting the need for clear communication and support throughout the trial process. Patients may not be satisfied with the compensation offered for their time.

Advice for Younger Patients

Question: What advice would you give to younger patients navigating hemophilia treatment?

Answer: For younger patients navigating hemophilia treatment, I would offer several pieces of advice based on my own experiences. Firstly, for parents of teenagers with bleeding disorders, I would emphasize the importance of exploring available free trial programs and financial support options, including insurance coverage and copay assistance programs, to alleviate the financial burden associated with treatment. To adolescents, I strongly urge you to sign up for hemophilia camps as soon as you become eligible (usually at 8 years old). Connecting with specialty pharmacies can also streamline the medication procurement process and provide access to additional support services. Building a strong support network, including healthcare providers, patient advocacy groups such as NBDF (National Bleeding Disorder Foundation), HFA (Hemophilia Federation of America), Hope for

Hemophilia, and peers living with hemophilia, can offer invaluable guidance and emotional support. Finally, I would encourage younger patients who are college-bound to prioritize their self-care and advocate for their needs within the healthcare system, empowering them to take control of their health and well-being.

Caregiver + Patient Advocate, Goa, India   Rohan Lepps
Aurora, IL, USA   P. M. Murali
Patient Advocate, Mumbai, India   Somesh Sambhani

Reference

[1] Chirmule N. Good genes gone bad: a short history of vaccines and biologics: failures, successes, controversies. Penguin Random House India Private Limited; 2021.

Foreword

Approved: The Life Cycle of a Drug and a Career

In the ever-evolving landscape of healthcare and medicine, the journey of a drug from conception to approval is a testament to the confluence of science, dedication, and innovation. *Approved: The Life Cycle of a Drug and a Career* by Naren Chimule provides a comprehensive exploration of this intricate process, offering invaluable insights into the multidimensional aspects of drug development.

I am so glad that Naren has taken on this herculean task of editing a book on drug development. His experience in this area started in academic applied research in 1999, when he was at the University of Pennsylvania, where I had started my work on CAR-T cells. He then worked in industry and contributed to drug development for two decades. As I reflect on my own journey in drug development, I am reminded of the importance of mentorship and collaboration.

This book arrives at a pivotal time when the need for rapid, yet thorough, drug development has never been more apparent, underscored by global health challenges and the push for medical advancements. Through detailed chapters and expert contributions, the book demystifies the complex processes behind pharmacology, toxicology, clinical trials, and more, all underpinned by the stringent requirements set by regulatory bodies like the FDA.

Drug development requires perseverance and relentless commitment to innovation, quality, and compliance. Over the past four decades, the landscape of drug development has been reshaped by the advent of biotherapeutics, offering new hope for patients and new challenges for researchers. The understanding of regulatory requirements, to ensure safe and effective drugs, has co-evolved with the drug development process. Central to the drug development approach is the principle of risk-based decision-making—a philosophy that emphasizes the importance of assessing and mitigating risk at every stage of the drug development lifecycle. Drawing upon a wealth of case studies, project reports, and peer-reviewed literature, the book highlights real-world implications of this approach, offering readers a nuanced understanding of its application in practice.

For aspiring professionals and entrepreneurs, this book serves as both a primer and a deep dive into the nuances of the industry. It illuminates the path drugs take from the lab to the clinic, highlighting the critical role of risk assessment and mitigation, and is peppered with case studies that bring theoretical concepts to life.

Moreover, "Approved" extends beyond the technical facets of drug development to consider the human element—the careers and lives of those who dedicate themselves to this field. It provides a framework not only for understanding drug development but also for appreciating the career trajectories that such work can foster.

As the world stands on the cusp of medical innovation, with cutting-edge therapies like gene editing and biologics reshaping treatment paradigms, this book is a crucial resource. It equips readers with the knowledge to navigate and contribute to the future of medicine, emphasizing that behind every breakthrough are the scientists, regulators, and healthcare professionals committed to improving human health.

Thank you, Naren Chimule, for assembling this guide. It promises to be an essential read for anyone looking to contribute to or understand the vital field of drug development, reflecting the journey of countless individuals who strive to push the boundaries of science for the greater good. I extend my heartfelt gratitude to the authors of this textbook for documenting their experiences in drug development and for the shared understanding of this complex and ever-evolving discipline. This book will inspire future generations of researchers and entrepreneurs to continue pushing the boundaries of scientific inquiry and to never lose sight of the transformative power of their endeavors.

W. Vague Professor in Immunotherapy,   Carl H. June
Professor of Medicine, Director, Center for Cellular
Immunotherapies, Director, Parker Institute
for Cancer Immunotherapy
University of Pennsylvania Perelman School of Medicine,
Philadelphia, PA, USA

Preface

Taking on a project to write a textbook on the entire processes involved in drug development is a daunting task. Over the past four decades, I have realized the importance of understanding the multi-disciplinary complexity of drug development to work on any single aspect of the process. It was only when I had the role of head of R&D at Biocon, one of India's largest biotechnology companies, that I got the "30000 foot-level view" of the entire drug development process—from science to finance, people management, and most importantly, the culture of the organization. The idea for this book has been brewing in my mind for more than a decade. My first book, intended for the lay-person, *Good Genes Gone Bad* [1] was an attempt to write stories about drugs that failed. The motivation to write the first book was to highlight that lessons learnt through failures were critical to subsequent magnificent discoveries and successful approvals. I first intended to call this textbook, "Failures." But this idea was squashed. When I talked to my friends and colleagues about writing more detailed stories of failures, many of them declined, since it is very difficult and could be controversial to talk about failures. So, I settled to focusing on editing this textbook to at least describe all the steps involved in drug development, which led to regulatory approvals. *Approved: The Life Cycle of Drug Development*. No single person can have the breadth and depth of this complex subject. Thus, writing this book, just as drug development itself, is indeed a collaborative endeavor.

This textbook delves into the intricate world of drug development, exploring its multifaceted journey from conception to regulatory approval. Through the expertise of experienced contributors, it offers a comprehensive guide that navigates the complexities of drug (vaccines, biologics, cell and gene modality) design, process development, pharmacology, biomarkers, toxicology, clinical trials, rare diseases, statistics, risk assessment, intellectual property considerations, and regulatory strategies. The story of the n-of-1 trial, which described the struggles and tribulations of one family to treat their son with an ultra-rare genetic disorder, humbles me. From the inception of innovative compounds to the meticulous orchestration of clinical trials, this book serves as a roadmap for both novices and seasoned professionals embarking on the journey of drug discovery.

As co-editor of this compendium, Vihang Ghalsasi has lent his expertise and vision, shaping the narrative with precision and insight. His insight into flow of the story has been instrumental in shaping this endeavor into a coherent textbook. The culmination of this project would not have been possible without the collective efforts of countless colleagues and students who have contributed with their shared experiences, transcribed by the authors with extensive subject matter expertise.

As you embark on this exploration of drug development, the information contained within these pages serves as both a guide and an inspiration, through stories of eminent scientists who have contributed to drug development. Whether you are a researcher seeking to unravel the mysteries of molecular design or a regulatory expert navigating the intricacies of compliance, this book may empower you to make meaningful contributions to the field. Together, let us continue to push the boundaries of pharmaceutical innovation, where every patient has access to the right treatments they need to thrive.

Philadelphia, PA, USA   Narendra Chirmule

Reference

[1] Chirmule N. Good genes gone bad: a short history of vaccines and biologics: failures, successes, controversies. Penguin Random House India Private Limited; 2021.

Preface

The journey of drug development is a fascinating and complex process that bridges the gap between scientific discovery and patient care. This book aims to demystify the intricate steps involved in bringing a new drug from the laboratory bench to the pharmacy shelf.

From the initial stages of drug discovery, through preclinical and clinical trials, to regulatory approval and post-market surveillance, each chapter delves into the critical phases of this journey. We explore the scientific, ethical, and regulatory challenges faced by researchers and developers, highlighting the collaborative efforts required to ensure safety and efficacy.

In my professional journey, I have served in various roles: molecular biology researcher, copy-editor, biotech industry professional, university professor, and biotech entrepreneur. Viewing drug development through this blend of experiences, I am continually intrigued and fascinated.

As a molecular biologist, I have always appreciated the beauty of the central dogma of life. The universal process of DNA to RNA to protein forms the foundation for designing biologics and biosimilars. Molecular biology and genetic engineering tools open vast avenues for manipulating genes, vectors, and cell lines, ultimately enabling researchers to develop new drug candidates.

As a copy-editor, I encounter numerous manuscripts emerging from the drug development process. These range from original research articles demonstrating proof of concept for novel drugs to clinical trial reports detailing statistical information on drug safety and efficacy, and case studies on individual patient responses to therapies.

As a university professor, I am excited to see the enthusiasm of students learning about the drug development process. Teaching them about the complexities and intricacies of this field is a unique experience. It is crucial for students to understand this process well, as their careers will likely be influenced by it.

As a biotech entrepreneur, although not directly involved in drug development, I often analyze case studies from the industry. These serve as templates to evaluate my venture's progress in terms of product properties, technical advancements, funding, and regulatory approvals.

The authors of this book bring diverse perspectives, making the compilation and editing of its chapters a deeply enriching experience. Whether you are a student, mid-career professional, entrepreneur, or medical writer, this book offers valuable insights to cater to your interests and curiosity.

Bajhol, Himachal Pradesh, India   Vihang Vivek Ghalsasi

Overview

The Life Cycle of Drug Development delves into a pivotal moment in human medicine, where groundbreaking advancements in biotherapeutics are reshaping the landscape of treatment options and patient care. Over the past four decades, these innovations have catalyzed a new era of medicinal breakthroughs, yet the journey from discovery to delivery is layered with complexity and high stakes. This comprehensive volume, authored by seasoned experts in the field, explores the multifaceted nature of drug development, emphasizing the critical roles of efficiency, regulatory compliance, quality assurance, and the balance between innovation and affordability.

Structured to serve as both an in-depth guide and a practical reference, the book presents key concepts across pharmacology, toxicology, clinical trial design, and the nuances of product development. Additionally, it addresses essential components of the drug development process, including intellectual property, regulatory frameworks, and statistical rigor, all while adhering to the stringent requirements set by the U.S. Food and Drug Administration (FDA). Unique to this volume are specialized discussions on emerging pathways such as those for rare diseases and personalized, n-of-1 trials, which represent cutting-edge approaches in targeted treatment.

Through a rich collection of case studies, project reports, and expert journal reviews, readers gain valuable insights into risk assessment and mitigation strategies that are crucial for navigating the challenges of drug development. Targeted especially toward early-career scientists, entrepreneurs, students, and educators, this book offers a thorough understanding of the scientific, regulatory, and operational intricacies in developing new medicines. It is an indispensable resource for anyone seeking to comprehend the dynamic ecosystem of modern drug development and its profound implications for global healthcare.

Contents

About the Editors

Narendra Chirmule has over three decades of experience in drug development. He has held senior leadership positions at Biocon (Bangalore, India), Amgen (Thousand Oaks, CA), and Merck (Philadelphia, PA). He is an expert in the area of immune responses to biologics and vaccines. Dr. Chirmule has a Ph.D. from the University of Mumbai, post-doctoral training at Cornell University Medical College, New York, and teaching and research experience University of Pennsylvania.

Vihang Vivek Ghalsasi is a molecular biologist and science educator. With over nine years of industry and academic experience, including roles at Syngene International (Bangalore, India) and Shoolini University (Solan, Himachal Pradesh, India), he is also a BELS-certified copy-editor who supports researchers in scientific writing. Dr. Ghalsasi holds a Ph.D. in Molecular Biology from Heidelberg University, Germany, and Master's in Biotechnology from the Indian Institute of Technology Bombay. He is a passionate explorer, travel blogger, and science communicator.

The Life Cycle of Drug Development

Narendra Chirmule

Abstract Drug development is a complex yet fascinating process. Having spent my entire career in this process, I feel honored to edit this book on challenges and pitfalls associated with developing a novel drug. This introductory chapter begins with a story and explains my motivation behind compiling this book. Further, it outlines how different chapters of this book are organized and which stories they foretell. For each chapter, I have provided the biography of the author(s) and explained how their personal journeys have left long-lasting impressions on the history of drug development. Thereafter, I have provided an overview of what each chapter includes and how each chapter highlights the importance of each step in drug development. Finally, I have provided biographies of stalwarts in each field. Their stories and accomplishments will surely serve as motivation for the readers.

Keywords Drug development · Author biographies · Book structure · Discoveries · Personal journeys

1 Introduction

1.1 Drug Development. Why the Book?

I have been given advice by writing coaches and friends who are highly published authors to always start a book with a story. So here are some stories to introduce the book.

Amgen was developing a monoclonal antibody drug for the treatment of psoriasis. It is a debilitating autoimmune disease with an extremely painful rash throughout the body. The mechanism of the drug action involved blocking the cytokine-induced inflammatory responses that led to the clinical symptoms. The

N. Chirmule (✉)
SymphonyTech Biologics, University of Pennsylvania, Philadelphia, PA, USA
e-mail: Narendra.Chirmule@symphonytech.com

pathway had shown a lot of promise, since there was genetic evidence that the genes involved in the inflammatory pathway were highly associated with psoriasis. The early phase studies demonstrated extremely good efficacy wherein the treatment resulted in near amelioration of psoriatic skin lesions. These studies were expanded to full-scale phase III trials, which also showed extremely effective clinical outcomes in the treated group, compared to the controls. However, an unexpectedly high rate of suicidal ideation behavior was observed after the completion and unblinding of the double-blind studies. The development of the drug had to be abandoned despite expensive investment [1].

Merck was developing a vaccine using adenoviral vectors to prevent HIV infection. Here again, early studies showed unprecedented levels of protection against live viral challenge in preclinical trials in non-human primates. Phase I studies showed that the vaccine was safe and did not cause any unanticipated side effects. However, the phase II proof-of-concept trial failed; not just failed to protect vaccinated human volunteers but made them more susceptible to HIV infection [2].

In clinical trials of gene therapy in the 1990s, for treatment of rare genetic diseases using adenoviral vectors and retroviral vectors, there were severe adverse events including insertional mutagenesis-mediated leukemia [3] and inflammation-induced death [4]. These, and many more colossal failures in drug development, have plagued the pharmaceutical and biotechnology industry for decades [5]. In fact, most drugs don't come anywhere close (to complete efficacy), since we have very little knowledge of the highly complex biology of humans. "It's a miracle, every time we find a drug that works" (quote by Roger Perlmutter, MD, a highly respected leader in drug development, in an interview with Andrew Dunn for *Endpoint News*, 2023). There is NO drug that is one hundred percent efficacious and has NO side effects. Even drugs, which we take for granted, such as paracetamol, have major toxicities in some individuals. So, the motivation for this book is to share the knowledge of experts in the field of drug development, such that lessons learned can be used to develop drugs accurately (to the mechanism of the pathogenesis of the disease), precisely (faster, by doing exactly the experiments required to prove that the drug works in the intended indication), and thereby with less cost. In my own experiences, while working in the pharmaceutical industry, there are very few individuals who have a complete understanding of all the complexities of drug development. No one knows everything. It takes a village to develop a drug.

I started teaching a curated course in drug development to various departments within the companies I worked in. I did this voluntarily, more for my own learning about functions and roles of the various departments that I worked with (the stakeholders). Over the past decade, I have been teaching a course on drug development at various universities including the University of Pennsylvania, Philadelphia; Indian Institute of Technology, New Delhi and Mumbai, India; Shoolini University, Solan, India; and many others. I have discussed the concepts of strategic drug development in academia, industries, investors, and even regulatory agencies. The book is addressed to the students of drug development across all areas of the sciences. All the chapters provide references and regulatory requirements for all the steps in the process. The book will also be useful for entrepreneurs developing drugs, diagnostics, and devices. The book provides a roadmap and some details on each step, with case studies and references. What the book does not cover are small molecule drug design, finances required for drug development, detailed aspects of pricing [6, 7, 8], requirements by the investors [9, 10], and soft skills necessary (such as decision-making, public speaking, maneuvering organizations, and attaining scientific eminence) [11, 12, 13]. There are excellent books on some of these topics. *Textbook of Drug Design and Discovery* edited By Kristian Stromgaard, Povl Krogsgaard-Larsen, Ulf Madsen; *Basic Principles of Drug Discovery and Development* by Benjamin E. Blass; *Fundamentals of Drug Development* by Jeffrey S. Barrett. In addition, the reader can refer to several peer-reviewed articles on financial aspects of drug development, such as net present value, cost of failures, and many others [14, 15, 16].

The book comprises chapters that follow the lifecycle of drug development: Design, (Biologics, Vaccines, Cell Therapy, Gene-and-Gene Editing Therapies), Process Development and Manufacturing, Pharmacology, Biomarkers, Toxicology, and Clinical Trials. Two special chapters, one on rare diseases, and another on an n-of-1 study, describe the progress and challenges in the desperate need for drugs for more than 10,000 rare and orphan diseases. Additional concepts involved in drug development are described in the chapters that review statistics aspects and regulatory strategies. Intellectual Property is the bedrock of the entrepreneurial process, and a chapter is dedicated to case study-based examples. Finally, drug development is about reducing risks (of failure of efficacy and resulting in toxicity). A chapter on risk management, a central concept in drug development, describes the processes involved in identification, assessment of causes and effects, ranking, mitigation, and communication of risks.

Merging of technologies in the fields of biological sciences, engineering, and mathematics has made it possible to generate, analyze, and interpret very large amounts of data, and search for needles in haystacks. Artificial intelligence (AI) and machine learning (ML) are in everyday vocabulary. Harnessing the power of these technologies has become critical in the drug development process. All the chapters in the book will be addressing the use of AI/ML in various applications.

The Appendix has brief articles that are important aspects of drug development, written by experts in the field. These subjects include the organizational structure of pharmaceutical companies (written by Amitava Saha), the importance of data integrity (written by Francis Crawley, and Perihan Elif Ekmekci), a road map for an entrepreneurial journey (written by Samir Mitragotri), and pricing of drugs (written by Bhavna Desai). The narration below provides a high-level overview of each chapter. To put the human behind the science of drug development, the biography of the lead author of the chapter and brief biographies of stalwarts in the field are listed. These biographies, listed at the back, are being provided for the intended readers to be inspired by the journeys of these accomplished authors.

2 Drug Design

2.1 Vaccines

The chapter on vaccines is written by Amitabh Gaur and his colleagues. Amitabh headed the Biologics Research Division at Becton Dickenson for more than two decades, where he collaborated with all major institutions and industries across the world on vaccines and biologics. He started his career by doing his PhD with Dr. G.P. Talwar, at the All India Institute of Medical Sciences, on development of a contraceptive vaccine. Following his PhD, he worked at Standford University on immunological mechanisms involved in immune tolerance and autoimmunity with C. Garrison Fathman.

Vaccine development stands at the forefront of biotechnological advancements, playing a pivotal role in global health by preventing and mitigating the impact of infectious diseases [17]. The history of vaccines dates back centuries, with early inoculation practices paving the way for modern vaccine development. Understanding the immunological principles, including innate and adaptive immunity, antigen recognition, and immune memory. Design of vaccines involves defining the nature of the protective immune response to the pathogen and identifying the protective antigenic epitopes. Vaccines can either be preventive or be used as therapies as immune modulators. The stages in vaccine development involve target identification, antigen selection, and pre-clinical testing for efficacy and safety. Clinical development phases evaluate the safety, efficacious dose, maximum tolerated doses, and effectiveness of the vaccine. Once a vaccine is approved and enters the market, ongoing monitoring is crucial for detecting rare side effects and ensuring long-term safety.

Several platform technologies are utilized, based on the requirement to induce innate, cell-mediated, or humoral immune responses. Some of the platforms include live attenuated, killed viruses, proteins, glycans, viral vectors, DNA, and recently mRNA. Computational approaches using machine learning-based algorithms using omics large data sets and synthetic biology have enabled accurate antigen prediction and epitope mapping in vaccine design. These approaches have enabled the design of personalized vaccines, especially in oncology. Challenges in vaccine development include antigen variations, on- and off-target adverse effects, and global equitable access. Fig. 1 lists the steps involved in the vaccine development process.

Each step requires extensive evaluation of the appropriate platform technology that addresses the mechanism of immune activation required for the vaccine. The iterative step(s) are required to refine the development process.

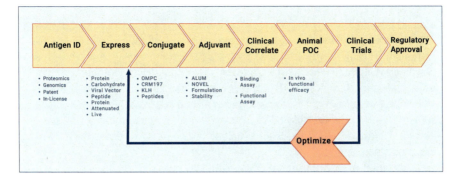

Fig. 1 Steps involved in vaccine development

Stalwarts in Vaccine Development Each section lists key individuals who have significantly contributed to the field. There will surely be many more individuals that the reader is encouraged to explore. Learning the career trajectories of such individuals can provide inspiration for one's own journey.

Maurice Hilleman: (1919–2005). Merck, West Point, PA

Maurice Hilleman stands as one of the most influential figures in the field of vaccination, credited with saving millions of lives through his groundbreaking work in vaccine development [18]. Born in a farming family in Montana, his journey was marked by perseverance and relentlessness in pursuit of innovation. Hilleman's career spanned over four decades at the pharmaceutical company Merck, where he developed more than 40 vaccines, including those for measles, mumps, rubella, chickenpox, hepatitis A and B, and meningitis. Perhaps his most notable achievement came in 1963 when he swiftly developed a vaccine for mumps in response to an outbreak, using his daughter's throat swab sample as a source of the virus. This rapid response saved countless lives and prevented widespread suffering. He received numerous accolades, including the National Medal of Science. His life's work continues to resonate, reminding us of the profound impact that dedication, innovation, and scientific ingenuity can have on humanity's well-being.

Title: Vaccinologist extraordinaire, developed the mumps vaccine from virus isolated from his daughter.
Education: Montana State University (undergrad), University of Chicago (Microbiology).
Accomplishments: Developed 40 vaccines and saved millions of lives.
Inflection point: Joining Merck, and access to resources to development of life-saving vaccines.

Paul Offit: (1951)

Paul Offit is a pediatrician and immunologist and is renowned for his contributions to vaccine development and advocacy. Serving as the Director of the Vaccine Education Center at the Children's Hospital of Philadelphia, son of a shirt-maker in Baltimore, MD, Offit has authored several influential books debunking vaccine myths and promoting evidence-based medicine. He played a pivotal role in the creation of the rotavirus vaccine, preventing severe diarrhea in children globally. Despite facing criticism and controversy, Offit remains a staunch advocate for vaccination, emphasizing the critical role of immunization in safeguarding public health and saving lives.

Title: Vaccinologist, Vaccine Science communicator, Rotavirus Vaccine, Educator.

Education: Tufts University (undergrad), University of Maryland (Medicine), Children's Hospital of Philadelphia (Pediatrics).

Accomplishments: Developed the rotavirus vaccine and has contributed to vaccine advocacy.

Inflection point: Finding Maurice Hilleman as a mentor.

2.2 Biologics

Taruna Arora and Sanjay Khare and their colleagues have written this chapter on Biologics. Taruna is Vice President of Biotherapeutics at Bristol Myers Squibb, leading analytical, translational, and biotherapeutic discovery functions. Her experience spans protein engineering, development, and approval for a diverse range of biologic modalities including bi-specifics, T cell engagers, ADCs, agonistic antibodies, and enhanced cell-killing modalities. Taruna earned her PhD in Immunology from the Mayo Clinic. Her interests lie in disruptive thinking, accelerated advancement of therapeutics for patients' benefit, and organizational strategies that connect science, people, functions, and cultures. Sanjay founded a start-up ImmunGene and developed several innovative bi-functional molecules. He successfully capitalized the company from Angel investors, venture capital fundings, nonprofit organization (Leukemia Lymphoma Society), and the NIH-SBIR (Small Business Innovation Research) grant. Sanjay has extensive experience in development of drugs with companies such as Amgen and Coheris. He has co-discovered key immune co-stimulatory molecules such as B7RP1 and B cell activating factor (e.g., BAFF). Sanjay received his PhD in Immunology at the All-India Institute of Medical Sciences, New Delhi, India, and did his postdoctoral fellowship in the Immunology Department at the Mayo Clinic and Medical School.

Biologic drugs have been highly effective in treatment of diseases for more than three decades [16]. The interactions of proteins and peptides through diverse intermolecular interactions result in activation of intracellular signaling pathways. Thus, recombinantly expressed growth factors (growth hormone, insulin), cytokines (interferon gamma), and neurotransmitters (dopamine), which induced agonistic signals, were the first generation of biologics. With the development of Chinese-Hamster-Ovary (CHO) cells to express the heavy and light chains of antibody molecules, recombinant monoclonal antibody therapies became the workhorse of biologic drugs. This technology enables the design of antagonist drugs which could effectively block signaling pathways (TNFa). The use of tagging payloads to antibodies, which could be specifically delivered to cells, resulted in development of killer-pathway-inducing drugs. Over the past decade, there has been an explosion of gene-based drugs for gene therapy, gene editing, and the "living drug: cell therapy." A systematic structure-function-based search using machine-learning-based systems (such as Deep-Mind Alpha-Fold [19]) will enable the discovery of an avalanche of drugs that will be extremely specific, effective, and safe.

The design of biologic drugs is a complex and dynamic process that draws upon a diverse array of platforms, measurement systems, discovery processes, and statistical considerations. As our understanding of biology deepens and technological capabilities expand, the potential for developing innovative and high effective therapies will continue to grow. This chapter provides a comprehensive exploration of various facets involved in designing biologic drugs, highlighting the interdisciplinary nature of this field and the collaborative effort required to drive therapeutic innovation,

Biologic drugs are being designed by leveraging diverse platforms. Protein therapeutics, particularly monoclonal antibodies, are specialized due to their specificity and long half-life. Targeting glycosylated residues of target proteins and developing drugs using lipid-nanoparticles have harnessed these molecular entities in drug design. Nucleic acid-based drugs (siRNA, RNAi, anti-sense oligonucleotides) have been developed with new mechanisms of action. Major advances have been made in gene therapy for treatment of monogenic diseases, and gene editing technologies enable precise modifications for new therapeutic innovations.

Fu-Kuen Lin: (1941)
Title: Molecular Biologist, Innovator, cloned recombinant erythropoietin.
Education: National Taiwan University (undergrad, and Master's in Plant Pathology), PhD (University of Illinois at Urbana-Champaign).
Accomplishments: Cloned and developed recombinant erythropoietin.
Inflection point: Graduate studies with David Gottleib.

Robert Swanson: (1947–1999)
As a founder of Genentech in 1976, he is regarded as an instrumental figure in launching the biotechnology revolution.

Title: Venture Capitalist, founded Genentech.
Education: BS (Chemistry); Management (MIT Sloan).
Accomplishments: Founding Genentech, one of the first biotech companies.
Inflection point: Meeting and convincing Herb Boyer to co-found Genentech.

Herbert W. Boyer: (1937)
A biochemist at the University of California at San Francisco, Herb Boyer developed the recombinant DNA technology for broad applications to developing new medicines such as human insulin, interferons, human growth hormone, and thrombolytic agents.

Title: Biochemist and Genetic Engineer who demonstrated the usefulness of recombinant DNA technology to produce commercial medicines.
Education: BS (St. Vincent College, PA); PhD (University of Pittsburg); Postdoc (Yale).
Accomplishments: Pioneering use of gene technology in medicine, Albert Lasker Basic Medical Research Award.
Inflection point: Meeting Robert Swanson and understanding the business of science.

2.3 Cell Therapy

Rahul Purwar and his colleagues have written a chapter on cell therapy. Rahul is a professor at the Indian Institute of Technology, Mumbai, and founder of the start-up ImmunoACT. His team at ImmunoACT, IIT, and Tata Memorial Hospital, Mumbai, have developed the CAR-T cell therapy for treatment of acute lymphoblastic leukemia in India. His team is also working in the field of skin immunology and examining the role of effector T cells and its related cytokine in skin inflammatory disorders (vitiligo, atopic dermatitis, and psoriasis) and T cell lymphoma with skin involvement. Rahul completed his PhD from Hannover Medical School, Hannover, Germany, and Postdoctoral studies from Harvard Medical School, Boston.

Cell therapy involves the administration of living cells to replace or repair damaged tissues, restore lost function, or modulate the immune system. These cells can be derived from various sources, including embryonic stem cells, adult stem cells, induced pluripotent stem cells (iPSCs), and even certain differentiated cells. The choice of cell type depends on the therapeutic goal and ethical considerations. Among the various cell therapies, chimeric antigen receptor (CAR) T-cell therapy has gained remarkable attention for its success in treating certain types of cancer, particularly hematological malignancies like leukemia and lymphoma. CAR-T cells are engineered to express synthetic receptors that recognize specific antigens on cancer cells, enabling targeted destruction. Recent advancements in CAR-T therapy involve refining the design of CAR constructs, enhancing T-cell persistence and function, and expanding its application to solid tumors.

Stem cell-based approaches show promise in treating neurodegenerative disorders such as Parkinson's disease, Alzheimer's disease, and amyotrophic lateral sclerosis (ALS). Several approaches have been utilized to develop specialized neural cells from pluripotent stem cells and transplant them into affected brain regions to replace lost or damaged neurons. Mesenchymal stem cells (MSCs) possess immunomodulatory and regenerative properties, making them attractive candidates for various therapeutic applications. Recent studies have explored the use of MSCs in treating conditions such as osteoarthritis, myocardial infarction, graft-versus-host disease (GVHD), and inflammatory disorders. Looking ahead, several future directions hold promise for advancing cell therapy, in areas of novel biomaterials, biomarker-driven precision medicine approaches, and delivery systems.

Donall Thomas and Robert Good, pioneers of bone marrow transplantation which was the origin of cell therapy.

Donall Thomas: (1920–2012)
Education: BS and MS (in Chemistry and Engineering, University of Texas, Austin); MD (Harvard Medical School); Postdoc at MIT, and trained on kidney transplantation at Bringham Hospital, Boston.
Accomplishments: Participated in the first kidney transplantation, Nobel Prize for bone marrow transplantation.
Inflection points: Association with Sidney Farber, who funded his first lab, being recruited to Seattle.

Robert Good, MD PhD: (1922–2003)
Education: MD, PhD from University of Minnesota.
Accomplishments. President and Director of the Sloan-Kettering Institute for Cancer Research from 1972 to 1982; conducted groundbreaking studies that helped identify the central role that the thymus plays in immunity; and performed the first successful bone marrow transplant, Laskar Award.
Inflection points: Training in human anatomy, opportunity to head the premier cancer research center, with access to resources.

Carl June, MD, PhD (University of Pennsylvania, Philadelphia) (1953)
Title: Immunologist, Pioneer in CAR-T cell therapy, cyclist.
Education: BS (US Naval Academy); MD (Baylor College of Medicine).
Accomplishments: Pioneered chimeric antigen receptor T cell therapy; Breakthrough Prize.
Inflection point: Learned adoptive cell therapy under the guidance of Donald Thomas, Nobel Laureate for bone marrow transplantation.

2.4   Gene Therapy

The chapter on gene therapy is written by Susan D'Costa and her colleagues. She is a molecular virologist with extensive experience in viral vector analytics, process development, manufacturing, commercial strategy, and building successful teams. She is currently the Chief Technical and Commercial Officer at Genezen, a leading viral vector CDMO. Susan led teams at Thermo Fisher Scientific, Viral Vector Services, and its predecessor companies—Brammer Bio and Florida Biologix, working with different viral vectors, liaising with diverse biotech clients, and building teams with scientific and operational excellence. Susan holds a PhD in Biology, specializing in molecular virology, from Texas Tech University; an MS in Biochemistry from Mumbai University (Grant Medical College); and a BS in Microbiology/Biochemistry also from Mumbai University (St. Xavier's College).

Gene therapy involves the introduction, alteration, or replacement of genetic materials to treat or prevent disease. This transformative approach addresses conditions caused by genetic conditions, by providing a functional copy of the mutant gene or modifying genetic regulation, to restore cellular functions. The efficiency of introducing transgenes into cells is achieved using (i) viral vector systems, such as retroviruses, lentiviruses, adenoviruses, adeno-associated viruses, herpes simplex viruses, (ii) non-viral systems such as lipid nanoparticles, liposomes, and physical and chemical methods such as electroporation, and chemicals. These vector systems have been used either by ex vivo or in vivo methods. In the former, cells are removed from the patient, genetically modified outside the body, and then reintroduced. In in vivo gene therapy, the therapeutic genes are delivered directly to the patient's tissues or organs. The choice between these approaches depends on the nature of the disease and the target cells. This chapter will discuss the processes and case studies of several gene therapies that have been approved for treatment of monogenic diseases, such as muscular dystrophy and hemophilia, among others.

R. Michael Blease, MD (1939)
Clinical Gene Therapy Branch in the National Human Genome Research Institute

Title: Immunologist, Pediatrician.
Accomplishments: Developed treatment for the first gene therapy using retrovirus vectors to express the adenosine deaminase gene in a retrovirus, expressed in the bone marrow.
Inflection Points: The collaborative and deeply scientific environment at NIH for doing research on gene therapy.

James M. Wilson, MD, PhD (University of Pennsylvania, Philadelphia)
Title: Gene therapy designer, inventor of several viral vectors, avid cyclist.
Education: Albion College (undergrad, Chemistry); University of Michigan (MD, PhD), residency and Postdoctoral work at Massachusetts General Hospital, and Whitehead Institute, Boston.
Accomplishments: Developed several viral-based gene therapies for orphan genetic diseases, which have been the basis of several recent approvals; elucidated host immune responses to viral vectors.
Inflection point: Death of a patient in a clinical trial that he led, which resulted in changing his research focus on AAV vectors.

Katherine High

Kathy grew up in a loving family in North Carolina. She was exposed to science and chemistry through her grandfather, who worked in chemistry at Lockheed Martin. She got her chemistry set at 10 years, and her dad, her inspiration, took her to visit MIT during his trips to Boston, every year. An early decision she made was to do chemistry at Harvard, instead of aeronautical engineering at MIT. Throughout her career, many decisions were balanced on a knife edge (could go either way). Her decision to work on gene therapy could well have been influenced by her dad's genetic heart condition. After the pioneering work in gene therapy in academia, she transitioned to industry, at Spark, to develop the first gene therapy, Luxturna®, for the treatment of genetic eye disease.

Title: Pioneering Physician-Scientist in gene therapy for hemophilia and other genetic diseases.

Education: Chemistry (Harvard), MD (University of North Carolina), Hematology Fellowship (Yale).

Accomplishments: Howard Hughes investigator; member of the National Academy of Sciences; developed gene therapies for hemophilia, and several genetic diseases; led the regulatory approval commercialization of the first FDA-approved gene therapy for genetic disease of the eye. Led the successful Phase 1/2 trial for Beqvez® for hemophilia B.

Inflection points: Hematology training at Yale with Edward Benz; working with a great group of scientists and collaborators at the Children's Hospital of Philadelphia and the University of Pennsylvania; leaving academia to lead R&D at Spark Therapeutics.

2.5 Gene Editing Technologies: Precision at the Molecular Level

Gene editing technologies enable precise modifications of the DNA sequence, offering unprecedented control over genetic material. These technologies have transformative implications for both gene therapy and various aspects of biologic drug design. The platforms for gene editing are expanding exponentially, and currently include CRISPR-Cas9, Cas12, Cas13, base editing, Zinc-finger nucleases, and TALENS. The challenges in the use of gene editing technologies include off-target effects and delivery methods. As these technologies continue to evolve, overcoming challenges, their impact on the landscape of medicine is poised to expand. The convergence of cell therapy and gene editing, in particular, holds tremendous potential for addressing genetic disorders, cancer, and other conditions at the molecular level. The ongoing synergy between scientific innovation, clinical research, and ethical considerations will shape the future of these transformative fields.

Jennifer Doudna and Emanuella Chapentier Known for their discovery of a molecular tool known as clustered regularly interspaced short palindromic repeats (CRISPR)-Cas9. The discovery of CRISPR-Cas9, made in 2012, provided the foundation for gene editing, enabling researchers to make specific changes to DNA sequences in a way that was far more efficient and technically simpler than earlier methods.

Jennifer Doudna: (1964)

Title: Gene Editor

Education: BS (Chemistry), Ponoma College, CA, Harvard University (PhD), Postdoc in University of Colorado

Accomplishments: Solved the elusive structure of ribozyme; developed the CRISPR-Cas9 gene editing technology

Inflection point: Training with Nobel Laureates Thomas Cech and Thomas Steitz

Emanuela Chapentier: (1968)

Title: Gene Editor

Education: Graduated in Biochemistry, Molecular Biology, and Genetics from Pierre and Marie Curie University; Postdocs at Instutut Pasteur, on antibiotic resistance; and Rockefeller University

Accomplishments: Developed the CRISPR-Cas9 gene editing technology

Inflection point: Training on bacterial genetics and molecular and biochemistry

3 Process Development and Manufacturing

Ankur Bhatnagar and Dhananjay Patankar have written the chapter on process development and manufacturing of biologics.

Ankur Bhatnagar is the Head of Process Sciences at Biocon Biologics Limited's Research and Development facility located in Bangalore, India. He was an integral part of the team involved in getting global approvals for biosimilars like Trastuzumab, PEG-GCSF, and insulin-glargine. He has Master's degree in Biochemical Engineering and Biotechnology from the Indian Institute of Technology (IIT) Delhi and an Executive MBA program at the Indian Institute of Management (IIM) Bangalore. Ankur has been recognized internally on various instances for his contributions in setting up a robust platform for the development of Biosimilars and as a people manager. An interview with him on YouTube can be seen in this link: https://youtu.be/wxwitIn84pQ?si=1pEvZ1DUa2dPGCG1

Dhananjay Patankar is an independent biopharmaceutical professional and advisor to several biopharmaceutical companies and research institutes. He has extensive experience in the development and manufacturing of novel biologics and biosimilars. Over his career, he led teams that developed India's first biosimilar therapeutic product (EPO), and developed EU and US-GMP-certified biologics manufacturing facilities. He has served as a Biologics Expert Committee member at the US Pharmacopeia. He earned his Bachelor's in Chemical Engineering from IIT Mumbai and PhD from the University of Utah in the United States. He is an avid traveler and historian.

Chemistry, Manufacturing, and Controls (CMC) modules of regulatory submissions involved a detailed description of the process and product development, quality control, and commercial manufacturing of drugs. Once a promising biological drug candidate is identified, process development begins. This stage involves optimizing the production process to ensure consistent quality, high yield, and scalability. The process involved developing the most suitable expression system, (e.g., mammalian cells, year, bacteria), optimizing grown conditions, and purification techniques. This iterative process aims to maximize productivity while minimizing costs and ensuring compliance with regulatory requirements. The first step is establishing a stable cell line capable of producing the desired therapeutic protein at high levels. This process involves transfecting host cells with the gene encoding the protein of interest and selecting clones with optimal expression characteristics. Cell line development typically employs techniques such as gene editing, screening assays, and clone selection to identify high-producing cell lines that can be scaled up for commercial production. The next step of upstream processing involves the cultivation of cells or microorganisms to produce the product. This stage includes cell culture, fermentation, and/or recombinant DNA technology to generate large quantities of the desired protein. Factors such as cell density, nutrient availability, oxygenation, and bioreactor design are carefully controlled to optimize protein expression and maintain cell viability. Upstream processing also involves monitoring and controlling various parameters to ensure reproducibility and consistency across batches. The next step is downstream development in which purification and isolation to remove impurities and obtain a highly pure and active product. These processes include cell harvest, clarification, filtration, chromatography, and formulation. Each purification step selectively separates the target protein from other cellular components, such as DNA, RNA, host cell proteins, and media components. Finally, the drug product undergoes formulation to ensure stability and compatibility with its intended delivery route (e.g., injection, infusion). Formulation may involve adding excipients, adjusting pH, or freeze-drying to preserve the biologic's activity during storage and administration. After formulation, the biologic is filled into vials, syringes, or other containers under sterile conditions (fill-finish).

Throughout the development and manufacturing process, rigorous quality control measures are implemented to assess the safety, purity, potency, and consistency of the biologic. Analytical techniques such as chromatography, mass spectrometry, electrophoresis, and bioassays are used to characterize the product and detect any impurities or deviations from specifications. Quality control testing is performed at various stages, including raw materials, in-process intermediates, and final product, to ensure compliance with regulatory standards and specifications.

Commercial manufacturing involves scaling up the production process to meet market demand while maintaining product quality and consistency. Commercial manufacturing facilities are equipped with state-of-the-art equipment, automation, and quality control systems to ensure efficient and compliant production. Continuous monitoring and process improvements are implemented to optimize productivity, minimize costs, and ensure supply chain reliability.

Jim Thomas, PhD
Process Development Scientist, and highly inspirational leader, and mentor who promoted innovation.

Education: PhD (Purdue); postdoctoral training (MIT), with Charles Cooney.
Accomplishments: Led, supported, and contributed to development of processes for manufacturing biologics that have transformed the field; elected member of the National Academy of Engineering.
Inflection points: Experiences at high-performing academic institutions and industries that provided the support for innovation and highly authentic leadership style.

Ann Lee, PhD
Chemical Engineer whose experience across early process development, to commercial manufacturing of vaccines, and mentor to many process-engineering scientists.

Education: BS (Cornell); PhD (Yale), Business (Harvard).
Accomplishments: Contributed to innovation in vaccine, biologics, and cell and gene therapies; elected member of the National Academy of Engineering.
Inflection Points: Courageous decision-making, self-confidence.

4   Pharmacology

The pharmacology chapter is written by Vibha Jawa. She is an Executive Director at Bristol Myers Squibb, leading biotherapeutic and cell/gene therapy bioanalytical function. She has led predictive and clinical immunogenicity groups and has more than two decades of experiences in development of vaccines, biologics, and gene therapy products, supporting 20 + IND, BLA, and MAA filings. She is an elected Fellow of the American Association of Pharmaceutical Scientists. Vibha volunteers for Steampark a nonprofit promoting STEM-based learning in underserved communities.

Pharmacology is the study of the effects of body on the drug (pharmacokinetics) and effect of the drug on the body (pharmacodynamics) (Fig. 2). *Pharmacokinetics*

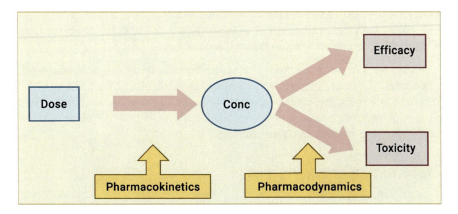

Fig. 2 Pharmacology: the study of the effects of body on the drug (pharmacokinetics) and effects of the drug on the body (pharmacodynamics)

involves understanding the processes by which drugs enter the bloodstream, reach their target sites of action, and are eventually eliminated from the body. The four main processes involved in pharmacokinetics are absorption, distribution, metabolism, and excretion (ADME). *Pharmacodynamics* involves how drugs interact with their target receptors or enzymes, and how these interactions lead to therapeutic or adverse effects. Pharmacodynamics can be influenced by factors such as drug concentration, receptor affinity, and downstream signaling pathways. The factors involved in pharmacodynamics include potency, efficacy, and selectivity. Finally, all biologic drugs have the potential to induce immunogenicity. Immunogenicity is the ability of a drug to induce an immune response in the body. When the immune system recognizes a biologic drug as a foreign substance, it can produce antibodies against the drug, leading to reduced efficacy and potential adverse effects. Immunogenicity can vary among individuals and can be influenced by factors such as the drug's structure, dose, and route of administration. Vaccines, on the other hand, require inducing the optimal immune response, and immunogenicity to vaccines will be discussed in the chapter on vaccines.

In order to understand the concepts of pharmacology, let's examine an example of a drug.

Herceptin in HER2-Positive Breast Cancer A 45-year-old woman is diagnosed with HER2-positive breast cancer, a subtype of breast cancer associated with aggressive tumor growth. Herceptin (trastuzumab) is a monoclonal antibody targeting the HER2 protein. Herceptin binds to the HER2 protein on the surface of cancer cells, blocking its signaling pathway and inhibiting tumor growth. The patient undergoes chemotherapy in combination with Herceptin, resulting in a significant reduction in tumor size and improved overall survival. In this case, binding of Herceptin to HER2 results in two actions: i) blocking the signaling pathway, and inhibiting tumor cell proliferation, and ii) directly killing the tumor by antibody-dependent-cellular-cytotoxic pathways (i.e., the effect of the drug on the body), is

pharmacodynamics. The dose, and number of times Herceptin has to be administered (since each dose will be cleared by the body in time), to achieve efficacy (i.e.,
the effect of the body on the drug) is pharmacokinetics.

This chapter covers the methods to measure the drug levels (pharmacokinetics),
and biomarkers that are indicators of clinical endpoints (pharmacodynamics). The
methods to quantitate the efficacious dose and determine the lowest effective dose
and maximum tolerated doses involve mathematical modeling and more recently
machine learning (quantitative pharmacology).

Gertrude Belle Elion: (1918–1999)
An American biochemist and pharmacologist, who shared the 1988 Nobel
Prize in Physiology or Medicine with George H. Hitchings and Sir James
Black for their use of innovative methods of rational drug design for the
development of new drugs. This method focused on understanding the target
of the drug rather than simply using trial and error. Her work led to the creation of the anti-retroviral drug AZT, which was the first drug widely used
against AIDS. Her well-known works also include the development of the first
immunosuppressive drug, azathioprine, used to fight rejection in organ transplants, and the first successful antiviral drug, acyclovir (ACV), used in the
treatment of herpes infection [4].

Title: Biochemist, Pharmacologist.
Accomplishments: Discovered AZT, azathioprine, ACV through rational
    drug design.
Inflection point: When she was 15, her grandfather died of stomach cancer,
    and being with him during his last moments inspired Elion to pursue a
    career in science and medicine in college.

Binodh DeSilva, PhD
Title: Pharmacology Scientist.
Education: Bachelor's in Analytical Chemistry from the University of
    Colombo, Sri Lanka; PhD in Chemistry (Bioanalytical) in 1994 from the
    University of Kansas; Postdoctoral Research Associate in the
    Pharmaceutical Chemistry Department at the University of Kansas.
Accomplishments: Responsible for leading the regulated analytical (small and
    large molecule), microbiology, and industrial hygiene laboratories as well
    as the Chemistry, Manufacturing, and Controls (CMC) regulatory docu-
    ment management functions in Product Development President of the
    American Association of Pharmaceutical Scientists (AAPS).
Inflection point: Leadership opportunities in major multinational companies;
    ability for mentorship through the AAPS.

5 Biomarkers

This chapter is written by Pradip Nair and Bindhu. **Pradip Nair** heads Discovery Science for Bicara at Syngene International in Bangalore. He has 19 years of experience in the Pharma industry associated with the development of novel and biosimilar biologics. BIOMAB™ (Nimotuzumab), an anti-EGFR monoclonal antibody approved for head and neck cancers, and the anti-CD6 monoclonal antibody ALZUMAB™ (Itolizumab) are novel molecules. He has played a significant role in the development of Bicara's lead asset, BCA101, which has successfully completed Phase 1 clinical trials in head and neck cancers under a US IND and is now beginning Phase 2 clinical trials. **Prof (Dr) Bindhu OS** is the Deputy Director at the Centre for Researcher Training and Administration, Jain (Deemed to be University), Bangalore. With more than 22 years of teaching experience, her research experience has focused on the expression of molecular markers such as matrix metalloproteinases and their regulators (TIMPs and NFkB) to comprehend the neoplastic transformation and progression biology of oral cancer.

Biomarkers are measurable indicators of biological processes and have become a critical step in the development of drugs. The integration of biomarkers in the early stages, during discovery of pathogenesis of disease and mechanism of drug actions, can contribute to designing drugs for more precise clinical indications. The use of biomarkers can enable confirmation and support validation of targets in disease processes, and potential therapeutic effects.

Once validated, biomarkers can become instruments for screening candidates, determining safety adverse events in preclinical studies, and can be used as proof-of-concept and dose determining in clinical trials. Biomarkers can be utilized for understanding patient heterogeneity and determining the right patients for the drug. In clinical trials, biomarkers are utilized as surrogate clinical endpoints to predict clinical benefit. During process development, biomarkers have been useful to establish ranges of critical quality attribute parameters and used to determine the impact of process and product-related changes. And finally, the use of biomarkers in post-market surveillance helps identify long-term adverse effects and potential alternate targets. Biomarker, many a time, leads to diagnostic tests for disease pathogenesis and determining levels of therapeutic effects of drugs.

A biomarker strategy for clinical development starts with studies conducted in healthy individuals to evaluate the levels of the targets of the drug in various conditions [Note: these studies are "phase 0" studies]. This process is to provide guidance on dose and schedule of the treatment for complete coverage of the target. These studies can be useful to understand if the drug was interacting with the target, or if the interaction did not have the desired effect on the disease. A second part of the strategy is to evaluate reversion of the dysregulated state of the biomarkers to normal ranges (established in phase 0 studies). Examples of such biomarkers are HbA1c and cholesterol levels. The third part of the strategy is to understand the heterogeneity of the clinical population. Thus, the biomarker strategy involves identifying subjects most likely to respond to the drug prior to drug administration (patient stratification), and those that the drug influences its effectiveness on efficacy or adverse effect (patient responders and non-responders).

With the advent of omics science (genomics, proteomics, metabolomics, etc.), the amount of data available has revolutionized biomarker discovery and validation. Combined with the use of machine learning-based data analytics, the patterns of multi-modal signatures of biomarkers are providing more precise and accurate indicators of endpoints for safety and efficacy. The integration of these technologies requires extensive standardization and validation. Regulatory considerations of biomarkers have been established through detailed guidelines, which provide a framework for the evidence required to support biomarkers during the entire drug development lifecycle.

Mary Claire King: (1946)
Title: Geneticist.
Education. Carleton College (undergrad in mathematics); University of California at Berkeley and San Francisco.
Accomplishments: First to show that breast cancer can be inherited due to mutations in the gene she called BRCA1; teaching in University of Chile, actively participated in the anti-war movement.
Inflection Points: Advisor Allan Wilson persuaded her to switch from mathematics to genetics.

Scott Patterson
Head of biomarkers at Gilead Sciences. Scott has over 30 years of industry experience in biomarker/translational sciences of drugs at Amgen, Celera, and Cold Spring Harbor, NY.

Education: BS and PhD in Pharmacology (University of Queensland, Australia).
Accomplishments. Utilized the largest mouse expressed sequence tag (EST) database to identify more than 1000 novel proteins [20].
Inflection points: Opportunity to work at Amgen, on functional genomics.

6   Toxicology

This chapter is being written by Padma Kumar Narayanan. He is Vice President and Head of Preclinical Safety at Wave Life Sciences. He has extensive experience in the field of preclinical safety, toxicology, and pathology spanning over two decades in companies including GlaxoSmithKline, Amgen, Ionis Pharmaceuticals, The Janssen Pharmaceutical Companies of Johnson & Johnson, and Los Alamos National Laboratory. He holds a Doctor of Veterinary Medicine (DVM) from Kerala Agricultural University, a Master of Science in small animal surgery from Tamil

Nadu Veterinary and Animal Sciences University, and a PhD in Immunophysiology/Immunotoxicology from Purdue University.

All drugs have the potential to have toxic effects in the body. The mechanisms of toxicity associated with biologics can be broadly categorized into immunogenicity, off-target effects, and non-specific immune activation. Biologics, being derived from living organisms, can elicit an immune response in the human body. This immune response can manifest as the production of antibodies against the biologic drug, leading to reduced efficacy and potential adverse effects. Biologics can interact with unintended targets in the body, leading to off-target effects. These effects can be due to the structural similarity of the biologic drug to endogenous molecules or the presence of shared receptors or pathways. Off-target effects can result in adverse events such as organ toxicity, systemic inflammation, or altered physiological functions. Some biologics, particularly those designed to modulate the immune system, can activate the immune response in a non-specific manner. This non-specific immune activation can lead to excessive inflammation, cytokine release syndrome, or autoimmune reactions. These immune-related toxicities can range from mild to severe, depending on the individual's immune status and the specific biologic drug.

Monoclonal antibodies that inhibit TNFa (infliximab, adalimumab) are extremely effective in ameliorating autoimmune diseases such as rheumatoid arthritis. However, treatment with these antibodies increases the risk of tuberculosis and fungal infections. Similarly, treatment of patients with B cell leukemia with anti-CD20 antibodies (rituximab) increases the risk of viral infections. These adverse events are on-target effects since this cytokine is essential for protective immune responses against this infection. Hypersensitivity immune reactions to drugs occur in a subset of subjects and can result in activation of innate immune responses or MHC-restricted adaptive immunity. These Hypersensitivity reactions (HSR) are classified into type I (IgE-induced anaphylaxis), type II (complement-induced systemic inflammation), type III (immune-complex induced kidney diseases), and type IV (delayed type, T-cell-mediated systemic reactions). Immune checkpoint inhibitors, such as pembrolizumab and nivolumab, have revolutionized cancer treatment by enhancing the immune system's ability to recognize and destroy cancer cells. However, these drugs can also cause on-target immune-related adverse events, including inflammation of various organs, autoimmune reactions, and even life-threatening conditions such as pneumonitis and colitis. Anti-vascular endothelial growth factor (VEGF) agents (bevacizumab and ranibizumab) are used in the treatment of various cancers and retinal disorders. These agents inhibit the growth of blood vessels, which is important for tumor growth and neovascularization in the retina. However, they can also cause hypertension, proteinuria, and impaired wound healing. In rare cases, severe gastrointestinal perforation and hemorrhage have been reported. Erythropoiesis-stimulating agents (ESAs) are used to treat anemia by stimulating the production of red blood cells. Treatment of patients with ESAs has the potential to increase the risk of cardiovascular events, such as heart attacks and strokes, in patients with chronic kidney disease. In addition, immunogenicity to ESAs has been shown to result in autoimmune anemia. Viral vectors used for transducing gene therapies cause severe systemic inflammation (e.g., adenoviral vectors) and can result in random insertional mutagenesis, resulting in leukemias (e.g., retroviral vectors for treatment of severe combined immune deficiency).

In order to predict and mitigate the risks of potential toxicities to therapies, ICHQ9 guidance by the FDA provides a systematic approach to mitigate these risks. A chapter is dedicated to discuss the process of risk assessment in this textbook. ICHQ9 states that risks are managed by the following steps: establishing a team; identification of failure modes; characterizing the severity, probability, and detectability of causes and effects; defining actions; defining residual risk; and making a risk register and risk communication.

Ruth Lighfoot Dunn, PhD FRCPath DVetMed(hc): (1951)
Ruth Lightfoot-Dunn vividly remembers her first day at vet school, as it was "a huge milestone" in her ambition to be a practicing equine vet. Had someone told her at the time that her veterinary degree would lead to the role of vice-president of R&D in the largest healthcare biotechnology company in the world, she would have insisted this would never happened. She did and made enormous contributions to the field.

Title: Veterinary Scientist, Equestrian
Education: Royal Veterinary College, London
Accomplishments: Leadership roles in large pharmaceutical companies, Chair, PhRMA Preclinical Safety Leadership Group
Inflection point: Met with Peter Bedford, one of the founders of modern veterinary ophthalmology

7   Clinical Trials

This chapter is written by Prajakt Barde, MD, and his colleagues. Prajakt is the Senior Medical Director at Med Indite Communications Private Limited, in Mumbai, India. He has more than a decade's experience in clinical operations including strategy planning, strategic alliances, licensing, negotiations and contracts, clinical development, medical writing and pharmacovigilance, NRA licensure, WHO pre-qualification, international registration, and product commercialization. He has contributed to the development of two H1N1 vaccines.

Clinical trials play a pivotal role in the drug development process, serving as a systematic and rigorous method to assess the safety and efficacy of new medical intervention. Clinical development, clinical operations pharmacovigilance, and medical affairs are distinct areas in drug development process. The *clinical development* processes involve defining the hypothesis, patient populations, trial design, clinical end points, and statistical analysis plan to establish the safety and efficacy of the drug. *Clinical operations* involve site selection and initiation, inclusion and exclusion criteria of patient recruitment and enrollment, randomization, blinding, data collection, monitoring and reporting, and safety monitoring. *Medical affairs* departments are responsible for post-approval life cycle management of the drug. The activities include medical communication, competitive intelligence, and pharmacovigilance. *Pharmacovigilance* involves detection, assessment, understanding, and prevention of adverse events. It includes data collection, management and reporting of adverse events, signal detection, and risk management. Regulations require reporting of periodic safety update reports (PSURs) at defined intervals. In some cases, the deficiencies in the BLA filing require post-market commitments.

Clinical trials are performed in phases in order to minimize risk to patients. Phase I trials focus on determining the safety, tolerability, and pharmacokinetics in a small group of healthy volunteers. Phase II studies establish the efficacious dose and frequency of dosing for the target disease indication. Phase III studies are trials to confirm the efficacy and safety of the drug, in comparison with the standard of care treatment or placebo. These studies validate the hypothesis, using pre-defined statistically defined acceptance criteria. Phase IV studies involve post-marketing surveillance that continues after regulatory approval to monitor long-term safety, effectiveness, and any rare adverse events.

Clinical trials are performed under strict ethical considerations. Informed consent is a fundamental ethical requirement, ensuring that participants are fully informed about the trial's purpose, procedures, potential risks, and benefits before deciding to participate. Institutional Review Boards (IRBs) play a crucial role in safeguarding participants' rights and well-being. An independent multi-disciplinary data safety monitoring board (DSMB) team is involved in real-time monitoring of a clinical trial.

Future trends in clinical trials include adaptive clinical trial designs, which allow for modifications of the study protocol, based on accumulating data. This approach enhances efficiency and flexibility in response to emerging insights. Real-world Evidence (RWE) is derived from routine clinical practice and complements traditional trial data. RWE provides a broader understanding of a treatment's effectiveness and safety in diverse patient populations.

Brian Kotzin, MD

Dr. Kotzin has a unique experience of translational science research with more than 25 years of academic research and 15 years of executive leadership for life science companies. In his academic career, he has made seminal discoveries in the field of autoimmunity. In his career in industry, in drug development, his leadership and regulatory acumen was critical in development of several drugs for immunological diseases.

Title: Clinical Immunologist, educator, clinical-development scientist and leader

Education: University of Southern California (undergrad, Mathematics), Stanford University (Doctorate in Medicine)

Accomplishments: Developed therapeutics for autoimmune and inflammatory diseases; elected Master of the American College of Rheumatology

Inflection point: Finding his mentors Phillipa Marrack and John Kappler

8   Rare Diseases

This chapter is written by Harsha Rajahansa, who is the founder member of the Organization for Rare Diseases in India and on the scientific expert panel for Rare Genomics Institute. He is academically affiliated as a faculty in the School of Systems Biology at George Mason University, Fairfax, VA.

Rare diseases, sometimes known as orphan diseases, present unique challenges in drug development. However, addressing the unmet medical needs of individuals affected by rare diseases is paramount. This chapter explores the importance of rare disease drug development from target identification to regulatory approval. Despite their rarity, the collective burden of rare diseases is substantial, with more than 100 million individuals worldwide facing limited treatment options and inadequate support.

In order to address the unmet medical need, the Orphan Drug Act in the United States and similar regulations worldwide incentivizes rare disease drug development and facilitates patient access to treatments. The Act provides accelerated paths to the approval process, in which there are opportunities to obtain frequent advice from regulators. The first step in the development pathway is the diagnosis of the rare disease, which has been reported to take as long as seven years (on average). There needs to be significant progress in early diagnosis and potential mechanism of action of the gene abnormalities. The next steps involve identifying specific targets through genetic studies, omics technologies, and mechanistic insights into disease pathology. These targets require to be validated through in vitro and in vivo pharmacology studies. Preclinical studies assess the safety, efficacy, pharmacodynamics, and pharmacokinetics of potential drug candidates using animal models and cellular assays. In parallel, appropriate drugs need to be designed and optimized using formulation development to enhance drug potency and selectivity. The investigational new drug (IND) application submission contains the following five chapters—(i) mechanism of action of the disease, and drug intervention; (ii) process of manufacturing the drug, detailed analytics of the potency and purity of the drug, and processes for control the processes for consistency (this section of the IND is called Chemistry-Manufacturing-Control [CMC]); (iii) pharmacology; (iv) toxicology; and (v) the protocol for the phase I clinical trial. The design of clinical trials is tailored to the unique characteristics of rare diseases, considering small patient populations, heterogeneous disease manifestations, and limited natural history data. The trials require close collaboration with patient advocacy groups and rare disease registries to facilitate patient recruitment, engagement, and retention in clinical trials.

In the past decade, there has been a significant advancement in development and approval of drugs for some rare diseases: e.g., enzyme replacement therapies for lysosomal storage disorder, such as Gaucher disease and Pompe disease, by genetically engineering recombinant enzymes, recombinant protein therapies, such as Factor VIII and IX for hemophilia, gene therapies, and oligonucleotide therapies for diseases with gene mutations such as Duchenne muscular dystrophy and cystic fibrosis.

The regulatory strategy for drug development for rare diseases as the potential of precision medicine in for all diseases by improving clinical outcomes for patients. Afterall, no drug works precisely the same way in every individual. In addition, another major lesson from development of drugs for rare diseases is the extraordinary benefit of collaboration among academia, industry, patient advocacy groups, and regulatory agencies, which is essential for overcoming the challenges associated with drug development and achieving meaningful progress in the field.

Stalwarts in the Field

Abbey Meyers, MD, is the founder and past President of the National Organization for Rare Disorders (NORD), a coalition of national voluntary health agencies and a clearinghouse for information about little known illnesses. She is known as the mother of the movement.

Title: Rare Disease Crusader.

Education: Meyers holds an Honorary Doctorate from Alfred University in New York.

Accomplishments: Established the rare disease advocacy movement, National Organization for Rare Disorders (NORD), the first national nonprofit to advocate and represent the voices and needs of all individuals and families affected by rare diseases; *FDA Commissioner's Special Citation for Exceptional Dedication and Achievements on Behalf of All People Afflicted with Rare Disorders* (1988), and the Department of Health and Human Services *Public Health Service Award for Exceptional Achievements in Orphan Drug Development* (1985).

Inflection point: Crusade started as a determined parent for whom "no" has never been an acceptable answer, especially when searching for treatment for her son's Tourette's syndrome in the 1970s.

Ishwar Chander Verma, MD.

Head of Genetic Medicine Department at Sri Ganga Ram Hospital, New Delhi. He was earlier Professor of Pediatrics and Genetics at the All India Institute of Medical Sciences (AIIMS), New Delhi; Fellow of the Royal College of Physicians (FRCP), London; the American Academy of Pediatrics (FAAP); and the National Academy of Medical Sciences (FAMS), New Delhi.

Title: Genetic Medicine Specialist

Education and work: After completing his MBBS, he worked in East Africa. He then trained in Genetics at London, and then worked in All India Institute for Medical Science from 1967 to 1996

Accomplishments: Set up the first Genetic Centre in India at AIIMS Delhi, Padma Shri (1983), and BC Roy National Award (1985) for his research in rare diseases

Inflection Points: Training in genetics in London; opportunity to lead genetics in AIIMS

9 N-of-1 Trial

Yiwei She has written this chapter, with writing assistance from Sneha Kedkar. Yiwei, who is the founder and CEO of the TNPO2 Foundation, is trained as a mathematician and worked as an Assistant Professor at Columbia University. She was enjoying the choice of being a stay-at-home parent when her second child Leo was diagnosed with a de novo ultra-rare neurodevelopmental genetic disease. Faced with no other options and unwilling to give up hope on a 4-month-old baby, Yiwei took on the project of attempting a pharmaceutical rescue of quality of life for her child. As parent, caregiver, advocate, and ad hoc drug developer, Yiwei brought together a team of basic and translational scientists from academia and industry to put a precision ASO into the clinic before Leo's second birthday.

N-of-1 trials are the ultimate goal of precision medicine. It involves designing and development of a specialized form of clinical trial to assess the effectiveness of a treatment or intervention for an individual patient. Unlike traditional clinical trials that involve groups of patients, n-of-1 trials focus on the unique response of a single patient to treatment over time. These trials are particularly useful in situations where there is uncertainty about the best course of treatment for a specific patient, or when standard treatments have failed to produce the desired results. By tailoring the treatment to the individual patient and closely monitoring their response over time, these trials can provide valuable insights into the effectiveness of different treatments for specific patients.

The process of conducting an n-of-1 trial typically begins with the identification of a patient who is experiencing symptoms that have not responded well to standard treatments or where there is uncertainty about the best course of action. Typically, it is an ultra-rare disease with single-digit numbers of patients worldwide. The first n-of-1 study was developed using the drug Milasen, to treat a patient, named Mila, who had a rare and fatal neurodegenerative disease with a novel mutation of neuronal ceroid lipofuscinosis 7 (CLN7, a form of Batten's disease). The patient was treated with an anti-sense oligonucleotide drug, with an anti-sense oligonucleotide drug within one year [21].

The US FDA has created guidelines for n-of-1 studies. These studies require the entire process of drug development to be conducted - from designing the drug to treat the pathway (mechanism) of disease pathogenesis, conduct pharmacology, and toxicology studies using a drug manufactured using good manufacturing processes. The clinical trial is performed through a defined clinical protocol. Throughout the trial, the patient's symptoms and response to treatment are closely monitored using various outcome measures, such as pain scores, symptom severity scales, or quality of life assessments. At the end of the trial, the data collected from each treatment cycle are analyzed to determine the effectiveness of the different treatments for that individual patient. The results of n-of-1 trials can provide valuable information to guide treatment decisions not just for the individual patients, but can contribute to the broader body of medical knowledge by providing insights into the effectiveness of different treatments for specific conditions. Aggregated data from multiple n-of-1 trials can be used to conduct meta-analyses, which can help to identify patterns and trends in treatment responses across different patients and settings. Despite their many advantages, n-of-1 trials also have some limitations. Because they focus on individual patients, the results may not always generalize to other patients with the same condition. Additionally, conducting n-of-1 trials can be time-consuming and resource-intensive, which may limit their feasibility in certain settings. Overall, n-of-1 trials represent a valuable tool for personalized medicine, allowing for the systematic evaluation of treatments for individual patients. Yiwei has described the entire workflow of the development of the anti-sense oligonucleotide drug to treat her son, in an n-of-1 clinical trial. This study is so inspirational and emphasizes the importance of patient advocacy groups, in supporting the development of drugs.

Timothy Yu
Associate professor of Pediatrics, Harvard Medical School. His pioneering work with Milasen set the stage for n-of-1 studies in clinical trials.

Education: Harvard (undergrad); UCSF (grad, medical school); fellowship (Mass General)
Accomplishment: Developed the first n-of-1 study through the regulatory process
Inflection point: Trained appropriately and being at an empowered institution to innovate

10   Intellectual Property

This chapter is written by Joanna Brougher who is a patent attorney focusing her practice on all aspects of services related to patents in the areas of biotechnology, pharmaceuticals, and medical devices, including patentability opinions, patent drafting, domestic and foreign patent prosecution, development and management of patent portfolios, and general client counseling during all phases of a product's lifecycle, from concept to commercialization. She is an Adjunct Professor at Cornell Law School and at the University of Pennsylvania School of Medicine. Previously, she was an Adjunct Lecturer at the Harvard School of Public Health and has published two books: *Intellectual Property and Health Technologies: Balancing Innovation and the Public's Health* and *Billion Dollar Patents: Strategies for Finding Opportunities, Generating Value, and Protecting Your Inventions* [22].

Intellectual property (IP) plays a pivotal role in the realm of biological drug development, where innovation is the cornerstone of progress. This chapter delves into the significance of IP in this domain and elucidates key steps involved in managing intellectual property effectively. Robust IP strategies are essential for fostering innovation, securing investments, and ensuring the continued advancement of biopharmaceuticals.

Incentivizing innovation by providing exclusivity is the cornerstone of IP laws. This exclusivity enables companies to recoup their investments and generate profits, driving further innovation. Investors seek assurance that their capital will be protected through patents and other IP rights, thereby facilitating funding for drug development endeavors. Patents and other forms of IP protection grant market exclusivity, allowing companies to commercialize their biological drugs without competition for a specified period. This exclusivity is crucial for recovering development costs and realizing profits before generic competition enters the market. IP rights enable biopharmaceutical companies to engage in collaborations, licensing agreements, and technology transfers, facilitating the exchange of knowledge and resources to expedite drug development.

Tom Irwing
Tom is a mentor and educator and has more than 48 years of experience in the field of intellectual property law. He is in the IAM Life Sciences Hall of Fame and a recipient of lifetime achievement awards.

Education: Chemistry, University of Utah (undergrad), J.D. (Duke University)
Accomplishments: Contributed to litigation of major drugs Kalydeco®, Crestor®, Allegra®, and many others

The initial steps of intellectual property management involve conducting a comprehensive IP landscape analysis. These activities include evaluating existing patents and publications to identify potential gaps and opportunities for innovation, and assessing patentability and freedom-to-operate to ensure the novelty and non-infringement of new inventions. The next steps involve filing patent applications early in the drug development process to secure priority rights and establish a strong IP foundation. Through the drug development life cycles the patent portfolio is optimized to cover novel aspects of the biological drug. Regularly review and update the IP portfolio to align with evolving business objectives, technological advancements, and regulatory changes.

Following assignment of patents, protecting the IP rights involves "offensive" and "defensive" strategies. This process requires monitoring third-party patents and intellectual property developments to identify potential infringement risks and formulate risk mitigation strategies. Defensive measures, such as filing defensive patents or negotiating cross-licensing agreements, can protect against litigation threats. Offensive measures include enforcing IP rights through litigation, cease-and-desist letters, or administrative proceedings to deter infringement and protect market exclusivity. Alternative dispute resolution mechanisms can include arbitration or mediation, to resolve IP disputes efficiently and cost-effectively.

11 Statistics

The chapters on Statistics and Risk are written by Ravi Khare who is the CEO of Symphony Technologies, a data analytics company. His focused areas of work have been Quality Engineering, Design, and Manufacturing in the Pharma, Engineering, and Automotive sectors. He is a mechanical engineer with expertise in data science and engineering.

Statistical concepts are integral to the development of biological drugs, ensuring robust study design, accurate data analysis, and reliable interpretation of results. This chapter explores the importance of statistical principles in biopharmaceutical research and elucidates key steps involved in their application throughout the drug development process. From study design to regulatory submission, statistical rigor plays a critical role in advancing safe and efficacious biological therapies. Variability is integral to nature. Understanding, measuring, and reporting variability requires several statistical principles including design of experiments, ensuring appropriate sample size, randomization, and blinding to minimize bias and maximize the validity of study findings. Several factors are described in the target product profile of the drug, such as effect size, acceptable biological variability, alpha level, and desired power.

In order to develop a systematic approach, the drug development process requires outlining a statistical analytical plan. Statistical methods, including hypothesis testing, regression analysis, and survival analysis, are employed to analyze complex datasets generated from preclinical and clinical studies. Validation of measurement methods is performed to ensure parameters such as precision, accuracy, linearity, and robustness. During the manufacturing processes, statistical methods involve statistical process controls, process development, process characterization, process validation, and continued process monitoring throughout the life cycle of the drug. The ultimate goal of statistical rigor is to assess the risks and benefits associated with biological drugs, including the likelihood of adverse events, treatment response rates, and overall clinical outcomes.

Ronald Fischer (1890–1962)
Ronald Fischer is the father of "design of experiments." His work on applying statistical methods for agricultural applications has been the subject of landmark books *Arrangement of Field Experiments* (1926) and *The Design of Experiments* (1935). Born in London, he was the son of a successful auctioneer. Lifelong he had poor eyesight, but clear vision of mathematics.

Title: A polymath, the father of DOE and neo-Darwinism
Education: Gonville and Caius colleges, Cambridge (undergrad in Mathematics)
Accomplishments: In 1918, he coined and proposed the word "analysis of variance (ANOVA)" and contributed to understanding genetic and population variance. He also proposed the "p-value" of significance of 0.05,

which is a 1:20 probability of being exceeded by chance alone. His work is the basis of genome-wide association studies (GWAS) today

Inflection points: Not accepting a position at the University of London, and taking on a temporary position in Rothamsted to study the vast amount of agricultural data there

Joseph Heyes: (1952–2019)

Joe Heyes was one of the most influential leaders in statistics of clinical trials. He worked at Merck from 1976 (for 43 years). He contributed to the development of the rotavirus, human papillomavirus, and varicella zoster vaccines, each of which required extremely large, complicated clinical studies to establish safety and efficacy. His work was instrumental in the availability of vaccines that have saved hundreds of thousands of lives.

Title: Statistician, Rotavirus-clinical trial designer, empathy

Education: Master's (Villanova University); MBA (Temple University); PhD (Temple University)

Accomplishments: His work on statistical design of the rotavirus vaccine trial, to ensure safety from intussusception, was instrumental to the availability of vaccines that have saved hundreds of thousands of lives

Inflection point: Joining Merck and participating and contributing to the development of vaccine clinical trials

12   Risk Analysis in Drug Development

Risk management plays a pivotal role in the development of biological drugs, where complex processes and inherent uncertainties pose significant challenges. This chapter explores the importance of risk management in biopharmaceutical research and elucidates key steps involved in identifying, assessing, and mitigating risks throughout the drug development lifecycle.

Ensuring Patient Safety is paramount to drug development. Risk management helps identify and address potential safety concerns early in the development process, minimizing the likelihood of adverse events in clinical trials and post-market use. One of the first steps in developing a risk management strategy is to engage cross-functional teams, including scientists, clinicians, regulatory experts, and quality assurance professionals, to leverage diverse perspectives in identifying risks. Risk assessment involves evaluation of the severity, likelihood, and detectability of identified risks to prioritize them based on their potential impact on patient safety, product quality, and project objectives. ICHQ9 regulatory guidance recommends utilizing qualitative and quantitative risk assessment tools, such as failure mode and effects analysis (FMEA) and probabilistic risk assessment (PRA), to systematically evaluate and prioritize risks. Risk mitigation strategies and action plans can reduce the likelihood or severity of identified risks to acceptable levels. Risk controls, such as process improvements, formulation adjustments, or enhanced monitoring measures, mitigate identified risks and enhance product safety and quality. Life cycle management of the risks involved establishing mechanisms for ongoing risk monitoring and surveillance throughout the drug development lifecycle.

13   Regulatory Strategy

There is no road map, checklist, or to-do list for regulatory strategy. It is a combination of multi-dimensional aspects of drug development. It requires experience, a lot of case study analyses, and planning. However, a few concepts are useful to know about. In the past (a few decades ago), one strategy to develop approvable, successful drug was to try many drugs and "hope" that one of them succeeds. This strategy in soccer analogy is "*shots on goal*." This strategy is generally expensive to support many molecules, most of which are doomed to fail. The more recent strategy has moved to "*pick the winner*," which has emerged in the face of the ability to collect, analyze, and interpret very large amounts of data with the advent of new analytical technologies. Another aspect of regulatory strategy for developing approvable drugs is "*biology first.*" Here, it is important to consider the mechanism of the biology of the disease and the drug action before determining the platform of the drug (e.g., small molecule or monoclonal antibody or gene therapy). Another strategy is to choose drugs that have a "*large effect size.*" The first few monoclonal antibodies approved targeting tumor targets had minimal efficacy, e.g. 10% responders had an increase in progression free-survival for a few months. To have higher market potential, therapeutic drugs had to achieve larger effect sizes, e.g., monoclonal antibodies targeting PCSK9 target had efficacies in the rage of 60–70% responders [23, 24]. Finally, some comments overheard during company motivational talks from leaders in the field include the following remarks: "An effective method" often deceives us in thinking that it is the only effective one, and most effective.

- The more you know, the more you know how much you don't know
- Ironically, the way to see clearly is to stand at a distance
- Most improved things can be improved
- R&D does not assure success, but lessens the risk

Roger Perlmutter, MD, PhD
Headed R&D for Amgen and Merck and contributed to development of many iconic drugs such as EPO, Enbrel, and Keytruda, among many others. He is also an accomplished pianist, and I have seen him perform in a company management meeting.

Education: Reed College (undergrad); MD and PhD from Washington University in St. Louis, clinical training in Internal Medicine at Massachusetts General Hospital and University of California, San Francisco, and California Institute of Technology (CalTech).

Accomplishments: Supervised the discovery and development of numerous lifesaving medicines including Keytruda, which continues to transform cancer care throughout the world; elected member of the American Academy of Arts and Sciences, a fellow of the American Association for the Advancement of Science, and both a distinguished fellow and past president of the American Association of Immunologists.

Inflection points: Experiences with legendary scientists including Leroy Hood, confident-clear-visionary decision-making at critical junctures in drug development (to progress and kill drugs).

Peter Marks, MD, PhD
Director of the Center for Biologics Evaluation and Research (CER) within the US FDA.

Education: Columbia University (undergrad), New York University (MD, PhD), Internal Medicine training in Bringham Women's Hospital, Boston; worked at Yale University, New Haven, CT.

Accomplishments: Member of the White House Coronavirus Task Force, and established Operation Warp Speed for COVID-19 vaccines.

Inflection points: Diverse experience in academia and industry; the COVID-19 vaccine experience; and understanding sense of urgency in drug review and approval processes for patients to have access to transformational therapies.

We return to the regulatory strategy:

Start with the end in mind. The strategy to develop a drug is an intricate journey, usually involving at least a decade of time, millions of dollars, intensive scientific rigor, and extremely talented people. Since no drug is 100% efficacious, and has no adverse effects, the development process is a delicate balance between efficacy, safety, ethical considerations, and legal requirements. A thorough evaluation of risks at various stages can anticipate potential failures and enable planning for mitigation strategies. Regulatory agencies oversee every aspect of drug development life cycle, with patient safety being at the center. Thus, considering what the label claims can be obtained can focus the teams to generate the appropriate data. The term used in the strategy is *"label as the driver* (of drug development)."

Collaboration between academia, industry, and regulatory authorities enables navigating the complex regulatory requirements. Regulatory frameworks, such as expedited review programs and orphan drug designation, facilitate the development of therapies for rare diseases or conditions with limited treatment options. Furthermore, regulatory agencies engage in initiatives to streamline regulatory processes, enhance regulatory science, and incorporate patient perspectives into decision-making, thereby promoting efficiency, transparency, and inclusivity in drug development efforts. As the landscape of drug development continues to evolve, regulatory strategies will remain essential tools for navigating the complex intersection of science, ethics, and public health, ensuring that innovative therapies reach those in need in a timely, cost-effective, and sustainable manner.

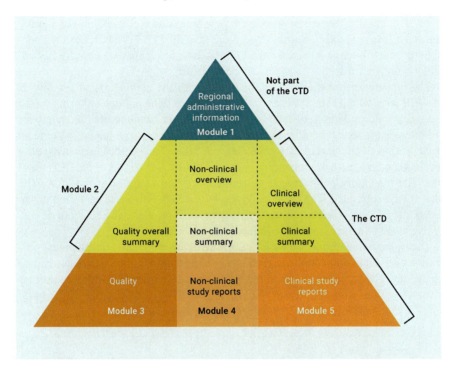

The five modules of the BLA are: (1) Administrative. (2) Overall summary. (3) Chemistry Manufacturing, and Controls. (4) Non-clinical. (5) Clinical. Each department in the company is responsible for their section of their module. In fact, each person in the company "works for the BLA." The BLA can comprise about a million pages. These pages comprise extremely great details of all the work done, written in specified formats, supporting documentation, such as SOPs, validation reports, training records, deviation reports, etc., need to be ready for audits. Digitization of documentation is vital to improve efficiency in the BLA preparation and filing processes. [Note: 20 years ago, BLA filing was done where hundreds of thousands of paper was printed, multiple copies made, and shipped in trucks, literally transported in trucks to the FDA offices in Washington DC].

The BLA is submitted. Celebration. Parties. Awards. The submission is just the beginning of the review process. All aspects of the BLA are secret to the company and FDA. No one outside the company knows the contents of the BLA. No one is supposed to share any information. The medical writing teams very carefully write manuscripts, and prepare for presentations in conferences, with oversight of legal and commercial teams, on the scientific results of the BLA filing, for dissemination to stakeholders. There is a systematic and strategic communication process that is planned and executed.

The regulatory agency will take precisely 18 months to review the documents, governed by the PDUFA Act. The PDUFA (Prescription Drug User Fee Act) is a US law passed in 1992 and subsequently renewed in 2017. PDUFA enables the FDA to collect fees from drug manufacturers to fund the drug approval process. These fees are used to expedite the review process for new drugs and biologics, aiming to bring safe and effective medications to market more quickly. PDUFA has been instrumental in reducing the time it takes for the FDA to review and approve new drugs, benefiting both pharmaceutical companies and patients.

The preparation for the BLA filing begins with the completion of the pivotal phase III trial, which shows successful data of meeting the pre-defined primary and secondary criteria. Management approves the team to commit to file (i.e., approves and releases the funds required for the BLA filing). The project management team maps the plan of all the activities, timelines, and costs. Each of the teams in the drug development process has a functional team and writes the sections of the BLA modules. The regulatory publishing teams collate the five modules of the BLA. The BLA is then submitted. The FDA, governed by the PDUFA Act, takes 18 months to review the BLA. It is a complicated, muti-functional process, which cannot be easily explained in a book, or lecture, it has to be experienced!

The figure below shows the various steps involved in the drug development process, the key interaction points of regulatory submissions, and communication with the agency. The reader can refer to the US FDA website for a step-by-step process (https://www.fda.gov/patients/learn-about-drug-and-device-approvals/drug-development-process).

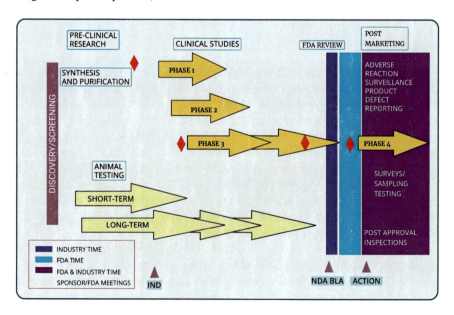

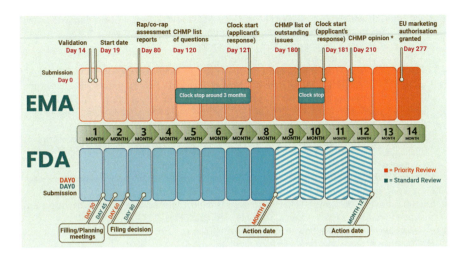

The figure above shows the precise steps involved in the review of the BLA submission by the EMA and FDA. More details can be found in guidance documents and in the websites of these agencies.

The future of drug development will be exciting with the advancement of new analytical biotechnologies, such as highly efficient and cost-effective sequencing, especially at the single-cell level, imaging technologies of spatial transcriptomics, nano-materials for drug delivery, and many more. Enhancement in the field of machine learning-based data analytics, quantum computing, and merging of engineering and biological fields will have an exponential impact on all aspects of drug development from discovery of new targets and pathways to highly efficient operations across the life cycle.

One of the aims of this book is to provide a glimpse of "what happens in the industry" during the drug development process. Of course, just reading this book, and references alone does not prepare you for a career in drug development, you must experience the process. It is a slow learning process. The chapters of the book (written by highly experienced professionals in the industry) are curated to give you an overview of all the major steps in the drug development life cycle.

Let's begin.

References

1. Schmidt C. Suicidal thoughts end Amgen's blockbuster aspirations for psoriasis drug. Nat Biotechnol. 2015;33:894–5.
2. Gray G, Buchbinder S, Duerr A. Overview of STEP and Phambili trial results: two phase IIb test-of-concept studies investigating the efficacy of MRK adenovirus type 5 gag/pol/nef subtype B HIV vaccine. Curr Opin HIV AIDS. 2010;5:357–61.
3. Fischer A, Hacein-Bey-Abina S. Gene therapy for severe combined immunodeficiencies and beyond. J Exp Med. 2020;217

4. Raper SE, Chirmule N, Lee FS, Wivel NA, Bagg A, Gao GP, Wilson JM, Batshaw ML. Fatal systemic inflammatory response syndrome in a ornithine transcarbamylase deficient patient following adenoviral gene transfer. Mol Genet Metab. 2003;80:148–58.

5. Chirmule N. Good genes gone bad: a short history of vaccines and biologics: failures, successes, controversies. Penguin Random House India Private Limited; 2021.

6. Tordrup D, van den Ham HA, Glanville J, Mantel-Teeuwisse AK. Systematic reviews of ten pharmaceutical pricing policies—a research protocol. J Pharm Policy Pract. 2020;13:22.

7. Lee KS, Kassab YW, Taha NA, Zainal ZA. Factors impacting pharmaceutical prices and affordability: narrative review. Pharmacy (Basel). 2020;9

8. Gronde TV, Uyl-de Groot CA, Pieters T. Addressing the challenge of high-priced prescription drugs in the era of precision medicine: a systematic review of drug life cycles, therapeutic drug markets and regulatory frameworks. PLoS One. 2017;12:e0182613.

9. Wouters OJ, McKee M, Luyten J. Estimated research and development investment needed to bring a new medicine to market, 2009–2018. JAMA. 2020;323:844–53.

10. Candio P, Frew E. How much behaviour change is required for the investment in cycling infrastructure to be sustainable? A break-even analysis. PLoS One. 2023;18:e0284634.

11. N. Chirmule, A process on how to ask questions, 2020.

12. N. Chirmule, A process of make decisions, 2020.

13. N. Chirmule, What the investor in biotechnology looks for, 2022.

14. Sun D, Gao W, Hu H, Zhou S. Why 90% of clinical drug development fails and how to improve it? Acta Pharm Sin B. 2022;12:3049–62.

15. Conti RM, Frank RG, Gruber J. Regulating drug prices while increasing innovation. N Engl J Med. 2021;385:1921–3.

16. Ghilardi N, Pappu R, Arron JR, Chan AC. 30 years of biotherapeutics development-what have we learned? Annu Rev Immunol. 2020;38:249–87.

17. Piot P, Larson HJ, O'Brien KL, N'Kengasong J, Ng E, Sow S, Kampmann B. Immunization: vital progress, unfinished agenda. Nature. 2019;575:119–29.

18. Offit P. Vaccinated: one man's quest to defeat the world's deadliest diseases. Harper Collins; 2008.

19. Jumper J, Evans R, Pritzel A, Green T, Figurnov M, Ronneberger O, Tunyasuvunakool K, Bates R, Žídek A, Potapenko A, Bridgland A, Meyer C, Kohl SAA, Ballard AJ, Cowie A, Romera-Paredes B, Nikolov S, Jain R, Adler J, Back T, Petersen S, Reiman D, Clancy E, Zielinski M, Steinegger M, Pacholska M, Berghammer T, Bodenstein S, Silver D, Vinyals O, Senior AW, Kavukcuoglu K, Kohli P, Hassabis D. Highly accurate protein structure prediction with AlphaFold. Nature. 2021;596:583–9.

20. Patterson SD. Decoding disease: Scott Patterson's perspectives on the power of biomarkers in drug development. BioTechniques. 2024;76:9–13.

21. Kim J, Hu C, Moufawad El Achkar C, Black LE, Douville J, Larson A, Pendergast MK, Goldkind SF, Lee EA, Kuniholm A, Soucy A, Vaze J, Belur NR, Fredriksen K, Stojkovska I, Tsytsykova A, Armant M, DiDonato RL, Choi J, Cornelissen L, Pereira LM, Augustine EF, Genetti CA, Dies K, Barton B, Williams L, Goodlett BD, Riley BL, Pasternak A, Berry ER, Pflock KA, Chu S, Reed C, Tyndall K, Agrawal PB, Beggs AH, Grant PE, Urion DK, Snyder RO, Waisbren SE, Poduri A, Park PJ, Patterson A, Biffi A, Mazzulli JR, Bodamer O, Berde CB, Yu TW. Patient-customized oligonucleotide therapy for a rare genetic disease. N Engl J Med. 2019;381:1644–52.

22. Brougher JT. Billion Dollar patents: strategies for finding opportunities, generating value, and protecting your inventions. JTB Publishing LLC; 2019.

23. Sabatine MS, Giugliano RP, Keech AC, Honarpour N, Wiviott SD, Murphy SA, Kuder JF, Wang H, Liu T, Wasserman SM, Sever PS, Pedersen TR. Evolocumab and clinical outcomes in patients with cardiovascular disease. N Engl J Med. 2017;376:1713–22.

24. Schwartz GG, Steg PG, Szarek M, Bhatt DL, Bittner VA, Diaz R, Edelberg JM, Goodman SG, Hanotin C, Harrington RA, Jukema JW, Lecorps G, Mahaffey KW, Moryusef A, Pordy R, Quintero K, Roe MT, Sasiela WJ, Tamby JF, Tricoci P, White HD, Zeiher AM. Alirocumab and cardiovascular outcomes after acute coronary syndrome. N Engl J Med. 2018;379:2097–107.

Vaccines as Prophylactics and Therapeutics

Priyal Bagwe, Sharon Vijayanand, and Amitabh Gaur

Abstract Vaccination has been a cornerstone of public health, markedly reducing morbidity and mortality associated with infectious diseases. This chapter offers an in-depth examination of vaccines, beginning with a historical overview that chronicles the progression of vaccination practices from their inception to contemporary advancements. It systematically addresses vaccines targeting viral, bacterial, and parasitic pathogens, elucidating their profound impact on global health.

Advances in vaccine development are thoroughly explored, with a focus on novel vaccine design methodologies and the optimization of delivery systems and administration routes. The chapter categorizes various vaccine types, including whole organism vaccines (live attenuated, inactivated), subunit vaccines (comprising proteins, peptides, and carbohydrates), and the revolutionary DNA and mRNA vaccines. Additionally, it discusses the utilization of virus-like particles (VLPs) for their efficacy in inducing robust immune responses.

A dedicated section on oncological vaccines investigates the advancements and obstacles in developing vaccines that target neoplastic cells, highlighting this as a promising area of immunotherapy. The chapter culminates with an exploration of the future trajectory of vaccines, reviewing emerging technologies and novel solutions for both infectious diseases and non-communicable diseases such as cancer.

Keywords Immunization · COVID-19 · mRNA vaccines · Vaccine development · Cancer vaccine

P. Bagwe · S. Vijayanand
Mercer University College of Pharmacy, Atlanta, GA, USA

A. Gaur (✉)
Innovative Assay Solutions LLC, San Diego, CA, USA
e-mail: agaur@neobiotechnologies.com

© The Author(s), under exclusive license to Springer Nature Switzerland AG 2025
N. Chirmule, V. V. Ghalsasi (eds.), *Approved: The Life Cycle of Drug Development*,
https://doi.org/10.1007/978-3-031-81787-8_2

1   Introduction

A vaccine is a biological preparation that provides acquired immunity to infectious agents or other diseases. Currently, there are vaccines to prevent over 20 life-threatening diseases, and immunization today prevents millions of deaths every year from diseases like diphtheria, pertussis, measles, tetanus, and influenza [1]. A recent study revealed that global immunization efforts have averted 154 million deaths over the past 50 years, including 146 million among children under the age of five [2]. Of these, most of the lives saved—101 million—have been those of infants under the age of one.

The origin of such immunization dates back to the late eighteenth century, when Edward Jenner pioneered the concept of vaccines and created the world's first successful vaccine. Jenner, an English physician and scientist commonly called the father of vaccinology, developed a vaccine for smallpox. Based on the knowledge that those who were infected with cowpox did not contract smallpox, Jenner inoculated an eight-year-old boy with pus, termed "matter" at that time, collected from a human cowpox sore. Although the boy was unwell for a few days following the inoculation, he made a full recovery, and did not fall sick upon being exposed to a smallpox sore. The "matter" had to be transported from England to the United States to inoculate people in America.

In the late 1800s, Louis Pasteur and his colleagues injected 14 daily doses of progressively inactivated rabies virus into a nine-year-old boy who had been severely bitten by a rabid dog. This ushered in the modern era of immunization. The early 1900s saw vaccines developed for outbreaks such as yellow fever, pertussis or whooping cough, and influenza [3]. Calmette and Guerin developed the Bacillus Calmette-Guerin vaccine containing a live attenuated form of the bacterium *Mycobacterium bovis* that prevented tuberculosis and other mycobacterial infections including leprosy [4]. This vaccine was first administered in 1921. The benefits of this vaccine in leprosy were observed for the first time in 1939 [4]. An immunotherapeutic and prophylactic modulatory vaccine was eventually approved for leprosy in 2016 [5]. The early 1920s saw researchers learning how to inactivate toxins such as those that cause tetanus. The inactivated toxins, or toxoids, still produce an immune response when introduced into the body, and a tetanus toxoid vaccine was developed on this basis in 1924. It was widely used during the Second World War.

By this time, the world had seen frequent epidemics of polio, making it the most feared disease in the world. Outbreaks killed thousands of people at a time, and those who survived were left with severe deformities. After a breakthrough where scientists successfully cultivated the poliovirus in human tissues, American physician Jonas Salk created the first successful vaccine against polio. This experimental killed-virus vaccine was tested on millions of people, and the inactivated poliovirus vaccine was eventually licensed in 1955. By 1957, the number of cases in the United States had dropped from 58,000 to 5600, and by 1961, only 161 cases remained [6].

A second type of polio vaccine, to be given orally was developed by Albert Sabin. The vaccine used a live attenuated or weakened form of virus, and was widely approved in 1961 [6].

The discovery of the hepatitis B virus by Baruch Blumberg in 1965 eventually led to his collaboration with microbiologist Irving Millman. Together, they developed the first hepatitis B vaccine using a heat-treated form of the virus in 1969. An inactivated vaccine was approved for commercial use from 1981 to 1990, and a recombinant DNA vaccine that was developed in 1986 is currently used for vaccination [7].

In 1971, Maurice Hilleman combined the measles, mumps, and rubella vaccinations developed in 1963, 1967, and 1969, respectively [8]. In 1995, a team led by Anne Szarewski established the role of human papillomavirus (HPV) in cervical cancer, causing researchers to begin work on an HPV vaccine. The first vaccine was approved in 2006, and since then vaccinations against HPV have been an integral part of the global fight against cervical cancer [9].

Aside from protecting against infectious and non-infectious diseases, vaccines have also been developed for contraception, and are undergoing clinical trials [10]. Such vaccines work by either inhibiting the production of gametes, generating immune responses against spermatozoa or oocytes to impair their function, or targeting the human chorionic gonadotropin (hCG), a hormone crucial for pregnancy [11, 12].

The 2019 coronavirus pandemic led to the approval of the first-ever mRNA vaccine that uses messenger RNA molecules, coding specific protein epitopes, to elicit an immune response. Several different pharmaceutical companies all over the world developed different types of vaccines against the SARS-CoV-2 to fight the COVID-19 pandemic. About 13.5 billion of these vaccines have been given, with almost 71% of the world's population, or about 5.6 billion people receiving at least one dose.

2   Vaccines Against Infectious Diseases

Vaccines stand as great defenders in the dynamic scope of infectious diseases, offering protection from various pathogens. They cover a wide spectrum of microbials, with every organism providing unique challenges in the journey toward immunization. Over the years, vaccines have been developed against bacteria, viruses, and parasites. Scientists have contributed immensely to combating infectious agents by developing innovative strategies. In the battle against infectious diseases, vaccines are indispensable offering protection against viruses, bacteria, and parasites. The journey from eradicating smallpox to controlling COVID-19 pandemic highlights the transformative power of vaccines in protecting human health (Fig. 1).

Fig. 1 The World Health Organization (WHO) poster commemorating the eradication of smallpox in October 1979, which was officially endorsed by the 33rd World Health Assembly on May 8, 1980. (Courtesy of WHO)

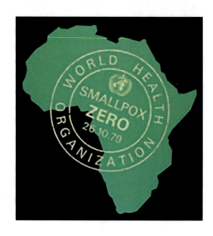

2.1   Vaccines Against Viral Diseases

Vaccines are an invaluable asset in protecting global health, playing a crucial role in preventing current and emerging viral threats to human health. They work by stimulating the immune system to recognize and combat pathogens, such as viruses, without causing the disease itself. This is achieved by introducing a harmless component of the virus, such as a protein or a weakened form of the virus, which trains the immune system to respond more effectively if exposed to the actual virus in the future. Vaccines have been instrumental in controlling and even eradicating diseases like smallpox and polio, and they continue to be vital in the fight against emerging viral threats like COVID-19 and influenza.

Viruses rapidly invade host cells and replicate, posing significant threats to human health. Viral vaccines have been a tremendous asset in combating these microscopic invaders. There are various types of viral vaccines, including live attenuated, inactivated whole cell, subunit, vector-based, and the recently developed mRNA vaccines. This chapter will delve into each type in detail.

The most common vaccine against measles, mumps, and rubella (MMR) is a live attenuated vaccine. It contains a weakened form of the virus that cannot cause the disease but can elicit a strong immune response. Live vaccines mimic natural infections, generating robust immunity.

The influenza vaccine is an example of an inactivated vaccine [13]. Since 2012 in Europe, the live attenuated influenza vaccine (LAIV) has been used as an alternative to traditional inactivated influenza vaccines (IIV). LAIVs mimic natural infections and have been found to provide broader clinical protection in children compared to IIVs. However, the detailed immunological mechanisms induced by LAIVs are not fully understood, and despite 14 years on the global market, there is no established correlation of protection. A better understanding of the immune responses after

LAIV administration may help achieve the goal of a future "universal influenza vaccine."

Inactivated vaccines utilize the whole killed virus, inactivated through physical or chemical processes. These vaccines do not pose an infection risk but can generate a strong immune response. They are relatively safer compared to live attenuated vaccines, though they often contain cellular debris apart from the critical viral components.

Hepatitis B vaccines [14, 15], for instance, contain purified protein subunit antigens that generate an immune response against specific viral components. By targeting key elements responsible for viral entry and replication, these vaccines minimize adverse effects and focus on generating targeted immunity.

The recent COVID-19 pandemic has spurred the development of novel vaccine delivery methods, such as mRNA vaccines by Pfizer-BioNTech and Moderna, which harness the power of mRNA, and adenovirus-based vaccines developed by several groups including Oxford-Astra Zeneca, Sputnik V, and Johnson & Johnson [16]. The adenovirus vaccine consists of a replication-incompetent recombinant adenovirus type 26 (Ad26) viral vector expressing the spike (S) protein of the coronavirus. Human embryonic cell lines—embryonic retinal cells or HEK293—permitting replication of the engineered virus are used for the production. The viral vector lacks the essential gene required for replication, thus eliminating the possibility of replication in individuals receiving the vaccine (Table 1).

Table 1 FDA-approved viral vaccines (1970–2021)

Vaccine	Disease	Type	Year approved	References
MMR	Measles, mumps, rubella	Live attenuated	1971	[17]
Varivax	Varicella (chickenpox)	Live attenuated	1995	[18]
Gardasil	Human papillomavirus	Recombinant	2006	[19]
Fluzone	Influenza	Inactivated	1980	[20]
HepA (Havrix)	Hepatitis A	Inactivated	1995	[21]
HepB (Engerix-B)	Hepatitis B	Recombinant	1986	[22]
Zostavax	Herpes zoster (shingles)	Live attenuated	2006	[23]
Shingrix	Herpes zoster (shingles)	Recombinant	2017	[24]
Comirnaty (Pfizer-BioNTech)	COVID-19	mRNA	2021	[25]
Moderna COVID-19 vaccine	COVID-19	mRNA	2021	[26]
J&J/Janssen COVID-19 vaccine	COVID-19	Viral vector	2021	[27]

2.1.1 Recent Successes in Viral Vaccine Development

SARS-CoV-2 Vaccines

The global COVID-19 pandemic called for a safe, effective, reliable, and permanent vaccine that would be accessible to a large population. The virus exhibited a high mutation frequency, which can render vaccines ineffective, highlighting the requirement for a vaccine that would be successful in curbing future outbreaks of SARS-CoV-2 variants. The most common approach to develop COVID-19 vaccines relied on eliciting an immune response that would produce antibodies against the spike protein of the virus by binding to the receptor-binding domain (RBD) [16, 28]. Some of the candidate vaccines developed include viral vector-based vaccines, nucleic acid-based RNA/DNA vaccines, vaccines that contain proteins purified from the virus, and live virus vaccines. Most of these vaccines were developed and approved in unprecedented timelines. The end of the pandemic can be achieved only by acquiring protection against severe disease and mortality, which requires the widespread availability of an effective vaccine [16, 28].

Aside from those already approved, there are over 330 different vaccine candidates against SARS-CoV-2. Among these is a pan-coronavirus vaccine candidate which can neutralize multiple strains including those that may emerge eventually. One such candidate is the spike ferritin nanoparticle (SpFN) vaccine, which displays spike proteins from up to two dozen variants on a ferritin nanoparticle [29]. This candidate vaccine has been shown to elicit potent and lasting neutralizing antibodies in non-human primates against known variants of concern. Here, we summarize the steps utilized for SARS-CoV-2 vaccine development, which helped condense the usual timeline of decades into 12–18 months. Various factors enabled these vaccines to be developed in record time. Firstly, the viral genes were sequenced, and the sequences were shared in early February 2020. Moreover, the urgency of the pandemic resulted in global collaborations between pharmaceutical companies like Pfizer-BioNTech and Astra Zeneca-Oxford-Serum Institute. In addition, the governments extended financial and regulatory support to various companies. The pandemic also saw an increase in speedy and open-access publications, with several manuscripts being published ahead of print on pre-print platforms like medRxiv and bioRxiv. Regulatory authorities were also involved in reviewing and providing guidance on the process. In addition, technological advancements in multi-omics, machine learning-based modeling and cryo-electron microscopy helped understand the viral life cycle and immune-pathogenesis mechanisms. The seriousness of the situation led to some changes in the vaccine manufacturing pipeline—planning and managing of at-risk manufacturing, scale-up, and supply chain logistics took place well before efficacy and safety results were obtained. The vaccines were also evaluated in an expedited manner by conducting the three phases of clinical trials in an overlapping manner. Synergizing public and private resources helped speed the timeline, and initiatives like those led by Coalition for Epidemic Preparedness Innovations (CEPI) and Operation Warp Speed helped obtain unprecedented levels of private and public funds. This minimized the risk for vaccine developers, which encouraged them to conduct concurrent cost-intensive studies [30].

Vaccine designing: In the initial steps of the viral life cycle, the RBD of the spike protein binds to the ACE2 and TMPRSS2 receptors on the host's cells. In the initial months of the pandemic, researchers experimentally showed that antibodies targeting the RBD of spike protein could effectively block viral entry and neutralize the virus. This observation significantly shortened the time required to identify the dominant protective antigen from several years to a few months. The widespread and immediate sharing of this data enabled several vaccine designers to begin developing the vaccine using various platforms.

A majority of the vaccines that were developed have utilized the spike protein or its RBD using mRNA, DNA, adenoviral vectors, and protein-based platforms. Additionally, Bharat Biotech developed killed-whole-virus vaccines, which were approved for emergency use. Researchers are currently generating data supporting the use of spike proteins to induce neutralizing antibodies and T-cell responses to elicit long-term protective immunity. Table 2 shows the different platforms of vaccines expressing spike proteins and their ability to elicit cell-mediated and humoral immune responses.

Respiratory Syncytial Virus (RSV) Vaccines

Respiratory syncytial virus (RSV) was discovered in 1956 from a laboratory animal with an upper respiratory tract infection. The virus was later isolated from children suffering from respiratory symptoms and found to be the cause of bronchiolitis. It is now known that RSV can infect people by the age of three and can cause repeated infections throughout life. The virus causes approximately 6000–10,000 deaths in the United States among people older than 65, and is responsible for 60,000–160,000 hospitalizations [31]. RSV is classified as antigenic subtype A or B based on the reactivity of the surface proteins to antibodies. Subtype A tends to be more prevalent globally, and is generally thought to be more virulent—with a higher viral load and faster transmission time—than subtype B.

Table 2 COVID-19 vaccines—ranked by reported efficacy

Platform	Company	Antigen	Efficacy
mRNA	Pfizer	Spike RBD	95%
mRNA	Moderna	Spike RBD	94%
mRNA	CureVAC	Spike	47%
Chimp-adenovirus	Oxford/AZ/serum	Spike	72%
Adenovirus 26	Janssen (J&J)[a]	Spike	68%
Human-adenovirus	Gamaleya institute	Spike	85%
Protein	Novavax	Spike	76%
Inactivated whole virus	Bharat biotech	All viral proteins	85%

[a]The use of J&J vaccine has been discontinued due to side effects with the FDA recommending the use of mRNA vaccines for general vaccination

Challenges in RSV Vaccine Development: Attempts to develop an RSV vaccine began in the 1960s, and four large studies in 1965 and 1966 were launched to test a whole inactivated vaccine. Infants younger than 6 months old were immunized before their first exposure to RSV. During the winter of 1966 and 1967, an outbreak of RSV in the cohort caused an infection in 20 infants out of the 31 who were vaccinated. Sixteen were hospitalized, and two children succumbed to the infection. This tragedy halted RSV vaccine development for decades [33].

A Structure-Based Design of a Fusion Glycoprotein Vaccine for RSV (The RSVPreF3 OA Vaccine)—The trimeric RSV F protein in its prefusion state was anchored on the viral envelope by a transmembrane domain at the C terminus. At the apex, an epitope of the prefusion F protein is targeted by antibodies with high neutralizing activity. However, when the F protein rearranges into the post-fusion form either spontaneously on the viral membrane or after creating a fusion pore with the host cell membrane, the epitope is lost. Stabilizing mutations were introduced on the interior of the protein to hold it in the prefusion conformation and preserve neutralization-sensitive epitopes at the apex for use as a vaccine antigen. The ectodomain of RSV F vaccines can be delivered as a soluble trimeric protein by constraining the C terminus or, if expressed by gene delivery, the membrane of the protein can be anchored by retaining the transmembrane domain.

Four clinical trials evaluated the effectiveness of three different single-dose RSV vaccines developed by various manufacturers. The GSK vaccine, tested in approximately 25,000 participants with an average age of 70, demonstrated a 94% efficacy against severe pneumonia caused by RSV and 72% efficacy against RSV-induced acute respiratory infection, with higher efficacy against the RSV subtype A (81–85%) compared to subtype B (70–72%). The Janssen vaccine trials enrolled 5800 participants aged 65 and older and showed an efficacy ranging from 70% to 80%, with 80% efficacy against the most severe forms of the disease. The largest trial, involving over 34,000 participants aged 60 and above, assessed the Pfizer vaccine, which displayed an efficacy of 67–86%, offering the highest protection against severe RSV pneumonia. Additionally, Pfizer conducted a separate trial with about 7000 pregnant women, where the vaccine showed an efficacy of 82% in preventing severe RSV in infants at 90 days post-birth, maintaining 69% efficacy 6 months after birth.

2.1.2 Vaccines Currently in Development for Viral Diseases

Dengue

Dengue virus (DENV) is an RNA virus of the Flaviviridae family that causes dengue fever, which affects about 400 million people globally. The virus is spread by mosquito vectors, with *Aedes* genus of mosquitoes responsible for transmitting four serotypes of the virus [34].

Several vaccines against DENV are in various stages of development, including Dengvaxia® which has been approved by the US FDA and other agencies. It is important to generate a balanced immune response against all the four serotypes to avoid more serious disease upon subsequent exposure to an alternate strain of the virus [35]. The severe disease is presumably due to antibody-dependent enhancement (ADE), where pre-existing antibodies after primary infection are ineffective against a different dengue serotype, and aid in viral replication, causing severe infection. The phenomenon of ADE also affects cross-viral vaccines that can derail vaccination efforts. Strategies that can eliminate or mitigate ADE are required to address this important caveat in an otherwise successful vaccine development process.

Dengvaxia® is a live attenuated tetravalent recombinant vaccine. It was designed by replacing genes encoding for envelope proteins of yellow fever virus with those coding for envelope proteins of dengue virus serotypes. The vaccine is recommended exclusively for those who have been exposed to a dengue infection at least once to avoid ADE-related adverse events [36]. Vaccines utilizing other approaches such as recombinant attenuated live virus and tetravalent vaccine-like particles are also being developed to control dengue fever [37, 38].

Chikungunya

Chikungunya is caused by the chikungunya virus (CHIKV), which is a mosquito-borne arthritogenic alphavirus. This enveloped RNA virus causes debilitating acute and chronic musculoskeletal disease all over the world.

Multiple vaccine candidates to prevent chikungunya are currently undergoing evaluation in clinical trials including a candidate vaccine utilizing a chikungunya virus-like particle (CHIKV VLP) developed by CEPI, which has shown significant neutralizing antibodies in a phase II clinical trial [39]. A French vaccine company called Valneva has reported encouraging results from a phase III clinical trial of a candidate live attenuated single-dose vaccine. The Valneva vaccine against chikungunya, Ixchiq, was approved in February 2024 by the US FDA based on a trial conducted in over 4000 adults. 98.9% of the subjects showed an antibody response capable of preventing infection after the vaccination was administered. Ixchiq contains a live, weakened version of the chikungunya-causing virus [40].

Zika

Zika virus (ZIKV), an arbovirus belonging to the Flaviviridae family, is a vector-borne virus that causes Zika fever. Although it is primarily spread by mosquitoes, it can also be transmitted by mixing bodily fluids. Transplacental transmission in pregnant women can result in congenital Zika syndrome, which manifests as fetal or infant microcephaly, other congenital malformations, and pregnancy complications like miscarriage. Some adults suffer from Guillain-Barre's syndrome following ZIKV infection.

Dozens of vaccines against ZIKV are being tested at preclinical stages. These vaccines employ a host of technological platforms including virus-like particle (VLP) technology as effective immunogens [41]. A couple of DNA vaccines in phase II clinical studies have demonstrated no adverse effects associated with concomitant Flaviviral infections [42].

Nipah

Nipah virus (NiV) and Hendra viruses (HeV) are members of the paramyxovirus family belonging to the genus *Henipavirus*. NiV can cause serious neurological and respiratory disease in humans and can cause death in 40–90% of infected individuals. The virus jumped species about 25 years ago from *Pteropus* bats (flying foxes) to horses and other animals including humans. Isolated cases of NiV have occurred in India, with a recent instance in 2018 that claimed the lives of 22 of the 23 infected individuals. Symptoms like encephalitis and meningitis may appear even up to 10 years after infection due to re-emergence of the virus in the central nervous system. The isolate from Bangladesh (NiV-B) and HeV is classified as a BSL-4 category C pathogen, requiring the highest biosafety level to be handled. Transmission from animals to humans and between humans has been recorded in the past decade.

Over ten studies have evaluated vaccines against NiV and HeV in animal models. The candidate vaccines employ vaccinia vector expressing NiV G and F proteins, VLP, mRNA, and recombinant protein subunits. The African green monkey model is widely accepted as an animal model that recapitulates the human NiV disease closely. Using this model, researchers have demonstrated the effectiveness of HeV soluble G protein (HeV-sG) recombinant subunit vaccine in inducing neutralizing antibodies. This helped controlled viremia and protected the animals from live challenge with a lethal dose of the NiV-B. This vaccine has been approved for use in horses and is undergoing tests for evaluation against NiV infection in a phase I clinical trial [43, 44].

Influenza: A Continuously Evolving Vaccine

Influenza is caused by Influenza A and B viruses. They are most commonly spread between people by respiratory droplets or fomite transmission. The genes encoding influenza virus surface proteins accumulate mutations which leads to the surface proteins undergoing antigenic drift. Moreover, the rearrangement of genes between human and animal influenza viruses causes strain A to undergo an antigenic shift resulting in completely new determinants on the surface proteins. Such phenomena decrease the effectiveness of pre-existing immunity—acquired either naturally or because of immunization—to combat a new strain of virus. This leads to the requirement of designing new vaccines each season to effectively prevent the infection. Approved vaccines used in developed nations are either a trivalent vaccine against two influenza A strains—H1N1 and H3N2 and one influenza B strain

(B-Victoria)—or a tetravalent vaccine that includes the two influenza A strains and two influenza B strains—Victoria and Yamagata.

These vaccines are tweaked for the Northern or Southern hemispheres based on data about circulating strains obtained from the WHO Global Influenza Surveillance Response System. Manufacturers develop variations of the following vaccines, and each formulation is tested in target groups for safety and efficacy [45]:

- Trivalent—inactivated and live attenuated vaccines (A Strains : H1N1 and H3N2; B Strain : Victoria)
- Quadrivalent—inactivated and live attenuated vaccines (A Strains : H1N1 and H3N2; B Strains : Victoria and Yamagata)

There are ongoing debates about the benefits of influenza vaccines in low- and middle-income countries given the cost of the vaccines. Interestingly, a recent retrospective study of about 75,000 subjects from over 50 healthcare centers revealed that influenza vaccine benefitted individuals who later went on to get infected with the SARS-CoV-2 virus [46, 47]. The researchers observed a significant reduction in medical complications like sepsis and deep vein thrombosis requiring emergency room or intensive care, suggesting potential benefits that should be explored.

HIV: mRNA Vaccines

Observations from prior vaccine trials provided insights about the effectiveness of certain antibodies in fighting against HIV. These antibodies, called broadly neutralizing antibodies (bnAbs), can bind to a site—the CD4 binding site—on the spiky protein envelope of the virus. This site is conserved among different variants of HIV since it is crucial for viral entry into host cells. A team at the Scripps Institute of Research discovered that they could induce the production of bnAbs by arranging several copies of an HIV glycoprotein critical for viral entry on the surface of a nanoparticle [48]. A phase I clinical trial for the vaccine resulted in exposure of participants' naïve B cell precursors to the protein-conjugated nanoparticle (called the eOD-GT8 60mer immunogen). The maturation of these B cells resulted in more antibody-producing cells, leading to the production of bnAbs by a vast array of immune cells [49].

The bnAbs are crucial to mount an effective immune response against HIV. A single bnAb is likely not effective in providing protection, hence requiring multiple bnAbs.

The eOD-GT8 60mer immunogen, a protein-based vaccine in a phase I clinical trial, found that almost all vaccine recipients did generate bnAb. Lack of any significant side effects suggested that the vaccine is safe. The next steps aim at transforming the eOD-GT8 60mer immunogen into an mRNA vaccine. This vaccine is undergoing a phase I clinical trial. Aside from this vaccine candidate, the teams at Scripps Research Institute and Moderna are testing three other mRNA vaccine candidates in a phase I clinical trial sponsored by the NIH. These candidates, similar to eOD-GT8, generate pieces of the HIV spike protein crucial for entry into host cells, but are genetically modified. Alternate versions that are being evaluated include a secreted kind, a version that is membrane bound, and another that has impaired binding to CD4 using a genetic deletion [50].

All four vaccine candidates present different versions of HIV envelope proteins; however, they are all mRNA-based. Once the genetic sequence of the target pathogen is known as was the case with SARS CoV2, an mRNA vaccine becomes a possibility. The mRNA vaccines offer advantages over protein-based vaccines in being easily customizable and faster to produce and amenable to multiple iterations.

2.2 Vaccines Against Bacterial Diseases

Bacterial vaccines are a critical component of public health, designed to protect against bacterial infections that can cause severe illness and death. These vaccines work by introducing killed or attenuated bacteria, or components of bacteria, into the body to stimulate an immune response without causing the disease. This process helps the immune system recognize and combat the bacteria if exposed in the future. Bacterial vaccines have been instrumental in controlling diseases such as tuberculosis, diphtheria, tetanus, and meningitis, significantly reducing morbidity and mortality rates worldwide.

Bacterial infections have long posed significant threats to human health due to their diverse array of virulent factors. The development and formulation of bacterial vaccines have made substantial progress in mitigating their impact on public health. Like viral vaccines, bacterial vaccines can be live attenuated or inactivated. Additionally, bacterial vaccines include toxoids and conjugates.

The oral polio vaccine, for example, is a live attenuated vaccine containing a weakened strain of the virus. The pertussis vaccine, on the other hand, is an inactivated vaccine. Current efforts are focused on utilizing different bacterial outer membrane components to develop subunit vaccine strategies.

Antimicrobial resistance, particularly in bacteria such as *Neisseria gonorrhoeae*, is an immediate concern [51]. As these bacteria become resistant to available antibiotics, vaccination remains the primary method of prevention. The tetanus toxoid vaccine targets bacterial toxins by inducing immunity against the harmful effects of the toxin. The *Haemophilus influenzae* type b (Hib) vaccine is a conjugate vaccine that combines bacterial polysaccharides with carrier proteins to enhance immunity [52].

Clostridium difficile is a bacterium that can cause severe diarrhea and more serious intestinal conditions, such as colitis. Rebyota, a medication approved by the FDA in 2022, is used to prevent recurrence of this infection. It is a fecal microbiota transplantation (FMT) product, where processed fecal material from a healthy adult donor is introduced into the patient's gastrointestinal tract [53].

Developing a vaccine against *Staphylococcus aureus* is particularly challenging due to the bacterium's ability to evade the immune system. *Staphylococcus aureus* Surface Antigen 2 (SA2Ag) is an engineered antigen derived from proteins on the bacterium's surface, representing a promising approach in this ongoing challenge [54].

2.2.1 Bacterial UTI (Urinary Tract Infections) Vaccines

Bacteria are a common cause of urinary tract infections (UTI), causing cystitis, prostatitis, and pyelonephritis. While patients rely on antibiotics to resolve UTIs, a potential alternative to these is a vaccine, especially in people who suffer from recurrent UTIs. Uromune, a bacterial vaccine also known as MV140, developed by Immunotek contains a suspension of whole-heat inactivated bacteria that cause recurrent UTIs. These are—*Escherichia coli*, *Klebsiella pneumoniae*, *Proteus vulgaris*, and *Enterococcus faecalis*.

MV140, the orally administered vaccine, induces immune responses in the systemic and genitourinary tracts. Studies have also shown that this vaccine can reduce the frequency, duration, and severity of UTIs. According to recent clinical trial data, 54% of the vaccinated participants remained free of UTIs for up to 9 years with no adverse effects [55, 56]. Some clinical studies have found that MV140 can reduce the total number of UTIs by 70%, increase the time to next UTI from 46 to 275 days, and reduce the symptoms, antibiotic use, and improve quality of life.

2.2.2 Leprosy Vaccine

Another bacterial infection that has affected humans since ancient times is leprosy, a chronic infectious disease caused by the bacterium *Mycobacterium leprae*. The BCG vaccine was found to be effective against leprosy in the 1900s [1, 4]. Several studies conducted in the 1960s showed that this vaccine offered limited protection against leprosy. Aside from socioeconomic issues, the technical challenges in culturing *Mycobacterium leprae* hindered leprosy research, and developing a successful and effective vaccination remained difficult [57]. The exact nature of immune responses responsible for the spectrum of symptoms of leprosy remains poorly understood. In the late 1990s and early 2000s, G. P. Talwar and team from India showed that the bacterium *Mycobacterium indicus pranii*, earlier known as *Mw*, has immunotherapeutic applications [58]. The *Mycobacterium indicus pranii* (MIP) vaccine is an immunomodulatory vaccine developed by Talwar for use in leprosy [5]. MIP is heat-killed *Mycobacterium w*, which is a non-pathogenic atypical mycobacterium belonging to Class IV of Runyon classification. MIP activates both innate and acquired immunity. It induces a Th1 and Th17 immune response while downregulating Th2 response. It shares epitopes with *Mycobacterium leprae* and *Mycobacterium tuberculosis* encouraging its use in leprosy and tuberculosis. MIP vaccine is safe with limited adverse effects including local site erythema, swelling, and occasionally fever and other systemic reactions. The MIP vaccine has received approval of the Drugs Controller General of India and the US FDA. The vaccination showed a protective efficacy of over 68% for 3 years and 39% after 9 years in people in contact with leprosy patients [59]. The National Leprosy Eradication Program has introduced MIP vaccine in India from the year 2016 in five endemic districts [58].

More recently a recombinant vaccine linking four *M. leprae* antigens: ML2531, ML2380, ML2055, and ML2028 (LEP-F1) is being investigated in mice. Mice immunized with Lepvax when challenged with *M. leprae* had reduced levels of the disease-causing bacteria [60] (Table 3).

2.3 Vaccines Against Parasitic Diseases

Protozoa and helminths have complex life cycles and immunomodulatory strategies. These present challenges to develop vaccines against them. Like viral and bacterial vaccines, parasitic vaccines have different types. The attenuated sporozoite malaria vaccine is a live attenuated vaccine. It contains a weakened form of the parasite. RTS,S malaria vaccine is a subunit vaccine that contains purified proteins, key components involved in targeting parasitic invasion. The Leishmania vaccine is a recombinant vaccine which uses genetic engineering technique to produce recombinant antigens derived from the parasite. The recombinant vaccines help focus the target without utilizing additional junk and reducing adverse effects. For parasitic

Table 3 FDA-approved bacterial vaccines (1970 onward)

Vaccine	Disease	Type	Year approved	References
DTaP (Daptacel, Infanrix)	Diphtheria, tetanus, pertussis	Inactivated toxoid and subunit	2002 (Daptacel), 1997 (Infanrix)	[61]
Tdap (Boostrix, Adacel)	Tetanus, diphtheria, pertussis	Inactivated toxoid and subunit	2005 (Boostrix), 2005 (Adacel)	[62]
Pneumovax 23	Pneumococcal disease	Polysaccharide	1983	[63]
Prevnar 13	Pneumococcal disease	Conjugate	2010	[64]
Menactra	Meningococcal disease	Conjugate	2005	[65, 66]
Menveo	Meningococcal disease	Conjugate	2010	[67]
Hib (PedvaxHIB, ActHIB)	*Haemophilus influenzae* type b	Conjugate	1989 (PedvaxHIB), 1993 (ActHIB)	[68]
BCG (Tice BCG)	Tuberculosis	Live attenuated	1991	[69]
Typhim Vi	Typhoid fever	Polysaccharide	1994	[70]
Vaxchora	Cholera	Live attenuated	2016	[71]
Biothrax (Anthrax Vaccine Adsorbed)	Anthrax	Inactivated	1970	[72, 73]

vaccines, adjuvants help enhance immune response to the antigen and provide immunity boost.

The proper understanding of intricate mechanisms by which vaccine confer immunity against the pathogen is essential. Vaccines generate immune response by presenting the antigens (which can be surface proteins, toxins, or other components of the target pathogen) to the cells of immunity such as T cells, B cells, and antigen-presenting cells. The process begins with the uptake of antigens by the innate immune cells, processing it, and presenting it to the cells of adaptive immunity which act by producing antibodies or activating cytotoxic T cells. This domino effect works to neutralize or kill the invading pathogen. Vaccines induce immunological memory, which means the immune system retains a ready-to-act heightened responsiveness upon re-exposure to the pathogen. By manipulating the antigen selection, vaccine formulation, and delivery methods, researchers optimize the efficacy, safety, and longevity of the vaccine-induced immunity, paving way for the development of next-generation vaccines, which would fight infections with great potency and precision.

2.3.1 Malaria Vaccine

Malaria is caused by protozoan parasites and has affected humans since ancient times. It is spread through the *Anopheles* mosquito, which is a major target to control the spread of the disease. The disease carries a significant socio-economic burden on societies where it is endemic. Over 220 million cases were estimated in 2019 from 89 different countries. The quest to develop a vaccine against malarial parasites for the past several decades has resulted in about 20 different vaccine candidates which are currently at various stages of clinical evaluation.

A WHO panel approved the first malaria vaccine in 2021 to reduce *P. falciparum*-induced clinical events in children in Africa. This vaccine, called Mosquirix®, requires three to four doses [74]. It is only about 40% effective against the disease, yet according to malaria researchers, its approval is promising in terms of obtaining prophylactic vaccines against parasitic diseases [75].

3 Advances in Vaccine Development

With time, significant advancements have been made in the field of vaccine formulation and development. The advances range from identifying the antigen, applying formulation strategies, and conducting preclinical and clinical testing. The use of adjuvants and carrier proteins has revolutionized the design of vaccines and helped to boost the immune response. Here we will be emphasizing on these aspects of vaccine development.

3.1 Antigen Selection and Development

The first and the most crucial step in vaccine development is antigen selection. Antigens can vary from whole cells of the pathogen to proteins, toxins, or nucleic acids. They differ in their virulence, and which determines their efficacy and specificity. The ever-evolving research advances have enabled applying molecular biology, protein chemistry, and genomic approaches to identify and characterize antigen targets, enabling the design of potent vaccine candidates. With the selection of antigens, depending on their virulence factor there may be a need for adjuvants or carrier proteins to enhance or boost the immune response. Moreover, recently, peptide-based vaccine development involves modifying the antigen by conjugating it with components that would help in targeting the antigen to right receptors.

Considering the most recent example of COVID-19 vaccine development, one approach to antigen selection involves the use of mRNA technology utilizing mRNA derived from the target pathogen. This approach provided precise control over antigen components, thereby reducing cellular junk avoiding unwanted adverse effects. Another strategy involved the use of spike proteins derived from coronavirus. The subunits helped focus on the antigenic determinants while reducing the risk of adverse effects.

3.2 Preclinical and Clinical Research

Preclinical testing of the vaccine candidates helps understand the safety, efficacy, and immunogenicity of the vaccine candidates before testing them in humans. For preclinical testing, animal models such as mice, rats, and non-human primates are used to assess the vaccine-induced immunity and evaluate the resistance to infection by challenge with the target pathogen. There are different parameters that need to be assessed such as the duration and magnitude of the immune response, the immunological memory and resistance to infection, and neutralizing ability. Moreover, safety is evaluated by understanding the adverse reactions—systemic and local upon vaccination [76].

Once the best preclinical candidate is chosen, it then advances to clinical evaluation, where they are tested in human subjects to assess immunogenicity, safety, and efficacy. Clinical trials are conducted in four phases. Phase I trial evaluates the safety and efficacy in a small population of healthy volunteers, followed by phase II trials where they assess the immunogenicity and dose-response in a relatively larger participant population. Phase III involves large efficacy trials to evaluate the effectiveness in preventing infection in the target population. This phase involves testing in thousands of people and includes study designs such as placebo, randomized, and double blind to minimize bias and ensure robustness of results. Phase IV is the post-marketing testing and observing the effectiveness of vaccine in the approved population [77].

3.3 Storage and Handling

Vaccine storage is a critical consideration as it would impact the vaccine stability, potency, and effectiveness. Proper vaccine storage includes appropriate temperature, humidity, and light, as that helps maintain the integrity of the components through their shelf life. Many vaccines need 2–8 °C of storage temperature. However, certain vaccines involving live pathogens attenuated require −20 °C to preserve the viability and efficacy. To ensure proper cold chain maintenance, vaccines are transported and shipped in refrigeration units or cold boxes with monitoring devices. The availability of cold chain for storage and transport is an important concern in developing and underdeveloped countries [78]. Thus, recent vaccine efforts are under way to develop candidates that do not need cold chain for storage. The advancements made in development of thermostable vaccines help in the accessibility and availability of the vaccine reducing dependence on cold chain infrastructure enhancing vaccine distribution and delivery in remote areas which are hard to reach.

3.4 Vaccine Delivery Systems

Vaccine routes refer to the pathways through which vaccines are administered to induce immune responses in the body. The choice of vaccine route depends on factors such as the characteristics of the vaccine, target population, desired immune response, and practical considerations. Here are some common vaccine routes.

3.4.1 Routes of Administration

Intramuscular (IM) Injection

Intramuscular injection involves delivering the vaccine into the muscle tissue, typically the deltoid muscle in the upper arm or the vastus lateralis muscle in the thigh. IM injection is one of the most common routes for vaccine administration, as it allows for rapid absorption and distribution of the vaccine antigen. Vaccines administered via IM injection include those against diseases such as influenza, measles, mumps, rubella, and tetanus [79].

Subcutaneous (SC) Injection

Subcutaneous injection involves delivering the vaccine into the tissue layer between the skin and the muscle. SC injection is commonly used for vaccines that require slower absorption and prolonged release of antigen, such as certain inactivated vaccines. Vaccines administered via SC injection include those against diseases such as hepatitis B, rabies, and varicella (chickenpox) [80].

Intradermal (ID) Injection

Intradermal injection involves delivering the vaccine into the dermal layer of the skin, just below the epidermis. ID injection is used for vaccines that require a small dose and induce strong immune responses with minimal reactogenicity. Vaccines administered via ID injection include those against diseases such as tuberculosis (BCG vaccine) and influenza (some formulations) [81].

Oral Administration

Oral administration is typically in the form of liquid drops, tablets, or capsules. Oral vaccines stimulate mucosal immunity in the gastrointestinal tract, providing protection against pathogens that enter the body through mucosal surfaces. Vaccines administered orally include those against diseases such as polio (oral polio vaccine), cholera, and rotavirus [82]. The oral polio vaccine (OPV) is a prime example of an orally administered vaccine. OPV contains weakened strains of the poliovirus and is typically administered as drops directly into the mouth. This delivery method stimulates immune responses in the gastrointestinal tract, providing protection against poliovirus infection. OPV has played a crucial role in global polio eradication efforts, particularly in regions where access to healthcare infrastructure is limited [83].

Intranasal Administration

Intranasal administration involves delivering the vaccine into the nasal passages, where it is absorbed by the mucosal tissues. Intranasal vaccines stimulate both systemic and mucosal immune responses, providing protection against respiratory pathogens. Vaccines administered intranasally include those against diseases such as influenza (intranasal flu vaccine) and COVID-19 (some experimental vaccines) [84].

Inhaled influenza vaccines, such as Flu Mist, offer an alternative delivery method to traditional injections. These vaccines are administered via nasal spray and deliver weakened or inactivated influenza viruses directly to the respiratory tract. Inhaled influenza vaccines stimulate mucosal immunity, providing protection against influenza infection. Flu Mist is approved for use in individuals aged 2–49 years and is particularly popular among children and individuals who prefer needle-free vaccination options [85]. Mucosal immunology is increasingly gaining attention as an area of great potential for the development of vaccines and immunotherapy. The mucosal immune system is even more complex than its systemic counterpart, both in terms of effectors and anatomy. It is instructive to divide the various tissue compartments involved in mucosal immunity into inductive sites and effector sites according to their main function. Inductive sites comprise lymphoid tissue, in which the triggering of naïve immune cells and the generation of memory–effector cells take place. A Phase II mechanistic clinical trial is under way to assess the systemic and mucosal immunogenicity of the multicomponent meningococcal serogroup B vaccine (4CMenB or Bexsero (R)) (group 1, 40 subjects) against *Neisseria gonorrhoeae*, using a placebo vaccine (normal saline) as a comparator (group 2, 10 subjects). There will be 50 participants, ages 18–49, both male and non-pregnant female subjects, enrolled at 1 site in the United States. The goal will be to ensure adequate representation of subjects by sex in both treatment groups. The enrollment will be stratified by both sex and treatment arm. The rectal mucosal biopsy cohort ("biopsy cohort," $N = 20$) will be enrolled first to address the primary objective of the study. During enrollment of the \"biopsy cohort," male and non-pregnant female subjects will be randomized 4:1 to either 4CMenB or placebo, up to a maximum of 10 male and 10 non-pregnant female subjects. Group 1 ($N = 40$) will receive two doses of 4CMenB on Day 1 and Day 29. Group 2 ($N = 10$) will receive two placebo injections on Day 1 and Day 29. Both groups will receive a single-dose prefilled syringe that is administered intramuscularly (0.5-milliliter each). The duration of each subject's participation is approximately 8 months, from recruitment to the last study visit, and the length of the study is estimated to be 14 months. The primary objective is to characterize the rectal mucosal Immunoglobulin G, IgG antibody response to *Neisseria gonorrhoeae* (GC) elicited by the 4CMenB vaccine as compared with the placebo vaccine (normal saline) in healthy adult subjects [86, 87].

Inhalable vaccines deliver antigens directly to the respiratory tract, where they stimulate immune responses in the lungs and airways. This delivery method is particularly relevant for vaccines targeting respiratory infections such as influenza and tuberculosis. Inhalable vaccines can elicit both mucosal and systemic immune responses, providing a comprehensive defense against respiratory pathogens [88].

Transdermal Administration

Transdermal administration involves delivering the vaccine through the skin using patches or microneedle arrays. Transdermal vaccines bypass the need for injections and can be self-administered, providing a convenient and painless alternative to traditional injection routes. Transdermal vaccination is being explored for a variety of vaccine candidates, including those against influenza, measles, and COVID-19 [89–91].

Intravenous (IV) Injection

Intravenous injection involves delivering the vaccine directly into the bloodstream through a vein [92]. IV injection is less commonly used for routine vaccination and is typically reserved for special circumstances, such as post-exposure prophylaxis or certain clinical trials. Vaccines administered via IV injection may include those against diseases such as rabies (post-exposure prophylaxis) or experimental vaccines in clinical research settings. Each vaccine route has its advantages and limitations, and the selection of the appropriate route depends on factors such as vaccine characteristics, target population, desired immune response, and practical considerations related to administration and logistics. By leveraging a variety of vaccine routes, healthcare providers can optimize vaccine delivery to maximize immunogenicity and protect against a wide range of infectious diseases. Vaccine delivery systems are important components of the health infrastructure. They ensure that vaccines reach their intended recipients safely and efficiently. From conventional methods like needle and syringe injections to novel technologies such as microneedle patches, these systems play a vital role in immunization programs all over the world.

3.4.2   Vaccine delivery devices

Needle and Syringe

The needle and syringe delivery method is the most conventional and widely used system for administering vaccines. An example is the seasonal influenza vaccination campaign. Healthcare professionals use sterile needles and syringes to inject vaccines into a patient's muscle tissue or into the skin layers. This method allows for precise dosage control and can accommodate a wide range of vaccines, from those requiring small doses to larger volumes. Seasonal influenza vaccination campaigns worldwide often rely on the needle and syringe delivery method. Healthcare professionals administer the influenza vaccine to individuals of all ages, typically targeting high-risk groups such as the elderly, young children, and individuals with underlying health conditions.

Auto-Disable Syringes

Auto-disable syringes are designed for single use, and they automatically become disabled after delivering a vaccine dose. This feature prevents syringe reuse, reducing the risk of needlestick injuries and the transmission of infections. Auto-disable syringes are particularly valuable in resource-limited settings where safe injection practices may be challenging to enforce.

Example: Immunization Programs in Developing Countries
Immunization programs in developing countries often utilize auto-disable syringes to enhance injection safety and prevent the transmission of bloodborne pathogens. Organizations such as UNICEF and the World Health Organization (WHO) support the procurement and distribution of auto-disable syringes to ensure safe immunization practices in resource-limited settings and appropriate disposal practices. These syringes are commonly used in mass vaccination campaigns targeting diseases such as measles, polio, and tetanus.

Microneedle Patches

Microneedle patches are an innovative vaccine delivery platform that involves applying a patch containing tiny, painless needles to the skin's surface. These needles, which are typically less than a millimeter in length, dissolve upon contact with the skin, delivering the vaccine into the underlying tissue. Microneedle patches offer several advantages over traditional injection methods, including ease of administration and the potential for self-administration.

Example: COVID-19 Vaccination Using Microneedle Patches
Microneedle patches have garnered significant interest for delivering COVID-19 vaccines due to their potential advantages, including ease of administration, reduced reliance on healthcare personnel, and simplified logistics. Several research groups and pharmaceutical companies are developing microneedle patch-based COVID-19 vaccines, which could revolutionize vaccination efforts by enabling self-administration and facilitating mass immunization campaigns [91, 93, 94].

Jet Injectors

Jet injectors use high-pressure streams of liquid to penetrate the skin and deliver vaccines without the need for needles. These devices can administer vaccines rapidly and efficiently, making them well-suited for mass vaccination campaigns. Jet injectors can be particularly valuable in settings where trained healthcare personnel are scarce or where needle reuse poses a risk [95].
Jet injectors have been deployed in various mass vaccination campaigns worldwide to deliver vaccines rapidly and efficiently. For example, during outbreaks of infectious diseases such as measles, jet injectors are used to immunize large populations quickly, particularly in emergency response situations. Organizations like Médecins Sans Frontières (Doctors Without Borders) have utilized jet injectors in humanitarian settings to reach displaced populations and communities affected by natural disasters.

4   Types of Vaccines

The first ever vaccine that was developed to treat smallpox in 1798 was a live attenuated vaccine, a weakened or "attenuated" form of the disease-causing pathogen that is capable of eliciting a robust immune response when administered. Following the development of the live attenuated vaccine platform, several other vaccine technology platforms have been developed. The live attenuated rabies vaccine was developed in the nineteenth century (1885) which was followed by the introduction of the killed whole organism vaccines developed for vaccination against Typhoid (1896), cholera (1896), and Plague (1897). The early twentieth century saw the development of the live attenuated BCG vaccine (1927), inactivated influenza vaccine (1936), and the introduction of purified protein or polysaccharide-based vaccines for diphtheria (1923) and tetanus (1926). More vaccines were developed in the second half of the twentieth century, notably the oral polio vaccine (1963), MMR vaccine (1963–1969), Varicella (1995), and the introduction of genetically engineered or recombinant vaccines. The hepatitis B surface antigen recombinant vaccine was introduced in 1986, and the cholera (recombinant toxin B) followed in 1993. In 2020, the rapidly-spreading, devastating COVID-19 pandemic required deployment of an accelerated vaccine development platform. The SARS COV2 vaccine became the first successful mRNA vaccine paving the way for the utilization of this versatile platform for other infectious diseases [96]. This section will delve into the details of the existing vaccine technology platforms (Table 4), their development,

Table 4 Types of vaccines

Vaccine formulation	Description
mRNA vaccines	Uses messenger RNA to instruct cells to produce a protein that triggers a response
DNA vaccines	Utilizes plasmid DNA to encode the antigen, which is then expressed in host cells
Viral vector vaccines	Uses a modified virus to deliver genetic material coding for an antigen into host cells
Virus-like particles	Mimics the structure of viruses but lacks viral genetic material, inducing a response
Recombinant protein	Uses genetically engineered proteins to stimulate an immune response
Peptide vaccines	Uses specific peptides (short chains of amino acids) that mimic epitopes of antigens
Liposome based	Uses lipid-based vesicles to encapsulate antigens, enhancing delivery and stability
Nanoparticle vaccines	Utilizes nanoparticles to improve antigen delivery and enhance immune responses
Conjugate vaccines	Links polysaccharides to proteins to enhance immunogenicity, especially in infants
Adjuvant systems	Combines different adjuvants to boost and modulate the immune response to vaccines

and how they generate an immune response when administered with examples cited for each type (3), (1), [97–99].

4.1 Live Attenuated/Weakened Vaccines

A live attenuated vaccine is the weakened form of the virulent organism which can generate an immune response against the same when administered. Typically, the virulent pathogen is weakened by repeated culturing in the laboratory to produce a non-virulent strain. For example, the measles virus used as a vaccine today was isolated from a child with measles disease in 1954. Almost 10 years of serial passage using tissue culture media were required to transform the wild virus into the attenuated vaccine virus. These vaccines produce an immune response by replicating in the host cell. Although these vaccines replicate in the host, they do not cause the disease as the wild type of organism. The immune response generated by the live attenuated vaccine is identical to that generated by natural infection as the host immune system cannot differentiate between the attenuated vaccine versus the wild type of organism. This is advantageous to produce a long-lasting immune memory. However, a small percentage of recipients do not respond to the first dose of an injected live, attenuated vaccine (such as measles, mumps, and rubella [MMR]) and a second dose is recommended to provide an extremely high level of immunity in the population. Orally administered live, attenuated vaccines require more than one dose to produce immunity. The live, attenuated viral vaccines currently available and routinely recommended in the United States are MMR, varicella, rotavirus, and influenza (intranasal). Other non-routinely recommended live vaccines include adenovirus vaccine (used by the military), typhoid vaccine (Ty21a), and Bacille Calmette-Guerin (BCG). These vaccines can possibly revert to their original pathogenic form and result in a fatal infection. People with a weak immune system are susceptible to infection upon live attenuated vaccine administration. For these reasons, the live attenuated vaccine platform is not widely preferred for vaccination with the exception of those vaccines mentioned above. Moreover, live attenuated vaccines require stringent storage conditions as they can easily be destroyed by light or heat [100].

4.2 Killed/Inactivated Vaccines

The late 1800s saw the development of inactivated whole-cell vaccines. In this type, the disease-causing pathogen was treated with gamma irradiation or chemical agents such as formaldehyde to kill the pathogen. Such treatments result in the killing of the pathogen and render it non-infectious but preserves the antigenic epitopes on the surface. Inactivated vaccines do not elicit an immune response as strong as the live attenuated vaccines as they are incapable of replication. Therefore, multiple doses may be required to generate a robust immune response in the host. In general, the first dose does not produce protective immunity, but "primes" the immune system. A protective immune response is developed after the second or third dose. The type of immune response generated is predominantly systemic humoral responses. Neutralizing antibodies against the pathogen are generated via the MHC Class II presentation pathway. Since the killed organism is extracellular, the antigen presentation occurs via phagocytosis and the MHC II pathway with little to no cross-presentation via the MHC I pathway. Therefore, inactivated vaccines typically induce very less cytotoxic T-cell responses. Antibody titers against inactivated antigens diminish with time. As a result, some inactivated vaccines may require periodic supplemental doses to increase, or "boost," antibody titers. These vaccines are more stable than the live attenuated vaccines but also require cold chain storage [6, 101].

4.3   Subunit Vaccine (Proteins, Peptides, and Carbohydrates)

Subunit vaccines incorporate only the antigenic portion of the pathogen that are known to best simulate the immune response. As these vaccines do not contain any live components, the immune response elicited by the vaccine is not as strong. Typically, adjuvants are used to enhance the immune response of subunit vaccines. Subunit vaccines to prevent bacterial infections utilize polysaccharides that form the outer coating of the bacteria. A polysaccharide vaccine by itself is not capable of eliciting a strong immune response as was previously observed in the first licensed vaccine against *Haemophilus influenzae* type B (Hib) which failed to produce a strong immune response in infants. This led to the development of a conjugate vaccine wherein the polysaccharide was linked to a protein antigen for improved protection. Conjugating a polysaccharide antigen to a protein molecule produces long-lasting protective immunity to the polysaccharide antigen. The immune response to a pure polysaccharide vaccine is typically T-cell-independent, which means these vaccines can stimulate B-cells without the assistance of T-helper cells. T-cell-independent antigens, including polysaccharide vaccines, are not consistently immunogenic in children younger than age 2 years, probably because of immaturity of the immune system. Attaching the polysaccharide antigen to a protein makes it possible to prevent bacterial infections in populations where a polysaccharide vaccine is not effective or provides only temporary protection. Toxins secreted by bacteria are also inactivated as used in vaccines which are known as toxoids [102]. These protein-based toxins are inactivated using heat, chemicals, or other methods. Some bacteria (e.g., tetanus, diphtheria) cause disease by producing toxins. The ability of the immune system to recognize and eliminate these toxins provides protection from the disease. Diphtheria and tetanus vaccines are toxoid vaccines. In case of the COVID-19 subunit vaccines, the spike protein or the receptor-binding domain (RBD) of the spike protein is used as the antigenic targets. The COVID-19 vaccine developed by Novavax is a protein subunit vaccine. Subunit vaccines are generally safe and have better storage conditions as compared to live attenuated vaccines [103].

4.4 Recombinant Vaccines

Recombinant vaccines refer to vaccine antigens that are produced in an expression system such as bacteria or yeast. Recently, mammalian cells are also being used as expression systems. Typically, genetic material that encodes for a desirable protein antigen is introduced into producer cells in culture. These producer cells serve as a bioreactor for the vaccine antigen production. Two examples of successful recombinant protein vaccines currently used by humans are the vaccines against hepatitis B and human papillomaviruses. Recombivax HB vaccine, was the first vaccine produced using recombinant DNA technology and approved for clinical use in 1986. Frazer and colleagues used the then relatively new technology of expressing genes in cell culture to create virus-like particles (VLPs) of HPV16, a key cancer-causing high-risk HPV type. Then in 2006, the first HPV vaccine (Gardasil), containing VLPs of four HPV types—high-risk HPV16 and HPV18 (which cause ~70% of cervical cancers) as well as low-risk HPV6 and HPV11 (which cause genital warts)—was approved for use in the United States in adolescent girls. This was followed shortly after by the approval of Gardasil or another VLP-based vaccine against HPV16 and HPV18 (Cervarix) in many other countries. In 2012, the US FDA approved the first mammalian cell-based influenza vaccine, Flucelvax. The vaccine was manufactured by Novartis using the Marlin-Darby canine kidney cell line [104].

4.5 Virus-Like Particles and Nanoparticle-Based Vaccines

Virus-like particles resemble the viral structure but are not capable of causing infection due to the absence of viral genetic material. VLPs mimic the viral structure thereby triggering an immune response in the host without producing any symptoms related to the disease. A VLP is typically constructed with one or more structural proteins that can be arranged in multiple layers. VLPs are also smaller in size and range between 20 and 200 nm, which means they can easily drain via the lymph nodes and activate an immune response. Their small size is also advantageous for antigen presentation as they can be easily recognized and engulfed by the antigen-presenting cells (APCs). The antigen presented via the APCs then triggers a cellular immune response which is desirable for vaccines. The core (HBc) and surface antigen (HBsAg) are the structural proteins of hepatitis B virus that were some of the earliest VLPs to be produced in heterologous expression systems. This led to the development of the first recombinant human vaccine against HBV in 1986, utilizing surface antigens. The human papillomavirus vaccine, commonly known as the cervical cancer vaccine, is administered to girls and boys aged 12–13 and is another example of a VLP vaccine. In the HPV vaccine, VLPs are composed of the main capsid protein, L1, which is the protein located on the surface of the virus. Nanoparticle-based VLP vaccines have the potential to eliminate the need for injections and could potentially be administered through intranasal vaccines or inhalers [105, 106].

4.6 Viral Vector Vaccines

Viral vectors were initially created for vaccine purposes nearly four decades ago, using vaccinia virus (VACV) as a vector to express the hepatitis B surface antigen HBsAg. When tested in chimpanzees for protective immunity, this vector proved effective in safeguarding the animals against hepatitis B infection. Unlike most traditional vaccines, viral vector-based vaccines do not contain antigens themselves, but instead utilize the body's own cells to produce them. This is achieved by employing a modified virus (the vector) to deliver genetic code for the antigen. The use of viral vectors results in a more robust immune response involving both B cells and T cells. However, prior exposure to the vector may diminish the vaccine's effectiveness. The viral vector serves as a delivery system, allowing for the invasion of the cell and the insertion of genetic code for the antigens of a different virus (the pathogen being targeted for vaccination). The virus itself is harmless, and by instructing the cells to only produce antigens, the body can mount an immune response safely, without developing the disease. Various viruses have been developed as vectors, including adenovirus (a cause of the common cold), measles virus, and vaccinia virus. These vectors have been stripped of any disease-causing genes and sometimes genes that enable them to replicate, rendering them harmless. The genetic instructions for producing the antigen from the target pathogen are integrated into the genome of the virus vector.

There are two primary categories of viral vector-based vaccines. Non-replicating vector vaccines are incapable of generating new viral particles; they solely produce the vaccine antigen. Replicating vector vaccines also generate new viral particles in the cells they infect, which then proceed to infect new cells that will also produce the vaccine antigen. Once introduced into the body, these vaccine viruses start infecting our cells and inserting their genetic material—including the antigen gene—into the cells' nuclei. Human cells produce the antigen as if it were one of their own proteins, and this is displayed on their surface alongside many other proteins. When the immune cells identify the foreign antigen, they initiate an immune response against it. One challenge of this approach is that individuals may have previously encountered the virus vector and develop an immune response against it, diminishing the effectiveness of the vaccine. Such "anti-vector immunity" also complicates the delivery of a second dose of the vaccine, assuming this is necessary, unless this second dose is administered using a different virus vector. A significant obstacle for viral vector vaccine production is scalability. Traditionally, viral vectors are cultivated in cells that are attached to a substrate, rather than in free-floating cells—but this is challenging to do on a large scale. Suspension cell lines are currently being developed, which would allow viral vectors to be cultivated in large bioreactors. Assembling the vector vaccine is also an intricate process, involving multiple steps and components, each of which increases the risk of contamination. Extensive testing is therefore necessary after every step, leading to increased costs. The Oxford-AstraZeneca vaccine utilizes the modified chimpanzee adenovirus ChAdOx1. Sputnik V uses human adenovirus serotype 26 for the first shot, and serotype 5 for the second [102, 107, 108].

4.7   Nucleic Acid Vaccines (DNA, mRNA)

Nucleic acid vaccines offer a safe, effective, and cost-effective approach. Additionally, these vaccines elicit immune responses that specifically target the chosen antigen in the pathogen. Nucleic acid-based vaccines, which include DNA (as plasmids) and RNA (as messenger RNA), show great promise in addressing a variety of conditions and illnesses. These vaccines utilize genetic material from a disease-causing pathogen to trigger an immune response against it. The genetic material, whether DNA or RNA, contains instructions for producing a particular protein from the pathogen, which the immune system recognizes as foreign. Once introduced into host cells, this genetic material is utilized by the cell's protein-making machinery to generate antigens that stimulate an immune response. In the case of DNA vaccines, a DNA segment encoding the antigen is initially inserted into a bacterial plasmid. Plasmids, circular DNA pieces used by bacteria to store and exchange beneficial genes, can replicate independently of the main chromosomal DNA and serve as a simple means of transferring genes between cells. DNA plasmids carrying the antigen are typically administered into the muscle, although a major challenge lies in facilitating their entry into human cells. This step is crucial because the machinery responsible for translating the antigen into protein is located within cells. Various technologies are under development to assist in this process, such as electroporation, which involves using brief electric current pulses to create temporary pores in patients' cell membranes; a "gene gun" that employs helium to propel DNA into skin cells; and encapsulating DNA in nanoparticles designed to merge with the cell membrane.

RNA vaccines contain the antigen of interest in messenger RNA (mRNA) or self-amplifying RNA (saRNA), which are utilized by cellular factories to generate proteins. Due to its temporary nature, there is no possibility of integration with our genetic material. The RNA can be administered alone, enclosed within nanoparticles (such as Pfizer's mRNA-based COVID vaccine), or transported into cells using similar methods used for DNA vaccines. Once inside the cell, the DNA or RNA initiates antigen production, which is then presented on the cell's surface for detection by the immune system, leading to a response that involves killer T cells, antibody-producing B cells, and helper T cells [109–111].

Recent advancements in vaccine development have resulted in several FDA approvals. Table 5 highlights the list of such FDA-approved vaccines. Tables 6 and 7 discuss different carrier proteins used in vaccine formulations and adjuvants used in vaccines, respectively.

Table 5 List of FDA-approved vaccines from 2014 to 2024 [112]

Vaccine	Manufacturer	Indication	Date of approval/ authorization
Gardasil 9 (9-valent HPV vaccine)	Merck	Human papilloma virus	December 10, 2014
Fluzone quadrivalent	Sanofi Pasteur limited	Influenza	December 11, 2014
Bexsero	GlaxoSmithKline	Serogroup B meningococcal disease	January 23, 2025
Quadracel	Sanofi Pasteur limited	Diphtheria, tetanus, pertussis and poliomyelitis	March 24, 2015
Fluad (adjuvanted, trivalent)	Seqirus	Influenza	November 24, 2015
Vaxchora	Bavarian Nordic A/S	Cholera	June 10, 2016
Shingrix	GlaxoSmithKline	Herpes zoster (shingles)	October 20, 2017
Heplisav-B	Dynavax technologies corporation	Hepatitis B	November 9, 2017
Vaxelis	MSP vaccine company	Diphtheria, tetanus, pertussis, poliomyelitis, hepatitis B, and invasive disease due to *Haemophilus influenzae* type b (Hib) in children 6 weeks through 4 years of age	December 21, 2018
Fluzone high-dose quadrivalent	Sanofi Pasteur Limited	Influenza (for >65 adults)	November 4, 2019
Ervebo (Ebola Zaire vaccine, live)	Merck	Ebola virus disease	December 19, 2019
Fluad quadrivalent (adjuvanted)	Seqirus	Influenza (for >65 adults)	February 21, 2020
MenQuadfi conjugate vaccine	Sanofi Pasteur, Inc.	Invasive meningococcal disease caused by *Neisseria meningitidis* serogroups A, C, W, and Y	April 23, 2020
Pfizer-BioNTech COVID-19 vaccine (EUA)	Pfizer-BioNTech	COVID-19	December 11, 2020
Moderna COVID-19 vaccine (EUA)	Moderna	COVID-19	December 18, 2020 2021
Janssen COVID-19 vaccine	Janssen (Johnson & Johnson)	COVID-19	February 27, 2021
Vaxneuvance (pneumococcal 15-valent conjugate vaccine)	Merck & Co	Invasive disease caused by *Streptococcus pneumoniae* serotypes 1, 3, 4, 5, 6A, 6B, 7F, 9 V, 14, 18C, 19A, 19F, 22F, 23F, and 33F	July 16, 2021

(continued)

Table 5 (continued)

Vaccine	Manufacturer	Indication	Date of approval/ authorization
Ticovac	Pfizer	Tick-borne encephalitis (TBE)	August 13, 2021
Comirnaty (first COVID-19 vaccine approved for use)	Pfizer-BioNTech	COVID-19	August 23, 2021
PreHevbrio	VBI Vaccines (Delaware) Inc	Hepatitis B	November 30, 2021
Spikevax	Moderna	COVID-19	January 31, 2022
Priorix (live)	GlaxoSmithKline	Measles, mumps, and rubella	June 3, 2022
Boostrix	GlaxoSmithKline	First vaccine specifically for use during the third trimester of pregnancy to prevent pertussis (whooping cough) in infants younger than 2 months old	October 7, 2022
Prevnar 20 (pneumococcal 20-valent conjugate vaccine)	Wyeth Pharmaceuticals LLC	Invasive disease caused by *Streptococcus pneumoniae* serotypes 1, 3, 4, 5, 6A, 6B, 7F, 8, 9 V, 10A, 11A, 12F, 14, 15B, 18C, 19A, 19F, 22F, 23F, and 33F	April 27, 2023
Arexvy (first RSV vaccine)	GlaxoSmithKline	Respiratory syncytial virus (>60 and older)	May 3, 2023
Abrysvo	Pfizer	Respiratory syncytial virus (>60 and older)	May 31, 2023
Cyfendus (anthrax vaccine adsorbed, adjuvanted)	Emergent BioSolutions	Anthrax (post-exposure prophylaxis; age 18–65)	July 20, 2023
Abrysvo	Pfizer	First vaccine for pregnant individuals at 32–36 weeks gestational age, to prevent RSV in infants up to 6 months old	August 21, 2023
Penbraya (pentavalent vaccine)	Pfizer	Invasive disease caused by *Neisseria meningitidis* serogroups A, B, C, W, and Y	October 20, 2023
IXCHIQ (live)	Valneva Austria GmbH	First chikungunya vaccine	November 9, 2023
MRESVIA (mRNA vaccine)	Moderna	Respiratory syncytial virus	May 31, 2024
CAPVAXIVE (Pneumococcal 21-valent conjugate vaccine)	Merck	*Streptococcus pneumoniae* serotypes 3, 6A, 7F, 8, 9 N, 10A, 11A, 12F, 15A,15B, 15C, 16F, 17F, 19A, 20A, 22F, 23A, 23B, 24F, 31, 33F, and 35B	June 17, 2024

Table 6 Carrier proteins used in vaccine formulations

Carrier protein	Source	Properties
Tetanus toxoid	Bacterium *Clostridium tetani*	Non-toxic variant of tetanus toxin, induces strong immune response due to its immunogenicity [113, 114]
Diphtheria toxoid	Bacterium *Corynebacterium diphtheriae*	Non-toxic variant of diphtheria toxin, elicits strong immune response, commonly used in combination vaccines [115]
CRM197	Genetically modified *Corynebacterium diphtheriae*	Non-toxic variant of diphtheria toxin, like diphtheria toxoid but produced through genetic engineering, safe for use [116]
Outer membrane protein complex (OMPC)	Bacterial outer membrane	Forms vesicles with potent immunogenicity, enhances immune response to co-administered antigens, used in conjugate vaccines [117]
Virus-like particles (VLPs)	Engineered to resemble viruses	Mimics the structure of viruses without genetic material, highly immunogenic, triggers strong immune response [105]
Recombinant protein antigens	Produced through genetic engineering	Allows for precise control over antigen structure, high purity, can be tailored to target specific pathogens or epitopes [118]
Recombinant protein nanoparticles	Engineered protein structures	Nano-sized particles with high surface area, enhance antigen presentation, improve stability and immunogenicity [119]

5 Cancer Vaccines

Although the deaths due to cancer have been declining, cancer incidence remains high [127]. Cancer vaccines aim to harness the body's immune system against cancer cells. Their primary goals include stimulating tumor-specific immune responses, promoting tumor regression, limiting spread to other organs, and eliminating minimal residual disease [128].

There are various challenges in the development of these vaccines [129]. Upon encountering foreign microbial invaders, the immune system initiates an innate and adaptive immune response. However, cancers, which are of "self" origin, fail to trigger the same innate immune response, thus evading immune detection. Moreover, the presence of an established cancer is indicative of the success of the tumor in overcoming any challenges posed by the host immune system. A desirable cancer vaccine by design must provoke the very immune system, that had failed initially to mount a potent and effective immune response against the tumor. The cancer vaccines additionally, are handicapped lacking the built-in advantages of therapies like immune checkpoint inhibitors or CAR-T cell therapy [133]. Despite this, they can target both surface and intracellular tumor antigens. Immune surveillance, exemplified by EBV-induced B cell lymphomas, underscores the potential of cancer vaccines as a potent and cost-effective strategy against tumors.

An encouraging result on the immunogenicity of tumor antigens was reported recently using samples from 13 patients with 8 different solid tumor types. The

Table 7 Adjuvants used in vaccine formulations

Adjuvant	Mechanism of action	Properties
Aluminum salts (e.g., aluminum hydroxide, aluminum phosphate)	Depot effect, enhances antigen uptake by antigen-presenting cells	Enhances immune response, widely used, generally safe, promotes Th2 response, example: Gardasil 9 (human papillomavirus vaccine) [120, 121]
MF59	Enhances antigen presentation and immune cell recruitment	Oil-in-water emulsion, promotes Th1 and Th2 responses, improves antibody production, example: Fluad (seasonal influenza vaccine) [122, 123]
AS03	Stimulates innate immunity, enhances antigen uptake	Squalene-based oil-in-water emulsion, promotes robust and durable immune response, example: Pandemrix (H1N1 influenza vaccine) [122]
AS04	Activates toll-like receptor 4 (TLR4) pathway	Combination of aluminum hydroxide and MPL (monophosphoryl lipid A), enhances antigen-specific antibody response, stimulates Th1 and Th2 immune responses, example: Cervarix (human papillomavirus vaccine), Fendrix (hepatitis B vaccine) [124]
QS-21	Stimulates antigen presentation by dendritic cells	Saponin extracted from *Quillaja saponaria* bark, enhances antibody and T-cell responses, often used in combination with other adjuvants, example: Shingrix (herpes zoster vaccine) [125]
MPLA (monophosphoryl lipid A)	Activates toll-like receptor 4 (TLR4) pathway	Derivative of bacterial lipopolysaccharide, enhances antigen presentation and cytokine production, promotes Th1-biased response, example: Shingrix (herpes zoster vaccine) [126]
CpG oligodeoxynucleotides	Stimulates toll-like receptor 9 (TLR9) pathway	Synthetic DNA sequences containing unmethylated CpG motifs, enhance antibody and T-cell responses, promote Th1 response [126]

initial data points toward a pre-existing natural immunity against the neoantigens—potentially the most immunogenic components of the tumor that can be mobilized and expanded with a tumor vaccine. In cases with highly mutated tumors such as lung cancers and melanomas, antigenically the tumors become different from normal cells to generate a T-cell response at least against a fraction of the mutations they carry. The challenge of a "cold tumor" thus becomes somewhat less challenging [130].

The road to effective cancer vaccines has been long and difficult. Although initial attempts to use a vaccine to treat cancer date back to the 1910s, the first effective therapeutic vaccine would not emerge until a hundred years later when Sipuleucel-T (Provenge) was approved to treat minimally symptomatic castrate-resistant prostate cancer (HRPC) in 2010 [131, 132]. T-VEC (Imlygic), approved in 2015 to treat certain types of advanced melanoma, is an oncolytic virus therapy—genetically modified herpes virus—that can also be classified as a therapeutic cancer vaccine as it releases antigens from tumors following their killing by the virus that stimulate the immune response to the tumor antigens [132]. Therapeutic cancer vaccines are being evaluated in more than 600 clinical trials [129].

Overcoming immune suppression in the tumor microenvironment and at central level remains a challenge. Breaking immune tolerance to the tumor at the central level would be extremely crucial. Use of the ubiquitous CMV as a vector could present a solution to this vexing problem as reported recently [32]. Co-existing viruses such as CMV, unlike cancers, escape immune destruction by eliciting a "benign" nondestructive T-cell response. Using CMV to transport cancer-related antigens could potentially elicit a long-lasting immune response to the tumor antigens. The nonpolymorphic major histocompatibility complex E (MHC-E) molecule is up-regulated on many cancer cells, thus contributing to immune evasion by engaging inhibitory NKG2A/CD94 receptors on NK cells and tumor-infiltrating T cells. To examine if MHC-E expression by cancer cells can be targeted for MHC-E–restricted T-cell response, rhesus macaques were immunized with rhesus cytomegalovirus (RhCMV) vectors genetically programmed to elicit MHC-E–restricted CD8+ T cells and to express candidate tumor-associated antigens (TAAs). The T-cell responses to TAAs were comparable to CMV antigen-specific responses, suggesting that CMV-vectored cancer vaccines can bypass central tolerance. Significantly, tumor antigen-specific MHC-E–restricted CD8+ T cells responded to PAP-expressing HLA-E+ prostate cancer cells, suggesting potential for T cell–based immunotherapies targeting the HLA-E/NKG2A immune response blocker.

5.1 Personalized Therapeutic Cancer Vaccines

Ideally, therapeutic vaccines train the immune system to recognize and destroy cancer cells by targeting tumor associated antigens—molecules present only on the surface of cancer cells or more abundant on cancer cells than on normal cells. However, tumor biology is complex, with a tumor microenvironment that suppresses immune activity and a vast molecular diversity that makes the identification of effective target antigens extremely difficult.

In the 1980s, researchers at Johns Hopkins University injected 20 people with colorectal cancer with a combination of their own cancer cells and the BCG vaccine. Originally developed to fight tuberculosis, the BCG vaccine is also a general immune system stimulator. The researchers' motivation was to elicit an immune response in the 20 subjects against their own tumors. After 2 years, all the people who received this treatment survived, while 4 people from the control group of 20 who did not receive this treatment succumbed to the disease.

Other trials testing such a technique showed similar positive results.

Recent technological advancements such as mass spectrometry, "omics" platforms for analyzing single-cell expression of genes and proteins, neoantigen prediction, and computational biology and machine learning have provided essential insights into the biology of cancer and the immune response. In addition, progress in creating sensitive, high-throughput platforms integrating various omics technologies, with the goal of understanding human T- and B-cell response to cancer and infection, and the behavior of immune cells in autoimmune disorders have been helpful. The use of "omics" platforms and machine learning has been helpful in characterizing the features of tumors and their microenvironments to improve vaccine strategies and antigen selection for the differing genetic landscapes of lung tumors in smokers and nonsmokers. Development of cancer mouse models like the CDX and PDX that help researchers define cancer-specific antigens and study tumor immune interactions are enabling the development of novel cancer vaccines.

Following the success of mRNA vaccines against COVID-19, researchers are exploring mRNA vaccines for *anti-tumor therapy*. mRNA cancer vaccines offer high specificity, better efficacy, and fewer side effects compared to traditional treatments [134, 135].

The process of developing personalized mRNA vaccines however is complex (Fig. 2). The use of algorithms as part of the manufacturing process presents unique regulatory challenges, while logistical difficulties, such as shipping materials and obtaining specimens from pathology labs, impact the speed of vaccine production.

Despite these challenges, personalized mRNA vaccines are delivering promising results in clinical trials [130]. Therapeutic mRNA cancer vaccines encode for the key components like the tumor-specific antigens (TSAs) and tumor-associated antigens (TAAs) with proven ability to induce anti-tumor immune responses [136].

Encouraging initial results were obtained from ongoing clinical trials of two personalized mRNA cancer vaccines that were tailored to the antigens presented by individual patient tumors. The first, developed by BioNTech in partnership with

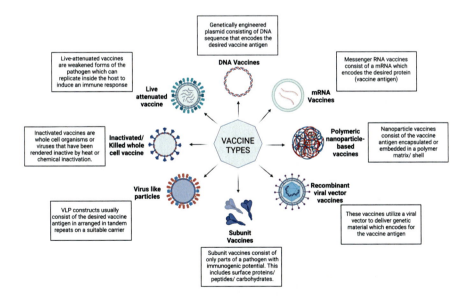

Fig. 2 Types of vaccines. Different types of vaccine platforms currently in use

Genentech for treating pancreatic cancer, was given alongside checkpoint blockade therapy after surgery, and resulted in almost no recurrence of tumors after 18 months of follow-up for patients who responded to the vaccine. The second vaccine, developed by Moderna in partnership with Merck, showed a 44% decrease in the recurrences of melanoma and a 65% decrease in distant metastases [137]. Therapeutic mRNA cancer vaccines have been reported to provide stronger cellular or humoral immunity than traditional inactivated pathogen or protein-based vaccines. Not only that, but it also has the advantages of low cost, rapid development, safety and flexibility, and potent immunogenicity. Over a dozen clinical trials are ongoing using mRNA cancer vaccines.

In addition to mRNA, other technologies that deliver cancer vaccines to lymph nodes for more potent activation of antitumor immunity showed positive results from an ongoing clinical trial (like those developed by Elicio Therapeutics). Designed to target cancers with mutation to KRAS, a mutation shared by many solid tumors, the vaccine was given to patients with pancreatic and colorectal cancers. Seventy-seven percent of patients experienced a reduction in biomarkers that show the persistence of tumors, and 32% cleared those biomarkers altogether.

5.2 Preventative Cancer Vaccines

In addition to therapeutic vaccines, there has been growing interest in developing preventative cancer vaccines. There are a couple of preventative vaccines against viruses associated with cancer:

Human Papillomavirus (HPV) Vaccine: The HPV vaccine is a preventative vaccine that targets the human papillomavirus. It has been highly effective in preventing cervical cancer and other HPV-related cancers. By vaccinating against specific HPV strains, we can significantly reduce the risk of cervical, anal, and oropharyngeal cancers. Several reviews on HPV vaccines have been published elsewhere [9]. These vaccines have had a significant impact on reduction of cervical neoplasia where the vaccine is administered [138].

Hepatitis B Virus Vaccine: Hepatitis B virus (HBV) is linked to liver cancer (hepatocellular carcinoma). The HBV vaccine prevents chronic HBV infection and reduces the risk of liver cancer. Widespread vaccination has led to a decline in liver cancer cases. Besides the above virus examples, vaccines that prevent cancer directly have proven to be elusive.

Recent analysis of healthy individuals who have never had cancer show immune responses to cancer-associated antigens. Some immune events, such as viral and bacterial infections or bone fractures, prime the immune system to recognize cancer cells. During these events, the body's cells may express antigens—for instance, to signal damage that needs to be cleared by the immune system—that cancer cells may also express in abundance. These findings led to the development of a preventative cancer vaccine, which in clinical trials showed a significantly lower rate of recurrence of advanced adenomas, the immediate precursor to colon cancer.

5.3   Antigen Spreading in Cancer Vaccines

Vaccines can also work as an adjunct to cell therapy. CAR-T cells engineered to target glioblastoma tumor antigens were found to lose their edge after initial success. It was discovered that some glioblastoma cells respond to the onslaught of CAR-T cells by halting production of the target antigen when attacked by the CAR-T cells as an escape mechanism rendering the therapy ineffective. In a 2019 study [139], the authors enhanced CAR-T cells' effectiveness against glioblastoma by delivering a vaccine to mice shortly after the engineered T cells were administered. This vaccine, which carries the same antigen targeted by the CAR-T cells, is taken up by immune cells in the lymph nodes, where they are the targets for the CAR-T cells. In that study, the researchers found that this vaccine boost not only helped the engineered CAR-T cells attack tumors, but it had another, unexpected effect: It helped to generate host T cells that now target other tumor antigens. This phenomenon, known as "antigen spreading," is desirable because it creates populations of T cells that, working together, can fully eradicate tumors and prevent tumor regrowth. This could help in addressing the challenge of antigen heterogeneity of solid tumors, the host T-cells that are now primed to attack other antigens may be able to kill the tumor cells that the therapeutic CAR-T cells could not. Other technologies like the cold atmospheric plasma (CAP) treatment, approved by FDA, could also help in exposing novel antigens from tumors [140–143]. Cold Atmospheric Plasma, made up of partially ionized gases, generates a variety of reactive oxygen (ROS) and nitrogen species that have been tried in cancer therapy to destroy cancer cells including solid tumors by influencing various mechanistic pathways at the same time. These include selective apoptosis of tumor cells, damage to the cellular DNA, control of the cell cycle of the tumor cells, among others. The tumor microenvironment (TME) becomes less immunosuppressive with increase in the cellular ROS levels and in the processing exposes more of the tumor antigens. CAP thus can have a role to play in facilitating development of cancer vaccines or as an adjunct to immunotherapies where it can reveal more neoantigens or more of the antigens targeted by immunotherapies.

5.4   Organ-Specific Cancer Vaccines

5.4.1   Pancreatic Cancer Vaccine

Personalized mRNA approach tailors the vaccine to an individual's cancer [144]. Until recently pancreatic cancer was thought to have very few mutations making it an unlikely candidate for a vaccine approach. However, focusing the antigen identification effort on the 10% of patients who survive the disease yielded targetable neoantigen candidates. Researchers can now customize and deliver personalized mRNA cancer vaccines using these target neoantigens to a patient in about 6 weeks. Promising results were reported from a Phase I trial using precision mRNA cancer vaccines to treat patients with one of the most fatal cancers, pancreatic cancer, which takes the lives of nearly 90% of its patients [145, 146]. Eight of the sixteen patients in the study generated neoantigen specific T cells that likely contributed to prolonged survival of these patients.

The next frontier for precision mRNA cancer vaccines will be figuring out which neoantigens are the best ones to target in each individual. Artificial intelligence would be of immense help in designing mRNA cancer vaccines, where efforts in training algorithms to identify the optimal target neoantigens for each patient would lead to a better therapeutic vaccine. The algorithms could be further refined by feeding back laboratory data into them. The self-learning reinforcement would be critical in developing a validated AI system for developing effective vaccines. BNT122/RO7198457-BioNTech and Roche-Genentech's messenger RNA (mRNA) cancer vaccine candidate Autogene cevumeran, continued to show promise 3 years into treatment of pancreatic ductal adenocarcinoma (PDAC) a particularly difficult-to-treat form of pancreatic cancer with a 5 year OSR of about 10%. A longer median recurrence-free survival in cancer vaccine responders was found to be associated with long-term tumor antigen(s)-specific T-cell response.

With more research, precision mRNA cancer vaccines may offer a breakthrough for pancreatic cancer, a disease with a low survival rate that has remained largely unchanged for nearly 60 years. Autogene cevumeran is currently being studied in three ongoing Phase II clinical trials in adjuvant PDAC, first-line melanoma and adjuvant colorectal cancer.

5.4.2 Breast Cancer Vaccine

An experimental vaccine in a decade-long Phase I human trial showed that about 80% of participants with late-stage HER2-positive breast cancer had prolonged life, with 40% experiencing no recurrence. Results from a Phase II study aimed at testing the vaccine's efficacy in a larger cohort of HER2-positive patients may confirm efficacy trends seen in the Phase I study (*Vaccine to Prevent Recurrence in Patients With HER-2 Positive Breast Cancer NCT03384914*) [147].

The recent studies on HER2 peptide vaccines, specifically the E75 and GP2 vaccines, have shown promising results in breast cancer patients:

E75 Vaccine Efficacy: The E75 vaccine demonstrated significant differences in clinical outcomes, such as lower recurrence rates and higher disease-free survival rates, compared to the control group. However, overall survival rates were not significantly different.

GP2 Vaccine Response: The GP2 vaccine elicited a strong immune response, indicated by changes in CD8+ T-cell numbers. More trials are needed to confirm its clinical efficacy. Both vaccines showed minimal side effects, indicating a favorable safety profile [148].

These findings suggest that HER2 peptide vaccines could be a safe and effective treatment option for patients with breast cancer, with the potential to improve clinical outcomes and quality of life [149].

5.4.3 Head and Neck Cancer Vaccine

A neoantigen cancer vaccine TG4050 being developed by NEC is in clinical trials as an adjuvant treatment of HPV negative head and neck cancer (HNNC). In this strategy, immunogenic epitopes were identified in all patients to create individualized vaccines. Tumor-specific variants were identified using next-generation sequencing of tumor and normal samples. Immune-relevant mutations were selected individually in 30 patients using a machine learning algorithm employing parameters such as MHC binding linked to immunogenicity. Following the vaccination regimen, TG4050 elicited persistent immune responses against multiple targets in several patients.

Poly-epitopic responses were induced over the course of vaccination in 16 out of 17 patients as determined by in vitro assays. The T-cell responses were durable lasting over 30 weeks [150]. Moreover, vaccinated patients did not suffer from any relapses during the 18-month observation period as compared to the control unvaccinated patients [151].

5.4.4 Glioblastoma Vaccine

Glioblastoma (GBM) is a devastating tumor of the central nervous system for which median survival remains 14–18 months. A dendritic cell vaccine DOC 1021 prepared by employing homologous antigenic loading of DCs improved survival compared to standard of care in a Phase I trial in glioblastoma multiforme (GBM) patients.

Stimulation of monocyte-derived DC following a specific protocol eliminates secretion of CTLA-4+ micro vesicles that helps in subsequent development of CD161int CD8+ T-cells. Such T cells are found to be effective in serial killing of tumor cells, resistance to exhaustion, and a tissue-homing capacity ideal for the treatment of solid tumors as described in animal models. The loaded DCs were administered to newly diagnosed and relapsed patients bilaterally in the vicinity of the deep cervical lymph nodes. Interim analysis of the open-label study showed that DOC1021 "substantially increased survival," well beyond the expected median overall survival of 12.7 months for patients receiving the standard of care [152, 153].

5.4.5 Lung Cancer Vaccine

An off-the-shelf dendritic cell vaccine platform is being developed by a European biotechnology company (PDC*line Pharma). The vaccines target various cancers by loading the plasmacytoid dendritic cell line pDCs with tumor-specific neoantigens. Initial results from a vaccine(PDC*lung01) targeting non-small cell lung cancer (NSCLC) revealed an ORR of 63%. The nine-month progression-free survival (PFS) was 52.1%, with a median PFS of 10.9 months. The vaccine when given without the anti PD-1 antibody was also able to generate a tumor-specific immune response in over two-thirds of the patients (AACR Abstract).

5.4.6 Liver Cancer Vaccine

A Phase I/II trial by Geneos Therapeutics investigating a therapeutic liver cancer vaccine candidate - a DNA plasmid PTCV (GNOS-PV02) encoding up to 40 neo-antigens - given with plasmid-encoded interleukin-12 and PD-1 inhibitor achieved the primary endpoints of safety and immunogenicity.

While overall, 11 of the 36 patients showed some response, 3 achieved complete response. Vaccine-induced antigen-specific T-cell responses were confirmed in 86.4% patients and were associated with clinical responses. The results support the PTCV's mechanism of action based on the induction of antitumor T cells and show that a therapeutic vaccine can be used to improve outcomes in advanced hepatocellular carcinoma in combination with immune checkpoint inhibitors.

The report is the first definitive demonstration of a personalized cancer vaccine improving clinical response to ICI therapy by inducing new, neoantigen-specific immune cells which traffic to the tumor. ClinicalTrials.gov identifier: NCT04251117 [136].

5.5 The Quest for a Universal Cancer Vaccine

An off-the-shelf cancer vaccine even as a therapeutic would be in great demand. However, developing one universal vaccine for a specific cancer is difficult, as *tumors with different antigens are different for different individuals*. Also, the tumor antigens may be similar to the body's own antigens; therefore, cancer vaccines should have the ability to distinguish between the tumor and the body's antigens.

5.5.1 KRAS as a Promising Target

Among the oncogenes that have been discovered in the past decades, RAS oncogenes are the most commonly mutated oncogenes and are associated with approximately one-third of all tumors. Over 1.9 million new cases of cancers were reported in 2023 by the American Association of Cancer Research [154]. Based upon mutation frequencies in the RAS oncogene family, of the 1.9 million new cases, it is likely that hundreds of thousands of patients present with RAS-mutated cancers each year in the United States. Over 85% of the RAS hot spot mutations associated with multiple cancer types are found in the Kirsten rat sarcoma viral oncogene homolog (KRAS) also associated with poor prognosis. KRAS mutations are found across a wide range of cancers and have been considered to be a prime target for universal therapies (Fig. 3) [155]. For over four decades, KRAS has been considered undruggable due to lack of small-molecule binding sites. Recent technological advances, however, have made direct KRAS targeting possible, yielding two approved drugs that target *KRAS* p.Gly12Cys (G12C) in non-small cell lung cancer.

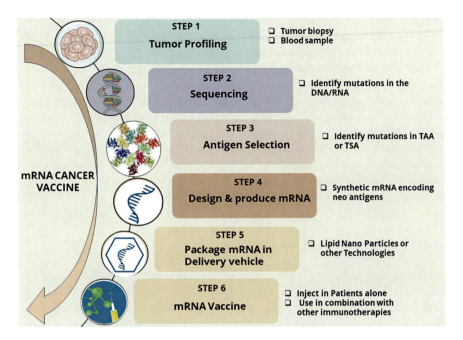

Fig. 3 Steps involved in the development and production of an mRNA vaccine for tumor antigens

Immunotherapeutic approaches including vaccines targeting mutations in the KRAS family offer an exciting possibility of developing off-the-shelf vaccines that may qualify as pan-cancer vaccines.

KRAS-G12C is a mutation common to lung cancers but is present in only 1% of pancreatic cancers. A new vaccine may be able to activate immune cells that in turn can target different KRAS mutations called *KRAS*-G12D and *KRAS*-G12R, which drive about 90% of pancreatic cancers and 40% of colon cancers. A vaccine made of synthetic peptides targeting the KRAS mutations in pancreatic cancers is currently being evaluated in clinical trials [156]. In the Phase I study, patients received the peptide vaccine in two phases—multiple injections as part of primary and secondary immunization, a few months apart. As the vaccine was found to be safe and well tolerated, a Phase II study has been initiated that involves 135 patients with KRAS mutation-positive tumors that are at a higher risk of relapse. The Phase II vaccine targets multiple KRAS mutations in the hopes of generating a broader immune response targeting more tumors.

Another therapeutic vaccine that targets tumors with mutations in the KRAS gene showed encouraging early results in a Phase I trial in 25 patients as a potential off-the-shelf treatment for certain pancreatic or colorectal cancer patients [157]. The vaccine was given after surgery to prevent or delay the recurrence in high-risk patients. The vaccine induced antigen-specific cells with 84% of patients showing increased activated T cells targeting KRAS mutations. The circulating tumor DNA in the blood was reduced in 84% of patients. Remarkably, in about 24% of patients, the tumor DNA was completely absent. On the clinical side, patients with a stronger T-cell response also experienced a longer relapse-free survival suggesting a link between the immune response and the clinical benefit.

Another trial with a multiepitope therapeutic vaccine targeting KRAS mutations was recently reported [158]. The investigational vaccine was given along with two immune checkpoint inhibitors, anti CTLA-4 and anti PD-1. Twenty shared neo-antigens were included in the vaccine construct of chimpanzee adenovirus (ChAd68) and self-amplifying RNA. Most patients (18 of 19) had a KRAS mutation associated with their tumors and tolerated the vaccine well. However, the multi-epitope vaccine demonstrated immunodominance of one epitope over another leading to a new KRAS vaccine design that includes only KRAS mutation determinants. The vaccine is currently in a Phase II clinical trial (NCT03953235).

5.5.2 Telomerase-Based Cancer Vaccine

Telomerase, a reverse transcriptase enzyme commonly associated with cancer, has emerged as a potential target for vaccination that may compensate for the lack of efficacy seen in patients being treated with immune checkpoint inhibitors. Using the telomerase enzyme as a target, a therapeutic vaccine, UV1 an off-the-shelf, peptide-based vaccine is being developed in combination with immune checkpoint inhibitors and is currently undergoing evaluation in five randomized Phase II trials. Following promising results from Phase I trials, UV1 has recently shown nearly double the objective response rate and improved overall survival in a randomized Phase II trial involving patients with malignant pleural mesothelioma [159, 160].

6 Future Directions in Vaccine Development

6.1 Role of Artificial Intelligence

Artificial intelligence and the iterative process of predication and validation it brings can provide a great boost in designing the next generation of vaccines. The first step of vaccine design consists of identifying the protein or peptide antigens that stimulate immunity. AI techniques are more likely to efficiently and presumably accurately identify the candidate antigens [161]. Vaxijen was one of the first companies to use AI-driven methods for antigen prediction. It operated on the premise that antigenicity is inherently encoded within the protein sequence and can be deduced from the chemical properties of amino acid residues. Researchers have trained the prediction model using the chemical properties of known bacterial antigens to achieve reliable results for antigen identification and quantification. The process begins with a known bacterial genome and in a short time (about 48 h) can identify candidate antigens accelerating the development process. This contrasts with the reverse-vaccinology approach that could take years. The validation of the AI-driven prediction comes from the successful creation of leads for some of the bacteria tested [150].

Evaxion is another clinical-stage company using AI in vaccine development [162]. From the target bacterial proteome, their AI platform for bacterial infectious diseases EDEN (Efficacy Discriminative Educated Network) - generates a ranked list of novel protective proteins, with the top 20–30 proteins selected for vaccine design and structural iterations. Animal models and functional assays determine the protection and immunogenicity potential, feeding that data back into the bioinformatics team to iteratively refine the leads based on preclinical model results. This leads to optimization of the antigens, moving it forward into testing with fusion proteins, and adjuvants. These compounds are subjected to chemistry, manufacturing, and controls (CMC). Based on these results, researchers define the best product candidate. This pipeline allows researchers to have candidates within 48 h in most cases.

Other AI platforms by Evaxion include RAVEN (Rapidly Adaptive Viral Response) and PIONEER (Personalized Immuno Oncology using NeoEpitope Efficacy Ranking) that can identify vaccine candidates against any existing, emerging, and mutating viruses. These vaccines can target both B and T cells to provide a robust and long-lasting humoral and cellular response. This approach has the potential to create more reliable and broader vaccines that are not specific to any future variants. This extends to creating reliable vaccines against viruses for which vaccine development has not been successful so far. Vaccines using these antigens can be developed as peptides, proteins, mRNA, or DNA depending upon other factors. Insights on the identification of B cell antigens from EDEN are combined with PIONEER module's ability to select T cell epitopes. The Bayesian Inference for Fragments Of protein Structures (BIFROST) is used finally to combine the two without altering the protein structure [163].

6.1.1   AI in Developing Vaccines for Infectious Diseases

The EDEN pipeline's leading candidate, EVX-B1, is a vaccine targeting *Staphylococcus aureus*. It addresses various virulence factors and toxins across multiple strains [164]. Validated through in silico analyses and preclinical studies using lethal sepsis and skin-abscess infection models, this vaccine aims to prevent skin and soft-tissue infections, as well as other *S. aureus*-related infections.

A multi-component prophylactic gonorrhea vaccine, EVX-B2, is also in the pipeline. With the rise in *Neisseria gonorrhoeae* infections worldwide linked with increasing antibiotic resistance, an effective vaccine would help in controlling the disease. Gonorrhea complications include infertility, ectopic pregnancy, blindness in newborns, septic infections, and heightened HIV transmission risk. The World Health Organization has reported cases of super-gonorrhea, which is resistant to antibiotics like ceftriaxone, azithromycin, penicillin, sulfonamides, tetracycline, fluoroquinolones, and macrolides. Evaxion using the EDEN system identified B cell antigens and tested multiple candidates. The final vaccine, a fusion of two novel protein antigens, EVX-B2, demonstrated high immunogenicity. Anti-EVX-B2 antibodies exhibited bactericidal activity against most of the 50 clinically relevant strains in vitro. In vivo testing in mice showed that immunization reduced vaginal colonization by various *Neisseria gonorrhoeae* strains. The vaccine is currently being evaluated in late-stage preclinical studies [162].

6.1.2 AI in Developing Vaccines for Cancer

In the area of therapeutic vaccines, Evaxion is evaluating therapeutic candidates. EVX-01, a liposomal/peptide immunotherapy, is in a phase II trial combined with the PD-1 inhibitor pembrolizumab for treating checkpoint inhibitor-naïve patients with metastatic or inoperable melanoma. Phase I trials indicated a higher response rate compared to placebo.

EVX-02, a personalized DNA therapeutic vaccine for adjuvant melanoma, is in phase 1/2a clinical trials for excised disease. Preliminary results showed T-cell responses in all patients without significant adverse effects, confirming the delivery technology's proof-of-concept.

EVX-03, a targeted DNA therapeutic vaccine optimized with an antigen-presenting cell (APC)-targeting unit, is being developed for non-small-cell lung cancer. Preclinical studies showed tumor reduction even at low doses, with a dose-dependent enhancement in response.

Evaxion and collaborators showed preclinical research results on delivering tumor neoantigens identified using the PIONEER platform. The lipid-nanoparticle mRNA platform used to deliver the vaccine blocked the growth of the tumor.

Evaxion's latest AI platform, ObsERV, aims to identify patient-specific virus targets for personalized cancer vaccines, paving the way for new cancer treatment paradigms targeting tumors that typically do not respond to immunotherapy.

6.2 Trained Immunity

Trained immunity refers to a functional adaptation of innate immune cells, wherein they acquire immunological memory (Fig. 4). Following exposure to specific stimuli, these cells modify their response to subsequent infections, whether related or unrelated, resulting in an enhanced immune reaction. MV130, an inactivated bacterial vaccine developed by INMUNOTEK, has demonstrated efficacy in protecting against viral infections, including those associated with childhood bronchitis. Notably, MV130 induces both cellular and humoral immune responses against the influenza virus, making it a trained immunity-based vaccine [165, 166] (Fig. 5).

Frequency of KRAS Mutations in Various Cancers

Cancer Tissue		KRAS (%)
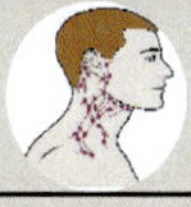	Head and neck	KRAS 2%
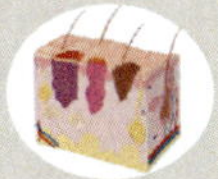	Melanoma	KRAS 1%
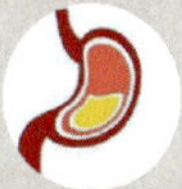	Esophagogastric	KRAS 3%
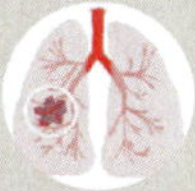	Lung adenocarcinoma	KRAS 32%
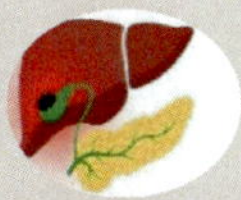	Gallbladder, cholangiocarcinoma	KRAS 21%
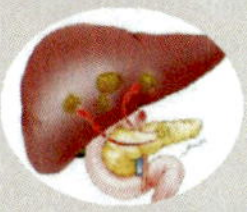	Pancreas	KRAS 88%
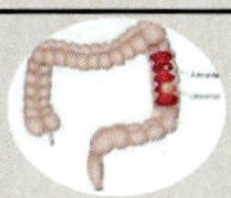	Colorectal	KRAS 50%
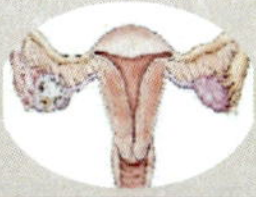	Endometrial	KRAS 17%
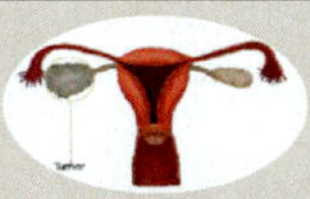	Ovarian	KRAS 9%
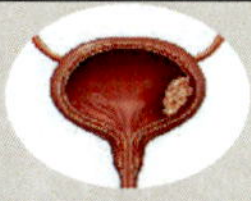	Bladder	KRAS 5%
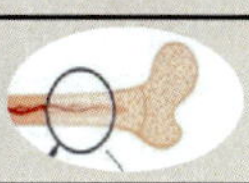	Myeloid leukemia	KRAS 5%

Fig. 4 Frequency of KRAS mutations in various cancers. (Figure adapted from: Singhal et al. [155])

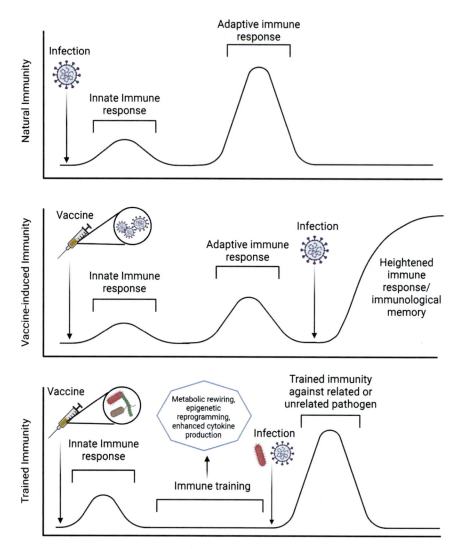

Fig. 5 Trained immunity. Comparative pathways of generating an immune response between natural infection, vaccine-induced, and trained immunity

6.3 Cryptic Tumor-Specific Antigens (TSA)

Scientists are leveraging a novel class of wild-type tumor-specific antigens (TSAs) identified through mass spectrometry, originating from non-coding DNA regions. Comprehensive mass spectrometry and bioinformatic analyses suggest these aberrantly expressed antigens present a significant opportunity for T cell vaccine against the tumors. An experimental mRNA vaccine using mouse antigens formulated into a lipid nanoparticle (LNP) was found to inhibit tumor growth in mouse models.

Immunogenicity studies demonstrate that numerous TSAs induce the expansion of specific T cell clones. These TSAs are being integrated into highly potent cancer vaccines, with each Class I TSA confirmed to be presented using mass spectrometry [167, 168].

In addition, machine learning approaches are being investigated to identify TSAs encoded by the non-coding sequences of the genome. And are being validated in the laboratory opening up possibilities of developing off-the-shelf vaccines for various cancers [169, 170].

6.4   Anti-allergy Immunizations

Anti-allergy shots are utilized for reducing the reactivity of patients to allergens. Allergen-specific immunotherapy requires vaccination with the disease-causing allergens. To ensure efficiency, the vaccines must be adapted to the sensitivity of each patient [171, 172]. As such, they are developed for individual patients. Some of the allergies that are relieved by such vaccines are those caused by pollen, mites, fungi, epithelium, and insects [173].

6.5   Vaccines for Non-infectious Diseases

6.5.1   Vaccine for Alzheimer's Disease

Alzheimer's disease (AD) is a neurodegenerative disease believed to occur due to the buildup of abnormal amounts of certain forms of beta-amyloid and tau proteins in the brain. Although the exact role of these proteins in the disease is poorly understood, they are strongly considered the causative agents of the disease.

Scientists are developing vaccines against both beta-amyloid and tau proteins, with six trials underway. The vaccine consists of an active ingredient intended to stimulate the immune system to generate antibodies against the proteins. These antibodies then traverse the blood-brain barrier and bind to the target proteins. This initiates a white blood cell response to clear the proteins. This process is aimed at slowing the progression of AD.

However, both beta-amyloid and tau are proteins normally expressed in the brain, and hence not recognized as foreign particles by the immune system. However, the disease pathophysiology is due to misfolded forms of these proteins, which present a novel target. Immune response directed against the misfolded proteins with vaccine additives called adjuvants can alert the immune system to danger.

Vaccines that use this mechanism of action are being developed by the Swiss biopharmaceutical company AC Immune. The two vaccine candidates—one against beta-amyloid and one against tau protein—are in early-stage human trials. Vaccinating individuals at the earliest stages of AD, even before developing significant symptoms, may be crucial to slow or halt disease progression. A blood test developed recently can detect the early signs of tau-related pathology before any apparent cognitive deterioration.

The US FDA recently granted a fast-track designation to a vaccine, UB-311 from the biotechnology company Vaxxinity, that targets amyloid protein. The vaccine has shown promising results in human trials, with 98% people showing a response. UB-311 is progressing toward a larger trial that will include around 3000 individuals.

A nasally administered vaccine that would stimulate the immune system in the lymph nodes of the neck, prompting monocytes to migrate into the brain to clear amyloid buildup, is being tested. The vaccine has successfully cleared accumulated beta-amyloid in mouse models of AD and is progressing toward early human trials [174].

6.5.2   Vaccine for Hypertension

A receptor in the arteries primarily regulates blood pressure. Its activation causes blood vessels to constrict, increasing the blood pressure. Although drugs called alpha-1-blockers can deactivate the receptor, they lack specificity and have a short half-life in the blood. This prevents them from being used as front-line drugs. To resolve this challenge a vaccine approach that targets arterial receptors is being tried. Mouse studies where antibodies blocking the arterial receptor have shown promising results [175–177]. One notable method involves using antibodies to block specific receptors involved in blood pressure regulation. For example, researchers have explored vaccines targeting the renin-angiotensin system, which plays a crucial role in regulating blood pressure. These vaccines aim to induce the production of antibodies that can block the effects of angiotensin II, a hormone that constricts blood vessels and raises blood pressure.

6.5.3   Vaccine for Aging and Senescence

Senescent or biologically old cells have undergone irreparable damage and ceased to divide yet remain viable and do not undergo cell death. The host immune system clears these cells, but the process falters with age, causing senescent cells to build up in tissue. This accumulation can result in the secretion of a cocktail of toxic inflammatory molecules which can cause damage to surrounding tissues that can extend into the bloodstream. Recent reports suggest that such buildup of senescent cells is a contributing factor for several age-associated diseases.

Some antigens, called seno-antigens, are exclusively present in senescent cells, which can be targeted by vaccines to boost the immune system to clear senescent cells.

In 2020, a team of researchers led by Hironori Nakagami and Osaka University Graduate School of Medicine developed a vaccine against a seno-antigen CD153 [178]. This antigen is found in senescent cells that build up in the visceral fat found on and around organs. When this vaccine was tested on mice fed with an obesity-inducing diet, vaccinated mice showed lower levels of senescent cells than unvaccinated mice and had improved glucose metabolism and reduced insulin resistance.

Shortly afterward, a team led by Tohru Minamino at Juntendo University Graduate School of Medicine conducted similar research in aged mice [179]. Senescent cells in the lining of blood vessels of aged mice are associated with a risk of atherosclerosis, which causes blood vessels to narrow, blocking blood flow. Vaccinated mice had a longer lifespan as well as health span—the length of time they were healthy—than placebo-treated mice.

In recent research from a team at Seoul National University, scientists isolated tumor cells from mice, induced senescence in these, and passed them through tiny, sieve-like membranes to generate miniscule particles. When they injected these particles into the mice, they observed a strong immune response that inhibited the growth and spread of tumors.

6.5.4 Vaccine for Preventing Pregnancy (hCG Vaccine)

A vaccine that induces antibodies against the pregnancy-related hormone human chorionic gonadotropin (hCG) can help prevent unwanted pregnancy [180]. This genetically engineered recombinant vaccine when tested in clinical trials was found to be effective in all women. All vaccinated women generated antibodies and were administered booster doses to maintain antibody titers. These antibodies did not intervene with normal reproductive functions—all women continued to ovulate and have regular menstrual cycles. The hCG vaccine development presented unique challenges from the beginning stages of the concept as developed by G.P. Talwar in the 1970s wherein a well-tolerated "self" molecule from the inception needed to be reconfigured so as to evoke an immune response that will neutralize the bioactivity of the target molecule. Various strategies were employed over the years including use of carrier proteins and adjuvants to break this immune tolerance [98, 174, 180–183].

Most recently due to the perseverance of Dr. Talwar, the recombinant vaccine is being manufactured to undergo more rounds of clinical trials. Besides being used as a contraceptive, the hCG vaccine would be of use in the treatment of cancers that involve the expression of the hCG hormone [184].

References

1. A Brief History of Vaccination [Internet]. [cited 2024 Aug 3]. Available from: https://www.who.int/news-room/spotlight/history-of-vaccination/a-brief-history-of-vaccination.
2. Shattock AJ, Johnson HC, Sim SY, Carter A, Lambach P, Hutubessy RCW, et al. Contribution of vaccination to improved survival and health: modelling 50 years of the expanded Programme on immunization. Lancet. 2024;403(10441):2307–16.

3. Montero DA, Vidal RM, Velasco J, Carreño LJ, Torres JP, Benachi OMA, et al. Two centuries of vaccination: historical and conceptual approach and future perspectives. Front Public Health. 2024;11:1326154.
4. Fritschi N, Curtis N, Ritz N. Bacille Calmette Guérin (BCG) and new TB vaccines: specific, cross-mycobacterial and off-target effects. Paediatr Respir Rev. 2020;36:57–64.
5. Talwar GP. An immunotherapeutic vaccine for multibacillary leprosy. Int Rev Immunol. 1999;18(3):229–49.
6. History of polio vaccination [Internet]. [cited 2024 Aug 3]. Available from: https://www.who.int/news-room/spotlight/history-of-vaccination/history-of-polio-vaccination.
7. Gerlich WH. Medical virology of hepatitis B: how it began and where we are now. Virol J. 2013;10:239.
8. Tulchinsky TH. Maurice Hilleman: creator of vaccines that changed the world. Case Stud Public Health. 2018:443–70.
9. Felsher M, Shumet M, Velicu C, Chen YT, Nowicka K, Marzec M, et al. A systematic literature review of human papillomavirus vaccination strategies in delivery systems within national and regional immunization programs. Hum Vaccin Immunother. 2024;20(1):2319426.
10. Talwar GP. A unique vaccine for control of fertility and therapy of advanced-stage terminal cancers ectopically expressing human chorionic gonadotropin. Ann N Y Acad Sci. 2013;1283:50–6. https://doi.org/10.1111/j.1749-6632.2012.06776.x. Epub 2013 Jan 9.
11. Naz RK. Antisperm contraceptive vaccines: where we are and where we are going? Am J Reprod Immunol N Y N 1989. 2011;66(1):5–12.
12. Gupta SK, Bansal P. Vaccines for immunological control of fertility. Reprod Med Biol. 2009;9(2):61–71.
13. Mohn KGI, Smith I, Sjursen H, Cox RJ. Immune responses after live attenuated influenza vaccination. Hum Vaccin Immunother. 2018;14(3):571–8.
14. Bull JJ, Smithson MW, Nuismer SL. Transmissible Viral Vaccines. Trends Microbiol. 2018;26(1):6–15.
15. Hepatitis B Vaccine – an overview | ScienceDirect Topics [Internet]. [cited 2024 Aug 16]. Available from: https://www.sciencedirect.com/topics/immunology-and-microbiology/hepatitis-b-vaccine.
16. Kale A, Gaur A, Menon I, Chirmule N, Bagwe P, Jawa R, et al. An overview of current accomplishments and gaps of COVID-19 vaccine platforms and considerations for next generation vaccines. J Pharm Sci. 2023;112(5):1345–50.
17. Research C for BE and. Measles, Mumps and Rubella Virus Vaccine Live. FDA [Internet]. 2023 Aug 22 [cited 2024 Aug 3]; Available from: https://www.fda.gov/vaccines-blood-biologics/vaccines/measles-mumps-and-rubella-virus-vaccine-live.
18. Research C for BE and. VARIVAX (refrigerated and frozen formulations). FDA [Internet]. 2023 Aug 23 [cited 2024 Aug 3]; Available from: https://www.fda.gov/vaccines-blood-biologics/vaccines/varivax-refrigerated-and-frozen-formulations.
19. Research C for BE and. GARDASIL 9. FDA [Internet]. 2024 Mar 15 [cited 2024 Aug 3]; Available from: https://www.fda.gov/vaccines-blood-biologics/vaccines/gardasil-9.
20. Research C for BE and. Fluzone Quadrivalent, Fluzone High-Dose Quadrivalent, Fluzone, Intradermal Quadrivalent, Fluzone Quadrivalent Southern Hemisphere, Fluzone High-Dose Quadrivalent Southern Hemisphere. FDA [Internet]. 2024 Jul 3 [cited 2024 Aug 3]; Available from: https://www.fda.gov/vaccines-blood-biologics/vaccines/fluzone-quadrivalent-fluzone-high-dose-quadrivalent-fluzone-intradermal-quadrivalent-fluzone.
21. Research C for BE and. HAVRIX. FDA [Internet]. 2023 Oct 30 [cited 2024 Aug 3]; Available from: https://www.fda.gov/vaccines-blood-biologics/vaccines/havrix.
22. Research C for BE and. ENGERIX-B. FDA [Internet]. 2023 Oct 30 [cited 2024 Aug 3]; Available from: https://www.fda.gov/vaccines-blood-biologics/vaccines/engerix-b.
23. Research C for BE and. Zostavax. FDA [Internet]. 2023 Mar 13 [cited 2024 Aug 3]; Available from: https://www.fda.gov/vaccines-blood-biologics/vaccines/zostavax.

24. Research C for BE and. SHINGRIX. FDA [Internet]. 2023 May 23 [cited 2024 Aug 3]; Available from: https://www.fda.gov/vaccines-blood-biologics/vaccines/shingrix.
25. Research C for BE and. COMIRNATY. FDA [Internet]. 2023 Dec 7 [cited 2024 Aug 3]; Available from: https://www.fda.gov/vaccines-blood-biologics/comirnaty.
26. Research C for BE and. Moderna COVID-19 Vaccine. FDA [Internet]. 2023 Nov 1 [cited 2024 Aug 3]; Available from: https://www.fda.gov/vaccines-blood-biologics/coronavirus-covid-19-cber-regulated-biologics/moderna-covid-19-vaccine.
27. Research C for BE and. Janssen COVID-19 Vaccine. FDA [Internet]. 2023 Jun 2 [cited 2024 Aug 3]; Available from: https://www.fda.gov/vaccines-blood-biologics/coronavirus-covid-19-cber-regulated-biologics/janssen-covid-19-vaccine.
28. Chirmule N, Khare R, Nair P, Desai B, Nerurkar V, Gaur A. Predicting the severity of disease progression in COVID-19 at the individual and population level: a mathematical model. Clin Exp Pharmacol. 2021;11(5):283.
29. Shepherd BLO, Scott PT, Hutter JN, Lee C, McCauley MD, Guzman I, et al. SARS-CoV-2 recombinant spike ferritin nanoparticle vaccine adjuvanted with Army Liposome Formulation containing monophosphoryl lipid A and QS-21: a phase 1, randomised, double-blind, placebo-controlled, first-in-human clinical trial. Lancet Microbe. 2024;5(6):e581–93.
30. Kuter BJ, Offit PA, Poland GA. The development of COVID-19 vaccines in the United States: why and how so fast? Vaccine. 2021;39(18):2491–5.
31. Jha A, Jarvis H, Fraser C, Openshaw PJ. Respiratory syncytial virus. In: Hui DS, Rossi GA, Johnston SL, editors. SARS, MERS and other viral lung infections [Internet]. Sheffield (UK): European Respiratory Society; 2016. [cited 2024 Aug 11]. (Wellcome Trust–Funded Monographs and Book Chapters). Available from: http://www.ncbi.nlm.nih.gov/books/NBK442240/.
32. Iyer RF, Verweij MC, Nair SS, Morrow D, Mansouri M, Chakravarty D, et al. CD8+ T cell targeting of tumor antigens presented by HLA-E. Sci Adv. 2024;10(19):eadm7515.
33. Blanco JCG, Boukhvalova MS, Shirey KA, Prince GA, Vogel SN. New insights for development of a safe and protective RSV vaccine. Hum Vaccin. 2010;6(6):482–92.
34. Harapan H, Michie A, Sasmono RT, Imrie A. Dengue: a minireview. Viruses. 2020;12(8):829.
35. Tully D, Griffiths CL. Dengvaxia: the world's first vaccine for prevention of secondary dengue. Ther Adv Vaccines Immunother. 2021;9:25151355211015839.
36. Shukla R, Ramasamy V, Shanmugam RK, Ahuja R, Khanna N. Antibody-dependent enhancement: a challenge for developing a safe dengue vaccine. Front Cell Infect Microbiol. 2020;10:572681.
37. Thoresen D, Matsuda K, Urakami A, Ngwe Tun MM, Nomura T, Moi ML, et al. A tetravalent dengue virus-like particle vaccine induces high levels of neutralizing antibodies and reduces dengue replication in non-human primates. J Virol. 2024;98(5):e0023924.
38. Aggarwal C, Ramasamy V, Garg A, Shukla R, Khanna N. Cellular T-cell immune response profiling by tetravalent dengue subunit vaccine (DSV4) candidate in mice. Front Immunol. 2023;14:1128784.
39. Cherian N, Bettis A, Deol A, Kumar A, Di Fabio JL, Chaudhari A, et al. Strategic considerations on developing a CHIKV vaccine and ensuring equitable access for countries in need. Npj Vaccines. 2023;8(1):1–8.
40. Ly H. Ixchiq (VLA1553): the first FDA-approved vaccine to prevent disease caused by chikungunya virus infection. Virulence. 2024;15(1):2301573.
41. Cimica V, Galarza JM, Rashid S, Stedman TT. Current development of Zika virus vaccines with special emphasis on virus-like particle technology. Expert Rev Vaccines. 2021;20(11):1483–98.
42. Wang Y, Ling L, Zhang Z, Marin-Lopez A. Current advances in Zika vaccine development. Vaccine. 2022;10(11):1816.
43. Amaya M, Broder CC. Vaccines to emerging viruses: Nipah and Hendra. Annu Rev Virol. 2020;7(1):447–73.

44. Geisbert TW, Bobb K, Borisevich V, Geisbert JB, Agans KN, Cross RW, et al. A single dose investigational subunit vaccine for human use against Nipah virus and Hendra virus. NPJ Vaccines. 2021;6:23.
45. Stierwalt EES. Scientific American. [cited 2024 Aug 7]. How Are Seasonal Flu Vaccines Made? Available from: https://www.scientificamerican.com/article/how-are-seasonal-flu-vaccines-made/.
46. Chung JR, Flannery B, Ambrose CS, Bégué RE, Caspard H, DeMarcus L, et al. Live attenuated and inactivated influenza vaccine effectiveness. Pediatrics. 2019;143(2):e20182094.
47. Examining the potential benefits of the influenza vaccine against SARS-CoV-2: A retrospective cohort analysis of 74,754 patients I PLoS One [Internet]. [cited 2024 Aug 7]. Available from: https://journals.plos.org/plosone/article?id=10.1371/journal.pone.0255541.
48. Leggat DJ, Cohen KW, Willis JR, Fulp WJ, deCamp AC, Kalyuzhniy O, et al. Vaccination induces HIV broadly neutralizing antibody precursors in humans. Science. 2022;378(6623):eadd6502.
49. A Phase 1, Randomized, First-in-human, Open-label Study to Evaluate the Safety and Immunogenicity of eOD-GT8 60mer mRNA Vaccine (mRNA-1644) and Core-g28v2 60mer mRNA Vaccine (mRNA-1644v2-Core) in HIV-1 Uninfected Adults in Good General Health [Internet]. Infectious Diseases. [cited 2024 Aug 7]. Available from: https://lsom.uthscsa.edu/infectious-diseases/clinical-trial/a-phase-1-randomized-first-in-human-open-label-study-to-evaluate-the-safety-and-immunogenicity-of-eod-gt8-60mer-mrna-vaccine-mrna-1644-and-core-g28v2-60mer-mrna-vaccine-mrna-1644v2-core-in-hiv-2/.
50. NIH Launches Clinical Trial of Three mRNA HIV Vaccines I NIAID: National Institute of Allergy and Infectious Diseases [Internet]. 2022 [cited 2024 Aug 7]. Available from: https://www.niaid.nih.gov/news-events/nih-launches-clinical-trial-three-mrna-hiv-vaccines.
51. Gaur A, Arunan K, Singh O, Talwar GP. Bypass by an alternate "carrier" of acquired unresponsiveness to hCG upon repeated immunization with tetanus-conjugated vaccine. Int Immunol. 1990;2(2):151–5.
52. Agrawal A, Murphy TF. Haemophilus influenzae infections in the H. Influenzae type b conjugate vaccine era ∇. J Clin Microbiol. 2011;49(11):3728–32.
53. Kitchin N, Remich SA, Peterson J, Peng Y, Gruber WC, Jansen KU, et al. A phase 2 study evaluating the safety, tolerability, and immunogenicity of two 3-dose regimens of a Clostridium difficile vaccine in healthy US adults aged 65 to 85 years. Clin Infect Dis. 2020;70(1):1–10.
54. Creech CB, Frenck RW, Fiquet A, Feldman R, Kankam MK, Pathirana S, et al. Persistence of immune responses through 36 months in healthy adults after vaccination with a novel Staphylococcus aureus 4-antigen vaccine (SA4Ag). Open forum. Infect Dis. 2020;7(1):ofz532.
55. Pérez-Sancristóbal I, de la Fuente E, Álvarez-Hernández MP, Guevara-Hoyer K, Morado C, Martínez-Prada C, et al. Long-term benefit of Perlingual Polybacterial vaccines in patients with systemic autoimmune diseases and active immunosuppression. Biomedicines. 2023;11(4):1168.
56. Saz-Leal P, Ligon MM, Diez-Rivero CM, García-Ayuso D, Mohanty S, Viñuela M, et al. MV140 mucosal vaccine induces targeted immune response for enhanced clearance of Uropathogenic E. coli in experimental urinary tract infection. Vaccines. 2024;12(5):535.
57. Talwar GP, Zaheer SA, Mukherjee R, Walia R, Misra RS, Sharma AK, et al. Immunotherapeutic effects of a vaccine based on a saprophytic cultivable mycobacterium, Mycobacterium w in multibacillary leprosy patients. Vaccine. 1990;8(2):121–9.
58. Dogra S, Jain S, Sharma A, Chhabra S, Narang T. Mycobacterium Indicus Pranii (MIP) vaccine: pharmacology, indication, dosing schedules, administration, and side effects in Clinical practice. Indian Dermatol Online J. 2023;14(6):753.
59. Sharma P, Mukherjee R, Talwar GP, Sarathchandra KG, Walia R, Parida SK, et al. Immunoprophylactic effects of the anti-leprosy Mw vaccine in household contacts of leprosy patients: clinical field trials with a follow up of 8-10 years. Lepr Rev. 2005;76(2):127–43.
60. Ali L. Leprosy vaccines—a voyage unfinished. J Skin Sex Transm Dis. 2021;3(1):40–5.

61. Research C for BE and. DAPTACEL. FDA [Internet]. 2022 Jul 22 [cited 2024 Aug 3]; Available from: https://www.fda.gov/vaccines-blood-biologics/vaccines/daptacel.
62. Research C for BE and. BOOSTRIX. FDA [Internet]. 2023 Oct 27 [cited 2024 Aug 3]; Available from: https://www.fda.gov/vaccines-blood-biologics/vaccines/boostrix.
63. Research C for BE and. PNEUMOVAX 23 - Pneumococcal Vaccine, Polyvalent. FDA [Internet]. 2023 Feb 28 [cited 2024 Aug 3]; Available from: https://www.fda.gov/vaccines-blood-biologics/vaccines/pneumovax-23-pneumococcal-vaccine-polyvalent.
64. Research C for BE and. Prevnar 13. FDA [Internet]. 2023 Mar 2 [cited 2024 Aug 3]; Available from: https://www.fda.gov/vaccines-blood-biologics/vaccines/prevnar-13.
65. Research C for BE and. Menactra. FDA [Internet]. 2023 Feb 28 [cited 2024 Aug 3]; Available from: https://www.fda.gov/vaccines-blood-biologics/vaccines/menactra.
66. Huston J, Galicia K, Egelund EF. MenQuadfi (MenACWY-TT): a new vaccine for meningococcal serogroups ACWY. Ann Pharmacother. 2022;56(6):727–35.
67. Research C for BE and. MENVEO. FDA [Internet]. 2023 Feb 28 [cited 2024 Aug 3]; Available from: https://www.fda.gov/vaccines-blood-biologics/vaccines/menveo.
68. Research C for BE and. Haemophilus B Conjugate Vaccine (Meningococcal Protein Conjugate). FDA [Internet] 2022 Nov 7 [cited 2024 Aug 3]; Available from: https://www.fda.gov/vaccines-blood-biologics/vaccines/haemophilus-b-conjugate-vaccine-meningococcal-protein-conjugate.
69. Research C for BE and. BCG Vaccine. FDA [Internet]. 2022 Oct 4 [cited 2024 Aug 3]; Available from: https://www.fda.gov/vaccines-blood-biologics/vaccines/bcg-vaccine.
70. Research C for BE and. Typhim Vi. FDA [Internet]. 2023 Mar 13 [cited 2024 Aug 3]; Available from: https://www.fda.gov/vaccines-blood-biologics/vaccines/typhim-vi.
71. Research C for BE and. VAXCHORA. FDA [Internet]. 2024 Jun 26 [cited 2024 Aug 3]; Available from: https://www.fda.gov/vaccines-blood-biologics/vaccines/vaxchora.
72. Research C for BE and. Biothrax. FDA [Internet]. 2023 Jul 6 [cited 2024 Aug 3]; Available from: https://www.fda.gov/vaccines-blood-biologics/vaccines/biothrax.
73. Jansen KU, Gruber WC, Simon R, Wassil J, Anderson AS. The impact of human vaccines on bacterial antimicrobial resistance. A review. Environ Chem Lett. 2021;19(6):4031–62.
74. Laurens MB. RTS,S/AS01 vaccine (Mosquirix™): an overview. Hum Vaccin Immunother. 2019;16(3):480–9.
75. Draper SJ, Sack BK, King CR, Nielsen CM, Rayner JC, Higgins MK, et al. Malaria vaccines: recent advances and new horizons. Cell Host Microbe. 2018;24(1):43–56.
76. Khehra N, Padda I, Jaferi U, Atwal H, Narain S, Parmar MS. Tozinameran (BNT162b2) vaccine: the journey from preclinical research to clinical trials and authorization. AAPS PharmSciTech. 2021;22(5):172.
77. Kandi V, Vadakedath S. Clinical trials and clinical research: a comprehensive review. Cureus. 15(2):e35077.
78. Pambudi NA, Sarifudin A, Gandidi IM, Romadhon R. Vaccine cold chain management and cold storage technology to address the challenges of vaccination programs. Energy Rep. 2022;8:955–72.
79. Zuckerman JN. The importance of injecting vaccines into muscle. BMJ. 2000;321(7271):1237–8.
80. Usach I, Martinez R, Festini T, Peris JE. Subcutaneous injection of drugs: literature review of factors influencing pain sensation at the injection site. Adv Ther. 2019;36(11):2986–96.
81. Prins MLM, Prins C, de Vries JJC, Visser LG, Roukens AHE. Establishing immunogenicity and safety of needle-free intradermal delivery by nanoporous ceramic skin patch of mRNA SARS-CoV-2 vaccine as a revaccination strategy in healthy volunteers. Virus Res. 2023;334:199175.
82. D'Souza B, Bhowmik T, Shashidharamurthy R, Oettinger C, Selvaraj P, D'Souza M. Oral microparticulate vaccine for melanoma using M-cell targeting. J Drug Target. 2012;20(2):166–73.
83. Oral Poliomyelitis Vaccine – an overview | ScienceDirect Topics [Internet]. [cited 2024 Aug 17]. Available from: https://www.sciencedirect.com/topics/medicine-and-dentistry/oral-poliomyelitis-vaccine.

84. Alu A, Chen L, Lei H, Wei Y, Tian X, Wei X. Intranasal COVID-19 vaccines: From bench to bed. eBioMedicine [Internet]. 2022 Feb 1 [cited 2023 Mar 11];76. Available from: https://www.thelancet.com/journals/ebiom/article/PIIS2352-3964(22)00025-1/fulltext.

85. Research C for BE and. FluMist Quadrivalent. FDA [Internet]. 2023 Jul 12 [cited 2024 Aug 17]; Available from: https://www.fda.gov/vaccines-blood-biologics/vaccines/flumist-quadrivalent.

86. Trial | NCT04722003 [Internet]. [cited 2024 Aug 17]. Available from: https://cdek.pharmacy.purdue.edu/trial/NCT04722003/.

87. Insight into Prevention of Neisseria Gonorrhoeae: A Short Review – PMC [Internet]. [cited 2024 Aug 17]. Available from: https://www.ncbi.nlm.nih.gov/pmc/articles/PMC9692366/.

88. Brandtzaeg P, Pabst R. Let's go mucosal: communication on slippery ground. Trends Immunol. 2004;25(11):570–7.

89. Braz Gomes K, D'Sa S, Allotey-Babington GL, Kang SM, D'Souza MJ. Transdermal vaccination with the Matrix-2 protein virus-like particle (M2e VLP) induces immunity in mice against influenza A virus. Vaccine. 2021;9(11):1324.

90. D'Sa S, Braz Gomes K, Allotey-Babington GL, Boyoglu C, Kang SM, D'Souza MJ. Transdermal immunization with microparticulate RSV-F virus-like particles elicits robust immunity. Vaccine. 2022;10(4):584.

91. Bagwe P, Bajaj L, Menon I, Braz Gomes K, Kale A, Patil S, et al. Gonococcal microparticle vaccine in dissolving microneedles induced immunity and enhanced bacterial clearance in infected mice. Int J Pharm. 2023;642:123182.

92. Wong J, Brugger A, Khare A, Chaubal M, Papadopoulos P, Rabinow B, et al. Suspensions for intravenous (IV) injection: a review of development, preclinical and clinical aspects. Adv Drug Deliv Rev. 2008;60(8):939–54.

93. Menon I, Bagwe P, Gomes KB, Bajaj L, Gala R, Uddin MN, et al. Microneedles: a new generation vaccine delivery system. Micromachines. 2021;12(4):435.

94. Menon I, Kang SM, D'Souza M. Nanoparticle formulation of the fusion protein virus like particles of respiratory syncytial virus stimulates enhanced in vitro antigen presentation and autophagy. Int J Pharm. 2022;623:121919.

95. Jet Injector – an overview | ScienceDirect Topics [Internet]. [cited 2024 Aug 17]. Available from: https://www.sciencedirect.com/topics/pharmacology-toxicology-and-pharmaceutical-science/jet-injector.

96. Nature Milestones in Vaccines [Internet]. [cited 2024 Aug 5]. Available from: https://www.nature.com/immersive/d42859-020-00005-8/index.html.

97. Immunize.org [Internet]. 2024 [cited 2024 Aug 5]. Vaccine History Timeline. Available from: https://www.immunize.org/vaccines/vaccine-timeline/.

98. Philadelphia TCH of. Vaccine History: Developments by Year [Internet]. The Children's Hospital of Philadelphia; 2014 [cited 2024 Aug 5]. Available from: https://www.chop.edu/centers-programs/vaccine-education-center/vaccine-history/developments-by-year.

99. Policy (OIDP) O of ID and H. Vaccine Types [Internet]. 2021 [cited 2024 Aug 5]. Available from: https://www.hhs.gov/immunization/basics/types/index.html.

100. Kumru OS, Joshi SB, Smith DE, Middaugh CR, Prusik T, Volkin DB. Vaccine instability in the cold chain: mechanisms, analysis and formulation strategies. Biologicals. 2014;42(5):237–59.

101. Inactivated Vaccine – an overview | ScienceDirect Topics [Internet]. [cited 2021 Aug 16]. Available from: https://www.sciencedirect.com/topics/immunology-and-microbiology/inactivated-vaccine.

102. Toxoid – an overview | ScienceDirect Topics [Internet]. [cited 2024 Aug 5]. Available from: https://www.sciencedirect.com/topics/immunology-and-microbiology/toxoid.

103. What are protein subunit vaccines and how could they be used against COVID-19? [Internet]. [cited 2024 Aug 5]. Available from: https://www.gavi.org/vaccineswork/what-are-protein-subunit-vaccines-and-how-could-they-be-used-against-covid-19.

104. CDC. Centers for Disease Control and Prevention. 2023 [cited 2024 Aug 17]. Cell-Based Flu Vaccines. Available from: https://www.cdc.gov/flu/prevent/cell-based.htm.
105. Nooraei S, Bahrulolum H, Hoseini ZS, Katalani C, Hajizade A, Easton AJ, et al. Virus-like particles: preparation, immunogenicity and their roles as nanovaccines and drug nanocarriers. J Nanobiotechnol. 2021;19(1):59.
106. Leleux J, Roy K. Micro and nanoparticle-based delivery systems for vaccine immunotherapy: an immunological and materials perspective. Adv Healthc Mater. 2013;2(1):72–94.
107. Choi Y, Chang J. Viral vectors for vaccine applications. Clin Exp Vaccine Res. 2013;2(2):97–105.
108. Deng S, Liang H, Chen P, Li Y, Li Z, Fan S, et al. Viral vector vaccine development and application during the COVID-19 pandemic. Microorganisms. 2022;10(7):1450.
109. Kapoor D, Suryawanshi R, Patil CD, Shukla D. Chapter 13: Recent advancements and nanotechnological interventions in diagnosis, treatment, and vaccination for COVID-19. In: Rai M, Yadav A, editors. Nanotechnological applications in virology [Internet]. Academic; 2022. p. 279–303. (Developments in Applied Microbiology and Biotechnology). Available from: https://www.sciencedirect.com/science/article/pii/B9780323995962000157.
110. Liu Y, Ye Q. Nucleic acid vaccines against SARS-CoV-2. Vaccine. 2022;10(11):1849.
111. The Coming of Age of Nucleic Acid Vaccines during COVID-19 | mSystems [Internet]. [cited 2024 Aug 5]. Available from: https://journals.asm.org/doi/10.1128/msystems.00928-22.
112. Research C for BE and. Vaccines Licensed for Use in the United States. FDA [Internet]. 2024 Jun 17 [cited 2024 Aug 5]; Available from: https://www.fda.gov/vaccines-blood-biologics/vaccines/vaccines-licensed-use-united-states.
113. Liang JL, Tiwari T, Moro P, Messonnier NE, Reingold A, Sawyer M, et al. Prevention of pertussis, tetanus, and diphtheria with vaccines in the United States: recommendations of the advisory committee on immunization practices (ACIP). MMWR Recomm Rep. 2018;67(2):1–44.
114. Chang MJ, Ollivault-Shiflett M, Schuman R, Nguyen SN, Kaltashov IA, Bobst C, et al. Genetically detoxified tetanus toxin as a vaccine and conjugate carrier protein. Vaccine. 2022;40(35):5103–13.
115. Pichichero ME. Protein carriers of conjugate vaccines. Hum Vaccin Immunother. 2013;9(12):2505–23.
116. Mishra RPN, Yadav RSP, Jones C, Nocadello S, Minasov G, Shuvalova LA, et al. Structural and immunological characterization of E. coli derived recombinant CRM197 protein used as carrier in conjugate vaccines. Biosci Rep. 2018;38(5):BSR20180238.
117. Micoli F, MacLennan CA. Outer membrane vesicle vaccines. Semin Immunol. 2020;50:101433.
118. Recombinant Antigen – an overview | ScienceDirect Topics [Internet]. [cited 2024 Jul 27]. Available from: https://www.sciencedirect.com/topics/immunology-and-microbiology/recombinant-antigen.
119. Gifre-Renom L, Ugarte-Berzal E, Martens E, Boon L, Cano-Garrido O, Martínez-Núñez E, et al. Recombinant protein-based nanoparticles: elucidating their inflammatory effects in vivo and their potential as a new therapeutic format. Pharmaceutics. 2020;12(5):450.
120. Oleszycka E, Lavelle EC. Immunomodulatory properties of the vaccine adjuvant alum. Curr Opin Immunol. 2014;28:1–5.
121. Rimaniol AC, Gras G, Verdier F, Capel F, Grigoriev VB, Porcheray F, et al. Aluminum hydroxide adjuvant induces macrophage differentiation towards a specialized antigen-presenting cell type. Vaccine. 2004;22(23–24):3127–35.
122. Seubert A, Calabro S, Santini L, Galli B, Genovese A, Valentini S, et al. Adjuvanticity of the oil-in-water emulsion MF59 is independent of Nlrp3 inflammasome but requires the adaptor protein MyD88. Proc Natl Acad Sci USA. 2011;108(27):11169–74.
123. Lofano G, Mancini F, Salvatore G, Cantisani R, Monaci E, Carrisi C, et al. Oil-in-water emulsion MF59 increases germinal center B cell differentiation and persistence in response to vaccination. J Immunol (Baltim Md 1950). 2015;195(4):1617–27.

124. Didierlaurent AM, Morel S, Lockman L, Giannini SL, Bisteau M, Carlsen H, et al. AS04, an aluminum salt- and TLR4 agonist-based adjuvant system, induces a transient localized innate immune response leading to enhanced adaptive immunity. J Immunol (Baltim Md 1950). 2009;183(10):6186–97.

125. Complete biosynthesis of the potent vaccine adjuvant QS-21 | Nature Chemical Biology [Internet]. [cited 2024 Jul 27]. Available from: https://www.nature.com/articles/s41589-023-01538-5.

126. Diwan M, Elamanchili P, Cao M, Samuel J. Dose sparing of CpG oligodeoxynucleotide vaccine adjuvants by nanoparticle delivery. Curr Drug Deliv. 2004;1(4):405–12.

127. Siegel RL, Miller KD, Wagle NS, Jemal A. Cancer statistics, 2023. CA Cancer J Clin. 2023;73(1):17–48.

128. Liu J, Fu M, Wang M, Wan D, Wei Y, Wei X. Cancer vaccines as promising immunotherapeutics: platforms and current progress. J Hematol Oncol. 2022;15(1):28.

129. Sheikhlary S, Lopez DH, Moghimi S, Sun B. Recent findings on therapeutic cancer vaccines: an updated review. Biomol Ther. 2024;14(4):503.

130. Miller AM, Koşaloğlu-Yalçın Z, Westernberg L, Montero L, Bahmanof M, Frentzen A, et al. A functional identification platform reveals frequent, spontaneous neoantigen-specific T cell responses in patients with cancer. Sci Transl Med. 2024;16(736):eabj9905

131. Kawalec P, Paszulewicz A, Holko P, Pilc A. Sipuleucel-T immunotherapy for castration-resistant prostate cancer. A systematic review and meta-analysis. Arch Med Sci. 2012;8(5):767–75.

132. O'Donoghue C, Doepker MP, Zager JS. Talimogene laherparepvec: overview, combination therapy and current practices. Melanoma Manag. 2016;3(4):267–72.

133. Gaur A, Chirmule N. The promise of Immunotherapeutics and vaccines in the treatment of cancer. In: Sobti RC, Ganguly NK, Kumar R, editors. Handbook of oncobiology: from basic to clinical sciences [Internet]. Singapore: Springer Nature; 2023. p. 1–43. [cited 2024 Aug 3] Available from: https://doi.org/10.1007/978-981-99-2196-6_62-1.

134. Frontiers | Recent advances in mRNA cancer vaccines: meeting challenges and embracing opportunities [Internet]. [cited 2024 Aug 7]. Available from: https://www.frontiersin.org/journals/immunology/articles/10.3389/fimmu.2023.1246682/full.

135. Palmer CD, Rappaport AR, Davis MJ, Hart MG, Scallan CD, Hong SJ, et al. Individualized, heterologous chimpanzee adenovirus and self-amplifying mRNA neoantigen vaccine for advanced metastatic solid tumors: phase 1 trial interim results. Nat Med. 2022;28(8):1619–29.

136. Yarchoan M, Gane EJ, Marron TU, Perales-Linares R, Yan J, Cooch N, et al. Personalized neoantigen vaccine and pembrolizumab in advanced hepatocellular carcinoma: a phase 1/2 trial. Nat Med. 2024;30(4):1044–53.

137. Moderna And Merck Announce mRNA-4157 (V940) In Combination with Keytruda(R) (Pembrolizumab) Demonstrated Continued Improvement in Recurrence-Free Survival and Distant Metastasis-Free Survival in Patients with High-Risk Stage III/IV Melanoma Following Complete Resection Versus Keytruda at Three Years [Internet]. [cited 2024 Aug 17]. Available from: https://investors.modernatx.com/news/news-details/2023/Moderna-And-Merck-Announce-mRNA-4157-V940-In-Combination-with-KeytrudaR-Pembrolizumab-Demonstrated-Continued-Improvement-in-Recurrence-Free-Survival-and-Distant-Metastasis-Free-Survival-in-Patients-with-High-Risk-Stage-IIIIV-Melanoma-Following-Comple/default.aspx.

138. Falcaro M, Soldan K, Ndlela B, Sasieni P. Effect of the HPV vaccination programme on incidence of cervical cancer and grade 3 cervical intraepithelial neoplasia by socioeconomic deprivation in England: population based observational study. BMJ. 2024;385:e077341.

139. Ma L, Dichwalkar T, Chang JYH, Cossette B, Garafola D, Zhang AQ, et al. Enhanced CAR-T cell activity against solid tumors by vaccine boosting through the chimeric receptor. Science. 2019;365(6449):162–8.

140. Reactive oxygen species overload: a review of plasma therapy and photobiomodulation for cancer treatment [Internet]. [cited 2024 Aug 7]. Available from: https://www.jkslms.or.kr/journal/view.html?doi=10.25289/ML.22.047.
141. Živanić M, Espona-Noguera A, Lin A, Canal C. Current state of cold atmospheric plasma and cancer-immunity cycle: therapeutic relevance and overcoming Clinical limitations using hydrogels. Adv Sci Weinh Baden-Wurtt Ger. 2023;10(8):e2205803.
142. Brány D, Dvorská D, Strnádel J, Matáková T, Halašová E, Škovierová H. Effect of cold atmospheric plasma on epigenetic changes, DNA damage, and possibilities for its use in synergistic cancer therapy. Int J Mol Sci. 2021;22(22):12252.
143. Kumar Dubey S, Dabholkar N, Narayan Pal U, Singhvi G, Kumar Sharma N, Puri A, et al. Emerging innovations in cold plasma therapy against cancer: a paradigm shift. Drug Discov Today. 2022;27(9):2425–39.
144. Mizrahi JD, Surana R, Valle JW, Shroff RT. Pancreatic cancer. Lancet. 2020;395(10242):2008–20.
145. Rojas LA, Sethna Z, Soares KC, Olcese C, Pang N, Patterson E, et al. Personalized RNA neoantigen vaccines stimulate T cells in pancreatic cancer. Nature. 2023;618(7963):144–50.
146. Balachandran VP, Łuksza M, Zhao JN, Makarov V, Moral JA, Remark R, et al. Identification of unique neoantigen qualities in long-term survivors of pancreatic cancer. Nature. 2017;551(7681):512–6.
147. Disis MLN, Guthrie KA, Liu Y, Coveler AL, Higgins DM, Childs JS, et al. Safety and outcomes of a plasmid DNA vaccine encoding the ERBB2 intracellular domain in patients with advanced-stage ERBB2-positive breast cancer: a phase 1 nonrandomized Clinical Trial. JAMA Oncol. 2023;9(1):71–8.
148. Zhou Y. HER2/neu-based vaccination with li-key hybrid, GM-CSF immunoadjuvant and trastuzumab as a potent triple-negative breast cancer treatment. J Cancer Res Clin Oncol. 2023;149(9):6711–8.
149. Tobias J, Garner-Spitzer E, Drinić M, Wiedermann U. Vaccination against Her-2/neu, with focus on peptide-based vaccines. ESMO Open. 2022;7(1):100361.
150. Abstract CT093: Vaccine immunotherapy by homologous antigenic loading as adjuvant therapy for glioblastoma: Ongoing phase I analysis | Cancer Research | American Association for Cancer Research [Internet]. [cited 2024 Aug 7]. Available from: https://aacrjournals.org/cancerres/article/84/7_Supplement/CT093/742698/Abstract-CT093-Vaccine-immunotherapy-by-homologous.
151. Investigational Vaccine TG4050 Shows Early Clinical Benefit in HPV– Head and Neck Cancers [Internet]. [cited 2024 Aug 3]. Available from: https://www.onclive.com/view/investigational-vaccine-tg4050-shows-early-clinical-benefit-in-hpv-head-and-neck-cancers.
152. miRNAAssayMayLeadtoEarlyDetectioninPDAC[Internet].[cited2024Aug3].Availablefrom: https://www.cancernetwork.com/view/mirna-assay-may-lead-to-early-detection-in-pdac.
153. Tarasiuk A, Mackiewicz T, Małecka-Panas E, Fichna J. Biomarkers for early detection of pancreatic cancer – miRNAs as a potential diagnostic and therapeutic tool? Cancer Biol Ther. 22(5–6):347–56.
154. AACR Cancer Progress Report [Internet]. Cancer Progress Report. [cited 2024 Aug 3]. Available from: https://cancerprogressreport.aacr.org/progress/.
155. Singhal A, Li BT, O'Reilly EM. Targeting KRAS in cancer. Nat Med. 2024;30(4):969–83.
156. (17) Abstract LB197: A pooled mutant KRAS peptide vaccine activates polyfunctional T cell responses in patients with resected pancreatic cancer | Request PDF [Internet]. [cited 2024 Aug 7]. Available from: https://www.researchgate.net/publication/370042265_Abstract_LB197_A_pooled_mutant_KRAS_peptide_vaccine_activates_polyfunctional_T_cell_responses_in_patients_with_resected_pancreatic_cancer.
157. Lymph-node-targeted, mKRAS-specific amphiphile vaccine in pancreatic and colorectal cancer: the phase 1 AMPLIFY-201 trial | Nature Medicine [Internet]. [cited 2024 Aug 7]. Available from: https://www.nature.com/articles/s41591-023-02760-3.

158. Rappaport AR, Kyi C, Lane M, Hart MG, Johnson ML, Henick BS, et al. A shared neoantigen vaccine combined with immune checkpoint blockade for advanced metastatic solid tumors: phase 1 trial interim results. Nat Med. 2024;30(4):1013–22.
159. Ellingsen EB, Bjørheim J, Gaudernack G. Therapeutic cancer vaccination against telomerase: clinical developments in melanoma. Curr Opin Oncol. 2023;35(2):100–6.
160. March 2023 – Volume 35 – Issue 2: Current Opinion in Oncology [Internet]. [cited 2024 Aug 28]. Available from: https://journals.lww.com/co-oncology/toc/2023/03000.
161. Kaushik R, Kant R, Christodoulides M. Artificial intelligence in accelerating vaccine development – current and future perspectives. Front Bacteriol. 2023; [cited 2024 Aug 3];2. Available from: https://www.frontiersin.org/journals/bacteriology/articles/10.3389/fbrio.2023.1258159/full.
162. Evaxion biotech [Internet]. [cited 2024 Aug 7]. Available from: https://www.evaxion-biotech.com/.
163. Crooke SN, Ovsyannikova IG, Kennedy RB, Poland GA. Immunoinformatic identification of B cell and T cell epitopes in the SARS-CoV-2 proteome. Sci Rep. 2020;10:14179.
164. Travieso T, Li J, Mahesh S, Mello JDFRE, Blasi M. The use of viral vectors in vaccine development. Npj Vaccines. 2022;7(1):1–10.
165. Montalbán-Hernández K, Cogollo-García A, Girón de Velasco-Sada P, Caballero R, Casanovas M, Subiza JL, et al. MV130 in the prevention of recurrent respiratory tract infections: a retrospective real-world study in children and adults. Vaccine. 2024;12(2):172.
166. Brandi P, Conejero L, Cueto FJ, Martínez-Cano S, Dunphy G, Gómez MJ, et al. Trained immunity induction by the inactivated mucosal vaccine MV130 protects against experimental viral respiratory infections. Cell Rep. 2022;38(1):110184.
167. Laumont CM, Vincent K, Hesnard L, Audemard É, Bonneil É, Laverdure JP, et al. Noncoding regions are the main source of targetable tumor-specific antigens. Sci Transl Med. 2018;10(470):eaau5516.
168. Kina E, Laverdure JP, Durette C, Lanoix J, Courcelles M, Zhao Q, et al. Breast cancer immunopeptidomes contain numerous shared tumor antigens. J Clin Invest. 2024;134(1):e166740.
169. Improving T-cell mediated immunogenic epitope identification via machine learning: the neoIM model | bioRxiv [Internet]. [cited 2024 Aug 28]. Available from: https://www.biorxiv.org/content/10.1101/2022.06.03.494687v1.full.
170. Lybaert L, Thielemans K, Feldman SA, van der Burg SH, Bogaert C, Ott PA. Neoantigen-directed therapeutics in the clinic: where are we? Trends Cancer. 2023;9(6):503–19.
171. Ramírez W, Torralba D, Bourg V, Lastre M, Perez O, Jacquet A, et al. Immunogenicity of a novel anti-allergic vaccine based on house dust mite purified allergens and a combination adjuvant in a murine prophylactic model. Front Allergy. 2022;3:1040076.
172. Generation of a virus-like particles based vaccine against IgE – Gharailoo – 2024 – Allergy – Wiley Online Library [Internet]. [cited 2024 Aug 7]. Available from: https://onlinelibrary.wiley.com/doi/full/10.1111/all.16090.
173. A Phase 1 First-in-Human Study (B4901001) Evaluating a Novel Anti-IgE Vaccine in Adult Subjects with Allergic Rhinitis – Open Access Library [Internet]. [cited 2024 Aug 3]. Available from: https://www.oalib.com/paper/5825337.
174. Yu HJ, Dickson SP, Wang PN, Chiu MJ, Huang CC, Chang CC, et al. Safety, tolerability, immunogenicity, and efficacy of UB-311 in participants with mild Alzheimer's disease: a randomised, double-blind, placebo-controlled, phase 2a study. eBioMedicine. 2023; [cited 2024 Aug 17];94. Available from: https://www.thelancet.com/journals/ebiom/article/PIIS2352-3964(23)00230-X/fulltext.
175. Garay-Gutiérrez NF, Hernandez-Fuentes CP, García-Rivas G, Lavandero S, Guerrero-Beltrán CE. Vaccines against components of the renin–angiotensin system. Heart Fail Rev. 2021;26(3):711–26.
176. Wu H, Wang Y, Wang G, Qiu Z, Hu X, Zhang H, et al. A bivalent antihypertensive vaccine targeting L-type calcium channels and angiotensin AT1 receptors. Br J Pharmacol. 2020;177(2):402–19.

177. Azegami T, Itoh H. Vaccine development against the renin-angiotensin system for the treatment of hypertension. Int J Hypertens. 2019;2019:9218531.

178. Yoshida S, Nakagami H, Hayashi H, Ikeda Y, Sun J, Tenma A, et al. The CD153 vaccine is a senotherapeutic option for preventing the accumulation of senescent T cells in mice. Nat Commun. 2020;11:2482.

179. Katsuumi G, Shimizu I, Suda M, Yoshida Y, Furihata T, Joki Y, et al. SGLT2 inhibition eliminates senescent cells and alleviates pathological aging. Nat Aging. 2024;4(7):926–38.

180. Talwar GP, Singh O, Pal R, Chatterjee N, Sahai P, Dhall K, et al. A vaccine that prevents pregnancy in women. Proc Natl Acad Sci USA. 1994;91(18):8532–6.

181. Talwar GP, Gaur A. Recent developments in immunocontraception. Am J Obstet Gynecol. 1987;157(4 Pt 2):1075–8.

182. Talwar GP, Singh O, Singh V, Rao DN, Sharma NC, Das C, et al. Enhancement of antigonadotropin response to the beta-subunit of ovine luteinizing hormone by carrier conjugation and combination with the beta-subunit of human chorionic gonadotropin. Fertil Steril. 1986;46(1):120–6.

183. Bröker M, Berti F, Schneider J, Vojtek I. Polysaccharide conjugate vaccine protein carriers as a "neglected valency" – potential and limitations. Vaccine. 2017;35(25):3286–94.

184. A unique vaccine for birth control and treatment of advanced stage cancers secreting ectopically human chorionic gonadotropin [Internet]. [cited 2024 Aug 3]. Available from: https://www.explorationpub.com/Journals/ei/Article/100326.

Discovery of Biologics as Therapeutic Drugs: Past, Present, and Future

Taruna Arora and Sanjay D. Khare

Abstract In the last four decades, biologics have been recognized as an important class of drugs for the treatment of debilitating autoimmune, metabolic, neurological diseases, and cancer. We will review the evolution of biologics, current status, and considerations in their development as drugs. During the course of this scientific breakthroughs, technical and manufacturability challenges combined with commercialization and regulatory considerations for product approval have led to improved design and development course of the biologic drugs, which now incorporate product's end goals for high quality, optimal activity, improved safety, and delivery for a given indication at the very early phase of the drug discovery and design. This chapter will discuss how engineering of formats has enabled optimization of biologics to maximize their effectiveness and minimize potential safety issues. We will also briefly describe potential pharmacological and safety issues observed with experimental drugs such as (i) short half life, (ii) development of neutralizing anti-drug antibodies, (iii) preclinical false alarms and/or redundant target biology and (iv) undesired safety concerns.

Finally, we will discuss next generation therapeutics that uses antibody and its derivatives to develop novel classes of drugs (e.g., bispecifics, bifunctional, T cell engagers, ADCs, etc.).

Keywords Biologics · Antibodies · Antibody-Drug Conjugates · Bi-specifics · Antibody Engineering

T. Arora (✉)
Bristol Myers Squibb, Trenton, NJ, USA

Immunome, Bothell, WA, USA
e-mail: tarora@immunome.com

S. D. Khare (✉)
RST Biotech, Palo Alto, CA, USA
e-mail: skhare@me.com

© The Author(s), under exclusive license to Springer Nature Switzerland AG 2025   93
N. Chirmule, V. V. Ghalsasi (eds.), *Approved: The Life Cycle of Drug Development*,
https://doi.org/10.1007/978-3-031-81787-8_3

1 Introduction

Biologics, often referred to as large molecules to distinguish them from small molecule therapeutics have evolved from naturally derived products such as body tissues and fluids to recombinant expression, revolutionizing the biotech and pharmaceutical industries. The introduction of recombinant expression technologies has been central to this transformation, enabling the large-scale production of native sequence-based proteins in the early days of biotechnology. Over the following decades, advancements in protein engineering and process optimization accelerated the development of therapeutic proteins, leading to the approval of various biologics, including hormones, enzymes, growth factors, cytokines, and antibodies. A significant milestone was the development of hybridoma technology, which facilitated the generation of monoclonal antibodies. The 1986 approval of the OKT3 monoclonal antibody by the US Food and Drug Administration (FDA) for treating acute graft rejection set a precedent for subsequent monoclonal antibodies, which were later improved through chimeric, humanized, or fully human versions to reduce the immunogenicity associated with non-human sequences. Today, biologics are used to treat a wide range of diseases, including autoimmune disorders, cancer, metabolic conditions, neurological diseases, infectious diseases, and genetic disorders.

As discovery platforms advanced and the understanding of structure-function relationships improved, biologics-predominantly antibodies have been further designed and engineered for higher specificity, selectivity, and half-life modulation, as well as targeted functional modulation. Insights gained from monoclonal antibodies have driven the development of more complex therapeutic modalities, such as antibody-drug conjugates (ADCs), and radioligand conjugates which combine the specificity of antibodies with the cytotoxic power of small molecules, radioisotopes, or peptides. Additionally, immune cell engagers have been designed to bring T or NK cells into close proximity with cancer cells to enhance immune responses. Other innovations include probodies, single-chain variable (V) domain fusions, and chimeric antigen receptor T-cell therapies (CAR-T), all of which have expanded the horizons of targeted therapies aimed at addressing unmet medical needs. Moreover, bi- and multi-specific antibodies, as well as fusion proteins that combine antibodies with peptides, receptor domains, or cytokines, are being developed to simultaneously modulate multiple disease targets, thereby increasing therapeutic efficacy and selectivity.

Advancements in Fc engineering have enabled the modulation of effector functions of Fc containing biologics, enhancing immune responses and extending the half-life of therapeutic antibodies. The conformational flexibility of the Fc region, along with access and engineering for cysteines and glycosylation sites, has allowed for site-specific conjugation and insertions. As a result, biologics, particularly Fc-based modalities, have greatly benefited from these design considerations by incorporating desired therapeutic properties in the design and allowing therapeutic candidates to be fine-tuned for specific immune-modulatory functions through Fc engineering.

Today, biologics are more complex, with additional functionalities incorporated into their design to develop potential treatments for a broader range of diseases, including those previously considered undruggable by large molecules. This increasing complexity requires careful consideration of several key factors, including structure, potency, safety, and manufacturing. Safety concerns, such as immunogenicity, are often addressed through predictive analyses, including in silico and in vitro cell-based assays. Additionally, large-scale manufacturing is essential for managing costs, while formulation development ensures drug stability and convenient dosing, both of which are critical to producing high-quality therapeutics.

Throughout the evolution of scientific breakthroughs in biologics, numerous lessons have been learned from both successes and failures. These lessons, stemming from technical challenges, manufacturability hurdles, commercialization, and regulatory considerations, have significantly improved the design of biologic drugs. Today, biologics are developed with a focus on high quality, efficacy, safety, and product delivery aspects for patient convenience. This chapter will delve into the critical factors and considerations involved in designing and developing various classes of biologic modalities to ensure a holistic approach that aligns each modality with its mechanism of action and desired therapeutic outcome for optimal patient treatment.

We will explore three key classes of biologics: small protein therapeutics, Fc-fusion proteins, and antibody-based modalities.

2 Small Protein Therapeutics

The treatment of various diseases often requires the use of endogenous secreted proteins as therapeutics, such as insulin, GLP-1 derivatives, erythropoietin, and cytokines like IL-2. Generally, polypeptides shorter than 40 amino acids can be manufactured via chemical synthesis. However, peptides and proteins with more complex structures and post-translational modifications, such as glycosylation and disulfide bonds, require microbial or mammalian sources for production. In the mid-to-late 1980s, scientists focused on key endogenous protein molecules deficient in pathological conditions (e.g., hormones, enzymes, growth factors, and cytokines), initially isolating these proteins from natural sources. Later on, the advent of recombinant DNA technologies enabled these products to be produced at scale with more consistent quality and predictable attributes to provide effective treatments for patients' unmet needs.

One of the most monumental achievements in biologic drug discovery was the development of insulin, a peptide consisting of 51 amino acids. Insulin serves as an excellent example of how lessons learned from early use, including efficacy, safety, and quality considerations, shaped its development. Unlike single-chain peptide hormones, insulin has a more complex structure, consisting of two chains linked by interchain disulfide bonds. In the body, it is produced as a hexamer, an inactive form that dissociates in circulation to release active insulin [1]. Understanding this structure-function relationship was crucial for the successful development of insulin as a therapeutic which has evolved from crude animal pancreatic extracts to recombinant human insulin and insulin analogs. First isolated by Frederick Banting in 1921, insulin was further developed by Charles Best under the leadership of John Macleod at the University of Toronto. Banting and Macleod earned the Nobel Prize in 1923 for the discovery of insulin [2]. Initially animal-sourced insulin provided an active form of insulin, but challenges arose with immunogenicity, leading to lipoatrophy and insulin resistance in many patients. Chromatographic purification techniques were developed in the 1970s to reduce product heterogeneity, and semisynthetic human insulin was produced in the 1980s by substituting a single amino acid in porcine insulin [3]. The discovery of the insulin gene in 1977 and evolution of recombinant DNA technology [4, 5] supported large-scale production of human insulin to ensure consistent source of product supply for patients. The first recombinant human insulin, Humulin® R, was approved in 1982, followed by Novolin® R and Insuman® R [reviewed in 6]. While low levels of anti-insulin antibodies still appear in most patients using recombinant products, these are considered to have less clinical significance. The development of insulin is a prime example of product optimization through engineering, manufacturing, and formulation, aligning the drug's modality with its physiological mechanism and therapeutic profile. Latest advancements have led to the approval of rapid-acting, ultra-rapid-acting, and even inhalable insulin analogs (Fig. 1).

A few years after the approval of recombinant insulin, other biotherapeutics like erythropoietin [Epogen, reviewed in 7] and granulocyte colony-stimulating factor

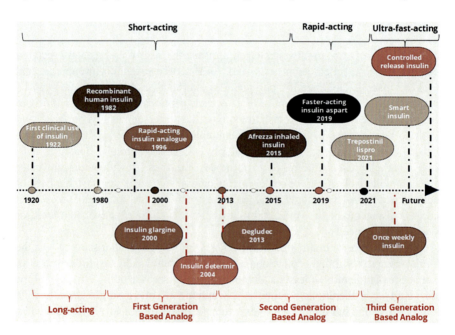

Fig. 1 Evolution of first biologic drug: insulin for diabetes

[G-CSF or Neupogen, reviewed in 8] were produced recombinantly. These drugs, used to stimulate red and white blood cell production respectively, were approved to treat anemia (1989, Epogen/Procrit, Amgen/JnJ) and neutropenia (1991, Neupogen, Amgen). In the next 5–10 years, important drugs like Activase (tissue plasminogen activator) and Infergen (interferon-a) were approved for treating heart attacks and viral infections, respectively. Recombinant human growth hormone (rhGH) was also approved by the FDA to treat growth hormone deficiency in children and adults (e.g., Somatropin, Genotropin, Humatrope, Norditropin). However, small proteins with a molecular weight of less than 150 kDa tend to have shorter pharmacokinetic and pharmacodynamic durations due to rapid renal clearance, necessitating frequent dosing for desired efficacy and exposure.

Short half-life and frequent administration requirements of small proteins were addressed by engineering small biologics for longer half-lives. For example, several GLP-1 receptor agonists have been developed in the form of protease resistance version of endogenous GLP-1 ligand (for example semaglutide), or as fusions of GLP-1 peptide variants with albumin or Fc (albglutide and dulaglutide respectively) (Table 1). These fusions were developed for once-weekly dosing for the treatment of diabetes [albiglutide; 9, 10, Dulaglutide; 11]. In 2017, GSK withdrew albiglutide from the market, citing economic reasons as its benefits did not outweigh those of semaglutide or tirzepatide (an analog of gastric inhibitory peptide GIP with dual agonism for receptors of GIP and GLP-1) . On a more positive note, the half-life

Table 1 A list of GLP-1 agonist drugs for the treatment of type 2 diabetes and for weight loss management

Drug name	Brand name	Administration	Primary use	Additional notes
Semaglutide	Ozempic	Injection (weekly)	Type 2 diabetes	May reduce risk of cardiovascular events; weight loss as a side effect
Semaglutide	Wegovy	Injection (weekly)	Weight loss	Higher dose of Ozempic; specifically approved for weight loss
Liraglutide	Victoza	Injection (daily)	Type 2 diabetes	Can be used for weight loss in higher doses (Saxenda)
Liraglutide	Saxenda	Injection (daily)	Weight loss	Higher dose of Victoza; specifically approved for weight loss
Dulaglutide	Trulicity	Injection (weekly)	Type 2 diabetes	Convenient once-weekly dosing
Exenatide	Byetta	Injection (twice daily)	Type 2 diabetes	Short-acting; taken before meals
Exenatide ER	Bydureon	Injection (weekly)	Type 2 diabetes	Extended-release version of Byetta; once-weekly dosing
Lixisenatide	Adlyxin	Injection (daily)	Type 2 diabetes	Short-acting; taken before meals
Semaglutide	Rybelsus	Oral (daily)	Type 2 diabetes	First oral GLP-1 agonist; taken once daily
Tirzepatide	Mounjaro	Injection (weekly)	Type 2 diabetes	Dual GLP-1/GIP receptor agonist; new class of medication

extension of products like Epogen and Neupogen, through engineered glycosylation (Aranesp) or pegylation (PEG-filgrastim), improved half life leading to less frequent dosing for patient convenience without compromising the efficacy of erythropoietin [12]. More recently, GLP-1 biologic drugs have shown efficacy in the treatment of obesity thus expanding the use of these anti-diabetic drugs for weight loss [13].

Cytokines have also been used as therapeutic drugs due to their immune-related effects. For example, recombinant IL-2 is used to treat melanoma, and interferon-a2b has been approved for treating diseases like hepatitis and follicular non-Hodgkin lymphoma. However, cytokines often exhibit pleiotropic effects due to the complexity of cytokine-receptor interactions, which can influence downstream signaling, adaptor protein recruitment, and immune modulation across different cell types. As a result, high-dose IL-2 treatment has been associated with severe toxicity. One hypothesis is that native IL-2 at high doses binds to the high-affinity IL-2Rabg receptor complex, which is expressed on both cytotoxic T cells and regulatory T cells. This binding leads to the expansion of regulatory T cells, thus limiting IL-2's efficacy. To overcome these limitations, protein engineering approaches have generated IL-2 muteins with selective binding to the low-affinity IL-2Rbg complex [15, and Table 2]. Additionally, recombinant cytokines are being explored as antibody fusions for tumor targeting, synergistic effects, and improved half-life. Several antibody-cytokine fusions are currently in clinical trials.

One of the known challenges with endogenous sequence-based protein therapeutics is their propensity for aggregation during manufacturing. In the body, physiological levels of cytokines, hormones, and growth factors are typically low in circulation. However, manufacturing and dosing considerations require these proteins to be produced and formulated at higher concentrations in buffers that are compatible with the route of administration and shelf stability. High concentrations can lead to aggregation, and both protein aggregates and certain excipients in the formulation can increase the risk of immunogenicity. Anti-drug antibodies (ADA) generated in response to recombinant proteins can also bind and neutralize endogenous proteins, potentially persisting in the body for extended periods and leading to serious adverse events. For instance, ADA elicited by Epogen administration have caused pure red blood cell aplasia (PRCA) [14]. A surge in PRCA cases was linked to a change in the excipients used in the Eprex formulation [16].

Proteins are also susceptible to clipping and proteolytic cleavage during production in host cells. To improve the solubility of proteins, glycosylation, pegylation, and fusions to albumin or Fc are commonly employed, while mutagenesis may be required to prevent cleavage while retaining the protein's activity [reviewed in 12]. Furthermore, advanced purification techniques and sophisticated analytical methods are essential to ensure consistent product quality across batches and mitigate risks associated with variable product attributes.

Table 2 Various forms of interleukin-2 in clinical development

Drug name	Company	Properties	Lead indications	Phase of development
Bempegaldesleukin	Nektar Therapeutics, BMS	IL-2 with six cleavable PEG groups	Melanoma, RCC, Bladder	Phase III
Nemvaleukin alfa	Alkermes	Circularly permuted IL-2v–IL2Rα fusion protein	Solid tumors	Phase II
SAR444245	Sanofi/Synthorx	IL-2 with one non-cleavable PEG group	Solid tumors	Phase I/II
RG6279	Roche	IL-2v–anti-PD1 mAb fusion protein	Solid tumors	Phase I
CUE-101	Cue Biopharma	IL-2–HLA complex–HPV16 E7 peptide fusion protein	Head and neck cancer	Phase I
NL-201	Neoleukin Therapeutics	IL-2 protein mimetic, computationally designed	–	IND
AU-007	Aulos Bioscience	Anti-IL-2 mAb, computationally designed	–	IND
STK-012	Synthekine	IL-2 partial agonist, targeting activated T cells	–	IND
KY1043	Kymab	IL-2v–anti-PDL1 mAb fusion protein	–	IND
BNT151	BioNTech	IL-2v, mRNA encoded	–	IND
MDNA11	Medicenna Therapeutics	IL-2 'superkine', albuminated	–	IND
WTX-124	Werewolf Therapeutics	Conditionally activated IL-2	–	Preclinical

3 Improving Half-life of Proteins Therapeutics

A variety of protein-engineering technologies are currently employed to enhance the circulating half-life, targeting, and functionality of therapeutic protein drugs, as well as to increase production yield and product purity. Techniques such as protein conjugation and derivatization—including Fc-fusion, albumin-fusion, and PEGylation—are commonly used to prolong a biologic drug's half-life in circulation. Extended in vivo half-lives are especially beneficial for patients undergoing factor, enzyme, or hormone replacement therapies, where frequent dosing can significantly impact patient well-being and compliance, particularly in young children.

Notable examples of albumin-fusion proteins include pegfilgrastim (agonist for GCSF-R), certolizumab pegol (antagonist, anti-TNFα Scfv-PEG), and pegintron (agonist for IFNα receptor). These fusion proteins leverage the long half-life of albumin to reduce dosing frequency. Receptor extracellular domains have proven effective as antagonists by binding and blocking their ligands from mediating pro-inflammatory effects through endogenous receptors. Fc fusion allows for dimeric configurations of the receptors, improving the activity, expression, and yield of these complex receptor ectodomains. Blockbuster antagonist drugs based on the receptor-Fc fusions include Etanercept (Enbrel, Amgen), which is used to treat rheumatoid arthritis, psoriasis, and spondyloarthropathies, and aflibercept (Eylea, Regeneron), approved for the treatment of wet macular degeneration and diabetic retinopathy. Etanercept is a fusion of the type 2 TNF receptor (TNFR2) ectodomain with IgG1 Fc and blocks TNFα and lymphotoxin alpha (LTα) from binding to TNFR1 and TNFR2 on inflammatory cells, thereby inhibiting inflammatory pathways. Aflibercept, a recombinant fusion protein comprising extracellular domains of human VEGF receptors 1 and 2 fused to the Fc portion of human IgG1, is formulated for intravitreal injection. It blocks VEGF ligands from binding to VEGF receptors on endothelial cells, thus inhibiting angiogenic pathways associated with retinopathies and wet age-related macular degeneration.

While PEGylation increases a protein's molecular weight and size, reducing rapid renal clearance, extended half-life for albumin and Fc fusion proteins relies on FcRn-mediated recycling. FcRn, present on endothelial and immune cells, binds to albumin and IgG Fc at acidic pH in endosomal vesicles, recycling these proteins to the cell surface, where they dissociate and re-enter circulation. FcRn knockout and transgenic mouse models have demonstrated the essential role of FcRn in extending the half-life of Fc fusion proteins. Some peptide-based therapies have also benefited from Fc fusions, which facilitate less frequent dosing and enhance peptide solubility and stability. For example, Romiplostim (Nplate, Amgen) is a thrombopoietin mimetic peptide-Fc fusion used to treat immune thrombocytopenia (ITP) with weekly dosing.

The Fc component of antibodies plays a dynamic and crucial role in both circulation and therapeutic efficacy. It can mediate immunomodulation and effector functions through both FcRn and Fc gamma receptors. When designing molecules with Fc fragments, it is vital to consider these biological effects, a topic that will be explored in greater detail in the following sections. A list of recombinantly expressed therapeutic proteins with -Fc and -Albumin is shown in Table 3.

4   Antibodies as Therapeutics

Over the past three decades, monoclonal antibody technology has become a powerful tool for developing a wide range of human therapeutics. More than 125 monoclonal antibodies (mAbs) are now FDA-approved for treating various inflammatory, neurological, metabolic diseases, and cancers (a partial list of therapeutic antibodies is shown in Table 4). The history of therapeutic antibodies is a fascinating journey, beginning with the pioneering observations of Emil von Behring and Shibasaburo Kitasato, who demonstrated that serum from infected animals could be used to treat and prevent infection in other animals. Subsequent research identified B cells as the primary source of immunoglobulins and led to the elucidation of the IgG1 antibody structure (Figs. 2 and 3).

Table 3 A list of therapeutic recombinant fusion proteins

Name	Brand name	rFusion protein	Indication	Approval year
Etanercept	Enbrel	TNFR2-Fc	Rheumatoid arthritis, psoriasis	1998
Alefacept	Amevive	LFA-Fc	Psoriasis	2003
Abatacept	Orencia	CTLA4-Fc	Rheumatoid arthritis	2005
Rilonacept	Arcalyst	IL-1 trap	Cryopyrin-associated periodic syndromes	2008
Belatacept	Nulojix	CTLA4-Fc	Organ transplant rejection	2011
Romiplostim	Nplate	TPO agonist	Chronic immune thrombocytopenia	2008
Aflibercept	Eylea	VEGF-trap	Macular degeneration	2011
Efmoroctocog alfa	Eloctate	Factor VIII – Fc	Hemophilia A	2014
Albiglutide[a]	Tanzeum	GLP-1 agonist	Type 2 diabetes	2014
Idelvion	Idelvion	Factor IX – Albumin	Hemophilia B	2016

[a]discontinued

Table 4 A partial list of therapeutic Antibodies for various clinical indication(s)

Antibody	Target	Brand name	Clinical indication(s)	Approval year
Rituximab	CD20	Rituxan/ Mabthera	Non-Hodgkin lymphoma, rheumatoid arthritis	1997
Infliximab	TNF-α	Remicade	Rheumatoid arthritis, Crohn's disease	1998
Trastuzumab	HER2	Herceptin	HER2-positive breast cancer	1998
Adalimumab	TNF-α	Humira	Rheumatoid arthritis, Crohn's disease	2002
Cetuximab	EGF receptor	Erbitux	Various cancer (e.g., colorectal, head, and neck)	2004
Bevacizumab	VEGF	Avastin	Various cancers (e.g., colorectal, lung)	2004
Eculizumab	Complement protein C5	Soliris	Paroxysmal nocturnal hemoglobinuria (PNH)	2007
Ustekinumab	IL-12 and IL-23	Stelara	Psoriasis, Crohn's disease	2009
Tocilizumab	IL-6 receptor	Actemra	Rheumatoid arthritis, cytokine release syndrome	2010
Denosumab	RANKL/ OPG-L	Prolia/ Xgeva	Post menoposal osteoporosis, bone metastatis	2010
Belimumab	BLyS/BAFF	Benlysta	Systemic lupus erythematosus	2011
Ipilimumab	CTLA-4	Yervoy	Various cancer (e.g., melanoma)	2011

(continued)

Table 4 (continued)

Antibody	Target	Brand name	Clinical indication(s)	Approval year
Pembrolizumab	PD-1	Keytruda	Various cancers (e.g., melanoma, NSCLC)	2014
Nivolumab	PD-1	Opdivo	Various cancers (e.g., melanoma, NSCLC)	2014
Vedolizumab	Integrin α4β7	Entyvio	Ulcerative colitis, Crohn's disease	2014
Blinatumomab	CD19 and CD3	Blincyto	Acute lymphoblastic leukemia	2014
Ramucirumab	VEGFR2	Cyramza	Various cancers (e.g., gastric, lung)	2014
Daratumumab	CD38	Darzalex	Multiple myeloma	2015
Secukinumab	IL-17A	Cosentyx	Psoriasis, ankylosing spondylitis	2015
Alirocumab	PCSK9	Praluent	Hypercholesterolemia	2015
Evolocumab	PCSK9	Repatha	Hypercholesterolemia	2015
Mepolizumab	IL-5	Nucala	Severe eosinophilic asthma	2015
Atezolizumab	PD-L1	Tecentriq	Various cancers (e.g., bladder, lung)	2016
Ixekizumab	IL-17A	Taltz	Psoriasis	2016
Reslizumab	IL-5	Cinqair	Severe eosinophilic asthma	2016
Dupilumab	IL-4 receptor α	Dupixent	Atopic dermatitis, asthma	2017
Ocrelizumab	CD20	Ocrevus	Multiple sclerosis	2017
Emicizumab	Factor IXa and X	Hemlibra	Hemophilia A	2017
Durvalumab	PD-L1	Imfinzi	Various cancers (e.g., bladder, lung)	2017
Sarilumab	IL-6 receptor	Kevzara	Rheumatoid arthritis	2017
Benralizumab	IL-5 receptor	Fasenra	Severe eosinophilic asthma	2017
Ibalizumab	CD4	Trogarzo	HIV-1 infection	2018
Cemiplimab	PD-1	Libtayo	Cancer (e.g., cutaneous squa cell carcinoma, NSCLC)	2018
Teprotumumab	IGF-1 receptor	Tepezza	Thyroid eye disease	2020
Aducanumab	Amyloid-β	Aduhelm	Alzheimer's disease	2021
Nivolumab+Relatlimab	PD-1 and LAG-3	Opdualag[a]	Melanoma	2022
Toripalimab	PD-1	Loqtorzi	Cancer (e.g., nasopharygeal carcinoma)	2023
Pemivibart	SARS-Cov-2 spike protein	Pemgarda	Covid-19 prevention	2024
Nipocalimab	FcRn	Pending	Myasthenia Gravis	
Bentracimab	Ticagrelor	Pending	Reversal of ticagrelor's antiplatelet effects	
Zenocutuzumab	Her2, Her3	Pending	Various cancer	
PF-06480605	TL1A		Ulcerative colitis	

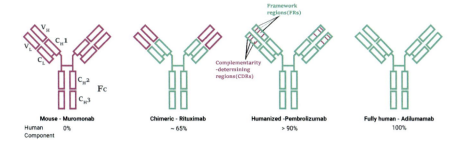

Fig. 2 Diagrammatic representation of various versions of therapeutic monoclonal antibodies (Mabs)

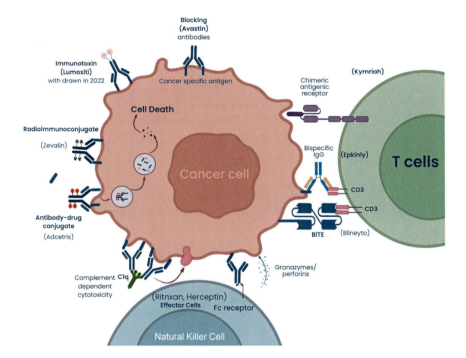

Fig. 3 Biological functions of various forms of antibodies, including therapeutics with antibody derivatives. Naked antibodies can effectively (i) block (e.g., Avastin, Keytruda), (ii) directly activate the biological function, in addition to (iii) targeted cell killing via ADCC/CDC/ADCP mechanisms (e.g., Rituxan) to ameliorate various diseases. Antibodies and its derivatives have also been used for targeted approaches in cell/gene/radioligand therapies (e.g. CAR-T, BiTE, ADCs, immunotoxins, radioimmunoconjugates, etc.)

It was the advent of monoclonal antibody technology, pioneered by César Milstein and Georges Köhler in 1975, which marked the era of development for monoclonal antibodies for therapeutic purposes. By fusing B cells with a myeloma cell line, they created hybridomas that produce antibodies with single specificity for a given target. This technology transformed the therapeutic application of monoclonal antibodies by enabling the production of a homogeneous source of clonal products. The first therapeutic antibody, a murine antibody against CD3 known as Muromonab-CD3 (trade name: Orthoclone OKT3), was approved by the U.S. Food and Drug Administration (FDA) in 1986. It has been used to prevent organ rejection in transplant patients, including those undergoing kidney transplants. While effective in preventing organ rejection, the murine nature of the antibody raised concerns about the development of anti-drug antibodies. This challenge led to the development of chimeric and fully human antibody technologies to reduce immunogenicity.

In 1969, Edelman et al. deciphered the amino acid sequence of IgG1, revealing its Y-shaped structure consisting of variable (V) regions and constant (C) regions of two polypeptide chains—heavy and light chains—joined by interchain disulfide bonds. They identified V regions as determinants of antigen specificity and the constant region of the heavy chain as playing a role in effector functions [19]. This structural and functional understanding paved the way for recombinant technologies to optimize antibody structures, reducing immunogenicity and enhancing desired effector functions through the Fc region. Thus a number of drugs were developed as chimeric, fully human, and later as multispecific antibodies. These classes of antibody drugs and Fc biology are discussed in detail in the following sections.

4.1   Chimeric Antibodies

Chimeric antibodies were developed to address the immunogenicity issues associated with murine-derived antibodies [20, 21]. These antibodies are created by fusing the variable domain of an antibody from a non-human species (such as mouse, rabbit, or llama) with the constant domain of an antibody from a human source. This fusion results in a chimeric antibody that maintains the specificity of the original antibody while incorporating the desirable properties of the human constant region (see Fig. 1).

Examples of successful chimeric antibodies include Infliximab [22] and Rituximab [23], which are widely used for treating autoimmune diseases and cancers. These chimeric antibodies revolutionized biologic treatments for conditions such as rheumatoid arthritis and non-Hodgkin lymphoma. Infliximab targets and neutralizes TNF-alpha, thereby suppressing its inflammatory effects in diseases like rheumatoid arthritis, psoriasis, and Crohn's disease [24]. Similarly, Rituximab binds to CD20 on the surface of lymphoma cells, leading to cell depletion through antibody-dependent cellular cytotoxicity (ADCC) and complement-dependent cytotoxicity (CDC). Other examples of chimeric antibodies include Cetuximab (an anti-EGFR IgG1) for treating colorectal cancer and head and neck cancer, and Basiliximab (an anti-IL2Ra) for preventing organ transplant rejection. Despite their success, the non-human variable regions in chimeric antibodies can elicit immune responses in patients, leading to the formation of anti-drug antibodies (ADAs). These ADAs can neutralize the therapeutic effects of the antibody or cause adverse reactions.

4.2   Humanized Antibodies

To address the immunogenicity issues associated with chimeric antibodies due to murine variable (V) regions, humanization technology was developed. This technology involves designing antibody sequences that more closely resemble human immunoglobulin frameworks in the variable region while preserving the complementarity-determining regions (CDRs) from the original source to maintain antigen specificity. The humanization process typically involves grafting CDRs onto an optimal human framework. Understanding the three-dimensional structure and prevalence of framework classes is crucial for designing humanized V regions. The result of this humanization process is the creation of antibodies that are predominantly human in their sequence, including both the V regions and the constant regions [25]. Humanized antibodies have enabled the industry to harness the therapeutic potential of antibodies while minimizing unwanted immune responses. Examples of marketed products developed using humanization technology include Pembrolizumab, Trastuzumab, and Bevacizumab.

Antibody Humanization and Engineering

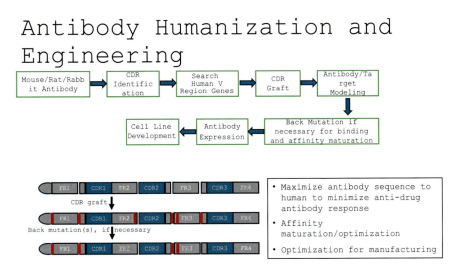

4.3 Fully Human Antibodies

With a deeper understanding of prevalent human framework classes and the ability to isolate sequences from the human B cell repertoire, display technologies like phage and yeast display have been developed to discover fully human antibodies. In phage display, antibody DNA sequences are inserted into a bacteriophage—a virus that infects bacteria—to create a library of antibody variants. The phage displays the antibody on its surface, allowing for the selection of those with desired binding properties. The library undergoes multiple rounds of selection and amplification: each round involves exposing the library to the target antigen, selecting antibodies that bind strongly, and then amplifying these selected antibodies for the next round. This iterative process enriches antibodies with enhanced binding affinity and specificity. Other methods, such as yeast display or ribosome display, also facilitate the directed evolution of antibodies, enabling the screening of extensive antibody libraries to identify optimized fully human antibodies with desirable properties. Traditionally, optimizing antibodies through directed evolution has been a time-consuming process, often involving specialized software tools and techniques to analyze and select the best antibody variants (figure display technologies)

In addition to display technologies, innovative human immunoglobulin (Ig) transgenic animal models have been developed to generate antibodies for therapeutic use. These models leverage in vivo mechanisms, such as hypermutation in germinal centers, to produce high-affinity antibodies with optimal biophysical properties. These transgenic mice contain randomly integrated human immunoglobulin heavy-chain (IGH) and kappa light-chain (IGK) transgenes, along with targeted loss-of-function alleles of the endogenous mouse Igh and Igk loci. However, the initial generation of these mice faced limitations, including position effects due to random transgene integration and constraints on the length of DNA that could be introduced via zygote injection, leading to incomplete human antibody repertoires. Additionally, these early models included both human variable and constant regions, resulting in suboptimal interactions between human constant regions and mouse signaling proteins (Igα and Igβ). This mismatch could impair antibody class switching, affinity maturation, and B-cell differentiation into mature antibody-secreting plasma cells. Subsequent improvements have addressed these issues by retaining mouse constant regions and inserting a complete human Ig variable repertoire. Modern advancements include common light chain mouse models and single VH domain mouse models for generating bispecific antibodies. Besides transgenic mice, fully human antibodies have also been developed using human Ig transgenic rats, rabbits, and chickens. Notable examples of antibodies generated from Ig transgenic mice that are FDA-approved and marketed include panitumumab (anti-EGFR) for cancer treatment, denosumab (anti-RANKL) for osteoporosis and cancer-related bone metastases, and nivolumab (anti-PD1) for immunotherapy. Kappa light chains were frequently used during the chimerization and humanization of wild-type rodent-derived antibodies. Many early display libraries were constructed using human kappa light chains, and antibodies containing kappa light chains have seen widespread success. However, antibodies with lambda light chains have also achieved clinical success. The use of both kappa and lambda light chain discovery technologies has expanded the scope of discovery programs, introducing novel binders. Lambda light chains may offer additional epitope diversity and unique functional activities that were not fully captured by kappa light chain discovery platforms.

Diversity of antibodies with these methods constitutes multiple attributes-diverse sequences of binders, different epitopes, species cross-reactivity, selectivity, affinity ranges, and intended applications (such as therapeutic, or reagents for IHC, and PK, etc.). Immunization and screening approaches both play an important role in obtaining antibodies of desired attributes against a target. For example, sequence conservation across mice and human would require methods that overcome immune tolerance in mice against the target of interest or the use of display methods. Multi-transmembrane proteins with limited extracellular region can be challenging targets for generating antibodies with desired profile. However, novel technologies such as mRNA and enriched membrane preps as immunogens have overcome some of these limitations.

4.4 AI/ML Application in Antibody Discovery

Artificial intelligence (AI) and machine learning (ML) are increasingly being integrated into antibody discovery processes to expedite discovery, optimization, and developability evaluation. By combining computational and biological sciences, these technologies are enhancing the design and development of protein and antibody drugs with the desired therapeutic profiles. AI and ML are employed in various aspects of antibody discovery, including sequence generation, analysis for liabilities and post-translational modifications, computational prediction of antibody structures and antigen interactions, and structural modeling. Many AI/ML and structural modeling tools are now available through open-access platforms.

ML tools are continually refined using experimental data, a practice known as "lab in the loop" or "design-make-test" cycles. While AI/ML tools are advancing sequence identification and some aspects of developability analysis, predictive design of desired sequences remains in its early stages. New AI techniques like generative models claim to create entirely new antibody sequences with desired properties. However, identifying target-specific candidates still involves an iterative process, and the comparative speed of AI approaches versus traditional methods in candidate discovery is yet to be fully determined. Nonetheless, AI/ML technologies clearly offer significant advantages in scaling analyses and reducing the time spent on manual evaluations. Biopharmaceutical companies and research institutions are utilizing AI/ML in their antibody discovery processes with the goal to accelerate the discovery of novel antibodies and optimize their characteristics for specific therapeutic applications.

There are several platforms that apply generative AI approach and use either (i) sequence-based or (ii) structure/geometry-based platforms. In addition, efforts are continuing to use the de novo approach to develop antibodies. These approaches are currently in nascent stages. Typically, AI/ML-based approaches are finding its applications for the following stages in biologic drug discovery [26].

(a) Affinity maturation: ML models can analyze vast datasets of antibody sequences and their corresponding antigen (target molecule) interactions using structure models. This allows computational scientists to predict how a new antibody sequence might bind to a specific antigen. This helps researchers in designing antibodies with targeting abilities in context of desired binding affinity.
(b) Predict immunogenicity and optimize humanization: AI/ML tools can also predict the potential immunogenicity of the humanized antibody by analyzing its sequence and structure. This can help in designing antibodies with reduced immunogenicity and improved safety profile. Accelerate humanization process by analyzing the sequence and structure of the mouse antibody and predicting the most suitable humanized variants. These tools utilize machine learning algorithms trained on large databases of antibody sequences and structures to make accurate predictions
(c) Support developability: AI/ML tools are useful in identifying amino acids or sequences within CDRs of antibodies or proteins such as cytokines that may be prone to modification during process development. These include prediction of sequence liabilities for deamidation, isomerization, cysteinylation, and tryptophan oxidation. Additionally, these tools can also identify hydrophobic patches in the solvent exposed regions of proteins and thus predict aggregation propensity [26, 27]

Currently, there are no AI/ML optimized antibodies in the clinical trials.

4.5 Fc Structure-Function Relationships and Its Impact on Design and Development Consideration

When designing Fc containing modalities, it is important to consider Fc biology and structure-function relationships. Following sections discuss the role of IgG Fc and its interactions with Fc receptors in modulating immune effector functions and half-life in circulation.

4.5.1 Role of Fc in Antibody-Mediated Functions

The Fc region of immunoglobulins plays a crucial role in mediating effector functions. There are five immunoglobulin isotypes found in serum, distinguished by the constant domain of the heavy chain: IgD, IgM, IgG, IgE, and IgA. The heavy chain constant domain is generally structured as CH1-CH2-CH3 for IgG, IgA, and IgD, with an additional CH4 domain present in IgM and IgE. Among these, IgG is the most prevalent isotype in circulation and is the backbone for the majority of approved therapeutic antibodies and Fc fusion proteins.

Both IgG and IgM isotypes can mediate effector functions such as complement-mediated cytotoxicity. Structurally, IgM is a pentamer, with each monomer joined by a J chain, while IgA can exist as both a monomer and a dimer. IgA is typically associated with mucosal immunity, whereas IgE is involved in immune responses to allergens, hypersensitivity, and parasitic infections.

The CH1 domain is located within the F(ab) region, while the remaining CH domains (CH2-CH3 or CH2-CH4) comprise the Fc fragment. The mobility or flexibility of the F(ab) and Fv portions of the antibody is primarily controlled by the CH1 domain and the hinge region. The Fc fragment defines the isotype and subclass of the immunoglobulin and mediates effector functions by binding to Fc receptors on effector cells or activating other immune mediators such as complement. Consequently, modifications to the Fc region can significantly impact the outcome of antibody-antigen interactions.

IgG has four subclasses, with the primary differences found in the hinge region of the constant heavy chain. These subclasses differ in their binding to Fcγ receptors and complement protein C1q, which affects their immune effector functions. These functions include antibody-dependent cellular cytotoxicity (ADCC), complement-dependent cytotoxicity (CDC), phagocytosis, opsonization, and immune modulation through signaling via ITAM and ITIM motifs in Fcγ receptors. Among the IgG subclasses, IgG1 is the most effective at mediating these functions, while IgG2 and IgG4 exhibit lower levels of effector function. Although IgG3 is a potent inducer of ADCC and CDC, it is less stable and prone to aggregation. IgG4 has a propensity for Fab arm exchange, a feature leveraged to generate bispecific antibodies, and is stabilized by substituting S228 with proline to ensure favorable manufacturing properties. Additionally, Fc glycosylation at N297 is critical for the ability of IgG subclasses to mediate effector functions.

Therapeutic antibodies are typically designed as IgG1, IgG2, or IgG4 (with S228P). The choice of IgG subclass is usually guided by considerations of the target type and the desired therapeutic outcome, particularly whether cell-killing functions should be retained or minimized. Researchers select the subclass that best aligns with these therapeutic goals.

The IgG1 subclass has proven to be highly stable, allowing for the production of homogeneous products that can be formulated at high concentrations. To achieve targeted therapeutic effects, protein engineers have developed effectorless IgG1 antibodies as an alternative to other subclasses. These effectorless IgG1 antibodies are engineered through mutations in the CH2 and CH3 regions. Early efforts focused on aglycosylation by mutating N297 to reduce effector functions. Recent advancements have identified mutations outside the glycosylation site to further reduce (LALA) or completely eliminate (LALAPG) binding to Fcγ receptors (FcγRs).

Conversely, the Fc region of IgG1 has also been engineered to enhance effector functions and immune modulation by introducing mutations that improve affinity for Fcγ receptors, such as FcγRIIIa and FcγRIIa.

Fcγ receptors are expressed on various immune cell types. There are three main types of FcγRs in humans: FcγRI, FcγRII, and FcγRIII. FcγRI, due to its high affinity for IgG1, is typically saturated by circulating IgG1. FcγRIIIa is associated with antibody-dependent cellular cytotoxicity (ADCC) mediated by NK cells. FcγRII has two isoforms with opposing functions due to their distinct signaling motifs: FcγRIIa, an ITAM motif-bearing receptor, is important for phagocytosis, while FcγRIIb, with an ITIM in its intracellular domain, acts as an immune inhibitory receptor (Fig. 4).

5   Role of FcRn in Recycling of Fc Proteins and IgG

The neonatal Fc receptor (FcRn) is a member of the extensive and functionally diverse family of MHC molecules. It plays a critical role in regulating the levels of IgG and albumin in circulation, contributing to their half-lives of 20–30 days. Nearly 120 years ago, Paul Ehrlich first described the protective effect of maternal antibodies on offspring, demonstrating their ability to cross the placenta and safeguard newborns from infections. Since then, it has been established that FcRn binds to IgG and albumin at low pH in endosomes and recycles them back to the bloodstream at neutral pH. This pH-dependent binding and recycling mechanism prevents IgG from being catabolized, thereby transferring protective immunity from mother to offspring across the placenta. This property of FcRn binding is also retained by Fc fusion proteins.

The FcRn-mediated diversion of IgG and albumin from lysosomal degradation as a result of Fc-FcRn interactions is fundamental to the development of new or modified drugs with enhanced FcRn binding, which can extend serum half-lives and improve pharmacokinetics. Several sites within the CH2-CH3 domains of Fc, specifically at residues like L253, H410, and H435, have been identified to significantly boost pH-dependent binding of IgG to FcRn [17, 18]. Notable examples of FDA-approved antibodies with enhanced FcRn binding include Nirsevimab (anti-RSV) and Satralizumab (anti-IL6R).

In summary, the Fc region is crucial in the design of therapeutic antibodies and Fc fusion proteins. Structural variations in the Fc region affect its interactions with Fc receptors, influencing IgG-mediated effector functions and sustaining IgG levels in circulation over extended periods.

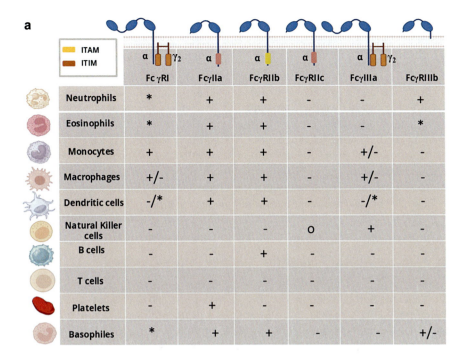

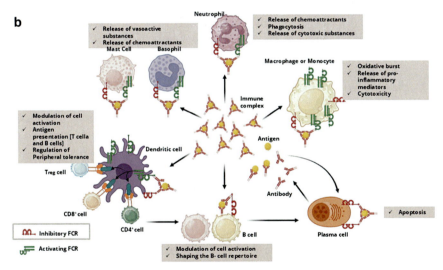

Fig. 4 Expression of inhibitory (FcγRIIb) and activating (e.g. FcγRI, FcγRIIa) Fc gamma receptors on immune cells (**a**) and their importance in cellular functions (**b**)

6 Next Generation of Biotherapeutics

The development of biologics has evolved significantly, driven by both successes, and desire to find better solutions for unmet need, leading to the engineering of more complex biotherapeutics aimed at enhancing potency through increased functionality. The flexible antibody structure has enabled innovative designs, including fusions and conjugations, to create more effective therapeutics. Advances in protein structure models, target characteristics, and disease biology have facilitated the development of bispecific and multi-specific antibodies and conjugates such as ADCs and radioconjugates. There are a number of these next-generation biologics progressing through clinical trials and some have been FDA approved.

Examples of classes of next-generation biotherapeutics are described below:

6.1 Bispecific Antibodies for Additive or Synergistic Functions

These include designs such as 1+1, 2+1, cross-mAbs, and duobodies. Such bispecific antibodies are engineered to bind to two different targets, improving tissue selectivity, mediating dual functions, or inducing cell cytotoxicity by engaging multiple cell types. The first bispecific antibody approved was Blinatumomab, a T-cell engager designed as an ScFv-ScFv fusion of anti-CD19 and anti-CD3 binding fragments. More bispecifics are currently in development. For example, Maritide, an antibody-peptide fusion, is being developed for obesity. This design involves an antibody targeting GIPR as an antagonist, combined with a GLP1 mimetic peptide conjugated to the antibody, which sends an agonistic signal via GLP1R.

In addition, bi-paratopic antibodies, (e.g. anti-Her2 - Zanidatamab), which recognize two epitopes on the same target and exhibit superior clustering of molecules, are being developed. Such bi-paratopics can result in better activity by simultaneously binding to two non-overlapping epitopes (sites) of the same target molecule. Bi-paratopics may also induce efficient internalization of a target which can be a useful property for ADC drug designs.

Bi-paratopic and bispecific antibodies can also enhance the ADCC response by clustering the target and Fc gamma receptors. It's important to note that optimizing ADCC function requires a comprehensive understanding of the specific target antigen, immune cell types involved, and the desired therapeutic outcome. Each optimization strategy should be carefully evaluated to ensure safety and efficacy.

6.2 T Cell Engagers

T cell engagers are a type of bispecific antibodies that are designed to bring target cells and T cells into close proximity, directing antigen-primed T cell-mediated killing activity specifically to tumor cells. T cell engagers typically consist of ScFv and/

or Fab domains that bind both the target cell and T cell. The simultaneous binding of both target and CD3 molecules leads to CD3 crosslinking, activating cytotoxic T cells. These cells then release cytotoxic granules, such as perforin, to kill the target cells. T cell engagers are highly potent and can mediate cell killing even against tumor targets with low expression density. More recently, some T cell engagers targeted to B cell antigens such as CD19 and BCMA are also showing promising efficacy in patients with systemic lupus autoimmune disease with manageable or less severe adverse events than cancer indications. These clinical data have led to increased clinical development activity of T cell engagers for diseases beyond oncology.

6.3   Antibody-Drug Conjugates (ADCs)

ADCs are a combination of antibodies with small molecules such as toxins or microtubule inhibitors. Traditional chemotherapy drugs are highly cytotoxic but often suffer from non-specific effects that limit their therapeutic index. To enhance the therapeutic index and target these potent drugs specifically to tumor cells, ADCs have been developed by leveraging the high affinity and specificity of antibodies for their targets.

ADCs consist of three key components: the antibody, the linker, and the payload. The design of an optimal ADC requires careful consideration of all three components:

- Antibody: Provides specificity and targeting to the tumor antigen.
- Linker: Crucial for the stability of the ADC. Early-generation ADCs faced challenges with premature payload release due to less stable linkers. Significant advancements in linker chemistry have improved serum stability and cell permeability.
- Payload: The cytotoxic drug attached to the antibody. The drug-to-antibody ratio (DAR) is vital for efficacy.

One notable example of ADC is Trastuzumab-DM1 (T-DM1), which combines the antibody trastuzumab with the maytansinoid DM1, achieving an average DAR of 4. Trastuzumab was used later for conjugation to a topoisomerase inhibitor (Deruxtecan, DxD) at a DAR ratio of 8 (fam-trastuzumab deruxtecan). DxD has mild cytotoxicity compared to DM1. In a head-to-head studies, fam-trastuzumab deruxtecan demonstrated superior efficacy compared to T-DM1, as measured by overall survival (OS) and durable activity. This difference in T-DM1 and fam-trastuzumab deruxtecan (marketed as Enhertu) is thought to be a result of payload choice, increased DAR, and bystander effect. The success of Enhertu has led to renaissance of ADCs as more industries are now pursuing this class of modality. More stable, hydrophilic linker designs have been designed to further improve ADC stability. New payload classes such as immune agonists and protein degraders are being explored to target intracellular proteins in cancer and immune cells. Moreover, another class of ADCs such as antibody oligo conjugates (AOCs) are also being explored for genetic disorders.

Despite the success in the treatment with ADCs, the dose-limiting toxicities due to free payload release in circulation is still a big challenge. Ongoing research by industry and academia aims to further refine ADC designs to enhance safety and efficacy (Table 5).

Table 5 FDA-approved antibody-drug conjugates (ADCs)

ADC name	Approval year	Target	Linker	Payload	Indication
Mylotarg (Gemtuzumab ozogamicin)	2000 (reapproved 2017)	CD33	Hydrazone	Calicheamicin	Acute myeloid leukemia (AML)
Adcetris (Brentuximab vedotin)	2011	CD30	Valine-citrulline dipeptide linker	Monomethyl auristatin E (MMAE)	Hodgkin lymphoma, sALCL
Kadcyla (Ado-trastuzumab emtansine)	2013	Her2	Non-cleavable thioether linker	DM1 (maytansine derivative)	HER2-positive breast cancer
Besponsa (Inotuzumab ozogamicin)	2017	CD22	Acid-cleavable hydrazone linker	Calicheamicin	Relapsed/refractory B-cell ALL
Polivy (Polatuzumab vedotin-piiq)	2019	CD79b	Valine-citrulline linker	MMAE	Relapsed/refractory DLBCL
Padcev (Enfortumab vedotin-ejfv)	2019	Nectin-4	Valine-citrulline linker	MMAE	Urothelial carcinoma
Enhertu (Fam-trastuzumab deruxtecan-nxki)	2019	Her2	Tetrapeptide-based cleavable linker	DXd (topoisomerase I inhibitor)	HER2-positive cancers
Trodelvy (Sacituzumab govitecan-hziy)	2020	Trop2	Cleavable CL2A linker	SN-38 (active irinotecan metabolite)	Triple-negative breast cancer (TNBC)
Zynlonta (Loncastuximab tesirine-lpyl)	2021	CD19	Cleavable dipeptide linker	Pyrrolobenzodiazepine (PBD) dimer	Relapsed/refractory large B-cell lymphoma
Elahere (Mirvetuximab soravtansine-gynx)	2022	FR alpha	Cleavable disulfide linker	DM4 (maytansine derivative)	Folate receptor-alpha positive ovarian cancer
Tivdak (Tisotumab vedotin-tftv)	2022	TF	Valine-citrulline linker	MMAE	Cervical cancer

6.4   Radioconjugates

Radioconjugates are targeted cancer therapies that combine a radioactive isotope with a molecule (usually an antibody or ligand) that specifically binds to cancer cells (Table 6). The targeting molecule delivers the radioactive payload directly to the tumor, allowing for precise radiation treatment of the cancer cells while minimizing damage to surrounding healthy tissue. This approach is used in both

Table 6 Some of the latest advancements in therapeutics that combine antibodies with radioligands

Therapeutic name	Target disease	Antibody	Radioligand	Mechanism of action	Approval status
Zevalin (Ibritumomab Tiuxetan)	Non-Hodgkin's lymphoma	CD20-targeting antibody	Yttrium-90 (Y-90)	Targets CD20 on B-cells and delivers radiation to destroy them	FDA Approved (2002)
Bexxar (Tositumomab)	Non-Hodgkin's lymphoma	CD20-targeting antibody	Iodine-131 (I-131)	Targets CD20 on B-cells and delivers radiation to destroy them	FDA Approved (2003)[a]
Xofigo (Radium-223 Dichloride)	Bone metastases from prostate cancer	N/A	Radium-223 (Ra-223)	Mimics calcium and targets bone metastases, delivering radiation	FDA Approved (2013)
Lutathera (177Lu-DOTATATE)	Neuroendocrine tumors	Somatostatin receptor-targeting peptide	Lutetium-177 (Lu-177)	Binds to somatostatin receptors on tumor cells and delivers targeted radiation	FDA Approved (2018)
Pluvicto (177Lu-PSMA-617)	Metastatic prostate cancer	PSMA-targeting antibody	Lutetium-177 (Lu-177)	Binds to PSMA on cancer cells and delivers targeted radiation	FDA Approved (2022)

These therapies represent significant advancements in the field of targeted cancer treatment, combining the specificity of antibodies with the destructive power of radioligands to effectively target and kill cancer cells while minimizing damage to surrounding healthy tissue diagnosis and treatment, providing a way to target and destroy cancer cells with high specificity. Examples include [Lutetium-177] Lu-DOTATATE (Lutathera), which targets somatostatin receptors in neuroendocrine tumors, and [Iodine-131] Ibritumomab Tiuxetan (Zevalin), used for treatment of non-Hodgkin's lymphoma by delivering radiation to CD20 + B-cells. Other notable radioconjugates include [Actinium-225] Ac-PSMA-617, an investigational treatment for prostate cancer that targets the prostate-specific membrane antigen (PSMA).

6.5 Conditionally Active Antibodies

To enhance the selectivity of complex multifunctional antibodies and minimize off-tumor effects while preserving on-target activity, researchers have focused on developing antibodies that are activated specifically in the disease-localized area. This approach involves the use of masking technology to prevent the antibody from binding to normal tissue at physiological pH. The mask is designed to be cleaved or removed by tumor-specific proteases present in the tumor microenvironment, thus enabling the antibody to become active and target the tumor site more precisely. This concept is currently being tested in clinical trials, such as with anti-EGFR T cell engagers.

7 General Safety Issues with Antibody-Based Therapeutics

Antibody therapies are a proven modality for the treatment in various diseases. Like any medical intervention, there are safety concerns that need to be carefully considered. Here are some of the key safety considerations related to antibody therapies:

A. Immunogenicity: Antibodies themselves can sometimes trigger an immune response in the body, leading to the production of anti-drug antibodies (ADAs). ADAs can reduce the effectiveness of the therapeutic antibody or cause adverse reactions. Currently, antibody developers take measures to minimize immunogenicity by engineering antibodies with human sequences and conducting rigorous preclinical and clinical testing to assess the risk of immune responses.

B. Off-Target Effects: Though antibodies have high specificity to their target, they can interact with unintended related and unrelated targets in the body, leading to off-target effects. This can result in unwanted side effects or complications. Extensive preclinical testing is conducted to evaluate the specificity of antibodies and minimize the risk of off-target effects.

C. Cytokine Release Syndrome: Some antibody therapies, particularly those that activate the immune system, can trigger a rapid and excessive release of cytokines, known as cytokine release syndrome (CRS). CRS can cause flu-like symptoms, fever, and in severe cases, organ dysfunction. Close monitoring and management of CRS are important to ensure patient safety.

D. Infusion-Related Reactions: Antibody therapies are often administered via intravenous infusion, which can sometimes lead to infusion-related reactions. These reactions can range from mild symptoms like fever and chills to more severe allergic reactions. Proper monitoring and management during infusions are necessary to mitigate these risks.

E. Long-Term Safety: Although antibody therapies may initially demonstrate safety in clinical trials, long-term safety data is often limited. It is important to continue monitoring patients receiving antibody therapies over an extended period to assess any potential long-term side effects or complications.

F. Drug Interactions: Antibody therapies, like other medications, can interact with other drugs a patient may be taking. These drug interactions can affect the efficacy and safety of both the antibody therapy and the other drugs. Careful consideration of potential drug interactions is essential when prescribing antibody therapies.

It is important to note that extensive preclinical testing and rigorous clinical trials are conducted to evaluate the safety and efficacy of antibody therapies before they are approved for widespread use. Regulatory authorities closely monitor the safety profiles of these therapies and require post-marketing surveillance to identify any previously unseen safety concerns.

8 Conclusions

Biologics have seen significant advances in discovery and design, moving from simple natural extracted products to recombinantly engineered therapeutic modalities like monoclonal antibodies and proteins. Designing the optimal format requires careful consideration of the therapeutic target profile, application goals, and indications of interest. Insights gained from preclinical and clinical experience have driven innovations, leading to more complex designs like antibody-drug conjugates (ADCs), T cell engagers (TCEs), and Chimeric Antigen Receptor T-cell (CAR-T) therapies. These advancements have expanded targeted treatment options for diseases with unmet needs. Today, biologic development focuses on enhancing specificity, safety, and efficacy while overcoming challenges in manufacturability and cost. By applying lessons from past successes and failures, biologics continue to advance medicine, providing targeted, effective treatments for increasingly complex conditions.

Box 1: Biophysical Characterization of Biologics

Biophysical characterization of therapeutic antibodies is a critical step in the development and evaluation of these biologic drugs. It involves the analysis and understanding of the physical and chemical properties of the antibodies, which helps to ensure their quality, stability, and efficacy. Here are some key aspects of the biophysical characterization process:

1. *Protein Structure*: The first step in characterizing therapeutic antibodies is determining their primary, secondary, and tertiary structures. Techniques such as mass spectrometry, peptide mapping, and nuclear magnetic resonance (NMR) spectroscopy are used to analyze the protein sequence and confirm its integrity.
2. *Molecular Weight and Purity*: Analytical techniques like size-exclusion chromatography (SEC) and mass spectrometry are used to measure the molecular weight of the antibody and assess its purity. This helps to ensure that the antibody is not degraded or contaminated by impurities.
3. *Binding Affinity*: The binding affinity of the antibody to its target antigen is an important parameter to evaluate its efficacy. Techniques like surface plasmon resonance (SPR) can measure the affinity and kinetics of the antibody-antigen interaction.
4. *Stability and Aggregation*: It is crucial to assess the stability of therapeutic antibodies under various conditions to ensure their shelf-life and effectiveness. Techniques like differential scanning calorimetry (DSC), circular dichroism (CD), and dynamic light scattering (DLS) are used to study the thermal stability, conformational stability, and aggregation propensity of the antibodies.
5. *Post-translational Modifications*: Therapeutic antibodies can undergo various post-translational modifications (PTMs) that can impact their stability and function. Techniques such as liquid chromatography-mass spectrometry (LC-MS) and capillary electrophoresis (CE) are used to identify and quantify PTMs like glycosylation, oxidation, and deamidation.
6. *Biophysical Characterization of Drug Product*: The biophysical characterization is not limited to the antibody itself but also extends to the final drug product, which includes formulation and stability studies, such as determination of pH, osmolality, viscosity, and particle analysis.

By conducting comprehensive biophysical characterization, researchers and developers can gain insights into the structural and functional attributes of therapeutic antibodies, ensuring their safety, efficacy, and stability throughout the drug development and manufacturing process.

Acknowledgments Authors would like to thank Dr. Ketki Kamthe and Dr. Riya Bhattacharya for their excellent assistance in drawing the figures.

References

1. Weiss M, Steiner DF, Philipson LH. Insulin biosynthesis, secretion, structure, and structure-activity relationships - Endotext - NCBI Bookshelf (nih.gov); 2014.
2. The Nobel Prize in Physiology or Medicine 1923 - NobelPrize.org.
3. Schernthaner G. Affinity of IgG-insulin antibodies to human (recombinant DNA) insulin and porcine insulin in insulin-treated diabetic individuals with and without insulin resistance. Diabetes Care. 1982;5(Suppl 2):114–8. [PubMed] [Google Scholar].
4. Goeddel DV, et al. Expression in Escherichia coli of chemically synthesized genes for human insulin. Proc Natl Acad Sci USA. 1979;76:106–10. [PMC free article] [PubMed] [Google Scholar].
5. Johnson IS. Human insulin from recombinant DNA technology. Science. 1983;219:632–7. [PubMed] [Google Scholar].
6. Ziegler AG, Danne T, Daniel C, Bonifacio E. 100 years of insulin: lifesaver, immune target, and potential remedy for prevention. Med. 2021;2(10):1120–37. https://doi.org/10.1016/j.medj.2021.08.003. Epub 2021 Sep 15. PMID: 34993499.
7. Adamson JW. The promise of recombinant human erythropoietin. Semin Hematol. 1989;26(2 Suppl 2):5–8. PMID: 2658101 Review.
8. Herman AC, Boone TC, Lu HS. Characterization, formulation, and stability of Neupogen (Filgrastim), a recombinant human granulocyte-colony stimulating factor. Pharm Biotechnol. 1996;9:303–28. https://doi.org/10.1007/0-306-47452-2_7. PMID: 8914196.
9. Davis PN, Ndefo UA, Oliver A, Payton E. Albiglutide: a once-weekly glucagon-like peptide-1 receptor agonist for type 2 diabetes mellitus. Am J Health Syst Pharm. 2015;72(13):1097–103. https://doi.org/10.2146/ajhp140260. PMID: 26092960.
10. Brønden A, Knop FK, Christensen MB. Clinical pharmacokinetics and pharmacodynamics of Albiglutide. Clin Pharmacokinet. 2017;56(7):719–31. https://doi.org/10.1007/s40262-016-0499-8. PMID: 28050889.
11. Jimenez-Solem E, Rasmussen MH, Christensen M, Knop FK. Dulaglutide, a long-acting GLP-1 analog fused with an Fc antibody fragment for the potential treatment of type 2 diabetes. Curr Opin Mol Ther. 2010;12(6):790–797. PMID: 21154170.
12. Strohl WR. Fusion proteins for half-life extension of biologics as a strategy to make biobetters. BioDrugs. 2015;29:215–39. https://doi.org/10.1007/s40259-015-0133-6.
13. Kusminski CM, Perez-Tilve D, Müller TD, DiMarchi RD, Tschöp MH, Scherer PE. Transforming obesity: The advancement of multi-receptor drugs. Cell. 2024;187(15):3829–3853. https://doi.org/10.1016/j.cell.2024.06.003. PMID: 39059360
14. Casadevall N, Nataf J, Viron B, Kolta A, Kiladjian J-J, Martin-Dupont P, Michaud P, Papo T, Ugo V, Teyssandier I, Varet B, Mayeux P. Pure red-cell aplasia and antierythropoietin antibodies in patients treated with recombinant erythropoietin. N Engl J Med. 2002;346:469–75. https://doi.org/10.1056/NEJMoa011931.
15. Raeber ME, Sahin D, Karakus U, Boyman O. A systematic review of interleukin-2-based immunotherapies in clinical trials for cancer and autoimmune diseases. eBioMedicine. 2023;90:104539.
16. Boven K, Stryker S, Knight J, Thomas A, van Regenmortel M, Kemeny DM, Power D, Rossert J, Casadevall N. The increased incidence of pure red cell aplasia with an Eprex formulation in uncoated rubber stopper syringes. Kidney Int. 2005;67(6):2346–53. https://doi.org/10.1111/j.1523-1755.2005.00340.x.

17. Tan H, Su W, Zhang W, Wang P, Sattler M, Zou P. Recent advances in half-life extension strategies for therapeutic peptides and proteins. Curr Pharm Des. 2018;24(41):4932–46. https://doi.org/10.2174/1381612825666190206105232.

18. Zaman R, Islam RA, Ibnat N, Othman I, Zaini A, Lee CY, Chowdhury EH. Current strategies in extending half-lives of therapeutic proteins. J Control Release. 2019;10(301):176–89. https://doi.org/10.1016/j.jconrel.2019.02.016. Epub 2019 Mar 5. PMID: 30849445.

19. Edelman GM, Cunningham BA, Gall WE, Gottlieb PD, Rutishauser U, Waxdal MJ. The covalent structure of an entire gammaG immunoglobulin molecule. Proc Natl Acad Sci U S A. 1969;63(1):78–85. https://doi.org/10.1073/pnas.63.1.78. PMID: 5257969.

20. Morrison SL, Johnson MJ, Herzenberg LA, Oi VT. Chimeric human antibody molecules: mouse antigen-binding domains with human constant region domains. Proc Natl Acad Sci USA. 1984;81(21):6851–5. https://doi.org/10.1073/pnas.81.21.6851. PMID: 6436822.

21. Shin SU, Morrison SL. Production and properties of chimeric antibody molecules. Methods Enzymol. 1989;178:459–76. https://doi.org/10.1016/0076-6879(89)78034-4. PMID: 2513467.

22. Knight DM, Trinh H, Le J, et al. Construction and initial characterization of a mouse-human chimeric anti-TNF antibody. Mol Immunol. 1993;30:1443–53.

23. Grillo-López AJ, White CA, Varns C, Shen D, Wei A, McClure A, Dallaire BK. Overview of the clinical development of rituximab: first monoclonal antibody approved for the treatment of lymphoma. Semin Oncol. 1999;26(5 Suppl 14):66–73.

24. Melsheimer R, Geldhof A, Apaolaza I, Schaible T. Remicade® (infliximab): 20 years of contributions to science and medicine. Biologics. 2019;13:139–78. https://doi.org/10.2147/BTT.S207246. PMID: 31440029; PMCID: PMC6679695.

25. Safdari Y, Farajnia S, Asgharzadeh M, Khalili M. Antibody humanization methods - a review and update. Biotechnol Genet Eng Rev. 2013;29:175–86. https://doi.org/10.1080/02648725.2013.801235. Epub 2013 Aug 2.

26. Tennenhouse A, Khmelnitsky L, Khalaila R, et al. Computational optimization of antibody humanness and stability by systematic energy-based ranking. Nat Biomed Eng. 2024;8:30–44. https://doi.org/10.1038/s41551-023-01079-1.

27. Hummer AM, Deane CM. Designing stable humanized antibodies. Nat Biomed Eng. 2024;8:3–4. https://doi.org/10.1038/s41551-023-01168-1.

28. Klein C, Brinkmann U, Reichert JM, Kontermann RE. Nature Review Drug Discovery 2024;23:301–319.

Cellular Therapies: From a Theoretical Concept to an Immuno-augmented Reality

Pratima Cherukuri, Aalia N. Khan, Karan Gera, Atharva Karulkar, Sweety Asija, Ankit Banik, and Rahul Purwar

Abstract Cellular therapies have emerged as a transformative approach in the treatment of various diseases, demonstrating their potential as a powerful tool in modern medicine. These therapies, which involve the transplantation of living cells to replace or repair damaged tissues and organs, have led to remarkable outcomes in conditions that were once deemed untreatable. From regenerative medicine to cancer immunotherapy, cellular therapies have provided new avenues for clinical interventions, showcasing significant success in improving patient outcomes. However, alongside these successes, there have also been notable setbacks.

In this chapter, we present a comprehensive overview of the historical development of cellular therapies and their progress over the years. We also highlight key breakthroughs, ongoing challenges, recent innovations, as well as the regulatory frameworks that ensure the safety and efficacy of these therapies.

Keywords Gene-modified cell therapies · Autologous cell therapy · Allogeneic cell therapy · Mesenchymal stem cells · Chimeric antigen receptor T-cell (CAR-T) · Natural killer (NK) cells · Tumour-infiltrating lymphocytes (TILs) · Dendritic cell (DC) vaccines · Vectors · US Food and Drug Administration (FDA) and European Medicines Agency (EMA) · Regulatory standards · Investigational New Drug (IND) · Good Manufacturing Practice (GMP) · Process development · Critical

P. Cherukuri
Genezen, Indianapolis, IN, USA

A. N. Khan · S. Asija · A. Banik
Department of Biosciences and Bioengineering, Indian Institute of Technology Bombay, Mumbai, Maharashtra, India

K. Gera · A. Karulkar
Immunoadoptive Cell Therapy Private Limited, Navi Mumbai, Maharashtra, India

R. Purwar (✉)
Department of Biosciences and Bioengineering, Indian Institute of Technology Bombay, Mumbai, Maharashtra, India

Immunoadoptive Cell Therapy Private Limited, Navi Mumbai, Maharashtra, India
e-mail: rahul.purwar@immunoact.com

process parameters (CPPs) · Critical quality attributes (CQAs) · Control strategy · Challenges · Technological innovations

1 Introduction to Cellular Therapies

As early as the late nineteenth century, investigators have attempted to leverage the potential of cells to treat disease. Over the last two decades however, researchers and clinicians have gained a much better understanding of the regenerative and disease-mitigative potential of cellular therapy, which involves enabling the grafting of autologous (from the same individual) or allogeneic (from a suitable donor) cells in to patients for the treatment of specific diseases (typically those which an otherwise unmet need, with poor clinical outcomes). Cellular therapies were once just a theoretical concept—today they have become the standard of care in the treatment of specific diseases. This chapter covers the historical and currently available cellular therapies—and the aspects involved from their design, research, process and clinical development, and regulatory pathways leading to usage in real-world settings.

This chapter explores successes and failures leading to the evolution of cellular therapies, along their unique manufacturing and quality control aspects. It also covers regulatory expectations that have formed the framework for their life-cycles and design considerations. It also touches on the probable directions in which these therapies will develop and evolve further—moving beyond a "simple" engraftment to "complex" gene-modified cellular therapies and next-generation multi-modal treatments in the field of cellular therapies.

2 The Past and Current Landscape of Cellular Therapies

Having existed in some form or another for almost a century, cellular therapies may involve a single cell type or multiple cell types. Depending upon the type of cells being used, they may be stem/progenitor cell-based therapies or those that use mature or differentiated cell subtypes. Furthermore, cellular therapies have now advanced to include targeted genetic modifications to certain cellular subtypes, with the aim of achieving a specific and durable response which may range from targeting a tumour-associated antigen (for cancer treatment) or to restore the function of a critical protein (for metabolic diseases).

2.1 Bone Marrow Transplantation

Cellular therapies are steadily advancing to the standard-of-care in the treatment of those respective diseases. The most widely accepted of these is bone marrow transplantation (BMT), owing to the potential of bone marrow cells to differentiate into, and reconstitute a variety of cellular populations [1]. Autologous (as early as 1939) and allogeneic (as early as 1957), BMTs have been performed worldwide for the treatment of multiple conditions such as blood and bone marrow cancers (such as acute myeloid leukemia, several forms of non-Hodgkins and Hodgkins lymphoma, chronic lymphocytic leukemia, acute lymphoblastic leukemia), neuroblastoma, non-malignant hematological conditions such as aplastic anemia, transfusion-dependent β-thalassemia, vaso-occlusive sickle cell disease, autoimmune diseases (such as Wiskott–Aldrich syndrome), as well as multisystem inherited metabolic disorders (such as the lysosomal storage diseases) [2]. While it is still regarded among clinicians as a complex process, over 30,000 successful annual procedures have been performed [3]. It involves the mobilization of hematopoietic stem/progenitor cells (HSPCs) from the bone marrow to the blood, so that a sufficient number of these can be extracted for eventual engraftment. CD34, a transmembrane phosphoglycoprotein, is ubiquitously expressed on HSPCs and serves as a common biomarker for this process [4]. A regimen of chemotherapy or chemoradiotherapy is followed to suppress the patient's immune system. Based on their intensity, these regimens may be either fully myeloablative in nature, middle/reduced intensity, or non-myeloablative chemotherapies. These options have established themselves as reliable (while risks may be varied with comorbidities) in patients where frontline treatments have failed or have been rendered inadequate. Outcomes in the treatment of these complex disease fronts may vary; however, in hematological conditions such as those mentioned above, significant improvements in overall survival have been reached, with a median of 54% in some countries for allogeneic transplants [5]. It is important to note that the applicability and effectiveness of bone marrow transplants face several critical caveats. In blood cancers, the option for autologous transplants is restricted by the availability of adequate HSPCs owing to proliferation and disease burden, as well as the potential for graft-vs-tumor response. Allogeneic transplants are dependent upon factors such as finding a suitable donor and restricting the post-procedural side effects—particularly graft-vs-host disease (GvHD). This occurs due to the presence of T-cells in the graft, which attack the patient's cells due to histological incompatibility. While standard practice in Allogeneic stem cell transplant involves matching the donor cells through human leukocyte antigen (HLA) typing, allo-transplant is not considered a viable treatment option if a patient has a declining performance status and/or other precipitating factors such as age and other which may not be able to handle this fallout such as comorbidities, etc. [6]. Allogeneic stem cell transplanted patients often may required hospitalization for upto 3 months for safety and to monitor the outcomes of the procedure, until their immune reconstitution is complete.

2.2 Beyond BMT: The Therapeutic Potential of Stem Cells

The potential of stem cells is just not restricted to BMTs. They can differentiate based on their developmental potency (totipotent stem cells rank the highest, followed by pluripotent options such as embryonic stem cells, and multipotent options such as HSPCs), and the processes to do so have been developed and optimized to commercial manufacturing scale. As technologies to induce pluripotency have evolved, stem cell therapies may no longer be restricted to conventional sources such as bone marrow, umbilical cord blood, and peripheral blood [7]. However, their therapeutic potential is still a subject of further investigation. Mesenchymal stem cells have been extensively explored clinically. Phase I/II trials in various forms of inflammatory bowel disease have shown initial complete remission rates as high as 83% at 24 weeks [8, 9], while outcomes in a trial on patients with multiple sclerosis showed a 40.4% response rate [10]. Though there are several other studies ongoing in mesenchymal stem cells for regenerative medicine, they are yet to show consistent efficacy.

Depending on the source, the differential populations of stem cells may differ. For example, cord blood, which is an approved source by the US FDA for at least eight different products to enable hematopoiesis, is characterized by a notably higher proliferative population of CD34+ve cells [11].

2.3 Gene-Modified Cell Therapies

Since the advent of therapeutic monoclonal antibodies in the 1990s, it has long been established that tumor-associated antigen targeting holds significant promise in treating cancers and achieving a durable response. Chimeric antigen receptor T-cell therapies (CAR-T) have shown unprecedented success in the treatment of B-cell malignancies in refractory/relapsing settings. Named for the peculiarity of their design on two counts (joining the targeting potential of B-cells and the cytotoxic potential of T-cells, along with originally being designed using human and animal protein sequences), CAR-T cells therapies represent the best success stories under the broadest sample sizes of clinical and commercial evaluation for gene-modified cell therapies, with over 20,000 patients having been treated since 2017, when the first was commercially approved [12] (Fig. 1). As of 2023, four CD19-targeted CAR-T cell therapies that have been US-FDA approved have shown curative potential in certain subsets of B-cell lymphomas and B-cell precursor acute lymphoblastic leukemia either on their own or in combination with bone marrow transplant; while in multiple myeloma, two US-FDA approved BCMA-directed CAR-T cell therapies have shown high overall response rates and superior progression-free survival [13, 14]. More than ten CAR-T cell therapies have been globally approved as of 2023, all of which are autologous. Allogeneic CAR-T cell therapies are under various stages of Phase I/II trials, with the aim of developing a more affordable,

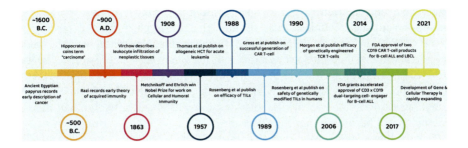

Fig. 1 The chronological development and milestones in the field of cellular therapies

scalable, and broadly deliverable treatment option. Furthermore, the gene-modified therapeutic potential of natural killer cells and macrophages is also being explored in Phase I/II trials in order to develop more robust off-the-shelf options [15, 16]. CAR-T cells have set a precedence for improvement of overall survival outcomes in the treatment of hematological malignancies in refractory/relapsing settings, while steadily showcasing better outcomes in earlier lines of treatment—prompting their eventual migration to the frontline, particularly in high-risk B-lymphomas and multiple myeloma [17–19]. However, the landscape is a more complex one for tackling solid tumors. For one, liquid tumors are generally more accessible compared to their solid counterparts, which are marked by an often histologically heterogenous territory that interlinks vascular tissues, the extracellular matrix, and connective tissue. The resultant immunological challenge is the exhaustion of T-cells by the time they have infiltrated the tumor or their stymieing at the metaphorical gates themselves.

The application of gene-modified cellular therapies extends beyond oncology and autoimmune conditions. HSPCs, particularly CD34+ve cells as mentioned above, are being increasingly used as a platform to deliver gene therapies for inherited metabolic diseases. Owing to their ubiquity of circulation and diverse differentiation capability, they possess restorative potential for several conditions. Thus, autologous CD34+ve cells have been successfully (in clinical and commercial settings) used as a platform for delivering gene therapies for metabolic diseases. By the end of 2023, lentiviral vector-modified as well as non-viral modification platforms (such as CRISPR-Cas9) received regulatory approval for their respective products from the UK MHRA, the EMA, and the US FDA to deliver corrective gene therapies for severe vaso-occlusive sickle cell disease and β-thalassemia. In these cases, the lentiviral method encodes a functional β-globin gene [20, 21], while the CRISPR/Cas9 method uses a gene-editing process driven by a guide-RNA that reactivates fetal hemoglobin at the erythroid-specific enhancer locus of BCL11A [22]. Both therapies result in the restoration of functional β-globin activity and a dramatic decrease in vaso-occlusive events, or dependence on transfusion, with progression-free survival rates remarkably improved in over 2 years of follow-up for each of these products. Other gene-modified cellular therapies on CD34+ve cells that have been approved for commercial use are treatments for diseases such as cerebral adrenoleukodystrophy [23] and metachromatic leukodystrophy, while various others are under clinical trials for rare diseases such as Hurler syndrome (NCT06149403). Gene-modified cellular therapies hold immense potential and have demonstrated their unique capability to bring about durable responses in treating diseases with a clinically unmet need. While there are several challenges in their accessibility and affordability, they also represent the shift of biopharmaceutical technologies toward "one-&-done" solutions and will play a critical role in transforming global healthcare in the next decade.

Cellular therapies involve developmental cycles and regulatory pathways that are notably different from conventional drugs. Their design (for modified as well as non-modified cellular therapies) and manufacturing considerations are complex, and as their development-to-approval cycles often follow an accelerated path, an emphasis is placed on generating enough chemistry, manufacturing, and controls (CMC) and pre-clinical data to prognose a favorable risk/benefit profile, which is backed up with applicable surrogate clinical endpoints, and later validated by post-approval clinical studies that compare these directly against the standard-of-care.

3   Regulatory Landscape in Cell-Based Therapies

Cell-based therapies hold immense potential for revolutionizing medical care by utilizing cells regenerative capacity to heal various diseases. However, bringing new therapies to market requires navigating complex regulatory systems to ensure safety, efficacy, and quality.

3.1   Overview of Global Regulatory Standards

Regulatory criteria for cell-based treatments varies by area, with each nation or union having their own set of norms. While the US Food and Drug Administration (FDA) and European Medicines Agency (EMA) are powerful regulatory bodies, worldwide regulatory standard alignment is critical for developing and marketing cell-based medicines in various markets. Organizations like the International Council for Harmonization of Technical Requirements for Pharmaceuticals for Human Use (ICH) and the World Health Organization (WHO) help establish harmonized guidelines and standards that ensure regulatory requirements are consistent worldwide. The FDA and the EMA are well-known regulatory authorities; other nations, such as India, regulate stem cell-based therapies. In India, the regulatory environment for stem cell-based therapies is still in flux, posing hurdles for industry experts. Preclinical and clinical investigations are critical for establishing batch uniformity, product stability, safety, and efficacy in locations with emerging regulatory frameworks [24, 25]. To successfully navigate global marketplaces, developers must be aware of regional regulatory nuances and modify their tactics accordingly.

3.2   Development and Optimization Pathways for Accelerated Approvals

Cell-based medicines are frequently unique approaches to disease treatment, demanding flexible regulatory procedures for rapid development and approval. The FDA and EMA provide options for rapid approvals, allowing patients with unmet medical needs quick access to promising treatments. These paths often include more efficient clinical trial designs, such as adaptive trial designs and surrogate outcomes, and faster review processes.

The FDA is crucial in regulating cellular therapies in the United States and has established two primary regulatory pathways for cellular therapy products: "361 products" and "351 products." The former, regulated under Section 361 of the Public Health Service Act, are not required to be licensed, approved, or cleared if they meet specific criteria. On the other hand, "351 products" require a premarket review of safety and efficacy data before being introduced into interstate commerce [24]. The FDA's regulatory framework focuses on preventing communicable diseases while guaranteeing the safety and efficacy of cellular and tissue-based products. The Center for Biologics Evaluation and Research (CBER) regulates stem cell-based treatments, approving drugs such as ApliGraf®, Carticel®, and Epicel®. Understanding the FDA's rules, including IND regulations, biologics regulations, and current Good Manufacturing Practice (cGMP), is critical for industry personnel navigating the research and approval procedure [26, 27]. In addition, the FDA has particular regulatory concerns for human cellular immunotherapy products used in cancer treatment, including genetically engineered cellular immunotherapies. These products, including autologous or allogeneic lymphocytes, CAR-T cells, antigen-presenting cells, or cancer cells modified or processed ex vivo, provide distinct manufacturing and validation hurdles compared to conventional immunotherapeutic [28].

In Europe, EMA oversees the regulatory framework for cell therapy products. Before cell-based medicines are licensed for clinical use, the EMA verifies that they meet high quality, safety, and effectiveness requirements. To receive clearance for their cellular products, industry experts must follow the EMA's requirements [29].

Overall, the role of the FDA and EMA is critical in assessing cell-based medicines' safety, effectiveness, and quality. Both authorities have developed recommendations and rules for cell therapy products, addressing manufacturing methods, product characterization, and preclinical and clinical testing needs. Developers can increase the possibility of successful regulatory submissions and clearances by aligning with regulatory expectations early in development. Furthermore, engaging with regulatory bodies through pre-submission meetings and scientific advice processes can give significant insights and help ease the regulatory review process.

3.3 Identification of Gaps and Guidance for Industry Professionals

Despite progress in regulatory frameworks, there are still gaps and obstacles in regulating cell-based treatments. These include uncertainty in regulatory expectations, variations in the interpretation and application of recommendations, and increasing scientific understanding of cell therapy products. Industry experts must handle difficulties related to standardized production procedures, disease progression monitoring, and safety concerns linked with particular products such as CAR-T cells. They may also help in building cooperation with regulatory agencies and other stakeholders. Additionally, sophisticated technology and analytical techniques can improve product characterization and assist regulatory compliance.

To launch breakthrough cellular therapies effectively, industry experts must traverse the complicated regulatory landscape, comply with global standards, and overcome regulatory gaps. Moreover, to solve quality control challenges, cell technology standards are crucial, and collaborative efforts are required to establish definite quality control standards that support scientific research and clinical translation [30].

In conclusion, manufacturing stages and product release are crucial aspects in preventing dangerous product usage. Before being released to patients, quality control tests, manufacturing practices, safety testing, and efficacy studies must be thoroughly reviewed. The regulatory framework for cell therapy products establishes manufacturing, clinical studies, and registration criteria, considering risk-benefit ratios [31]. Therefore, regulatory requirements are crucial to guarantee cell-based treatments' safety, effectiveness, and quality. Understanding and complying with regulatory standards allows developers to negotiate the complicated regulatory landscape and provide breakthrough cures to needy patients. Continued communication between industry stakeholders, regulatory bodies, and continuous scientific and technological breakthroughs will drive the creation of regulatory standards to aid in developing safe and effective cell-based treatments.

4 Design Considerations for CTs

Cell therapy holds immense promise in healthcare by offering personalized and targeted treatment options for various diseases. The commercialization of cell therapies has been a reality for the better part of the twenty-first century. However, the successful translation from bench to bedside requires meticulous attention to critical design considerations for cell therapies, as well as careful evaluation of logistics and technology.

4.1   Autologous Products

Autologous cell therapies, which use patient's own cells, have garnered significant interest for their potential to revolutionize patient care (Fig. 2). This approach minimizes risks associated with immune rejection, transmission of infections, and compatibility issues, making it one of the quickest and safest routes to bring new treatments to market [32]. Remarkably, some patients with advanced hematological cancers have achieved over a decade of remission following treatment [33]. Despite the vast potential of this emerging modality, which has sparked considerable interest and investment within the biopharma industry, the personalized nature of autologous therapies poses unique challenge that must be addressed to ensure their safety, efficacy, and delivery in a reproducible manner.

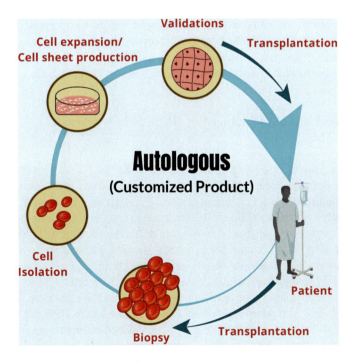

Fig. 2 An overview of autologous cellular therapy

4.1.1 Historical Milestone and Key Takeaways

Sipuleucel-T (Provenge) by Dendreon Corporation, an autologous cellular product against metastatic castration-resistant prostate cancer (mCRPC), is regarded as the world's first personalized cancer therapy, having been approved by the US FDA in 2010 [34]. The process involves exposing a patient's immune cells (which have undergone leukapheresis by a peripheral blood mononuclear cell protocol) ex vivo to prostatic acid phosphatase (PAP), a prostate tumor-associated antigen, and reintroducing these activated cells into the patient. The therapy is administered in three sessions over a period of one month, priced at $93,000 for a course of treatment. However, the handling and production of Sipuleucel-T is manual and time-consuming, presenting significant logistical challenges in terms of manufacturing and delivering the therapy to patients in a timely manner. It was also the first in a series of approved autologous cellular therapies which have a very high cost, when compared with off-the-shelf biologics.

The cascading issues of their affordability, reimbursement, and coverage under insurance, Medicare, and Medicaid programs are challenges that exist with most cellular therapies today, which may be rate-limiting factors in their prescription [35].

This example underscores the need for consensus guidelines on manufacturing and administration protocols to enhance reproducibility and reliability of results. Advanced bioprocessing technologies, including automated systems and closed-system bioreactors, are being explored to improve scalability, reproducibility, and cost-effectiveness of such products [36, 37]. However, in the context of lower- and middle-income economies, it is likely that a combination of these solutions may have to be implemented to arrive closer to equitization and expanded access.

4.1.2 Rapid Progress and Expanding Horizons in Autologous Cell Therapy

Since the approval of first autologous cell therapy, the field has progressed swiftly, with several therapies receiving FDA approval and numerous products in phase III clinical development [38]. For instance, autologous cell therapy for cardiovascular diseases, such as myocardial infarction and heart failure, has shown improvements in cardiac function and symptom relief. In a study by Hare et al. (2012), heart failure patients receiving autologous bone marrow-derived mesenchymal stem cells (MSCs) showed improvements in left ventricular ejection fraction and reduced adverse cardiac remodeling, indicating therapeutic promise in heart failure management [39]. Similarly, autologous cell therapy for neurological conditions like stroke, spinal cord injury, and neurodegenerative diseases has shown potential for neuroprotection and functional recovery (Fig. 3). Clinical studies involving patients with acute ischemic stroke who received intravenous infusion of autologous bone marrow-derived mononuclear cells demonstrated improvements in neurological function and reductions in infarct volume [40, 41]. Furthermore, transplantation of autologous olfactory ensheathing cells (OECs) in a rat model of spinal cord injury

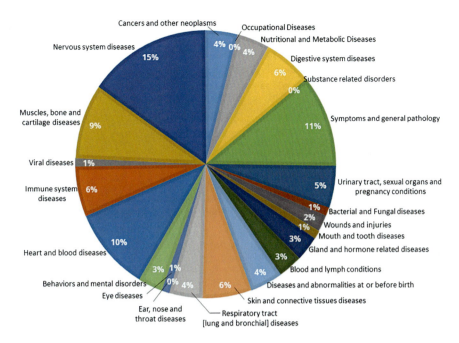

Fig. 3 Disease-specific classification of clinical trials involving mesenchymal stem cell-based therapies

has shown enhancements in locomotor function and axonal regeneration at the injury site [42].

Autologous cell-based cancer therapies have also shown encouraging results. While tumor-infiltrating lymphocyte (TIL) therapy and dendritic cell (DC) vaccines have demonstrated varied response rates across different cancer types, they remain promising modalities in the pursuit of effective cancer treatment. For instance, a study by Rosenberg et al. (2011) reported durable responses and long-term survival in a subset of patients with metastatic melanoma post the use of autologous TIL therapy [43]. While some patients were greatly benefited from these therapies, others experienced disease progression or limited clinical benefit. CAR-T cell therapies, particularly for hematological malignancies, have also demonstrated remarkable efficacy, with high response rates and durable remissions. For example, CAR-T cell therapy targeting CD19 in pediatric and young adult patients with relapsed or refractory B-ALL has shown notable rates of complete remission in a substantial number of patients [44]. Likewise, CAR-T cell therapy exhibited promising responses and long-lasting remissions in patients with relapsed or refractory DLBCL [45].

Despite notable successes, not all patients respond to CAR-T therapy, and some may experience disease relapse following treatment [46]. Treatment-associated toxicities and high costs continue to pose significant challenges to the widespread adoption of this therapy [47]. Efforts are underway to refine CAR design, aiming to bolster T cell activation, persistence, and tumor targeting [48–51].

Advances in synthetic biology have further facilitated the development of modular CAR designs, providing precise control over CAR-T cell functions, including fine-tuning antigen specificity, and signaling strength [52]. For this reason, the field is expanding beyond the currently approved second-generation CAR-Ts, which typically consist of an extracellular antigen-binding region (commonly a single-chain variable fragment [scFv] or VHH nanobody), a single co-stimulatory domain (typically 4-1BB or CD28), and a CD3ζ signaling domain. Efforts to develop more advanced CAR generations are actively underway, including designs with dual antigen-binding regions, membrane-bound or secretory cytokines, and even dual co-stimulatory domains. These advancements are being investigated in several early-stage clinical trials. Much like their predecessors, these next-generation CARs would seek to incorporate the elements of multiple types of immune cells into a singular or combinatorial adoptive cellular therapy strategy [53].

4.1.3   Cost and Scalability in Autologous Cell Therapy Production

Autologous cell therapies face significant challenges, notably in reducing vein-to-vein time, processing costs, and batch-to-batch variability (Fig. 4). Manufacturing processes play a pivotal role in deciding vein-to-vein time, which indicate the need to streamline operations and reduce batch failure rates that may be attributed to human error, contamination, and variability [54]. Efforts are therefore focused on automating manufacturing processes for cellular products to enhance efficiency and reduce variability through controlled cell expansion and modification, ultimately seeking to lower costs [55]. Companies like Miltenyi Biotec (CliniMACS Prodigy®) and STEMCELL Technologies (RoboSep™) offer automated cell separation systems for consistent isolation of target cells, while Cytiva's Xuri™ W25 and Terumo

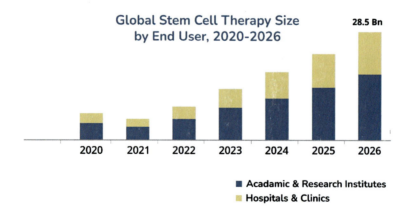

Fig. 4 Trends in global cell therapy market growth

BCT's Quantum Flex Cell Expansion System provide scalable solutions for cell therapeutics [56]. Despite these advancements, current processing equipment is often optimized for CAR-T cells and lack flexibility for other cell types such as natural killer cells.

While automated systems offer benefits in efficiency, scalability, and reproducibility, they come with significant costs. Current CAR-T cell therapies remain financially out-of-reach for many patients, as a single dose range from $300,000 to $500,000. Achieving equitable distribution and access to these therapies requires coordinated efforts to reduce costs and improve efficiency.

Innovations like ImmunoACT's NexCAR19™ (actalycabtagene autoleucel), which offers a cost-effective integrated manufacturing platform, represent a noteworthy stride toward rendering CAR-T cell therapy more accessible and affordable on a global scale [57, 58]. Comparably lower grades of cytokine-release syndrome and immune-effector cell-associated neurotoxicities may be attributed to the design of the antigen-binder CAR-T cell therapies such as NexCAR19® and Fucaso®, which use either a humanized or fully-human scFv [59].

Beyond cost considerations, logistical issues which arise from the intrinsic variability in patient cells pose challenges in managing manufacturing processes efficiently. Scalability remains a concern, as individualized processes may not be feasible for large-scale production, limiting accessibility and affordability. Furthermore, continued vigilance is essential to monitor long-term safety outcomes and mitigate risks associated with emerging technologies, including genome editing and cell reprogramming. These complexities emphasize the necessity for alternative approaches to overcome current limitations and broaden the therapeutic landscape in healthcare.

4.2 Allogeneic Products

While autologous therapies have advantages in patient safety, significant progress has also been made in developing allogeneic cellular therapies, which use cells sourced from a single donor for multiple patients [60] (Fig. 5). Over the past decade, research in allogeneic cell therapy has seen substantial evolution. Initially, much of the focus centered on establishing the feasibility and safety of allogeneic cell therapies, particularly in the context of hematopoietic stem cell transplantation (HSCT) for hematological malignancies. Given the potential for allogeneic cell-based therapies to provoke immune responses in recipients, which could result in graft rejection or adverse reactions, strategies to mitigate these responses are crucial for the successful application of allogeneic products.

Early studies aimed to address the risks of GvHD and immune rejection through various approaches, including donor-recipient HLA matching, T-cell depletion, and the use of immunosuppressive regimens [61, 62]. MSCs were identified to have benefit over other stem cell types due to their inherent immunoprivileged nature, indicating no requirement for immunosuppressant therapy with allogeneic MSC

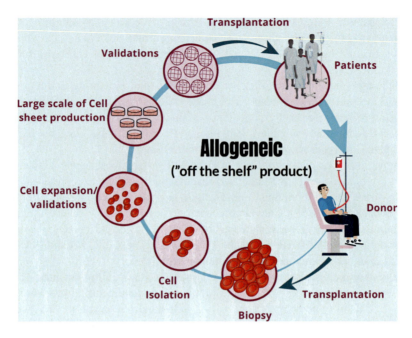

Fig. 5 Key highlights of allogeneic cell therapy

transplantation. This characteristic allowed for the potential use of MSCs from a single donor to treat many patients through the establishment of cell bank system [63].

Prochymal, developed by Osiris Therapeutics Inc., became the world's first approved therapy in 2012 for acute GvHD in pediatric patients. MSCs sourced from the bone marrow of healthy adult donors aged between 18 and 30 years were cultured to generate up to 10,000 doses of prochymal from a single donor [64]. However, setbacks emerged as subsequent clinical trials failed to consistently demonstrate efficacy across various indications and presented challenges in obtaining regulatory approval as well as achieving widespread adoption.

4.2.1 Therapeutic Potential of Allogeneic Therapies

As the field progressed, efforts focused on augmenting the therapeutic effectiveness of allogeneic cell products. MSCs derived from alternative tissues have emerged as promising candidates for allogeneic cell therapy. Umbilical cord blood-derived MSCs (UCB-MSCs) soon attracted attention owing to their abundance, immunomodulatory properties, and low immunogenicity. A few of researches have illustrated the therapeutic potential of UCB-MSCs in various disease models, including GvHD, inflammatory bowel disease (IBD), and neurological disorders. A study by Yang et al. (2020) demonstrated that UCB-MSCs infusion effectively ameliorated

GvHD symptoms in murine model and prolonged their survival. This was achieved by suppressing alloreactive T-cell responses and promoting regulatory T-cell expansion [65]. Similarly, adipose tissue-derived MSCs (AD-MSCs) have been extensively studied for their regenerative and immunomodulatory properties. AD-MSCs exhibit robust proliferation capacity and multilineage differentiation potential, making them attractive candidates for tissue repair and immune modulation. Researchers have also highlighted their therapeutic potential for conditions such as rheumatoid arthritis [66].

Some clinical studies have assessed the safety and efficacy of allogeneic NK cell therapies across various malignancies, including leukemia, lymphoma, and solid tumors. For instance, a Phase I trial investigating allogeneic NK cell therapy in patients with relapsed or refractory acute myeloid leukemia (AML) demonstrated feasibility, safety, and clinical activity [67]. Similarly, a Phase I/II trial evaluating the safety and efficacy of a CD19-directed CAR-NK that includes a membrane-expressed IL-15, and which uses cord blood as the starting material has shown promise as an off-the-shelf approach [15]. While CAR-NKs may lean toward superior safety compared with CAR-Ts, there may be concerns regarding their durability as they do not possess an immunological memory component.

Nevertheless, allogeneic CAR-based cell therapies offer advantages such as scalability, off-the-shelf availability, and the potential to target multiple antigens. Locke et al. (2023) demonstrated the clinical feasibility and efficacy of allogeneic CD19 CAR-T cell therapy in patients with relapsed or refractory B-cell lymphoma, achieving high response rates and durable remissions [68]. Additionally, advancements in genome editing technologies, such as CRISPR-Cas9, have enabled precise modification of allogeneic cells to enhance their functionality, reduce immunogenicity, and improve safety profiles. Efficacy assessments have revealed promising results, with allogeneic cell products demonstrating therapeutic benefits across various conditions, including hematological malignancies, solid tumors, and autoimmune diseases.

4.2.2   Challenges and Technological Innovations

Concerns about manufacturing scalability informed the need for standardized protocols to ensure product consistency [69, 70]. Challenges pertaining to cell expansion kinetics pose another area of concern in allogeneic cell therapy manufacturing. Expansion phase is pivotal for generating a sufficient quantity of therapeutic cells for clinical applications, yet achieving optimal cell growth kinetics while preserving cell identity and functionality remains a formidable challenge for certain cellular products. Careful optimization of culture media formulations and growth factors are paramount to attain consistent cell expansion kinetics and ensure product quality [71, 72]. Moreover, cryopreservation techniques play a critical role as they enable long-term storage and distribution of these allogeneic cellular products. Studies have investigated diverse cryopreservation strategies aimed at enhancing the post-thaw recovery and potency of allogeneic products [73].

In autologous therapies, the impact of batch failure is typically confined to that individual. Conversely, in allogeneic therapies, the loss of a manufacturing batch can have far-reaching consequences, affecting numerous intended recipients (Fig. 6). Studies have also illustrated the effectiveness of automated platforms in mitigating the risks associated with batch loss and variability in allogeneic cell therapy manufacturing [74]. Hassan et al. (2020) suggest that larger automated bioreactors are ideal for expanding allogeneic MSCs to maximize cell yield and enable production of thousands of therapeutic doses per batch [75].

Furthermore, by incorporating barcode labeling, electronic batch records, and real-time monitoring systems, manufacturers can effectively track the movement of raw materials, intermediates, and final products at each manufacturing stage (example: InstantGMP™ all-in-one software). This not only facilitates batch reconciliation and quality control but also enables rapid identification and containment of any deviations or discrepancies, thereby ensuring regulatory compliance. Industries can further explore the utilization of closed-system bioreactors, robotic platforms, and automated liquid handling systems to streamline allogeneic cell therapy manufacturing and reduce reliance on manual manipulation.

In future, allogeneic cell therapies will require continued advancements in ongoing manufacturing technologies, development of strategies to enhance cell engraftment and persistence, and refinement of regulatory frameworks to uphold safety and efficacy standard to further unlock their full potential in clinical application (Fig. 7).

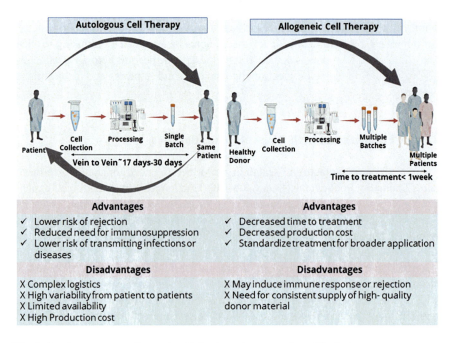

Fig. 6 Autologous versus allogeneic cellular therapies: strengths and limitations

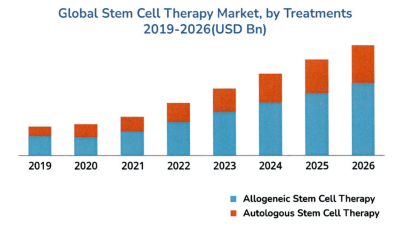

Fig. 7 Projections on the economic landscape shaped by cellular therapies

4.3 Risk Assessment and Considerations in Cell Therapy Development

Given that cell therapies involve manipulation and transplantation of living cells into patients, it is imperative to diligently evaluate and manage unique risks and challenges throughout the development process. A crucial aspect is to identify potential biological, technical, and regulatory risks linked to the specific cell type, manufacturing process, and intended clinical application. For example, in the case of MSC therapies, potential biological risks may encompass tumorigenicity, immunogenicity, and their unintended differentiation into undesired cell lineages [76]. Technical risks could include challenges related to cell expansion kinetics, genetic stability, and contamination during the manufacturing process [77]. Regulatory risks involve ensuring compliance with regulatory requirements, including adherence to Good Manufacturing Practice (GMP) guidelines, comprehensive product characterization, and thorough preclinical safety assessments. Through systematic evaluation of these risks, developers can implement effective risk mitigation strategies to ensure the development of safe and effective cell therapies.

Furthermore, considerations related to manufacturing scalability, reproducibility, and cost-effectiveness are critical in cell therapy development. The shift from small-scale research-grade production to large-scale commercial manufacturing presents a multitude of challenges, such as ensuring process scalability, sourcing raw materials, and meeting infrastructure requirements. Studies conducted by several researchers have highlighted the hurdles encountered during the transition from manual cell culture to automated bioreactor systems for the large-scale expansion of MSCs, emphasizing the importance of process optimization and scalability in cell therapy manufacturing [78, 79].

Fig. 8 Quality control strategies for maintaining standards in cell therapy

Moreover, ensuring product consistency and quality control is paramount in cell therapy development to minimize batch-to-batch variability and ensure reproducibility of therapeutic outcomes. Variability in cell source, culture conditions, and manufacturing processes can impact the safety, potency, and efficacy of cell therapies. Therefore, developers must establish rigorous quality control measures, encompassing in-process testing, release criteria, and stability testing, to uphold product consistency and adhere to regulatory standards (Fig. 8). Mastrolia et al. (2019) extensively discussed the importance of implementing standardized manufacturing protocols and quality control measures to attain consistent and reproducible MSC products for clinical use in their review [80].

Additionally, risk assessment in cell therapy development extends to various aspects including clinical trial design, patient selection, and post-marketing surveillance. Clinical trial protocols need to incorporate suitable safety monitoring, dose escalation strategies, and patient eligibility criteria to minimize potential risks and ensure patient safety. Post-marketing surveillance and pharmacovigilance efforts are crucial for monitoring the long-term safety, efficacy, and durability of cell therapies in real-world clinical settings [81]. By systematically evaluating and mitigating these risks, developers can ensure the safety, efficacy, and quality of cell therapies, facilitating their successful translation from the laboratory to the clinic. Thus, by drawing lessons from past failures and leveraging insights from successful endeavors, the field of cell therapy holds the promise of delivering transformative treatments for patients in need.

5 Manufacturing Considerations of CTs

Cell therapy product manufacturers are responsible for ensuring that their products adhere to stringent quality standards encompassing patient safety, efficacy, and regulatory compliance for both drug product (DP) and drug substance (DS). Critical quality attributes (CQAs), defined by the International Conference on Harmonization (ICH) Guideline Q8(R2), outline the indispensable physical, chemical, biological, or microbiological properties required to uphold product quality. These attributes play a pivotal role in maintaining patient safety, product effectiveness, and consistency across production batches. Manufacturers leverage CQAs to validate each product batch, generate clinical evidence, maintain batch-to-batch consistency, define meaningful release specifications, and evaluate product comparability post-manufacturing process alterations. Understanding the intricate relationships between CQAs, clinical outcomes, and critical process parameters (CPPs) is necessary for establishing a robust control strategy. This understanding ideally facilitates consistent operational performance, uniform batch quality, and the ability to anticipate the implications of planned process changes. The manufacturing control strategy characterizes a complete, risk-based methodology aimed at ensuring the utmost quality of the manufactured drug product. Its overarching objective is to mitigate variability across all factors and processes influencing the final quality and clinical efficacy of the product [82].

The expectations for demonstrating control over cell therapy manufacturing processes has evolved substantially within the regulatory communities of the United States, the United Kingdom, the European Union, and China [83]. The integrated control strategy, supported by data from process characterization and validation studies, is presented to regulatory agencies as part of filing documents. Any changes to product specifications or modifications to the control strategy post-approval must be communicated to and approved by regulatory bodies.

Incomplete characterization or insufficient validation of the manufacturing process can result in a lack of robustness to support commercialization, potentially leading to rejection of the strategy and further delays in the commercialization of the cell therapy drug product. Therefore, dedicating time to establish, monitor, and enhance the comprehensive control strategy is the most efficient approach for any company involved in cell-based therapy production.

5.1 Control Strategy: Ensuring Quality and Consistency

Control strategy is the critical element of the product development life cycle and is essentially all the parameters of your product and process, and it is expected to be evolved over the lifecycle of drug product development (Fig. 9).

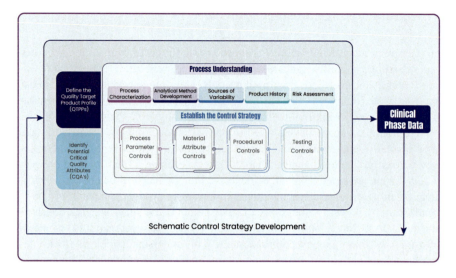

Fig. 9 Control strategy development

In preclinical and early stages, unified strategy needs to be constructed by integrating smaller-scale strategies that encompass all integral facets of the manufacturing system:

1. Process parameters
2. Raw material characteristics
3. Procedural controls
4. Quality management systems (QMS) and
5. Testing controls

As clinical data emerges, it should influence further characterization and refinement of all components of the control strategy. Long-term safety profiles should be established, and a deeper understanding of patient heterogeneity and variability needs to be attained. Patient-centric data should be used to validate criticality of various manufacturing CQA's.

5.1.1 Elements of Control Strategy

Specific control strategy elements influencing regulatory filings encompass various aspects of the integrated control strategy, including raw material controls, procedural controls, critical process parameters (CPPs), in-process controls (IPCs), and intermediate/release specifications (Fig. 10). These elements not only inform the overarching strategy across programs but also play a crucial role in facilitating regulatory approval. Furthermore, a robust holistic control strategy framework empowers cell therapy developers to uphold competitiveness and ensure patient safety in

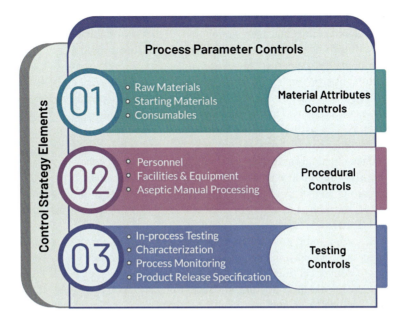

Fig. 10 Specific elements of control strategy

this rapidly evolving field. This becomes particularly significant as critical clinical questions pertaining to cell therapy are continually addressed.

To provide clarity and structure for readers, a methodology like that of the "Technical Report Portal TR 81 2018" will be employed. This approach will use a hypothetical scenario involving a CAR-T cell therapy product within the context of the existing regulatory framework and guidelines as necessary but will broadly cover other cell therapy products as deemed required. This will help illustrate key concepts and principles aiding in understanding the application of regulatory requirements to real-world scenarios for cell therapy development.

5.1.2   General Manufacturing Process of a Cell Therapy Product

To delve further into each of the control strategy elements and establish initial CQAs or potential CQAs, a comprehensive understanding of cell manufacturing process is essential. This involves analyzing the various steps in manufacturing, identifying critical points where quality attributes can impact the final product, and assessing how different parameters influence these attributes.

Cell therapy products encompass a wide range of products but can be mostly characterized as autologous or allogeneic therapies. Autologous products are customized to each patient and are derived from the individual's own cells. In contrast, allogeneic products are not specific to any particular patient and are typically sourced from cells obtained from donors. Allogeneic products may also undergo genetic modification to enhance their therapeutic properties or target specific diseases. Indeed, both allogeneic and autologous cell therapy products come with their own distinct set of challenges in manufacturing to ensure compliance with safety, efficacy, and quality standards.

For autologous products, the main challenge lies in process consistency and reproducibility, which requires a degree of flexibility and adaptability to accommodate individual variations. This can increase complexity and the risk of errors during production, as well as the potential for longer turnaround times to produce each dose. On the other hand, allogeneic products face challenges related to donor variability and the potential for immune rejection. Ensuring consistency in cell sourcing and its gene-modified consequences on safety, quality, and potency across different donor batches is crucial. Additionally, the genetic modification of allogeneic cells, if applicable, introduces additional complexities related to gene editing efficiency, off-target effects, and long-term safety considerations (Fig. 11).

Overall, navigating these challenges requires careful optimization of manufacturing processes, robust quality control measures, and adherence to regulatory guidelines to deliver cell therapy products that are safe, effective, and of high quality, regardless of whether they are allogeneic or autologous. On the other hand, regulatory approaches themselves would behoven by a degree of malleability as platforms for the manufacturing of cellular therapies are still quite heterogenous and are not as industry-normalized as those for manufacturing recombinant proteins. Once the approach is decided, the manufacturing process for both autologous and allogeneic processes follow a similar approach that can be split into five-unit operations as shown in Fig. 12:

1. Cell collection/isolation (leukapheresis)
2. Activation
3. In vitro genetic modification (transduction with viral vectors or other approaches)
4. Expansion

Fig. 11 Production challenges in manufacturing cellular therapies

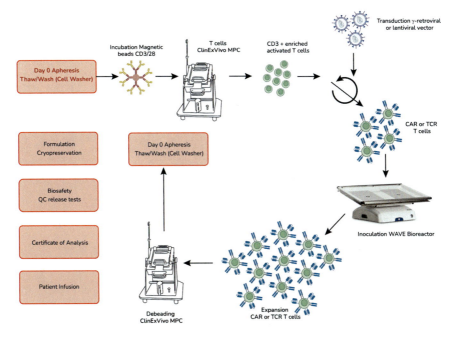

Fig. 12 Steps involved in the manufacturing process of cellular therapies

5. Harvest and formulation
6. Cryopreservation

5.1.3 Critical Quality Attributes (CQA's)

Once the manufacturing process is established, stage dependent CQA's with respect to patient safety and efficacy need to be established. CQAs are pivotal in delineating the desired product quality of a manufactured item. According to the ICH Guideline Q8(R2), CQAs encompass physical and biological characteristics [84]. These attributes should fall within specified limits, ranges, or distributions to guarantee the desired standard of product quality and are essential for:

1. Generating clinical data
2. Demonstrating lot-to-lot consistency
3. Assessing in-process attributes
4. Supporting meaningful specifications for product release
5. Demonstrating product comparability after manufacturing process changes

A representative list of CQA's for CAR-T genetically modified with LVV are listed in Table 1.

Initially, CQAs are defined during the drug development process, often starting with the creation of a Quality Target Product Profile (QTPP) that outlines the

Table 1 Critical quality attributes for gene modified cell-based product

CQA Category	Quality Attributes for Gene modified Cell based product
Safety	Endotoxins, Sterility, Mycoplasma, Adventitious viruses Replication competent lentivirus (RCL), vector copy number (VCN), off-target edits.
Identity	Surface marker expression specific to the intended cell phenotype (T-cell marker CD3+) and Confirmation of CAR expression
Residuals	Surface marker expression specific to the intended cell phenotype (T-cell marker CD3+) and Confirmation of CAR expression
Potency and Strength	Percentage Viable therapeutic cells, Intended Biological activity of cells or CAR transgene
Compendial	pH, Osmolality, appearance, visible particles

Therapeutic Attribute Category	DP Attributes for Gene modified Cell based product
Therapeutic Indication	Ocular, Immune, Cardiovascular etc.
Target population	Adult or pediatric
Cell type	PBMC's
Cell Origin and source	Autologous or Allogeneic
Volume per Dose	Cell numbers and volume in milliliters
Container closure System	Drug product in bags or vials
Stability and Storage Conditions	Cryopreserved or 2-8oC

product design criteria. The principal source of risk that may impact product attributes is the manufacturing process and process parameters identified during the drug product production. The process parameters that directly impact the CQA's are defined as the Critical Process Parameters (CPP's).

5.1.4 Process Parameters Considerations/Controls

Process parameters, including CPP's, play a pivotal role in determining the quality of the final product by impacting CQAs of the process (Fig. 13). Assessing the criticality of these parameters involves evaluating their potential impact on CQAs (Safety, Potency, Purity, and Identity).

Implementing Failure Mode and Effects Analysis (FMEA) early in process development will act as a valuable tool for assessing the CPPs impact on the CQA's. FMEA helps in determining whether the existing controls are sufficient and whether any potential failures can be detected before completing the process steps. This systematic approach aids in identifying and prioritizing potential failure modes, assessing their effects, and implementing appropriate preventive measures to mitigate risks and ensure product quality and safety. Implementing robust control strategies, including FMEA, contributes to regulatory compliance and patient safety. At the onset, it is advisable to develop a comprehensive overview of the production process. This can be achieved by employing various tools such as a simple flow diagram, deployment flow chart, SIPOC (Supplier, Input, Process, Output, Customer) diagram, or IPO (Input, Process, and Output) diagram. Each step of the production process should be meticulously described, outlining the product obtained after each stage. Furthermore, it is imperative to identify and delineate the specific parameters that govern the overall process (Table 2).

The quality of the product can be compromised at any stage of the manufacturing process, underscoring the importance of selecting appropriate process parameters. The criticality of these parameters is evaluated based on their potential impact on CQAs, which are closely monitored throughout manufacturing to uphold the Quality Target Product Profile (QTPP) [85]. While not all process parameters pose a significant risk to product quality, those identified as truly critical are incorporated into the manufacturing control strategy as essential components.

5.1.5   Material Controls

Materials utilized in biological manufacturing processes pose a significant risk to the CQAs of the final product. In this discussion, we will expand our focus beyond CAR-T cells to encompass other cell types used as starting materials. However, we will primarily concentrate on cells and viral vectors (starting materials), while also briefly mentioning other emerging gene editing tools. These materials play a crucial

Fig. 13 Critical Quality Attributes (CQA) and Critical Process Parameters (CPP) in cellular therapy manufacturing

Table 2 IPO diagram for CAR-T manufacturing

Controllable parameters		
• Number of single-use bioreactor bags • Cell seeding density • Media volume for each bag	• Culture duration • Incubation temperature • %CO_2 • Dissolved oxygen	• Rocker/agitation speed • Feed volume • Feed timing
Unit Operation Inputs	**Cell expansion unit operation steps**	**Unit Operation Outputs**
Starting materials		**Performance attributes**
• Cells from previous step (activation / gene editing)	1. Seed cells and media into bioreactor bag 2. Incubate for 5 days with a set agitation speed / profile	• Cell counts • Cell viability
Raw materials	3. Add media feed 4. Incubate for 5 additional days	**Quality attributes**
• Culture medium • Serum • Culture feed • Antibiotics	5. Obtain cell count / viability 6. Harvest; transfer cells to another bag for further processing Performance attributes	• Safety: sterility, endotoxin, mycoplasma • Purity: dead cells, contaminating B cells, process-related impurities (e.g., remaining beads)
Consumables		
• Single-use bioreactor bags • Transfer bottles / bags • Serological pipettes • Pipette tips • Filters • Sterile centrifuge tubes • Welder blades		• Identity: % CAR+ CD3+ • Potency: cytotoxicity, cytokine release
Equipment		
• Incubator • Biological Safety Cabinet (BSC) • Tissue culture pipettor • Rocking motion bioreactor/ stirred tank bioreactor • Aseptic welder • Grade B / Grade C room		

role in determining the quality, safety, and efficacy of the manufactured product, underscoring the importance of stringent selection, characterization, and control measures throughout the manufacturing process.

5.2   Starting Materials

5.2.1   Cells as Starting Materials

Quality of cell therapy products rely on the quality of their raw materials, primarily cells that serve as the foundation of the therapy. Ensuring the quality of these cells is extremely crucial for the CQA's of the final therapeutic product as they pose several inherent safety risks:

- Origination: Patients themselves (autologous) or healthy donors (allogeneic)
- Contamination with adventitious agents such as bacteria, viruses, fungi, or parasites
- Cells with tumorigenic potential
- Ex-vivo cell modification consequences, that pose physiological or anatomical consequences that can be hazardous
- Potential to elicit an immunogenic reaction (therapeutic gene expressed or in case of allogenic treatment)

Cell Collection Process

Various sources of cells for therapeutic use are broadly categorized into three groups based on their origin: autologous cells, allogeneic cells, and xenogeneic cells. The cell source is a critical factor affecting procurement, manufacturing, efficacy, and safety of therapeutic products, as it determines the patient's immune response to the transplanted cells (Table 3).

Autologous cells are derived from the patient and aim to treat diseases by leveraging the patient's own cells. This approach includes bone marrow-derived hematopoietic stem cells (HSCs), immune effector cells from peripheral blood, and induced pluripotent stem cells (iPSCs). While autologous therapies avoid immune responses and offer potential for re-administration, they face challenges such as variability in

Table 3   Advantages and disadvantages of various cell sources for cell-based therapeutics [86]

Cell Source	Cell Types	Effect On Immune Cells	Cell Engraftment	Durability Of Response	Dosing
Autologous	HSCs, T cells	Cells recognized as self, no requirement for immunosuppression, but some immune response to protein expression might be observed	Potentially permanent	Long term	Re-administration possible, variability in dosing
Allogeneic	MSCs, NK cells, B cells	Cells recognized as foreign, immunosuppression required	Transient engraftment	Short-term, variable	Re-administration possible, variability in dosing
Xenogeneic	Porcine pancreatic islet cells, choroid plexus cells	Cells and proteins recognized as foreign, immunosuppression required	Transient engraftment	Short-term, variable	Low feasibility for re-administration, limited control overdosing

cell quality due to the patient's health status, complex manufacturing processes, and high costs. Although immune responses to transgenes have been observed in preclinical studies, clinical manifestations remain absent.

Allogeneic cells, sourced from individuals distinct from the patient, offer scalability and simplification of manufacturing processes but may pose immunogenic reactions due to immune mismatch. Notable examples include NK cells and MSCs, which have shown minimal immunogenicity. However, most allogeneic therapies require immunosuppression regimens or engineering approaches to address immune mismatch challenges. Encapsulation of cells in biopolymer matrices is one approach to prevent immune recognition and enable re-dosing, although foreign body responses to encapsulation materials remain a challenge. Despite these challenges, encapsulated allogeneic cell therapies are under development for various conditions such as type 1 diabetes, hemophilia, and glaucoma.

5.2.2 Viral Vectors as Critical Starting Materials

Viral vectors play a significant role in transferring and integrating transgenes of interest into cells during the cell therapy manufacturing process and are considered as critical starting materials/raw materials in the CT manufacturing process, even when the modification occurs ex vivo, such as in CAR-T cell products. Although they are starting materials, regulatory agencies, like the EMA, suggest applying cGMP principles for vector manufacturing beginning with cell banks (master cell bank: MCB and working cell bank: WCB) used to produce the vector when used as a starting material for advanced therapy medicinal products (ATMPs). While EU regulatory agencies clearly define vectors (and all components used to make them) as starting materials, the FDA's language is less explicit, referring to vectors as "critical materials." However, expectations regarding vectors are similar across regulatory bodies, as vector attributes directly influence the final drug product. Tables 4 and 5 give examples of where GMP or GMP principles apply in the manufacturing process for CT's (ATMPs) [87].

Retroviral Vectors

Retroviruses (particularly gammaretroviruses and lentiviruses, which are widely used in gene-modified cell therapies) consist of an RNA genome that undergoes reverse transcription to form a double-stranded DNA molecule. This DNA is then integrated into the host cell genome by a viral pre-integration complex (PIC) and the integrase protein (IN). Once integrated, the provirus is formed, containing two long terminal repeats (LTRs) that flank the structural and accessory genes necessary for reverse transcription and packaging. The LTRs contain essential regulatory sequences, including a polyadenylation signal, primer sequence, and viral promoter/enhancer elements crucial for viral gene expression regulation.

Table 4 Application of cGMP principles in manufacturing of ATMP's

Example Products	Application of GMP to manufacturing steps is shown in dark grey. GMP Principles should be applied where shown in light gray. starting material - active substance - finished product				
In vivo gene therapy: mRNA	Plasmid, manufacturing and linearization	In vitro transcription		mRNA manufacturing and purification	Formulation, filling
In vivo gene therapy: non-viral vector (e.g. naked DNA)	Plasmid, manufacturing	Establishment of bacterial bank (MCB, WCB)		DNA Manufacturing fermentation and purification	Formulation, filling
In vivo gene therapy: viral vector	Plasmid, manufacturing	Establishment of a cell bank (MCB, WCB) and virus seeds when applicable		Vector Manufacturing and purification	Formulation, filling
Ex-vivo: genetically modified cells[3]	Donation. procurement and testing of tissues / cells[1]	Establishment of a cell bank (MCB, WCB) for plasmid and/or vector expansion and viral seeds when applicable	Plasmid manufacturing, Vector manufacturing	Genetically modified cells manufacturing	Formulation, filling

Table 5 Autologous CAR-T cell therapy ATMP manufacturing

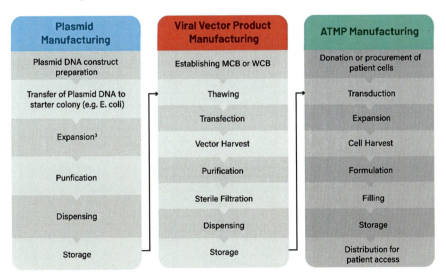

Gammaretroviruses have few fundamental genes—gag, pol, and env—involved in viral particle production and integration. These genes encode precursor proteins that are processed into mature proteins involved in particle assembly, reverse transcription, and envelope fusion with host cells. Lentiviruses, on the other hand, have additional accessory genes that regulate viral transcription, RNA processing, nuclear entry, and interactions with host cells.

From viruses to viral vectors Gene-transfer vectors derived from retroviruses and lentiviruses replace viral genes with a therapeutic gene expression cassette while retaining sequences essential for vector packaging, reverse transcription, and integration. This modification allows for the delivery of therapeutic genes into target cells while minimizing viral pathogenicity. It is crucial to keep in mind that the choice of viral vector for a cell and gene therapy (C&GT) program significantly impacts its viability and success. Each viral vector possesses unique traits that can render it suitable for specific applications. Consequently, it is essential to assess factors such as the vector's integrating capability, the length of the transgene, and whether it will be utilized for in vivo or ex vivo therapy right from the outset of the C&GT project. Table 6 outlines the fundamental characteristics of gammaretroviral vectors and lentiviral vectors.

Gammaretroviral vectors are potent tools extensively utilized in cell therapy applications, offering the advantage of transferring larger transgenes for stable integration. Like lentiviruses, they can integrate their genetic material into the target cell genome which will be passed down to target cell progeny. Although retroviruses cannot transduce non-dividing cells (only those that are dividing), their ability

Table 6 Characteristics of gammaretroviral vectors and lentiviral vectors

Characteristic	Lentivirus	Gamma retrovirus
Gene expression	Transient or stable	Transient or stable
Infect dividing cells	Y	Y
Infect non-dividing cells	Y	N
Integration into target cell genome	Y	Y
Relative immune response in target cells	Low	Low
Max cassette size	Up to 8.5 kb	Up to 10 kb
Particle size	Large	Large
Cell therapy	Y	Y
Envelope proteins	• Vesicular stomatitis virus (e.g., VSV-G protein) • Baboon retrovirus (e.g., Baboon retroviral envelope [BaEV] glycoprotein)	• Amphotropic MLV-A • Gibbon ape leukemia virus (GALV) • Feline endogenous virus (RD114) • Baboon retrovirus (e.g., Baboon retroviral envelope [BaEV] glycoprotein) • Vesicular stomatitis virus (e.g., VSV-G protein)

to carry a large transgene and high rates of transduction makes them a particularly attractive tool for C&GTs [88].

The pioneering product in this category, Strimvelis®, obtained marketing authorization in Europe in 2016. It was approved for treating patients with severe combined immunodeficiency caused by adenosine deaminase deficiency (ADA-SCID) who lack an HLA-matched related stem cell donor. Strimvelis® consists of CD34+ hematopoietic stem/progenitor cells (HSPCs) that have been transduced using a first-generation γ-retroviral vector. This vector expresses ADA cDNA under the control of the viral long terminal repeat (LTR) promoter/enhancer [89].

The current manufacturing approach for RVV (retroviral vector vectors) involves the use of stable producer cell lines during upstream production, with minimal downstream processing due to the inherent instability of RVV envelopes and the size of the vector (mature viral particles range from 80 nm to 120 nm). However, the overall vector manufacturing process and regulatory aspects are similar to the manufacturing process for lentiviral vectors, which will be discussed in more detail in the next section, which include the following:

1. Starting materials control
2. Process control
3. Defining process parameters
4. Testing and Characterization

Lentiviral Vectors

Lentiviruses belong to the broader family of retroviruses, as previously noted, and are derived from HIV-1. One of their key advantages is the ability to transduce both dividing and non-dividing cells, including T-cells with comparable efficiencies. Indeed, LVs are currently utilized in the manufacturing of several recently approved CAR-T cell therapies, such as Kymriah®, Breyanzi®, Carvykti®, and Abecma®.

Different generations of LV systems have been developed, and presently, third-generation LVs are the most widely employed for clinical applications. They are typically manufactured using a four-plasmid transient transfection system consisting of three packaging plasmids and one self-inactivating gene transfer plasmid or plasmid with gene of interest (GOI plasmid). Lentiviral vectors have versatile applications and can serve as either a DS or a DP depending on their application. However, for the purposes of this document, we will focus on discussing their usage as a DS in CAR-T.

LVV use for ex-vivo application requires development of separate QTPP and identify CQAs. Additionally, the QTPP of the final drug product should encompass considerations for vector manufacturing process residuals, impurities, and multiplicity of infection (MOI). This underscores the regulatory significance of the vector: while the EMA views lentiviral vectors as starting material, the US Food and Drug Administration (FDA) simultaneously categorizes vector used for ex vivo genetic modification of cells as bulk drug substance, an active pharmaceutical ingredient, and a critical component of the final drug product.

Regardless of classification, LVs must undergo appropriate characterization, and all assays used for quality control during manufacturing should be properly qualified and validated in a phase-appropriate manner. Within any Investigational New Drug application (IND), a separate drug substance section should be provided for vectors used for ex vivo modification of cells, as outlined in Module 3 of the Common Technical Document (CTD).

Lentiviral Vector Manufacturing

Manufacturing lentiviral vectors (LVV) involves distinct upstream and downstream processes. Upstream production (USP) focuses on generating high-titer vectors favorable to downstream purification. These processes vary based on the chosen cell line for production and whether cells undergo transient transfection or are engineered into stable producer cell lines. Adherent cell cultures are typically employed for smaller-scale needs during developmental stages, while suspension cultures are preferred for large-scale production. The supplementation of cell media, such as with serum or sodium butyrate, can enhance producer-cell viability and vector stability but may introduce challenges to product safety and downstream processing ease. Downstream processing aims to maximize viable vector recovery while minimizing impurities that could compromise the efficacy or safety of the final drug product. These steps generally include purification, enrichment/concentration, sterile filtration, and storage in a sequential order.

5.2.3 Starting Materials for Lentiviral Vector Manufacturing

For LVV manufacturing, the starting materials shall be the components from which the viral vector is obtained (i.e., the plasmids used to transfect the packaging cells and the MCB of the packaging cell line), and the principles of GMP shall apply from the bank system used to produce the vector onwards [87].

Cell Lines and Plasmids

The cell bank system utilized for LV production typically relies on HEK 293 or HEK 293T packaging cell lines. The 293T cell line is derived from the 293-cell line by introducing the gene encoding the SV40 T-antigen. This addition of the T-antigen enhances the promoter activity of plasmids transfected into these cells, thereby improving protein expression and facilitating lentiviral vector molecule formation. LVs commonly utilize a four-plasmid system approach to mitigate the risk of unwanted recombination events that could lead to the development of replication-competent lentivirus during viral vector manufacturing. This strategy helps to ensure the safety and integrity of the vector production process by separating essential viral elements across multiple plasmids, reducing the likelihood of recombination, and maintaining control over vector construction.

5.2.4 Regulations for Cell lines and Plasmids

During lentiviral vector (LV) production, all components must be thoroughly described in the manufacturing summary submitted as part of an Investigational New Drug (IND) application. It encompasses:

1. Detailed descriptions of plasmid construct generation, including the DNA sequence of the entire plasmid(s).
2. Appropriately qualification of the bacterial MCB, with sufficient information on its qualification, certificates of analysis (COA's) for cell banks, executed batch records, and evidence of passed release tests establishing the safety, identity, purity, and stability of the microbial cell preparation used in the bank.
3. The genotype and origin of the mammalian cell line (e.g., HEK 293, HEK 293T) utilized or Producer Master Cell Banks (MCBs) and Working Cell Banks (WCBs).
4. Charecterization testing of both the MCBs and WCBs to ensure identity, safety, and stability.

This comprehensive documentation and qualification process are crucial for ensuring the integrity and safety of LV production for human gene therapy applications.

5.2.5 Vector Manufacturing Process Characterization

Process characterization in LV production involves a series of studies aimed at understanding the process thoroughly. This includes process design, characterization, qualification, and ongoing monitoring. These steps should align with the clinical development phases of both LV and cellular drug product. Proper characterization of the LV manufacturing process is crucial for minimizing risk and ensuring consistent quality. Throughout clinical development, process design is refined, and CPPs are identified to achieve a consistent process with minimal variability. Process characterization requirements may vary depending on the intended application of the LV, with formal studies implemented once the final commercial process is established.

Design-of-experiment (DOE) approaches are applied to understand how different factors impact the process output (CQAs). Examples include [83]:

1. Raw material inputs
2. pH and total manufacturing time
3. Clarification (filter material and pore sizes)
4. Purification step: membrane versus column chromatography
5. TFF membrane cutoff size, pressure, size, and flow rates

Knowledge of CPPs from DOE studies helps define operating ranges for process parameters, ensuring consistent performance. Analytical methods used should be scientifically sound, providing reliable results. Stability assessments of manufacturing intermediates aid in establishing process limits and maximum hold times. Analytical data from process characterization inform the development of in-process controls (IPCs), which guide manufacturing. A well-characterized process reduces the risk of lost batches and facilitates qualification and validation efforts. A detailed description of the drug substance manufacturing process and controls must be included in regulatory submissions, with any changes or updates submitted as amendments.

5.2.6   Critical Quality Attributes of LVV

As a pivotal element of the final product, the CQAs of viral vectors are paramount considerations in the early stages of developing any CAR T-cell therapy. Examples of CQAs specific to LVV are mentioned in Table 7.

Table 7   Critical quality attributes of lentiviral vectors used in ex vivo therapy

CQA Category	Attributes for LVV used for CAR-T
Identity	Transgene presence
	Envelope
	Adventitious virus (human, bovine, and porcine if animal-derived materials used)
	Replication competent Lentivirus
Safety	Mycoplasma
	Sterility
	Endotoxin
Purity	Residual plasmid
	Residual host cell DNA, total DNA
	Host Cell Protein
	Residual serum/nuclease and transfection reagent
Potency/strength	Physical/genomes titer
	Infectious/functional titer
	Physical titer: infectiou titer ratio
	Functional/biological potency (transduced primary cells)
Other	Visible particulates
	pH
	Osmolality
	Appearance

5.2.7 Release Testing

Testing is conducted on every LV lot to verify its quality and adherence to specifications for each key attribute. The understanding of appropriate testing methods and corresponding limits evolves as LVs progress through clinical development stages and as the field matures. In early clinical development, a comprehensive set of assays for LV quality should be performed after each process step. This helps identify critical steps and determines which assays are most effective in detecting process changes that affect the final LV product. As the LV product advances toward commercialization, assay selection, timing, and specifications are refined based on enhanced product knowledge gained over time.

5.2.8 Other Delivery Systems: Emerging Technologies and Manufacturing Processes

Efforts are currently underway to explore alternatives to viral transduction in the field of CAR-T cell therapy. One promising endeavor involves engineering CAR-T cells using effective, transient messenger RNA (mRNA) delivery. This approach circumvents the challenges associated with viral manufacturing and optimization, potentially allowing for multi-dose administration instead of single-dose therapy. While murine models have shown equivalent efficacy, translating these findings to clinical settings is still pending, making this approach highly appealing.

Another potential approach to achieve stable expression of the anti-CD19 receptor on T cells is through site-specific knock-in of the transgene. Roth et al. have demonstrated site-specific knock-in in literature; however, this technique has yet to be applied in clinical settings. Despite these advancements, their translation into clinical applications remains incomplete.

Gene Editing

Developers have various options to enhance CAR performance, including transient or stable expression or inhibition of factors. One such therapeutic option is the inhibition of PD-1, frequently targeted for knock-out using gene-editing tools. Clinically validated gene-editing platforms, such as CRISPR/Cas9, Zinc finger nucleases, and megaTALs, offer developers options. CRISPR/Cas9, particularly, is popular due to its ease of use and flexibility. The CRISPR/Cas9 system is typically delivered without an integrating viral vector, often by electroporation using mRNA guides and Cas9 RNPs. As delivery is transient, the editing tool acts and dissipates, yet the editing results persist in the cell. This capability allows for entirely non-viral, site-specific integration of a persistent transgene, potentially disrupting current manufacturing processes [83].

Electroporation

Electroporation (EP) serves as the leading non-viral delivery platform for cell engineering. This technology involves passing electric current pulses through a solution containing cells. EP can deliver various cargos, including plasmids, mRNA, proteins, transposons, and gene-editing tools. Commercial EP units, such as MaxCyte, Lonza Nucleofection, and ThermoFisher Neon, are widely used in clinical trials to introduce transgenes into human primary cells. While EP offers efficient delivery, it can cause cell damage, scalability challenges, and lower efficiency and viability at larger scales. Additionally, most commercial EP systems are semi-continuous processing systems [83].

Lipid Nanoparticles

Lipid nanoparticles (LNPs) are biodegradable polymeric structures utilized for in vivo and ex vivo delivery of mRNA and RNPs to cells. LNPs are being evaluated for transient CAR-T therapies and gene editing, demonstrating comparable transfection efficiency to EP but with lower cytotoxicity. Scalability advantages of LNPs include no device component; however, challenges include optimizing LNP chemistry and improving process reproducibility.

5.3   Advanced Manufacturing Control (AMC) and Analytical Considerations

5.3.1   Procedural Controls

Implementing procedural controls that are clear and succinct are essential to ensure product quality and to support cGMP manufacturing operations. It enables cell therapy manufacturers to mitigate risks arising from personnel, facility design, equipment, and environmental conditions.

5.3.2   Aseptic Manual Processing

Indeed, modern cell therapy manufacturing processes often leverage closed and automated systems to ensure consistency, efficiency, and sterility. However, certain unit operations may still necessitate manual interventions and open manipulations. During these steps, rigorous aseptic techniques are imperative to minimize the risk of contamination, especially given that cell therapy manufacturing typically does not involve a final sterile end filtration step. Instead, the sterility of the final product relies on maintaining aseptic conditions throughout the manufacturing process. These manual operations may include cell isolation, culture expansion, transduction, and formulation steps, where individual operators or manufacturing personnel play a critical role in executing procedures aseptically. Strict adherence to standard operating procedures (SOPs), cleanroom protocols, and personal protective equipment (PPE) requirements is essential to prevent microbial contamination and ensure the safety and quality of the final cell therapy product. Moreover, ongoing training, monitoring, and quality assurance measures are vital to reinforce aseptic techniques and detect and address any deviations or breaches in sterile conditions promptly. By prioritizing aseptic practices during manual operations and maintaining a culture of quality and vigilance, cell therapy manufacturers can mitigate the risk of contamination and produce safe and effective therapies for patients.

5.3.3 Personnel

Developing and manufacturing CAR-T cell products requires a skilled and adaptable workforce capable of navigating the dynamic landscape of cell therapy. To ensure safe, effective, and consistent production, a strong quality culture must be established at the organizational level, supported by personnel with comprehensive product knowledge and robust training in aseptic manufacturing and product-specific processes. Personnel must possess the appropriate qualifications, practical experience, and understanding of their responsibilities, as recommended by regulatory authorities like the European Commission. Prior to engaging in routine manufacturing operations, personnel should undergo successful aseptic process simulation tests to demonstrate competency. Additionally, adherence to personnel hygiene levels and health conditions, along with procedures for disease prevention, is essential, as outlined by the World Health Organization. Ongoing training and retraining are crucial to maintain proficiency in assigned responsibilities, including cleaning and maintenance of aseptic processing areas. Manufacturing facilities must ensure redundancy in both personnel and organizational structure to facilitate thorough review of all manufacturing and testing activities [90]. This ensures a robust quality management system and minimizes the risk of errors or deviations that could impact product quality and patient safety. By investing in a skilled workforce and fostering a culture of quality and continuous improvement, cell therapy manufacturers can meet the challenges of a rapidly evolving field and deliver high-quality therapies to patients.

5.4 Manufacturing Strategy: Optimization and Scalability

Manufacturing strategy for cell therapies involves optimizing production processes and ensuring scalability to meet clinical and commercial demands. Here are some considerations:

1. *In-house Versus Contract Development and Manufacturing Organization (CDMO)*: Deciding whether to manufacture in-house or outsource to a CDMO depends on factors such as available resources, expertise, infrastructure, and timelines. In-house manufacturing offers greater control and flexibility but requires significant investment in facilities and expertise. Outsourcing to a CDMO provides access to specialized expertise, infrastructure, and scalability without the need for upfront capital investment.
2. *Early Clinical Versus Phase 3 and Commercial*: Manufacturing strategies may vary at different stages of clinical development. Early clinical stages may prioritize rapid production of small batches for testing safety and efficacy. As therapies progress through clinical phases and toward commercialization, manufacturing processes must be optimized for scalability, consistency, and cost-effectiveness to meet increased demand and regulatory requirements.
3. *Autologous (Auto) Versus Allogeneic (Allo)*: Manufacturing strategies differ between autologous and allogeneic therapies. Autologous therapies are patient-specific and typically involve customized manufacturing processes for each patient, focusing on rapid turnaround time and patient-specific quality control. Allogeneic therapies are derived from a single donor and require scalable, standardized manufacturing processes to produce large batches for multiple patients, with emphasis on consistency, quality, and cost-efficiency.
4. *Optimization and Scalability*: Manufacturing processes must be optimized to maximize efficiency, yield, and product quality while ensuring scalability to meet increasing demand. This involves identifying and optimizing critical process parameters, implementing robust quality control measures, and leveraging automation and technology to streamline production. Scalability considerations include facility design, equipment scalability, supply chain logistics, and regulatory compliance.

Ultimately, the manufacturing strategy for cell therapies should align with the specific characteristics of the therapy, stage of development, regulatory requirements, and business objectives to ensure successful translation from early development to commercialization.

5.5 Facility Designs: Meeting Regulatory Requirements and Ensuring Safety

The success of a cell therapy in the clinic relies on all the aforementioned control elements. Establishing new facilities to manufacture cell-based therapies in accordance with essential regulatory standards is a challenging undertaking. Ensuring proper layout and infrastructure is vital for safe and efficient production in the realm of cell therapy manufacturing [91].

5.5.1 Layout and Infrastructure

A distinguishing characteristic of cell-based therapy product manufacturing is the necessity for full aseptic processing, unlike the sterile filtration or final product terminal sterilization typical of traditional biologics. Hence, facility design must prioritize minimizing the risk of product contamination throughout the entire process, from incoming raw materials to final product storage, ensuring both product quality and patient safety. Autologous processes present design challenges due to the simultaneous manufacturing of multiple, small-scale, patient-specific batches. Conversely, mid- to large-scale allogeneic processes require facilities designed to accommodate processes at the boundary of manual processing capability, with provisions for future integration of automated technologies that are still in development.

In this section, we try to provide an overview of the considerations for facility design process, highlighting key regulatory requirements, design layout, operational considerations, and qualification, validation, and manufacturing start-up processes. These processes aim to transform initial facility requirements into a fit-for-purpose operational facility for cell-based therapies.

5.5.2 Fit for Purpose

Tailoring the facility design and infrastructure to meet the unique needs of the manufacturing process. A thorough grasp of the process and fundamental CQAs is essential for a facility to meet product manufacturing requirements and plays a vital role in ensuring consistent product delivery, influencing manufacturing operations, and managing costs. Cell-based therapy developers usually enlist architectural and engineering specialists to spearhead the design process, beginning with requirement definition and concluding with the final agreed design. Success hinges on collaboration across various functions within the developer's organization, encompassing manufacturing, quality, regulatory, engineering, project management, and business management.

Cell-based therapy developers have two primary options for manufacturing: In-House Manufacturing, which involves utilizing existing facilities or constructing new ones to conduct manufacturing internally; or outsourcing to a Contract Development and Manufacturing Organization (CDMO) where developers can opt to outsource manufacturing to a specialized CDMO, which offers expertise and infrastructure for cell therapy production. Each option presents distinct advantages and considerations, and the choice depends on factors such as available resources, expertise, scalability requirements, timelines, and strategic objectives. Table 8 provides a concise overview of key considerations for cell-based therapy developers when deciding between in-house manufacturing and outsourcing to a CDMO.

Table 8 Key factors deciding to keep manufacturing in-house vs outsource

Aspect	In-House Manufacturing	CDMO
Infrastructure	Requires establishment or modification of facilities: Time and Costs	Established infrastructure available
Expertise	Requires hiring/training specialized personnel	Access to specialized expertise
Control	Full control over manufacturing processes	Reliance on external partner for manufacturing
Flexibility	Greater flexibility to adapt to changing needs	Limited flexibility, subject to CDMO's capacity
Scalability	Capacity to scale production in alignment with demand	Limited by CDMO's capacity and availability
Cost	Upfront investment in facilities and personnel	Outsourcing costs based on services rendered
Time to Market	Longer lead time due to facility setup and validation	Potentially faster time to market with CDMO
Regulatory Compliance	Direct responsibility for compliance with regulations	Shared responsibility with CDMO
Risk Management	Full oversight of risk management	Shared risk management with CDMO
Focus	Focus on core competencies and proprietary processes	Allows focus on core development activities

5.5.3 In-House Building of Manufacturing Capacity or Choosing a CDMO Facility

Facility should meet the capabilities and expertise to meet the requirements of LV and CAR-T cell therapy manufacturing. This may involve assessing the facility's track record, capabilities, compliance with regulatory standards, and ability to scale up production as needed.

1. *User Requirement Specifications (URS)*: Clearly defining the functional and process requirements for the facility and equipment based on the specific needs of LV and CAR-T cell therapy manufacturing. This includes detailing the intended use, performance criteria, and regulatory requirements for each component of the manufacturing process.
2. *Design*: Ensuring that facility design and equipment selection prioritize safety, regulatory compliance, and efficient workflow. This includes adhering to relevant regulations and guidelines (e.g., cGMP, FDA regulations) and implementing appropriate layouts, containment measures, and environmental controls to minimize risks to product quality and personnel safety.

5.5.4 Cleanrooms Design and Classification

As discussed previously, cell-based therapy products cannot be sterile filtered or terminally sterilized and therefore must be manufactured in cleanrooms using aseptic processes. Cleanrooms are designed to mitigate the risk of contamination of products through control of the concentration of airborne particles and viable organisms that pose risks to product quality and sterility. The appropriate selection of the cleanroom environment to mitigate the risk of contamination is critical for successful manufacturing.

5.5.5 Flow of Personnel and Materials

The design of process rooms and support rooms in a facility is crucial for ensuring a smooth flow of personnel and materials to minimize the risk of contamination. Flow diagrams are generated during the facility design phase to convey the operational intent. Personnel flow diagrams illustrate the movement of personnel from entry to exit, including gowning and de-gowning stages, as well as cleaning activities. Material flow diagrams depict the entry of raw materials and consumables, their journey through manufacturing areas, and exit routes as waste or finished products. Intermediates and quality control samples are also accounted for. Specific emphasis is placed on tracking the movement of patient cellular materials. Unidirectional flows are often used to minimize contamination risk by ensuring a one-way flow of personnel, materials, and waste. This approach reduces the likelihood of confusion between products, cross-contamination, and errors in manufacturing or control steps.

When designing processing rooms for cell-based therapy manufacturing, various layout options should be considered to meet operational, quality, and regulatory requirements, particularly cGMPs, ensuring consistent production of compliant products. Layouts may be directly influenced by the nature of the manufacturing process, and vice versa. Since cell therapies cannot undergo sterile filtration or terminal sterilization, the layout must accommodate aseptic processing and minimize contamination risks.

The chosen layout should focus on segregating processes and controlling the flow of people, products, materials, and waste to prevent contamination and cross-contamination effectively. Ideally, facility design should align with well-defined operational criteria and Quality Risk Management (QRM) policies, even if processes are not finalized during the design phase. Developers should establish robust process definitions to guide the layout design, while maintaining flexibility to accommodate future process modifications. Several facility design approaches are commonly used in cell-based therapy manufacturing, each with its advantages and considerations. It is important to note that a facility design may incorporate elements from multiple approaches based on the specific needs and risks associated with the manufacturing processes. Here are some common approaches:

5.5.6 Dedicated Production Space

- This approach involves dedicating processing rooms to the production of a single patient lot, particularly common for autologous cell-based therapies where segregation of individual patient batches is crucial.
- Each room is isolated from others to maximize process segregation, minimizing the risk of cross-contamination or material mix-up.
- While effective for early clinical-stage production, dedicated production spaces can be inefficient in terms of space and equipment utilization, especially for autologous therapies with limited scalability.
- Dedicated rooms may be suitable for both open and closed processes, with open processes often conducted in Biological Safety Cabinets (BSCs) requiring higher cleanroom classifications.
- Considerations include the use of isolators for containment, capital and maintenance costs, and ease of operation, particularly for scaled-out production models.

The Ballroom Space

- The ballroom design concept involves a large, open manufacturing area that offers flexibility to accommodate various processes and equipment without extensive structural segregation.
- This design is ideal for processes and products expected to evolve over time.
- For autologous processes, multiple workstations within the ballroom can efficiently handle individual patient lots. However, this design increases the risk of mix-ups and cross-contamination due to concurrent operations within the same space.
- In allogeneic manufacturing utilizing campaign approaches, the ballroom provides flexibility for different equipment and processes. Cleaning activities between campaigns and equipment validation processes should be considered. Use of isolators or BSCs depends on the process requirements and associated cleanroom classifications, with attention to increased operational costs and cross-talk risks with BSCs.
- Ballroom designs require fewer facility airlocks compared to other designs, reducing construction and operational costs.

The Dance-Floor Design

Also known as segregated unit operations, this approach combines aspects of dedicated room segregation with the flexibility of a ballroom design.

- Unit operations are distributed across adjacent smaller spaces, allowing for specific process segregation and flexible manufacturing.

- Each room in the dance-floor layout can be connected through walls or common corridors, maintaining process segregation as guided by operational procedures and quality risk management.
- This design facilitates efficient space and equipment utilization by enabling decontamination and turnover between patient lots or processes.
- While offering advantages in process segregation, the dance-floor approach may require increased airlock usage and operational coordination for decontamination between lots compared to a single open ballroom space.
- It is best suited for processes utilizing the same facility, equipment, and platform technologies in different locations or for single-process operations. However, it may not offer the same level of segregation as fully dedicated processing rooms.

In summary, while dedicated production space is commonly used for autologous cell-based therapies, it may also be suitable for allogeneic approaches depending on scale and room size restrictions. Facility design decisions should not be limited to one specific approach. Instead, they should be based on an assessment of the intended processes, associated risks, and the site's risk-based management approach. Often, a combination of different approaches may be employed to create an effective and efficient manufacturing facility design.

5.5.7   Cleanroom Classification for Open and Closed Processes

Cleanrooms are meticulously designed to minimize the risk of product contamination by controlling the concentration of airborne particles and viable organisms that could compromise product quality and sterility. Selecting the appropriate cleanroom environment is crucial for successful manufacturing.

Regulatory authorities such as the FDA (U.S.), MHRA (U.K.), and EMA (EU) define standards that characterize different tiers of cleanroom environments, which are integral to CGMP guidelines (Table 9). These standards specify allowable particle concentrations across various size categories, ranging from 0.1 to 5 microns,

Table 9 Comparison of FDA versus EU cleanrooms terminology

Cleanroom Class FDA terminology	EU Terminology	Viable Particle Limit (per m³)	Non-Viable Particle Limit (per m³)
ISO 1	Grade A	< 1	10
ISO 2	Grade B	< 10	100
ISO 3	Grade C	< 100	1,000
ISO 4	Grade D	< 200	10,000
ISO 5	Grade D in operation	-	100,000

within a specified volume of airspace. Additionally, they offer guidance on the expected air quality for different types of processes. Although different regulatory authorities use varying terminology to describe cleanroom environment tiers, there are generally consistent standards and guidelines across jurisdictions.

6 Future Direction of CTs

Cell therapy has revolutionized the landscape of oncology, offering a personalized approach that harnesses the body's own cells to combat diseases and injuries. Recent years have witnessed remarkable advancements in this field, with the emergence of novel therapies holding immense promise for the future of healthcare. Engineering disciplines such as immune-modulatory approaches, synthetic biology, and genome editing are pivotal in advancing cell therapy research. These technologies are anticipated to enhance autologous and allogeneic cell therapy pipelines by improving potency, viability, and manufacturing processes.

The integration of engineering marvels like CRISPR-Cas9 gene editing and CAR-T cell therapy has propelled cellular therapeutics to new heights [86]. CRISPR-Cas9, with its precision in genome editing, offers a revolutionary approach to correcting genetic mutations underlying various diseases. Comparing CRISPR-Cas systems to other genome editing tools like zinc finger nucleases (ZFN) and transcription activator-like effector nucleases (TALEN), which require complex coding, makes it easier to precisely target specific genomic regions using guide RNA sequences [92]. This method promises to cure genetic illnesses by repairing the underlying genetic mutations. Clinical trials are being conducted to evaluate CRISPR-Cas9 for several genetic diseases, including sickle cell anaemia and cystic fibrosis. The approach has previously shown encouraging results in preclinical investigations [93].

CAR-T cell therapy, as described earlier, stands as another beacon of hope in the fight against cancer, which employ genetically engineered T cells to target and eliminate cancer cells. While offering unprecedented efficacy, challenges such as off-target effects and ethical considerations underscore the need for continued research and refinement. Moreover, clinical research has given rise to significant ethical concerns, such as patient informed consent in clinical trials—particularly when it comes to weighing the possible risks and benefits—and striking a balance between the interests of patients and researchers own scientific objectives [94].

Regenerative medicine, through innovative approaches like induced pluripotent stem cells (iPSCs) and organoids, offers a transformative paradigm for organ regeneration. This strategy might revolutionize transplantation and enhance results for individuals requiring organ replacement. Stem cells are often used in organ regeneration because of their unique ability to develop into various cell types.

1. *Induced Pluripotent Stem Cells (iPSCs)*: Adult cells reprogrammed to function like embryonic stem cells are known as iPSCs. It is possible to induce these cells

to develop into distinct cell types, such as kidney, liver, or heart cells. Since the new organ would be genetically identical to the patient, researchers may be able to prevent the issue of organ rejection by using the patient's cells to create iPSCs [95].

2. *Organoids*: Organoids are laboratory-grown, scaled-down, simplified replicas of actual organs. Organoids can be employed to study organ development and diseases. They are generally produced from stem cells. Organoids may be utilized in organ regeneration to develop more substantial, valuable organs for transplantation [96].

Researchers have made great strides in regenerative medicine by conducting successful lab experiments to develop organs like kidneys, livers, and hearts. Although these developments are encouraging, certain obstacles must be removed before organ regeneration becomes a standard therapeutic option. The intricacy of organ structure and function, the requirement for vascular networks to develop to deliver nutrients and oxygen to the developing organ, and the possibility of immunological rejection even when utilizing a patient's own cells are some of the difficulties associated with organ regeneration. Research studies are still ongoing to overcome these obstacles and improve the methods used in organ regeneration. Therefore, cell therapy for organ regeneration has enormous potential for advancing medicine. Organ regeneration may become a standard therapeutic option with more study and technical developments, saving many lives and enhancing the quality of life for patients everywhere [97].

While the promise of cell therapy is vast, it comes with its share of challenges. High costs, complex manufacturing processes, and ethical considerations pose formidable barriers to accessibility [98]. Addressing these challenges requires a multifaceted approach. From navigating insurance coverage to bridging the urban-rural divide in healthcare access, concerted efforts are underway to ensure that no patient is left behind. Innovative solutions, including financial assistance, telemedicine, and policy reforms, aim to democratize access to life-saving cell therapies.

Furthermore, fostering collaboration among stakeholders is paramount in advancing cell therapy research and development. Public-private partnerships, alongside global collaborations, can create dynamic ecosystems to accelerate progress and broaden therapeutic possibilities. Organizations such as the World Health Organization (WHO) and the International Society for Cellular Therapy (ISCT) are working to promote collaboration and knowledge sharing among researchers, clinicians, and policymakers worldwide to facilitate equitable access to cell-based therapies. Global collaborations also enable stakeholders to pool their resources and expertise to address regulatory challenges collectively, ensuring compliance and accelerating the translation of research findings into tangible therapies. As we understand the ethical and regulatory landscapes of next-generation therapies, it is imperative to prioritize patient welfare and ensure the secure and ethical application of emerging technologies.

By advocating for equitable access, fostering innovation, and navigating regulatory frameworks responsibly, we can work toward a future where cell therapy is universally accessible. Together, we can harness the transformative potential of cell therapy to advance healthcare into an era where the health of every patient is prioritized and no one is left behind.

Acknowledgments We extend our deepest gratitude to the patients and their caregivers whose courage and resilience have made cellular therapies some of the most advanced and hope-fulfilling treatments available for diseases with unmet clinical needs. Special thanks to the dedicated researchers and clinicians in the field of cellular therapies, whose relentless pursuit of knowledge, crucial research, and clinical innovations have led to the development of life-saving, breakthrough treatments. Their groundbreaking work laid the foundation for this chapter.

We also thank Dr. Narendra Chirmule for envisioning and supporting this extensive book. We are immensely grateful to SpringerNature for the opportunity to contribute this chapter and hope that it will serve as a valuable resource for researchers, clinicians, and industry professionals involved in the development and implementation of cellular therapies. Special thanks to Ketki Kamthe for the figures.

References

1. Henig I, Zuckerman TJRMmj. Hematopoietic stem cell transplantation—50 years of evolution and future perspectives. Rambam Maimonides Med J. 2014;5(4):e0028.
2. Tan EY, et al. Hematopoietic stem cell transplantation in inborn errors of metabolism. Front Pediatr. 2019;7:433.
3. De La Morena MT, Gatti RAJHOC. A history of bone marrow transplantation. Immunol Allergy Clin North Am. 2011;25(1):1–15.
4. Cottler-Fox MH, et al. Stem cell mobilization. Hematology Am Soc Hematol Educ Program. 2003;2003(1):419–37.
5. Zahid MF, et al. Outcome of allogeneic hematopoietic stem cell transplantation in patients with hematological malignancies. Int J Hematol Oncol Stem Cell Res. 2014;8(4):30.
6. Ferrara JL, et al. Graft-versus-host disease. Lancet. 2009;373(9674):1550–61.
7. Takahashi K, et al. Induction of pluripotent stem cells from adult human fibroblasts by defined factors. Cell. 2007;131(5):861–72.
8. Vieujean S, et al. Mesenchymal stem cell injection in Crohn's disease strictures: a phase I–II clinical study. J Crohns Colitis. 2022;16(3):506–10.
9. Ko JZ-H, Johnson S, Dave MJB. Efficacy and safety of mesenchymal stem/stromal cell therapy for inflammatory bowel diseases: an up-to-date systematic review. Biomolecules. 2021;11(1):82.
10. Islam MA, et al. Mesenchymal stem cell therapy in multiple sclerosis: a systematic review and meta-analysis. J Clin Med. 2023;12(19):6311.
11. Hordyjewska A, Popiołek Ł, Horecka AJC. Characteristics of hematopoietic stem cells of umbilical cord blood. Cytotechnology. 2015;67:387–96.
12. Benjamin B. CAR-T cell cancer immunotherapy gets personal. 2023; Available from: https://seas.harvard.edu/news/2023/02/car-t-cell-cancer-immunotherapy-gets-personal.
13. Zhang X, et al. CAR-T cell therapy in hematological malignancies: current opportunities and challenges. Front Immunol. 2022;13:927153.
14. Cappell KM, Kochenderfer JNJNRCO. Long-term outcomes following CAR T cell therapy: what we know so far. Nat Rev Clin Oncol. 2023;20(6):359–71.
15. Marin D, et al. Safety, efficacy and determinants of response of allogeneic CD19-specific CAR-NK cells in CD19+ B cell tumors: a phase 1/2 trial. Nat Med. 2024:1–13.

16. Reiss KA, et al. A phase 1, first-in-human (FIH) study of the anti-HER2 CAR macrophage CT-0508 in subjects with HER2 overexpressing solid tumors. American Society of Clinical Oncology; 2022.
17. Neelapu SS, et al. Axicabtagene ciloleucel as first-line therapy in high-risk large B-cell lymphoma: the phase 2 ZUMA-12 trial. Nat Med. 2022;28(4):735–42.
18. Kambhampati S, et al. Real-world outcomes of brexucabtagene autoleucel (brexu-cel) for relapsed or refractory (R/R) mantle cell lymphoma (MCL): a CIBMTR subgroup analysis by prior treatment. American Society of Clinical Oncology; 2023.
19. San-Miguel J, et al. Cilta-cel or standard care in lenalidomide-refractory multiple myeloma. N Engl J Med. 2023;389(4):335–47.
20. Locatelli F, et al. Betibeglogene autotemcel gene therapy for non–β0/β0 genotype β-thalassemia. N Engl J Med. 2022;386(5):415–27.
21. Kanter J, et al. Lovo-cel gene therapy for sickle cell disease: treatment process evolution and outcomes in the initial groups of the HGB-206 study. Am J Hematol. 2023;98(1):11–22.
22. Frangoul H, et al. Exagamglogene autotemcel for severe sickle cell disease. N Engl J Med. 2023;142:1052.
23. SKYSONA. Available from: https://www.fda.gov/vaccines-blood-biologics/skysona.
24. George BJPicr. Regulations and guidelines governing stem cell based products: clinical considerations. Prespect Clin Res. 2011;2(3):94–9.
25. Regulatory for cellular therapies. Available from: https://www.aabb.org/regulatory-and-advocacy/regulatory-affairs/regulatory-for-cellular-therapies.
26. Parson AJC. The long journey from stem cells to medical product. Cell. 2006;125(1):9–11.
27. El-Kadiry AE-H, Rafei M, Shammaa RJFiM. Cell therapy: types, regulation, and clinical benefits. Front Med. 2021;8:756029.
28. Duraiswamy J, et al. Navigating regulations in gene and cell immunotherapy. In: Gene and cellular immunotherapy for cancer. Springer; 2022. p. 141–64.
29. Cellular & Gene Therapy Guidances. Available from: https://www.fda.gov/vaccines-blood-biologics/biologics-guidances/cellular-gene-therapy-guidances.
30. Cao J, et al. Developing standards to support cell technology applications. Cell Prolif. 2022;55(4):e13210.
31. Kondo AT, Ribeiro AAFJJOBMT, C. THERAPY. Regulatory considerations for cellular therapy. J Bone Marrow Transplant Cell Ther. 2022;3(1):166.
32. Kazmi B, Inglefield CJ, Lewis MPJW. Autologous cell therapy: current treatments and future prospects. Wounds. 2009;21(9):234–42.
33. Melenhorst JJ, et al. Decade-long leukaemia remissions with persistence of CD4+ CAR T cells. Nature. 2022:1–7.
34. Anassi E, Ndefo UAJP. Sipuleucel-T (provenge) injection: the first immunotherapy agent (vaccine) for hormone-refractory prostate cancer. P T. 2011;36(4):197.
35. Jaroslawski S, Toumi MJB. Sipuleucel-T (Provenge®)-autopsy of an innovative paradigm change in cancer treatment: why a single-product biotech company failed to capitalize on its breakthrough invention. BioDrugs. 2015;29(5):301.
36. Heathman TR, et al. The translation of cell-based therapies: clinical landscape and manufacturing challenges. Regen Med. 2015;10(1):49–64.
37. Challener CAJBI. Process analytical technologies for manufacturing cell and gene therapies timing is everything, and it might be ideal for acceleration of real-time monitoring solutions. Duluth: Advanstar Communications Inc; 2021. p. 10–4.
38. Kochenderfer JNJMT. FDA approval of the first cellular therapy for a solid (non-hematologic) cancer. Cell Press; 2024.
39. Hare JM, et al. Comparison of allogeneic vs autologous bone marrow–derived mesenchymal stem cells delivered by transendocardial injection in patients with ischemic cardiomyopathy: the POSEIDON randomized trial. JAMA. 2012;308(22):2369–79.
40. Savitz SI, et al. Intravenous autologous bone marrow mononuclear cells for ischemic stroke. Ann Neurol. 2011;70(1):59–69.

41. Prasad K, et al. Autologous intravenous bone marrow mononuclear cell therapy for patients with subacute ischaemic stroke: a pilot study. Indian J Med Res. 2012;136(2):221–8.
42. Assinck P, et al. Cell transplantation therapy for spinal cord injury. Nat Neurosci. 2017;20(5):637–47.
43. Rosenberg SA, et al. Durable complete responses in heavily pretreated patients with metastatic melanoma using T-cell transfer immunotherapy. Clin Cancer Res. 2011;17(13):4550–7.
44. Maude SL, et al. Tisagenlecleucel in children and young adults with B-cell lymphoblastic leukemia. N Engl J Med. 2018;378(5):439–48.
45. Schuster SJ, et al. Long-term clinical outcomes of tisagenlecleucel in patients with relapsed or refractory aggressive B-cell lymphomas (JULIET): a multicentre, open-label, single-arm, phase 2 study. Lancet Oncol. 2021;22(10):1403–15.
46. Gu T, et al. Relapse after CAR-T cell therapy in B-cell malignancies: challenges and future approaches. J Zhejiang Univ Sci B. 2022;23(10):793–811.
47. Khan A, et al. Immunogenicity of CAR-T cell therapeutics: evidence, mechanism and mitigation. Front Immunol. 2022;13:886546.
48. Yang J, et al. Advancing CAR T cell therapy through the use of multidimensional omics data. Nat Rev Clin Oncol. 2023;20(4):211–28.
49. Ghorashian S, et al. Enhanced CAR T cell expansion and prolonged persistence in pediatric patients with ALL treated with a low-affinity CD19 CAR. Nat Med. 2019;25(9):1408–14.
50. Sakuishi K, et al. Targeting Tim-3 and PD-1 pathways to reverse T cell exhaustion and restore anti-tumor immunity. J Exp Med. 2010;207(10):2187–94.
51. Locke FL, et al. Tumor burden, inflammation, and product attributes determine outcomes of axicabtagene ciloleucel in large B-cell lymphoma. Blood Adv. 2020;4(19):4898–911.
52. Eyquem J, et al. Targeting a CAR to the TRAC locus with CRISPR/Cas9 enhances tumour rejection. Nature. 2017;543(7643):113–7.
53. Gumber D, Wang LDJE. Improving CAR-T immunotherapy: overcoming the challenges of T cell exhaustion. EBioMedicine. 2022;77:103941.
54. Abou-el-Enein M, et al. Scalable manufacturing of CAR T cells for cancer immunotherapy-clinical production of CAR T Cells. Blood Cancer Discov. 2021;2(5):408–22.
55. Lock D, et al. Automated, scaled, transposon-based production of CAR T cells. J Immunother Cancer. 2022;10(9):e005189.
56. Song HW, et al. Scaling up and scaling out: advances and challenges in manufacturing engineered T cell therapies. Int Rev Immunol. 2022;41(6):638–48.
57. Karulkar A, et al. Making anti-CD19 CAR-T cell therapy accessible and affordable: First-in-Human Phase I clinical trial experience from India. Blood. 2022;140(Supplement 1):4610–1.
58. Jain H, et al. High efficacy and excellent safety profile of actalycabtagene autoleucel, a humanized CD19 CAR-T product in r/r B-Cell malignancies: a phase II pivotal trial. Blood. 2023;142:4838.
59. Mallapaty SJN. Cutting-edge CAR-T cancer therapy is now made in India—at one-tenth the cost. Nature. 2024;627(8005):709–10.
60. Karantalis V, et al. Allogeneic cell therapy: a new paradigm in therapeutics. Am Heart Assoc; 2015. p. 12–5.
61. Gladstone DE, Bettinotti MPJH, the American Society of Hematology Education Program Book. HLA donor-specific antibodies in allogeneic hematopoietic stem cell transplantation: challenges and opportunities. Hematology. 2017;2017(1):645–50.
62. O'Reilly RJ, et al. T-cell depleted allogeneic hematopoietic cell transplants as a platform for adoptive therapy with leukemia selective or virus-specific T-cells. Bone Marrow Transplant. 2015;50(2):S43–50.
63. Le Blanc K, et al. HLA expression and immunologic propertiesof differentiated and undifferentiated mesenchymal stem cells. Exp Hematol. 2003;31(10):890–6.
64. Kurtzberg J, et al. Allogeneic human mesenchymal stem cell therapy (remestemcel-L, Prochymal®) as a rescue agent for severe refractory acute GVHD in pediatric patients. Biol Blood Marrow Transplant. 2013;20:225–31.

65. Albu S, et al. Clinical effects of intrathecal administration of expanded Wharton jelly mesenchymal stromal cells in patients with chronic complete spinal cord injury: a randomized controlled study. Cytotherapy. 2021;23(2):146–56.
66. Baharlou R, et al. Human adipose tissue-derived mesenchymal stem cells in rheumatoid arthritis: regulatory effects on peripheral blood mononuclear cells activation. Int Immunopharmacol. 2017;47:59–69.
67. Ahmadvand M, et al. Phase I non-randomized clinical trial of allogeneic natural killer cells infusion in acute myeloid leukemia patients. BMC Cancer. 2023;23(1):1090.
68. Locke FL, et al. Phase 1 results with anti-CD19 allogeneic CAR T ALLO-501/501A in relapsed/refractory large B-cell lymphoma (r/r LBCL). American Society of Clinical Oncology; 2023.
69. Liu Y, et al. Optimizing the manufacturing and antitumour response of CAR T therapy. Nat Rev Bioeng. 2023;1(4):271–85.
70. Murphy MB, et al. Mesenchymal stem cells: environmentally responsive therapeutics for regenerative medicine. Exp Mol Med. 2013;45(11):e54.
71. Koh SK, et al. Natural killer cell expansion and cytotoxicity differ depending on the culture medium used. Ann Lab Med. 2022;42(6):638.
72. Moseman JE, et al. Evaluation of serum-free media formulations in feeder cell–stimulated expansion of natural killer cells. Cytotherapy. 2020;22(6):322–8.
73. Yao X, Matosevic SJB. Cryopreservation of NK and T cells without DMSO for adoptive cell-based immunotherapy. BioDrugs. 2021;35(5):529–45.
74. Morrow D, et al. Addressing pressing needs in the development of advanced therapies. Front Bioeng Biotechnol. 2017;5:55.
75. Hassan MNFB, et al. Large-scale expansion of human mesenchymal stem cells. Stem Cells Int. 2020;2020:9529465.
76. Barkholt L, et al. Risk of tumorigenicity in mesenchymal stromal cell–based therapies—bridging scientific observations and regulatory viewpoints. Cytotherapy. 2013;15(7):753–9.
77. Herberts CA, Kwa MS, Hermsen HPJJotm. Risk factors in the development of stem cell therapy. J Transl Med. 2011;9:1–14.
78. Jankovic MG, et al. Scaling up human mesenchymal stem cell manufacturing using bioreactors for clinical uses. Curr Res Transl Med. 2023;71:103393.
79. Fitzgerald JC, et al. GMP-compliant production of autologous adipose-derived stromal cells in the NANT 001 closed automated bioreactor. Front Bioeng Biotechnol. 2022;10:834267.
80. Mastrolia I, et al. Challenges in clinical development of mesenchymal stromal/stem cells: concise review. Stem Cells Transl Med. 2019;8(11):1135–48.
81. Abou-el-Enein M, et al. Evidence generation and reproducibility in cell and gene therapy research: a call to action. Mol Ther Methods Clin Dev. 2021;22:11–4.
82. Association, P.D. Technical report no. 81. Cell-based therapy control strategy. Bethesda: PDA; 2019.
83. A-Cell ARM. Available from: https://alliancerm.org/manufacturing/a-cell-2022/.
84. International Conference on Harmonization (ICH) Guideline Q8(R2). 2009.
85. Kwilas A. CMC challenges during accelerated development of human cell & gene therapy products. 2020; Available from: https://www.casss.org/docs/default-source/cgtp/2020-cgtp-speaker-presentations/kwilas-anna-us-fda-2020.pdf?sfvrsn=6ddcba7e_6.
86. Bashor CJ, et al. Engineering the next generation of cell-based therapeutics. Nat Rev Drug Discov. 2022;21(9):655–75.
87. European Medicines Agency. Questions and answers on the principles of GMP for the manufacturing of starting materials of biological origin used to transfer genetic material for the manufacturing of ATMPs. 2021; Available from: https://www.ema.europa.eu/en/documents/other/questions-and-answers-principles-gmp-manufacturing-starting-materials-biological-origin-used-transfer-genetic-material-manufacturing-atmps_en.pdf.
88. Aiuti A, Roncarolo MG, Naldini LJEm. Gene therapy for ADA-SCID, the first marketing approval of an ex vivo gene therapy in Europe: paving the road for the next generation of advanced therapy medicinal products. EMBO Mol Med. 2017;9(6):737–40.

89. Williams DA, Thrasher AJJSctm. Concise review: lessons learned from clinical trials of gene therapy in monogenic immunodeficiency diseases. Stem Cells Transl Med. 2014;3(5):636–42.
90. Pimpaneau Valérie SJ. Blackton Michael PDA Technical Report No. 81 (TR 81) Cell-Based Therapy Control Strategy.
91. Roger A. TR-43 Revised: identification and classification of nonconformities in molded and tubular glass containers, for pharmaceutical manufacturers. 2013.
92. Akram F, et al. CRISPR/Cas9: a revolutionary genome editing tool for human cancers treatment. Technol Cancer Res Treat. 2022;21:15330338221132078.
93. Sharma G, et al. CRISPR-Cas9: a preclinical and clinical perspective for the treatment of human diseases. Mol Ther. 2021;29(2):571–86.
94. Ren S-S, et al. Ethical considerations of cellular immunotherapy for cancer. J Zhejiang Univ Sci. 2019;20(1):23.
95. Singh VK, et al. Induced pluripotent stem cells: applications in regenerative medicine, disease modeling, and drug discovery. Front Cell Dev Biol. 2015;3:2.
96. Arjmand B, et al. Advancement of organoid technology in regenerative medicine. Regen Eng Transl Med. 2023;9(1):83–96.
97. Mao AS, Mooney DJJPotNAoS. Regenerative medicine: current therapies and future directions. Proc Natl Acad Sci. 2015;112(47):14452–9.
98. Dodson BP, Levine ADJBB. Challenges in the translation and commercialization of cell therapies. BMC Biotechnol. 2015;15:1–15.

Hope Unlocked: Gene Therapy with Viral Vectors and Gene Editing

Jennifer J. Thiaville, Mariana Santana Dias, Sayuri E. M. Kato,
Shashwati Basak, Srinivas Rengarajan, Hilda Petrs Silva, Natasha Rivas,
and Susan M. D'Costa

Abstract Gene therapy is a novel approach that involves changing the expression of a person's genes to treat, cure, and/or ultimately prevent a disease. It may involve replacing a disease-causing gene with a healthy copy, adding a new gene to help treat a disease or deficiency, inactivating a gene that is not functioning properly, or permanently changing a gene on a patient's genome. Since its inception, much progress has been made in advancing gene therapy delivery and gene editing technology, and the field has seen unprecedented development with a wave of therapeutic platforms that have gained regulatory approval worldwide ranging from cancer therapies to treating monogenic diseases. One of the most critical factors in gene therapy is deciding the appropriate gene delivery method into the target tissue/cells, and viruses have been the top choice vectors due to their natural tropism to transduce and infect cells. This chapter will focus on the most common viral vectors used in gene therapy, their biology, clinical applications, and production strategies, highlighting approved platforms, adverse events, and regulatory impacts. In addition, we will review the most recent gene editing tools and their application on gene therapy, discussing the analytics required to ensure that the viral vector platforms can be validated and approved for the clinic, their challenges, and future outlooks.

Keywords Viral vector gene therapy · Adeno-associated virus · Adenovirus · Herpes simplex virus · Gene editing · Gene replacement · Gene regulation ·

J. J. Thiaville · S. E. M. Kato · N. Rivas · S. M. D'Costa (✉)
Genezen Laboratories, Indianapolis, IN, USA
e-mail: sdcosta@genezen.com

M. S. Dias · H. Petrs Silva
Universidade Federal do Rio de Janeiro, Instituto de Biofísica Carlos Chagas Filho,
Precision Medicine Research Center, Rio de Janeiro, RJ, Brazil

S. Basak
San Diego, CA, USA

S. Rengarajan
Kodo Lifescience Pvt. Ltd., Vapi, Gujarat, India

© The Author(s), under exclusive license to Springer Nature Switzerland AG 2025
N. Chirmule, V. V. Ghalsasi (eds.), *Approved: The Life Cycle of Drug Development*,
https://doi.org/10.1007/978-3-031-81787-8_5

171

CRISPR · Cas9 · Lentivirus · Gammaretrovirus · Retrovirus · AAV · AdV · LVV · RVV · ZFN · HSV · Talen · Oncolytic virus · Gutless adenovirus

1 Introduction

The technique of altering a patient's genetic material, either directly in vivo or by modifying human cells (patient or healthy donor) ex vivo, for the treatment of a particular disease is gene therapy. Gene therapy is particularly targeted toward diseases that cannot be cured with current conventional medications. These modifications can be in the form of a healthy copy of a non-functional gene (gene replacement) and/or regulating the control of gene expression (gene regulation). Gene regulation may include silencing or down-regulation to inactivate a mutated gene that produces a toxic product. Alternatively, it can be the upregulation of a gene that insufficiently expresses a protein by controlling gene regulatory sequences, or re-activation of a healthy copy on the chromosome that is silenced (for example, reactivating the silent X-chromosome in females). Additionally, gene regulation can be in the form of gene editing, a permanent modification of a patient's genetic material. There are various ways to elicit gene modification including viral and non-viral systems. Non-viral systems can include naturally occurring delivery systems like transposons or directly transferring DNA or mRNA using various techniques (transfection, RNP formation, nanoparticles). Non-viral systems with the exception of gene editing techniques will not be discussed in this chapter. Nature has devised efficient carrier delivery systems in the form of viruses. For viral vector gene therapy, these viruses are modified in the laboratory to lose their pathogenicity and carry therapeutics genes and/or regulatory elements as payloads while still maintaining their effectiveness as delivery and expression systems. Several viruses including adenoviruses (AdV), adeno-associated viruses (AAV), lentiviruses (LVV), gammaretroviruses (RVV), and herpes viruses (HSV) have been successfully used as gene therapy vectors. Some others including but not limited to poxviruses, Semliki Forest virus, reoviruses, and measles viruses have also been explored as viral vectors (Fig. 1). The development of gene therapy has focused on oncology and rare diseases (Fig. 2). Over the last decade, several of these gene therapies have been approved across the world for the treatment of various diseases from monogenic diseases like inherited blindness and muscular dystrophies to hemoglobinopathies like sickle cell disease as well as various cancers (Fig. 3).

The history of gene therapy is well over five decades with Dr. Stanfield Rogers, of Oak Ridge National Laboratory, TN, attempting to treat two sisters suffering from hyperargininemia with a papillomavirus that potentially contained arginase [4]. The trial failed to reverse the condition since the virus did not encode arginase production but rather affected cellular arginase levels. The first ex vivo therapy was a 1980 study, where Dr. Martin Clin, Professor Emeritus of Medicine at the University of California, Los Angeles, re-introduced patient bone marrow cells transfected with β-globin back into two β-thalassemia patients [5]. Unfortunately,

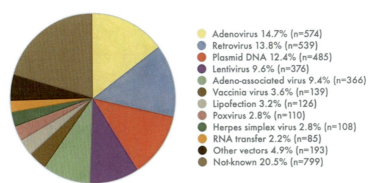

Fig. 1 Different types of vectors used for gene transfer in gene therapy clinical trials. Chart represents analysis of data entries on clinical trials undertaken in 46 countries from 1989 to March 2023 graphed by vector types (Republished with permission from Gene Therapy Clinical Trials Worldwide, provided by The Journal of Gene Medicine [225])

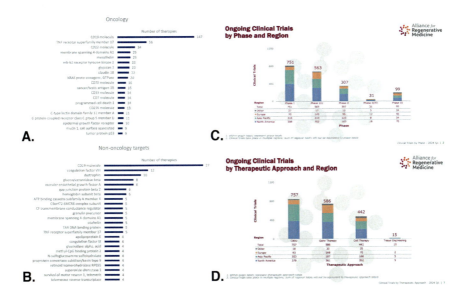

Fig. 2 Gene therapy pipeline: (**a, b**) The most commonly targeted therapeutic areas remain oncology and rare diseases as of Q1 2024 (Source: [1]). (**c, d**) Current clinical trials by phase as of Q1 2024, graphed by modality and region. (Source: [2])

this study was unsuccessful as well. Nevertheless, both these studies marked very important events in the history of gene therapy. It wasn't until the early 1990s that viral vector gene therapies saw their first success in a trial for the correction of adenosine deaminase (ADA) deficiency. The study used a recombinant retrovirus

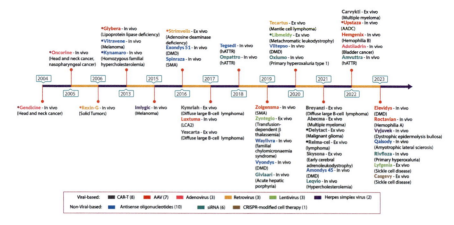

Fig. 3 Gene therapies have been approved across the world for the treatment of various diseases. Delivery platforms for gene therapy products are categorized into two groups: viral and non-viral. Viral-based gene therapy products use viruses, such as those described in this chapter, as gene delivery vectors. Non-viral-based gene therapy products include other mechanisms for gene delivery, such as antisense oligonucleotides, siRNAs, and CRISPR genome editing. * indicates the gene therapies approved outside the United States. (Adapted from [3])

carrying the ADA gene to transform T-cells ex vivo and infuse these transformed cells into a patient with ADA-SCID [6, 7]. Initial gene therapy studies focused only on repairing or replacing defective genes. As the field of functional genomics developed, other tools such as nucleases were incorporated into gene editing. First-generation nucleases were based on zinc-finger technology (ZFN), followed by transcription activation-like effector nuclease (TALEN) technology. The discovery of clustered regulatory-interspaced short palindromic repeats/CRISPR-associated protein 9 (CRISPR-Cas9) technology in 2012 accelerated and simplified the use of gene editing tools in gene therapy [8, 9]. The first approved therapy utilizing CRISPR/Cas9 for the treatment of sickle cell disease (SCD) was in 2023, just 11 years after this milestone discovery [10] (Fig. 3). Gene editing tools have been used in the clinic using both non-viral and viral delivery systems.

This chapter will elaborate on the most common viral vectors used in gene therapy—specifically adenoviruses (AdV), adeno-associated viruses (AAV), lentiviruses (LVV), gammaretroviruses (RVV), and herpes viruses (HSV). For each virus, a short description of their biology, their utility as a viral vector, and the different ways to manufacture these vectors will be discussed. A section of this chapter will also introduce gene editing tools and techniques and their utility in gene therapy. In addition, where possible, validation of these vectors in the clinic will be elucidated with special attention to regulatory impacts, adverse events, and approvals. Finally, a portion of this chapter will focus on the analytics needed to ensure that these vectors can be successfully used in the clinic.

2 Adenovirus (AdV)

Adenoviruses (AdV) were among the first viruses modified to be used as a vector for gene therapy [11]. Their first use for gene transfer studies dates to the early 1990s and was followed by an increasing number of clinical trials involving AdV vectors in gene therapy. In 1993, an AdV vector was first used for in vivo gene therapy in humans and administered to patients in a clinical trial for cystic fibrosis. This initial series of clinical trials showed that it was possible to achieve relevant therapeutic levels of expression, but also indicated some issues related to AdV vector immunogenicity and short-term effects [11–13]. In 1999, the gene therapy field experienced a significant setback following the tragic death of Jesse Gelsinger, an 18-year-old patient who received a systemic infusion of an early-generation AdV vector in a clinical trial for ornithine transcarbamylase deficiency (OTC) gene therapy [14]. In this trial, a high dose of 38 trillion particles was infused in the patient via a hepatic artery, which led to systemic inflammation resulting in a cytokine storm [15, 16]. This severe adverse event highlighted several concerns about the use of AdV vectors, and gene therapy in general, impacting the development of this field for almost a decade. In response to these safety concerns, the NIH focused on enhancing the safety and efficacy of AdV vectors, including a shift from single-gene therapies to applications such as vaccines and treatments for cancer, arthritis, and vascular dysfunction [11, 17]. Adstiladrin®, a non-replicating vector for the treatment of a type of bladder cancer [18], is the only AdV product that is currently approved by the FDA. AdV vectors have the advantages of a large packaging capacity and broad tropism, and they do not integrate into the host cell genome. Despite their extensive use, special safety concerns arise when utilizing AdV vectors for gene therapy, including their high immunogenicity and hepatotoxicity associated with strong liver targeting [19].

2.1 Overview of Adenovirus

Adenoviruses are commonly associated with infections of the upper respiratory and gastrointestinal tracts, with cold-like symptoms in humans, and are not related to severe disease in immunocompetent individuals, frequently being asymptomatic [20, 21]. There are seven human-infecting species, A–G, divided into more than 60 serotypes [21]. Group C adenoviruses type 2 (Ad2) and type 5 (Ad5) were the first commonly applied for gene delivery, due to the vast knowledge of their biology [22]. Adenoviruses are 90–100 nm in diameter with a non-enveloped, icosahedral protein capsid and a 26–48 kb linear, double-stranded DNA genome. The AdV genome also has a cis-packaging element (Ψ), which signals for DNA packaging into capsid, and is flanked by two inverted terminal repeat sequences (ITR), important for priming DNA replication [19, 23, 24]. Major capsid proteins include hexon, penton base, and fiber, which are important for AdV targeting and cell entry [25]. The viral genome encodes approximately 35 proteins, which are temporally expressed—early-phase (E) and late-phase (L) proteins. The immediate-early (IE), E1A protein, is expressed soon after infection and activates transcription of other early genes (E1B, E2, E3, and E4) responsible for DNA replication and cell- and immune modulation. Late-phase genes (L1, L2, L3, L4, and L5) are expressed after DNA replication and are associated with virus assembly, release, and cell lysis [19, 23, 24] (Fig. 4).

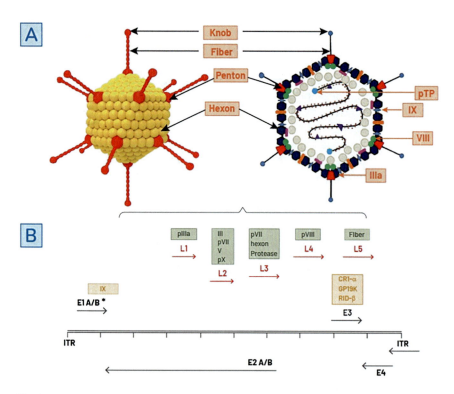

Fig. 4 Structure of adenovirus. (**a**) Adenovirus capsids are icosahedral and non-enveloped. Major components of the capsids are hexons, pentameric penton bases, and trimeric fiber proteins. Abbreviations: pTP, pre-terminal protein; IX, protein IX; VIII, capsid protein VIII; IIIa, protein IIIa. (**b**) AdV has a 26–48 kb linear, double-stranded DNA genome, which is divided into early (E) and late (L) transcriptional units. Abbreviations: IX, protein IX gene; pIIIa, capsid protein precursor pIIIa gene; III, gene-encoding penton base; pVII, protein precursor VII gene; V, protein V gene; pX, protein X gene; pVIII, protein precursor VIII gene; CRl-α, gene-encoding membrane glycoprotein E3 CRl-α; GP19K, membrane glycoprotein E3 gp19K gene; RID-ß, membrane protein E3 RID-ß gene; ITR, inverted terminal repeat. (Adapted from [26])

2.2   Engineering of Adenovirus for Gene Therapy: First and Second-Generation AdV Vectors

The first generation of AdV vectors for gene therapy was created by removing *E1A/ E1B*, and sometimes *E3*, from the original viral genome (Fig. 4). Deletion of E1genes generated replication-defective (RD) AdV that required propagation in complementing cells, with a transgene capacity of 4.5–8 kb [19, 23, 27–29]. RD AdV vectors cannot initiate DNA replication and persist in target cells as extra-genomic episomes [30]. Despite the deletion of E1, first-generation AdV vectors still showed leaky expression of several viral proteins that could activate immune responses and potentially mediate the destruction of transduced cells, affecting patient safety and efficacy [27, 31, 32]. Gene expression mediated by these AdV vectors was typically transitory which is not suitable for many gene therapy applications. Additionally, replication-competent particles could still be generated in the production process due to homologous recombination between viral and host cell genomes [23, 29].

Second-generation AdVs were generated by further deleting other viral genes, such as *E2A*, *E2B*, or *E4*. This provided a larger capacity of ~10 kb to accommodate a therapeutic sequence, decreased viral gene expression, diminished immune response, and extended transgene expression (Fig. 5). However, these vectors still

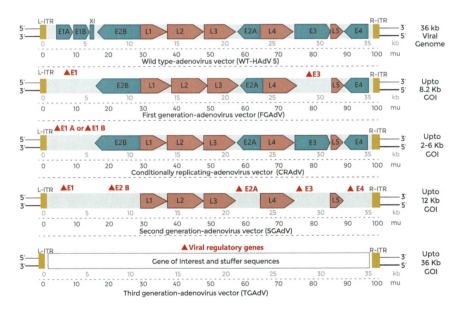

Fig. 5 Genome structure of the different types of AdV vectors. Comparison of the original AdV genome with the modifications made in the different vector versions. The red triangles represent genes that have been deleted. On the right, the size of the genome or the space available for insertion of the gene of interest is indicated. (Adapted from [33])

contained several AdV genes and did not present a robust improvement in reducing immunogenicity compared to first-generation vectors [23, 30].

2.3 Third-Generation AdV Vectors

Third-generation vectors, also known as "gutless," "high-capacity," or "helper-dependent" AdV vectors were developed by removing all viral genes (Fig. 5). The only remaining parts of the original AdV genome are the cis-packaging element (Ψ) and the ITRs. These vectors have a capacity of up to 36 kb, and a safer profile, with reduced immunogenicity, and prolonged gene expression [23, 30, 34]. However, these changes imply more complex manufacturing. Production process requires adenoviral helper sequences, which will allow replication and packaging, to generate full AdV vectors. Importantly, the helper genome is typically engineered to exclude the cis-packaging element (Ψ) from its sequence to prevent its packaging into the vector capsid [19, 35]. A special concern is the need to exclude the helper virus from the final product during vector manufacturing [36, 37]. This concern must be alleviated using purification strategies that minimize the amount of helper virus that co-purifies with the vector. This process assures a lower risk of production of replication-competent particles and results in AdV vectors with improved safety and lower immunogenicity than previous generations [30, 38, 39].

It is important to consider that even third-generation AdV vectors elicit an immune response since AdV capsid is by itself associated with strong immunogenicity [40, 41]. This feature has been limiting the application of AdV vectors for gene therapy, but it has been explored in the development of therapeutic strategies where it might be beneficial, such as cancer therapy (oncolytic viruses) and vaccines delivery.

2.4 Oncolytic Viruses

Apart from classical gene delivery strategies, AdV has also been largely studied for oncolytic therapies. In this strategy, recombinant viruses are used to infect and replicate in cancer cells, thus attacking the tumor, eliminating adjacent cancer cells, and inhibiting their growth [42, 43]. These oncolytic viruses are generated to preferentially replicate in cancer cells and are also known as selectively or conditionally replicating recombinant AdV (CrAdV) (Fig. 5). This function of the vector is achieved by engineering E1 genes with tumor-specific promoters [43, 44]. Oncolytic AdVs may also be modified to express adjuvant molecules, including immunomodulatory cytokines or checkpoint inhibitors to improve therapeutic outcome [45, 46].

2.5   Adenoviruses as Vaccine Vectors

The strong transgene expression and immunogenic profile of AdV vectors make them excellent tools for vaccine development. The vector, generally first- and second-generation RD AdV, is engineered to express an antigen from the targeted pathogen, inducing an immune response against it. The AdV itself, because of its high immunogenicity, can function as a vaccine adjuvant, inducing inflammation. The importance of AdV application as vaccine vectors was highlighted during the COVID-19 pandemic when different vaccines were developed and approved for emergency use all over the world [33, 47].

2.6   Adenovirus Vector Manufacturing

A few different methods have been described for the design of first-generation AdV vectors including homologous recombination of a shuttle vector containing the gene of interest (GOI) and the adenovirus genome in E1A-expressing cells (most commonly HEK293), in a way that E1 gene from AdV is replaced by the GOI [19, 27]. A more efficient but complex process is the pAdEasy method for homologous recombination in *E. coli*, where the plasmids are linearized and transfected into producer cell lines [48]. The simplest process with a good frequency of AdV vector rescue is the AdMAX system utilizing a Cre/LoxP-based recombination in HEK293 cells co-transfected with shuttle and Ad-genome plasmids [49, 50]. Another possible design involves directly cloning the GOI and regulatory sequences from the shuttle vector into the AdV backbone plasmid, a technically demanding process but one that avoids the uncertainty of homologous recombination. In this strategy, the final plasmid is also linearized and transfected into helper cells (Fig. 6) [19, 51].

In the case of third-generation AdV vectors, the production process relies on the co-supply of packaging-defective AdV helper sequences. In this process, helper cells are infected with AdV helper and transfected with the gutless plasmid containing the GOI. Manufacturing of these vectors is challenged by the low vector production efficiency and the risk of contamination with residual helper virus in the final vector product [19, 35]. Importantly, the cis-packaging element (Ψ) of the helper genome is flanked by loxP sites, and the packaging cell line expresses Cre to excise this packaging signal from the helper genome and prevent its encapsidation into AdV capsids.

Alternatives for the use of an AdV helper virus include a helper baculovirus-dependent system, in which infection with a baculovirus is used as a substitute to provide AdV genome sequences in a similar packaging method as the helper AdV-dependent one [53]. More recently, another method has been described which completely abolishes the use of helper viruses. In this case, producer cells are co-transfected with an expression cassette containing the GOI and a helper plasmid containing the AdV genome deleted for Ψ element, both ITRs, and the E3 gene [54].

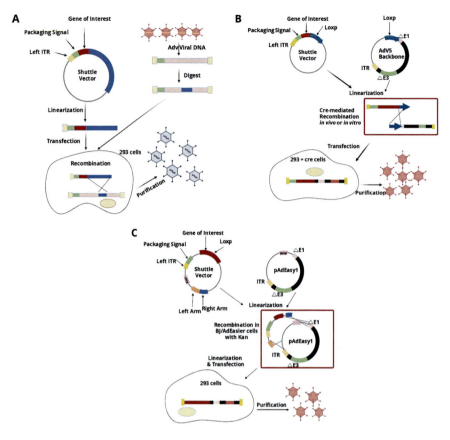

Fig. 6 Different methods for packaging of AdV vectors. (**a**) Homologous recombination in HEK-293 cells, where E1 gene from AdV is replaced by the GOI, previously cloned into a shuttle vector. (**b**) AdMAX system utilizing a Cre/LoxP-based recombination in HEK293 cells co-transfected with shuttle and Ad-genome plasmids. (**c**) pAdEasy method for homologous recombination in *E. coli*, after which the plasmids are linearized and transfected into producer cell lines. (Adapted from [52])

This method generated a good amount of gutless AdV vectors and excluded the presence of AdV and replication-competent adenovirus particles in the final preparation [19] (Fig. 7).

When producing AdV vectors, an adenovirus seed stock is produced and used to infect producer cells in subsequent amplification steps. AdV vector production can be done in adherent or suspension cells, the latter being more suitable for large-scale

Fig. 7 (continued) baculovirus-dependent gutless packaging system, a baculovirus is used as a substitute to helper AdV to provide necessary genome sequences. (**c**) The helper plasmid-based system does not use any helper virus. Producer cells are co-transfected with a plasmid containing an expression cassette with the GOI and a helper plasmid containing the AdV genome deleted for Ψ element, both ITRs, and the E3 gene. (Adapted from [19])

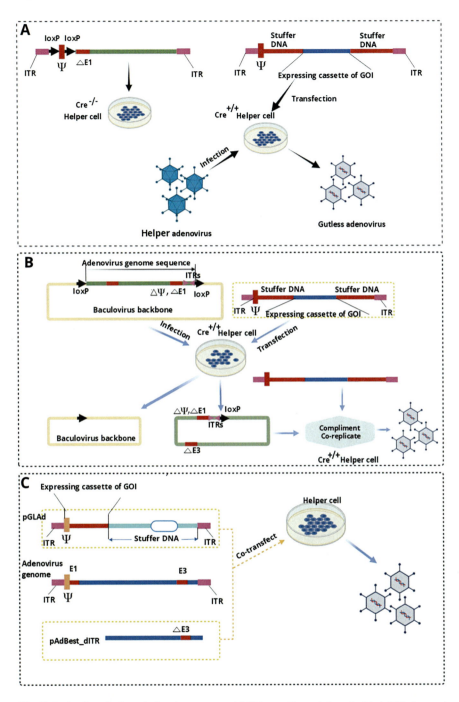

Fig. 7 Packaging of adenoviral gutless vectors. (**a**) Helper cells are infected with AdV helper and transfected with the gutless plasmid containing the GOI. The cis-packaging element (Ψ) of the helper AdV is flanked by loxP sites. The packaging cell line expresses Cre; therefore, the helper AdV Ψ is excised, preventing encapsidation of helper genome into capsids. (**b**) In the helper

manufacturing. Following a suitable incubation period, cells are lysed and the AdV vectors are purified. Lysis can be induced using freeze/thaw cycles, pressure-based methods, or detergents and protease treatment. The cell lysate is then clarified to remove cell debris, either by centrifugation or depth filtration, and treated with endonuclease to remove extraneous host-cell- and viral-DNA. Downstream vector purification utilizes ultracentrifugation (primarily for small-scale production), and/ or chromatography methods, such as anion exchange chromatography (AEX) and affinity chromatography, and tangential flow filtration (TFF) for concentration and diafiltration into the final formulation buffer [33] (Fig. 8).

When considering the manufacturing of AdV vectors for clinical use, some specificities need to be addressed. RD-defective preparations must be tested for the presence of replication-competent, parental, or wild-type viruses that might be generated during the production process. In the case of oncolytic viruses, which are adapted to replicate selectively in cancer cells, it is possible that recombination leads to the formation of particles with a parental genome. The FDA does not recommend the use of HEK293 for production of E1-modified oncolytic adenoviruses, due to the presence of E1 gene in this cell line [55]. Other possible impurities from the production process may include residual DNA and proteins from the host cell, viral proteins, as well as empty or immature particles [56, 57].

2.7 Clinical Applications of Adenovirus Vectors

The earliest adenovirus vector approvals were in China. In 2003, the China State Food and Drug Administration (SFDA now NMPA) approved Gendicine, a replication-defective adenovirus serotype 5 (Ad5) vector expressing the tumor suppressor protein p53, for the treatment of head and neck cancer. In 2005, the SFDA approved Oncorine, an oncolytic virus with E1B 55-kD and E3 deletions, that selectively replicates in cancer cells as a cancer treatment [3, 58]. However, neither of these treatments are approved in the United States or Europe. Currently, Adstiladrin, a non-replicating adenovirus serotype 5 (Ad5) vector containing the human interferon alfa-2b (IFNα2b) transgene, for high-risk Bacillus Calmette-Guérin (BCG)-unresponsive non-muscle invasive bladder cancer, is the only AdV gene therapy vector approved by the FDA (NMIBC) [18]. Apart from already approved products, new treatments are currently in clinical development, mostly for cancer therapy, as depicted in Table 1, and new approvals are expected in the following years. In this sense, the current clinical scenario of AdV vectors clearly depicts the shift in the field of AdV gene therapy toward applications for cancer, largely due to initial safety concerns regarding the high immunogenicity of this vector.

As mentioned previously, vaccines are another way to exploit AdV immune response in an advantageous way. The crisis scenario during the COVID-19 pandemic encouraged numerous efforts to develop SARS-CoV-2 vaccines as an effective way to contain the pandemic. Four different AdV vaccines were developed and approved for emergency use in different countries, with widespread application:

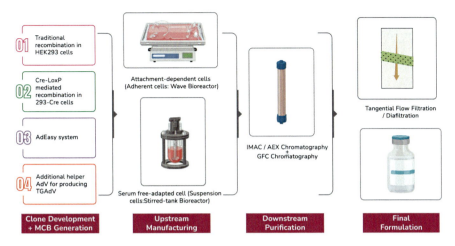

Fig. 8 Adenovirus vector production. AdV vector large-scale manufacturing starts with clone development and creation of a Master Cell Bank (MCB) and a Master Virus Seed. The most appropriate platform is selected, either for suspension or adherent cells, and the upstream process is carried out after cell seeding. Downstream purification involves mostly chromatography techniques, and the final formulation is achieved after tangential flow filtration/diafiltration. IMAC, metal-ion affinity chromatography; AEX, anion exchange chromatography; GFC, gel filtration chromatography. (Adapted from [33])

Covishield/Vaxzevria developed by AstraZeneca/University of Oxford in the UK, a recombinant chimpanzee adenovirus vector, ChAdOx1; Sputnik V created by Gamaleya Research Institute in Russia (rAd26-S and rAd5-S as first and second dose, respectively); Ad26.COV2-S from Johnson & Johnson; and Ad5-S by CanSino Biologics in China [19]. All of them express protein S, a surface glycoprotein that mediates the entry of SARS-CoV-2 into host cells. This scenario highlighted the potential and safety of AdV applications for vaccines and boosted the research and development of the field. Vaxzevria and Jcovden now have standard marketing authorization by EMA. Furthermore, an AdV vaccine for Ebola, consisting of a recombinant Ad26 encoding the glycoprotein (GP) of the Ebola virus Zaire (ZEBOV) Mayinga strain, was approved in Europe in 2020 [59] (Table 2). Several other AdV-based vaccines are currently under clinical development for infectious pathogens including not only for SARS-CoV-2 and Ebola, but also for HIV, Influenza, *Plasmodium falciparum* (malaria), *Mycobacterium tuberculosis*, and Zika virus [33].

Overall, AdV vectors are largely applied for gene delivery, but their use has changed over time from viral vectors for monogenic diseases to oncolytics and vaccine vectors. Their strong immunogenicity and transient gene expression motivated vector engineering strategies that led to improved safety and efficacy, and yet others leveraged their use in the development of therapies for cancer and vaccines, the last application being extensively successful during the COVID-19 pandemic. Today, AdV vectors are mostly applied in cancer therapies and vaccine production, comprising 80% of their usage.

Table 1 AdV vector-based gene therapy products

Gene therapy products currently approved

Product	Vector type	AdV vector construct	Target disease	Approval
Gendicine	Replication defective	rAd5-p53	Head and neck cancer	SFDA—2003
Oncorine (H101)	Oncolytic	rAd5.ΔE1B 55-kD	Head and neck cancer/ nasopharyngeal cancer	SFDA—2005
Adstiladrin	Replication defective	rAd5-IFNα2b	Bladder cancer	FDA—2022

Gene therapy vectors in clinical trial—Phase II or III

Clinical trial ID	Vector type	AdV vector construct	Target disease	Phase
NCT04452591	Oncolytic	rAd5-E2F-1. E1a + GM-CSF	Non-muscle invasive bladder cancer	Phase III
NCT06111235	Oncolytic	rAd5-E2F-1. E1a + GM-CSF	Intermediate risk non-muscle invasive bladder cancer (IR-NMIBC) following transurethral resection of bladder tumor (TURBT)	Phase III
NCT02630264	Replication defective	rAd5-hEndostatin	Squamous cell carcinoma of the head and neck	Phase III
NCT02928094	Replication defective	rAd5-FGF-4	Refractory angina due to myocardial ischemia	Phase III
NCT02798406	Oncolytic	rAd5-Δ24-RGD	Glioblastoma/gliosarcoma	Phase II
NCT06340711	Oncolytic	rAd5-hTERT. E1A/B	Esophagogastric adenocarcinoma	Phase II
NCT05419011	Replication defective	TRI-Ad5 (CEA/ MUC1/ Brachyury)	Lynch syndrome	Phase II
NCT03039751	Replication defective	rAd5-VEGF-D	Refractory angina pectoris	Phase II
NCT04739046	Oncolytic	rAd5-yCD/ mutTK(SR39) rep-ADP	Pancreatic cancer	Phase II
NCT04673942	Oncolytic	rAd5-TGF-β	Sarcoma and refractory solid tumors	Phase II
NCT06283121	Oncolytic	rAd5-TD-nsIL-12	Gastric cancer	Phase II

3   Adeno-Associated Virus (AAV)

The high immunogenicity of adenovirus vectors that led to the death of Jesse Gelsinger profoundly impacted the development of gene therapy by the end of the 1990s, bringing concerns and skepticism to the field. In the 2000s, the research of vectors for in vivo gene therapies began to focus on another virus, the adeno-associated virus (AAV) [61, 62]. Despite their name, they bear little similarity to

Table 2 AdV vector-based vaccines

Vaccines approved worldwide for emergency use during COVID-19

Product	Vector type	Developer	AdV vector construct	Target disease
Covishield/ Vaxzevria	Replication defective	AstraZeneca/University of Oxford	rChAdOx1-S	COVID-19
Sputnik V	Replication defective	Gamaleya Research Institute	rAd26-S and rAd5-S	COVID-19
Jcovden	Replication defective	Johnson & Johnson	rAd26-S	COVID-19
Convidecia	Replication defective	CanSino Biologics	rAd5-S	COVID-19

Vaccines with full approval for infectious diseases

Product	Vector type	AdV vector construct	Target disease	Approval
Zabdeno	Replication defective	rAd26.ZEBOV-GP	Ebola virus disease	EMA— 2020
Vaxzevria	Replication defective	rChAdOx1-S	COVID-19	EMA— 2022
Jcovden	Replication defective	rAd26-S	COVID-19	EMA— 2023

Source: [60]

adenovirus, and their low immunogenic profile and non-pathogenic nature brought a new hope for gene therapy.

Gene therapy vectors derived from AAV proved to be safe and have low toxicity in pre-clinical and clinical contexts [63]. The number of INDs and clinical trials with rAAVs has been in constant rise in the last decades (Fig. 9) [64–66]. In fact, nowadays it represents the most used vector for in vivo gene therapy clinical trials especially for monogenic diseases [67]. As of the beginning of 2025, FDA had approved seven gene therapy products that use AAV as a vector: Luxturna®, Zolgensma®, Hemgenix®, Elevidys®, Roctavian®, Beqvez®, and Kebilidi®. Furthermore, in addition to the above, the EMA had approved Glybera® in 2012 (since withdrawn from the market) and Upstaza® in 2022 [68]. However, some issues are of interest when considering gene therapy applications with rAAVs and need to be addressed during product development. Importantly, reported rAAV toxicities that require caution include hepatotoxicity, oncogenicity, thrombotic microangiopathies, and neurotoxicity [69].

3.1 Overview of AAV

AAVs are naturally found in the environment without being associated with any human disease. They were first discovered by Bob Atchison at the University of Pittsburg as a contaminant in adenovirus preparations and shown to replicate only in the presence of adenovirus, which is why he named them adeno-associated

A

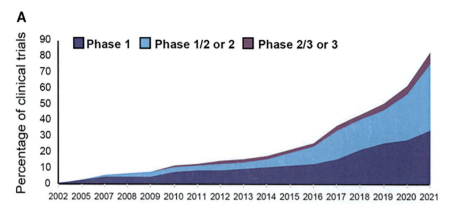

Fig. 9 Clinical trials with rAAV. Graph represents cumulative AAV-mediated gene therapy clinical trials from 2002 to 2021, either in Phase 1, 1/2–2, or 2/3–3. (Republished with permission [66])

viruses [70]. AAVs are single-stranded DNA viruses from the genus *Dependoparvovirus*. They are essentially replication-deficient; their replication depends on co-infection with a helper virus, such as adenovirus (or other large DNA viruses including HSV, poxviruses and baculoviruses). The AAV genome is small, 4.7 kb in size, and has two main genes, *rep* and *cap*, which encode proteins responsible for virus replication and capsid components, respectively [71]. Additional proteins, such as AAP and MAAP, have also been identified and shown to impact productivity, encapsidation, and stability [71, 72].

There are four replicase proteins that originate from promoters p5 and p19: Rep78, 68, 52, and 40 and are obtained by alternative splicing of their transcripts. Rep78 and Rep68, the two larger rep proteins, have helicase, endonuclease, and ATPase activities and are responsible for DNA replication, while Rep52 and Rep40 have helicase activity and mediate genome packaging into capsids. The three capsid proteins (VP1, 2, and 3) are expressed from the p40 promoter and generated by splicing. VP2 and VP3 proteins are generated as a result of alternative translation initiation codons from the same transcript. To compose the viral capsid, VP1, VP2, and VP3 are associated in approximately a 1:1:10 ratio. The AAP protein is also obtained from the p40 promoter, with an alternative open reading frame, and is important for capsid assembly and nuclear import in some AAV serotypes [71, 73]. The cap gene also encodes for protein X, described to enhance AAV replication, and the membrane-associated accessory protein (MAAP), which might be associated with capsid stability [72, 74]. The AAV genome is flanked by inverted terminal repeats (ITRs), 145 bases each, that contain signals for replication, transcription regulation, and packaging of DNA into the viral capsid [74] (Fig. 10).

Another characteristic of AAV is its variety of naturally occurring serotypes: at least 13 serotypes have been described with more than 100 variants discovered in primates. These variants have been divided into six clades, A–F, based on functional and serologic similarities [75, 76] (Fig. 11). AAV1 to 9 are the most studied and applied in the generation of vectors [77]. Viral serotypes are classified by their distinct antigenic characteristics, which reflect changes in amino acid sequences in exposed portions of their capsids that contain epitopes recognized by antibodies [78]. Structural differences between serotypes also lead to changes in the interaction with molecules present on the cell surface, which confers the capacity to infect

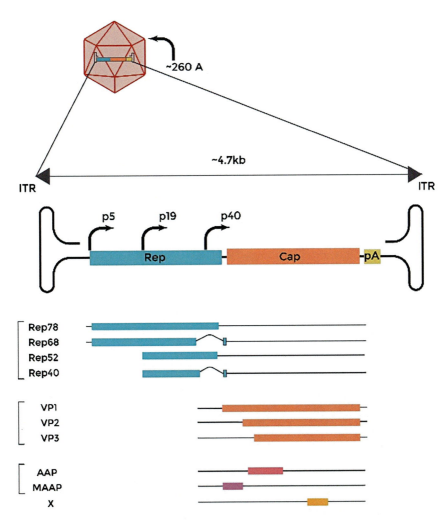

Fig. 10 Schematic of the AAV genome. AAV is a small virus, with a genome size of approximately 4.7 kb. There are two major genes: rep, coding for proteins important in virus replication, starting transcription from promoters p5 and p19; and cap, which originates the capsid proteins, expressed from the p40 promoter. Other proteins encoded from the p40 promoter are AAP, MAAP, and protein X and are important for several steps in viral infection, such as capsid assembly, nuclear import, encapsidation, and stability. (Adapted from [74])

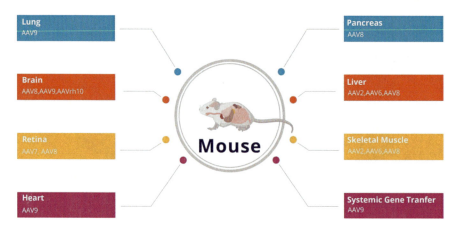

Fig. 11 Tropism of different AAV serotypes. Vector capsid variations have a big impact on tissue tropism. (Adapted from [81])

different tissues for each of them, diversifying the possible therapeutic applications with these vectors [79].

Naturally occurring capsids can be engineered to modify and/or improve vector tropism, creating novel capsids with high efficiency and/or specificity to target usually non-permissive cells, or to improve immune evasion. Specifically, rational design of AAV variants involves the modification of amino acids in specific locations in the capsid, with the exchange of known residues for a specific purpose. Alternatively, peptide ligands can be inserted in the capsid surface to modulate cell targeting. In directed evolution, a library of AAVs with several randomly modified capsids is used in many steps of infection and selection to generate a vector with high transduction efficiency in a cell type or tissue of interest. Recently, new capsids have also been generated in silico by computational design [80].

3.2 Manufacturing of AAV Vectors for Gene Therapy

Vectors derived from recombinant adeno-associated virus (rAAVs) are built after the removal of all viral genes, and substituted by the therapeutic gene and regulatory sequences that may include a tropism-specific promoter, intron(s), enhancer elements, and polyadenylation signals. The only remaining viral sequences in the vector genome are the inverted terminal repeats (ITRs), required for genome replication and packaging [66]. This reduces the possibility of harmful viral sequences in the final product, even though replication-competent AAV may be formed in the production process of rAAVs, as discussed below. rAAVs have a low frequency of genome integration, with their genome primarily persisting in host cells as episomes, and therefore, present a lower risk of oncogenesis as compared to other integrating vectors [82].

Considering that rAAV vectors are completely gutted, replication and capsid genes are provided in *trans* during production. rAAV production is possible by transient transfection using plasmids, stably transfected producer (or packaging) cells, insect cells (Sf9) infected with a baculovirus helper system, or mammalian cells infected with a herpesvirus helper system [83] (Fig. 12). In the case of transient transfection, HEK293 cells (adherent or suspension) are typically transfected at final cell mass with plasmids containing AdV helper (pHELP), AAV replication, and capsid genes (prep2capX) and a plasmid containing the transgene of interest with all its necessary regulatory sequences flanked by ITRs. The total number of plasmids used varies depending on the ability to co-locate these functions on the same plasmid such that up to a total of three plasmids may be used for transient transfection. In the case of packaging and/or producer cell lines, typically A549 or HeLa cells are stably transfected such that both AAV helper genes and the GOI, or only the AAV helper genes, are integrated into the cell's genome. AdV helper sequences are supplied by infection with wild-type (wt) or thermo-sensitive (ts) live AdV and, if necessary, AAV genes are supplied by a first-generation AdV vector (E1-deleted) [3, 83, 84].

Viral infection-based protocols usually provide high yield and scalability but carry the risk of the presence of helper viruses in the preparations. These systems do not require AdV helper genes, the helper function is provided by the herpesvirus or baculovirus. HSV-1 is a well-described helper virus for AAV productive replication [85–88]. A high-performing HSV-based system has been described using two

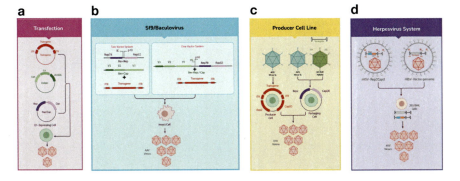

Fig. 12 Different Platforms for AAV Production. (**a**) Transient transfection. Here, production cells, usually HEK293, are co-transfected with three plasmids: transgene plasmid, helper plasmid containing AdV helper sequences, and Rep/Cap, which encodes for AAV viral proteins important for genome replication and capsid formation. (**b**) Sf9/Baculovirus system. In this system, helper functions are provided by baculovirus, which are used to infect insect Sf9 cells and deliver rep, cap, and transgene sequences. This can be a one- or two-vector system, differing if one or two baculovirus vectors are used to provide rep and cap. (**c**) Producer cell line. AAVs can also be packaged using producer cell lines stably expressing both the transgene and AAV viral sequences (Rep/Cap), or only Rep/Cap. Here, the remaining necessary sequences are provided by adenovirus. (**d**) Herpesvirus system, in which rep, cap, and the transgene are provided by replication-deficient HSV vectors. (Adapted from [83])

engineered recombinant replication-deficient HSVs lacking ICP27, carrying either the AAV vector genome containing the expression cassette with the gene of interest or the AAV rep/cap packaging functions. Simultaneous infection of HEK293 or BHK cells with both HSV stocks, in an adherent or suspension format, is needed for the production of AAV [87]. For the baculovirus production system, fall armyworm cells (*Spodoptera frugiperda*, Sf9) are similarly used and co-infected with baculoviruses containing AAV rep/cap functions and the ITRs flanking the transgene of interest [89].

Purification processes include five major phases: harvest of the cells producing AAV, cell lysis to liberate AAV particles (and/or collection of spent media), extraneous cellular and viral nucleic acid removal with endonuclease, capture (affinity resins) and polish (full particle enrichment) of particles by chromatography and/or density gradients, and concentration/formulation using tangential flow filtration followed by sterile filtration [3, 86, 90] (Fig. 13). Gradient ultracentrifugation can be used for rAAV purification and is usually achieved with cesium chloride (CsCl) or iodixanol gradients. Apart from removing impurities, this method allows for significant enrichment of full capsids but can be difficult to scale. Ion-exchange chromatography (AEX) is popularly used for separation of full and empty capsids, even though challenged by the similarity in charge of these particles [3, 90, 91].

The different production systems described above can produce AAV with high yields, potency, and full capsids; however, there are pros and cons for each process. The transient transfection system, a successful production system for Luxturna®, Zolgensma®, Upstaza®, and Elevidys®, requires expensive, highly pure plasmids as

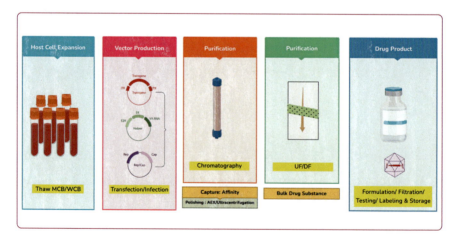

Fig. 13 Generic process flow diagram for AAV manufacture. Production of rAAV vectors starts with expansion of either adherent or suspension host cells, followed by transient transfection of the three plasmids for vector production. After cell lysis and clarification, the AAV vector particles are purified with different steps of chromatography, followed by ultrafiltration (UF)/diafiltration (DF) and final formulation.

starting material. The HSV-based system requires complex preparation of the upstream reagents (HSV and cell banks), and purification processes adapted to ensure complete removal of HSV-derived products and associated detection assays [87, 92]. Similarly, for producer cell systems, it is important to show complete removal of adenovirus-derived products. At least three different approved AAV products have successfully used the baculovirus system for production (Glybera®, Hemgenix®, and Roctavian®), yet it is important to understand and optimize the impact of post-translational modifications, capsid protein stoichiometry, and other aspects of baculovirus production system as it relates to potency and efficacy.

Known impurities in rAAV preparations represent a risk for immunotoxicity, and vector manufacturing under good manufacturing practices (GMP) is still a challenge. The safety and efficacy of rAAV products are largely dependent on purification processes to remove impurities after vector assembly in host cells, and quality control relies on robust assays for analysis of impurities [55]. Analytics are further explored in detail later in this chapter. Impurities in rAAV preparations may include residual DNA from the production process, which can elicit innate immune responses [93, 94]. Even though extra-capsid DNA is mostly degraded with nuclease treatment in the production process, non-therapeutic genetic materials, such as helper and *rep/cap* sequences or host cell-derived DNA, are packaged into rAAVs. The FDA recommends that care should be taken when selecting a host cell for production, due to the possibility of co-packaging of non-vector sequences and unwanted DNA sequences, especially tumorigenic ones. Packaged impurity DNA is difficult to remove or reduce in the final product, and adequate quality data, risk assessment, and details of the process should be provided [55].

The formation of replication-competent AAV (rcAAV) during the production process is possible due to recombination of AAV elements in the vector with the *rep* and *cap* sequences provided in trans during manufacturing. Even though AAV is non-pathogenic and replication deficient, FDA has expressed concern about immune responses to products of *rep* and *cap* genes, which may impact the safety of the product.

Another important concern related to the production process of rAAVs is the co-purification of gene therapy preparations with capsids lacking genetic material (empty). The proportion of full/empty capsids varies considerably in rAAV preparations, ranging from <10% to >70% depending on several factors such as genome size and transfection method [95]. These empty capsids represent a real challenge in rAAV manufacturing, since they are unable to deliver the therapeutic gene thus reducing the therapeutic load on one hand, but, on the other hand, are capable of triggering innate and adaptive immune responses [94, 96, 97] and competing with full particles for binding to target cells [98]. Therefore, the proportion of empty capsids impacts vector quality and safety and must be carefully assessed on a case-by-case basis per therapeutic product being developed.

3.3   Clinical Applications of AAV Vectors

Nine rAAV-based products have already been approved by the FDA and/or EMA for marketing (Table 3). Glybera®, developed for the treatment of lipoprotein lipase deficiency, was the first AAV gene therapy product approved by EMA in 2012. It consisted of an rAAV1 vector encoding the gene for the naturally occurring gain-of-function variant LPLS447X of the human lipoprotein lipase (LPL). However, its marketing authorization renewal was not requested due to very limited demand, and it was withdrawn [99, 100]. The first gene therapy rAAV approved by the FDA was Luxturna®, an rAAV2 vector expressing RPE65, for patients with biallelic RPE65 mutation-associated retinal dystrophy, subsequently approved also in Europe in 2018. Following that, Zolgensma® was authorized in 2019 by the FDA and in 2020 by EMA, an rAAV9 vector for the treatment of pediatric patients with spinal muscular atrophy (SMA). After 2022, the number of marketing authorizations in the United States and Europe began to rise, with four new approved products between 2022 and 2023, including applications for aromatic L-amino acid decarboxylase (AADC) deficiency, Duchenne muscular dystrophy (DMD), and hemophilia B and

Table 3 Gene therapy products based on AAV approved for marketing in the United States and Europe

Product	rAAV construct	Target disease	Marketing approval
Glybera® (Alipogene tiparvovec)	rAAV1-LPL^{S447X}	Lipoprotein lipase deficiency	EMA—2012 (withdrawn)
Luxturna® (Voretigene neparvovec-rzyl)	rAAV2-hRPE65v2	Leber congenital amaurosis type 2 (LCA2).	FDA—2017 EMA—2018
Zolgensma® (Onasemnogene abeparvoveque)	rAAV9-SMN1	Pediatric spinal muscular atrophy (SMA)	FDA—2019 EMA—2020
Upstaza® (Eladocagene exuparvovec)	rAAV2- DDC	Severe aromatic L-amino acid decarboxylase (AADC) deficiency	EMA—2022
Hemgenix® (Etranacogene dezaparvovec)	rAAV5-FIX	Hemophilia B	FDA—2022 EMA—2023
Elevidys® (Delandistrogene moxeparvovec)	AAVrh74-micro-dystrophin	Duchenne muscular dystrophy (DMD)	FDA—2023
Roctavian® (Valoctocogene roxaparvovec)	rAAV5-hFVIII-SQ	Hemophilia A	EMA—2022 FDA—2023
Beqvez® (Fidanacogene elaparvovec-dzkt)	rAAVrh74-FIX	Hemophilia B	FDA—2024 EMA—2024
Kebilidi® (eladocagene exuparvovec-tneq)	AAV2-DDC	Ornithine transcarbamylase (OTC) deficiency	FDA—2024

A. Furthermore, several therapies are in late clinical development and future approvals are expected (Table 4).

The overall low immunogenicity and safety profile of rAAV vectors is reflected in the number of products currently approved and in late clinical trials. Despite its clinical success, clinical development of AAVs has been fraught with headwinds. Major toxicity risks of rAAV vectors are hepatotoxicity and thrombotic microangiopathy (TMA) after intravenous injection, neurotoxicity after intrathecal or intraparenchymal delivery, and oncogenicity, as indicated by the FDA [69]. Strategies to mitigate adverse events due to immune-mediated toxicities and improve the safety

Table 4 Late-stage clinical trials with rAAV vectors

Clinical trial ID	rAAV construct	Target disease	Phase
NCT05203679	rAAV843-hFIX	Hemophilia B	Phase III
NCT05139316	rAAV8-G6Pase	Glycogen storage disease type IA	Phase III
NCT05345171	rAAV8-OTC	Ornithine transcarbamylase (OTC) deficiency	Phase III
NCT06297486	SPK200[a]-FVIII-SQ	Hemophilia A	Phase III
NCT03569891	rAAV5-hFIXco-Padua	Hemophilia B	Phase III
NCT03584165	rAAV2-REP1/ rAAV8-RPGR	Choroideremia/ X-linked retinitis pigmentosa	Phase III
NCT03496012	rAAV2-REP1	Choroideremia	Phase III
NCT04370054	rAAV6-h FVIII	Hemophilia A	Phase III
NCT03861273	rAAV-Spark100-hFIX-R338L	Hemophilia A	Phase III
NCT05568719	rAA6-FXIII	Hemophilia A	Phase III
NCT04281485	rAAV9-mini-dystrophin	Duchenne muscular dystrophy	Phase III
NCT03293524	rAAV2-ND4	Leber hereditary optic neuropathy	Phase III
NCT05926583 NCT04671433 NCT04794101	rAAV5-hRKp.RPGR	X-linked retinitis pigmentosa	Phase III
NCT05407636	rAAV8- anti-VEGF fab	Age-related macular degeneration	Phase III
NCT05335876	rAAV9-SMN1	Spinal muscular atrophy (SMA)	Phase III
NCT03566043	rAAV9-hIDS	Mucopolysaccharidosis type II (Hunter syndrome)	Phase II/ III
NCT03612869	rAAVrh10-h.SGSH	Mucopolysaccharidosis type IIIA	Phase II/ III
NCT04850118	rAAV2tYF-hRPGRco	X-linked retinitis pigmentosa	Phase II/ III
NCT03199469	rAAV8-Des-hMTM1	X-linked myotubular myopathy	Phase II/ III
NCT02716246	scAAV9.U1a.hSGSH	Mucopolysaccharidosis type IIIA	Phase II/ III
NCT04884815	rAAV9-ATP7B	Wilson disease	Phase I/II/ III

[a]Bioengineered capsid derived from AAV3

of the treatment may include the exclusion of patients with high anti-AAV antibody titers from clinical trials, and prophylactic or therapeutic immunosuppression. Avoiding systemic application, whenever possible, also reduces the risk of immune response.

Hepatotoxicity is a major concern for the use of AAV, and the most common adverse effect after intravenous administration. In a clinical trial for the treatment of X-linked myotubular myopathy (XLMTM), intravascular administration of AT132, an AAV8 expressing MTM1, led to severe hepatotoxic effects, with the death of four children, putting the clinical trial on hold. In this case, hepatotoxicity may have been aggravated by hepatobiliary disease associated with XLMTM. Zolgensma led to a high incidence of adverse effects associated with hepatotoxicity in clinical trials (one-third of patients had at least one adverse event) with one case of severe liver injury, managed with prednisolone. After approval, other cases of severe hepatotoxicity were also observed. In 2022, two children died due to liver failure 5–6 weeks after treatment with commercial Zolgensma. Hepatic alterations have also been observed in clinical trials for hemophilia, and in some cases, elevated aminotransferases coincided with loss of transgene expression [69, 101].

Cases of TMA, a hematological emergency with endothelial pathology and thrombosis of small vessels, have also been observed in clinical trials for spinal muscular atrophy (SMA) with Zolgensma, and for Duchenne muscular dystrophy (DMD) with SGT-001 and PF-06939926. In general, this event was associated with complement activation. In the case of DMD, the risk mitigation plan involved the prophylactic use of complement inhibitors and management of vector manufacturing to decrease the percentage of empty vectors. In other clinical trials, neurotoxicity was identified in the form of dorsal root ganglion (DRG) degeneration and brain MRI alterations. DRG toxicity was observed after intrathecal delivery in animal models, and also in clinical trials for giant axonal neuropathy (GAN) and amyotrophic lateral sclerosis (ALS), with neuronal loss in this structure. Intraparenchymal administration of AAV led to MRI alterations in the site of injection in a trial for late infantile Batten disease with AAVrh.10CLN2. However, there was no clear definition if the cause was related to the vector or to the delivery procedure [69, 102].

So far, experiences in clinical trials have not indicated oncogenicity as a major issue of rAAV. The genetic material of these vectors persists mostly as episomes, but integration is a possibility. Tumorigenesis has been shown in rodent models and large animals. In clinical trials, no tumor development has been associated with rAAV. In a single patient that developed hepatocellular carcinoma (HCC), no clear association to insertional mutagenesis of the vector was found [103].

A meta-analysis of AAV clinical trials from 2022 analyzed the toxicity of these vectors in 255 studies. It showed that 30.6% had treatment-emergent serious adverse events (TESAEs) and 60% of clinical holds were due to toxicity findings. As expected, the most frequent TESAEs were hepatotoxicity and TMA after systemic delivery, and neurotoxicity if delivery was to the central nervous system. Interestingly, the study did not find a difference in TESAEs due to AAV production methods, showing similar outcomes for transient transfection of HEK293 cell or baculovirus/Sf9 system. In total, 11 patient deaths in 8 clinical trials were identified, but in most cases the relation to the vector was unclear [104].

Overall, AAV vectors are considered safe for gene therapy applications, which is reflected in the increasing number of products in clinical trials and approvals for marketing. However, its production process can be very complex and still presents several points of improvement that should be addressed to increase the safety and efficacy of the final product, in particular issues related to immunogenicity and inflammatory toxicities.

4 Retroviruses (Lentivirus, LVV and Gamma Retrovirus, RVV)

The use of viral vectors from the family Retroviridiae has become increasingly popular in gene therapy applications. This virus family has the distinctive ability to reverse-transcribe the virion RNA into linear double-stranded DNA and then integrate the DNA into the genome of the host cell, making them desirable for therapeutic applications. Two members of this family, gammaretroviruses and lentiviruses, are most commonly used for ex vivo gene-modified cell therapy applications, including chimeric antigen receptor T-cell (CAR-T) therapies. The gammaretroviruses are simple retroviruses and include the murine leukemia virus (MLV) and gibbon ape leukemia virus (GALV). Gammaretroviruses were among the first viral vectors used for clinical gene therapy applications and, despite safety concerns surrounding their integration sites, have maintained popularity for certain therapeutic applications. The lentiviruses, including the human immunodeficiency virus types 1 and 2 (HIV-1, HIV-2), are complex retroviruses and possess characteristics that distinguish them from gammaretroviruses, such as the ability to infect and integrate into both dividing and non-dividing cells [105].

Retroviruses are enveloped RNA viruses that are approximately 80–100 nm in diameter and display viral glycoproteins on the outer lipid envelope [106]. Retroviral particles contain two sense-strand RNAs that are bound by nucleocapsid proteins (Fig. 14). Retroviruses can broadly be grouped as simple or complex viruses, based on their genome organization. All retroviruses contain three major coding domains, which are required for survival and function: *gag*, encoding the structural proteins that form the matrix, capsid, and nucleoprotein; *pol*, encoding the enzymes required for reverse transcription and integration and protease activity; and *env*, encoding the components of the viral envelope protein [106, 107]. Simple viruses, such as the gammaretroviruses, encode only these proteins; however, complex retroviruses, such as lentiviruses, encode additional accessory proteins involved in regulating viral replication or host cell response to the viruses (Fig. 15). For example, one of the most well-known lentiviruses, human immunodeficiency virus type 1 (HIV-1), encodes additional genes *tat* and *rev* to support transcriptional activation, RNA polymerase elongation, and orchestration of nuclear export of the viral RNA.

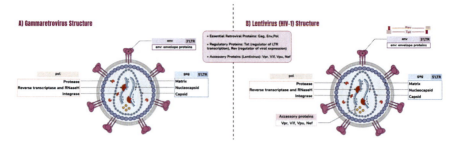

Fig. 14 Gammaretrovirus and lentivirus structure (adapted from [108]). All retroviruses (**a**. gammaretrovirus and **b**. lentivirus) contain two sense-strand RNAs (dimer) coated by nucleocapsid, NC, proteins which are a mature product of the Gag polyprotein. The viral core, also formed by the mature product of the Gag polyprotein, Capsid. The vector core contains integrase and reverse transcriptase enzymes (mature products of the Pol polyprotein) in addition to the genomic RNA-NC complex. The protease enzyme is present outside the core and within the matrix and is responsible for the maturation of the Gag polyprotein. This structure is enveloped by a lipid bilayer containing the envelope proteins, env. In addition, complex retroviruses, such as lentiviruses (**b**), encode additional genes *tat* and *rev* that support transcriptional activation, and accessory proteins involved in regulating viral replication (Vpr, Vif, Vpu, Nef)

Additional auxiliary genes contributing to viral production and pathogenesis present in HIV-1 and other lentiviruses include *vif*, *vpr*, *vpu*, and *nef* [23].

The retrovirus genome is flanked by long terminal repeats (LTRs) at the 5′ and 3′ ends. The LTRs can be divided into three elements: the U3 and U5 sequences are unique to the 3′ and 5′ termini and the R sequence is repeated at both ends. The transcription initiation site is between the U3 and R sequences, and the polyadenylation signal is between the R and U5 sequences. The U3 region contains most of the transcriptional control elements of the provirus, including the promoter and enhancer sequences responsive to both cellular and viral, in cases of complex viruses, transcriptional activator proteins. The enhancer sequence of the U3 region plays a critical part in determining the specificity of the viral replication and pathogenesis [106, 109].

Members of the retrovirus family begin their life cycle with the entry of the mature virus into the cell through either direct membrane fusion or receptor-mediated endocytosis via the viral glycoproteins binding to receptors on the cell surface (Fig. 16). Following entry, the *gag* subunits are dissociated from the viral core and the viral RNA is converted to a proviral double-stranded DNA by the reverse transcriptase. The dsDNA is imported into the nucleus where it can be integrated into the host genome by the pre-integration complex (PIC) forming the provirus, a process involving the virally encoded integrase along with host-cell transcription factors. The gammaretroviruses must access the host cell genome during mitosis when the nuclear envelope is disassembled. Lentiviruses, on the other hand, can access chromosomes by active transport of the DNA through the nuclear pore of an intact nuclear envelope. This ability to import via the nuclear pore allows

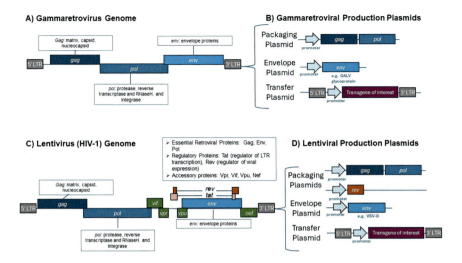

Fig. 15 Genomic organization of retroviral vectors. (**a**) Gammaretrovirus is a simple retrovirus containing the three essential retroviral genes (*gag*, *env*, and *pol*) flanked by the 5' and 3' LTRs. The genome size of the gammaretrovirus is approximately 8.3 kb. (**b**) Three plasmids are used for gammaretroviral vector production. The three essential genes are separated onto packaging and envelope plasmids, and the transfer plasmid contains the transgene of interest flanked by the LTRs. (**c**) Lentiviruses are examples of complex retroviruses containing the three essential retroviral genes, as well as the regulatory genes *tat* and *rev* plus additional accessory genes. The HIV-1 genome is shown here with the accessory genes *vpr*, *vif*, *vpu*, and *nef*. The genome size of the HIV-1 genome is approximately 9.75 kb. (**d**) Production of lentiviral vectors requires four plasmids under the third-generation production systems. *gag/pol* and *rev* are encoded on separate packaging plasmids and the envelope plasmid contains the *env* gene and can be modified for pseudotyping. The transfer gene contains the transgene flanked by the LTRs

the lentiviruses to integrate DNA in both dividing and non-dividing cells, whereas gammaretroviruses are limited to only dividing cells [106, 108].

Host cell machinery is leveraged to initiate and complete transcription of the integrated virus ("provirus"). Whereas simple retroviruses rely on the cellular machinery, HIV-1 and other complex retroviruses utilize both cellular transcription factors and trans-acting factors encoded by the virus (e.g., the HIV-1 Tat protein is a trans-activator of transcription) [110]. The integrated proviral genome relies on host machinery to initiate and complete the transcription and translation of viral proteins necessary to assemble infectious particles. Viral RNAs are transported out of the nucleus using cis-acting viral sequences. Viral progeny are released from the cell through a process called budding from the plasma membrane. During this process, endogenous membrane proteins within the host cell can be incorporated into the viral envelope [111, 112].

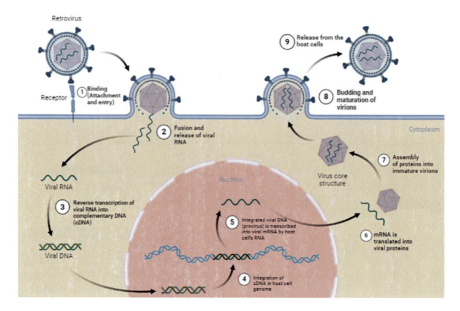

Fig. 16 Retrovirus lifecycle using HIV-1 as a model (figure from Biorender). 1. The Env glycoprotein of the virus binds to the cell surface receptor of a susceptible cell. 2. This causes fusion of the glycoprotein with the cell membrane and release of the viral RNA into the cytoplasm of the cell. 3. The ssRNA is reverse-transcribed into complementary DNA (cDNA) in the core starting in the cytoplasm, the core then enters the nucleus for completion of reverse transcription and uncoating. 4. The cDNA is then integrated into the host cell genome by the enzyme integrase. 5. The integrated viral DNA, called a provirus, is then transcribed into viral mRNA by the host RNA polymerase. 6. The viral mRNA is then translated to form capsid proteins that assemble to form immature virions (7). Budding and final maturation of virions takes place at the cell membrane (8) resulting in release from the host cells (9)

4.1 Gammaretroviruses

The interest in gammaretroviruses as a tool for gene delivery began in the 1970s, and a deeper understanding of the biology of these viruses and advancements in molecular engineering brought the idea of utilizing gammaretroviruses as a gene therapy tool to reality. The first use of gammaretroviruses in a human gene therapy trial was in 1990 to treat patients with adenosine deaminase-deficient severe combined immunodeficiency (ADA-SCID) by ex vivo transduction of the bone marrow cells [113, 114]. Retroviruses quickly emerged as a desirable tool for cell transductions, and their use expanded to treatments of diseases such as X-linked SCID and malignant glioma.

4.2 Production of Retroviruses

Gammaretroviruses encode only the three essential genes *gag*, *pol*, and *env* (trans-elements) flanked by the 5' and 3' LTRs (cis-elements) (Fig. 15). Production systems for recombinant retroviruses rely on providing these three genes and the transgene of interest (TOI) flanked by the LTRs either on separate plasmids or by stable expression in a producer cell line. Transient transfection of a naïve cell line (e.g., HEK293T) requires simultaneous co-transfection of the three plasmids: packaging (*gag/pol*), envelope (*env*), and transfer plasmids to produce RVVs. The transfer plasmid contains the TOI flanked by the LTRs that also includes the packaging signal to allow incorporation of the TOI in the retroviral particle. The envelope plasmid and packaging plasmids do not contain other retroviral elements such as a packaging signal. Because of the lack of the packaging signal, the *env*, *gag*, and *pol* genes are not incorporated into the retroviral particles produced, rendering the particles replication deficient. In lieu of transfection of multiple plasmids, stable packaging cell lines have been developed that express *env*, *gag*, and *pol*, and require only a single plasmid transfection with the transfer plasmid containing the TOI. These stable cell lines can be further modified and selected to incorporate the TOI generating a stable producer cell line (PCL), eliminating the need for plasmids and reducing production costs. While PCL can be time-consuming to isolate and characterize, the RVV manufacturing process using PCL is more cost-effective and easily scalable and reduces batch-to-batch variability compared to processes using transient transfection.

Gammaretroviruses are sensitive to a variety of physical and chemical conditions (e.g., temperature, pH, salt concentration, etc.); therefore, conditions must be maintained within suitable ranges during processing to maintain stability and potency. When developing manufacturing processes, considerations are also needed to maintain the reverse transcriptase and integrase enzymatic activities throughout the upstream and purification processes. The viral particles produced during the upstream process are secreted into the supernatant which is harvested, treated with endonuclease, and clarified via membrane filtration to remove extraneous host- and plasmid DNA, host cells, and cell debris. Additional downstream processing steps beyond clarification will be dependent on the properties of the virus, including the pseudotype, and these steps may impact recovery and infectivity. Additional purification steps may include tangential flow filtration (TFF) to remove host cell proteins and anion exchange chromatography (AEX). Most processes also include sterile filtration as the final step; however, this step can be challenging as up to 70% of viral titer can be lost due to the large size of the RVVs and vector clumping observed with certain pseudotypes. Some manufacturers may opt to perform sterile filtration during an earlier stage in a closed system to minimize loss; however, additional microbiological testing may be required to ensure safety [113]. In the absence of sterile filtration, the entire process will need to be validated for asepsis which makes for a very expensive endeavor.

One of the critical steps for the development of retroviral therapies is pseudotyping by selecting envelope proteins for expression in the retroviral vector. The glycoproteins expressed on the viral envelope are critical for the virus to gain access to the cells; therefore, expressed glycoprotein can impact the cell specificity and improve transduction [113, 115]. Pseudotyping can either broaden or narrow the range of cell types targeted. Other advantages of pseudotyping include improved stability and increased yields during manufacturing [115]. The pseudotype of the virus can impact cellular toxicity, since some viral proteins can be toxic when overexpressed which can impact the yield during the manufacturing process. Common glycoproteins used for pseudotyping include those from the amphotropic murine leukemia virus (Ampho MLV-A), Gibbon ape leukemia virus (GALV), feline endogenous virus (RD114), foamy virus, baboon retrovirus (BAEV), and vesicular stomatitis virus (VSV-G protein) [113].

4.3 Clinical Use of Retroviruses

Studies early on using the gammaretrovirus MLV hinted at the promise of these viruses as a therapeutic tool; however, these studies also revealed safety concerns related to the inherent nature of retroviral integration. Clinical studies for therapies using ex vivo retrovirus-mediated gene transfer into hematopoietic progenitor cells showed a preference to insert transgenes near oncogene promoters, enhancing risks of other malignancies. A clinical study for the treatment of SCID-X using a retrovirus-mediated gene transfer into bone marrow-derived CD34+ cells of patients was initially deemed a success with 9 out of 10 patients developing functional immune system. However, within 5 years of treatment, four of the patients developed uncontrolled clonal proliferation of mature T-lymphocytes [116, 117]. Insertional mutagenesis associated with retrovirus-mediated gene transfer was speculated, prompting reassessment of the risks associated with the use of gammaretroviruses.

Later clinical studies using a retroviral mediated ex vivo transduction for cell therapy were also plagued with concern. A gene therapy treatment for Wiskott-Aldrich syndrome showed resolution of disease indication, but several patients developed acute leukemia associated with integration sites and chromosomal translocations [118]. In 2016, a breakthrough was achieved with the EMA authorization of Strimvelis® in Europe for treatment of patients with severe combined immunodeficiency caused by adenosine deaminase deficiency (ADA-SCID). Strimvelis® consists of CD34+ hematopoietic stem cells transduced with gammaretrovirus to express the adenosine deaminase. Unfortunately, an occurrence of leukemia linked to insertion site was observed in a single patient within 5 years of the approval, yet the incidence of insertional oncogenesis was significantly lower than in previous studies described for the Wiskott-Aldrich and SCID-X therapies [119]. While dosing of patients was put on hold and ultimately discontinued by Orchard Therapeutics due to cost, Strimvelis® was recently revived by the Telethon Foundation [120] and remains authorized for use in Europe (EMA website). Despite the progress made with gammaretroviral vectors, safety concerns around the insertional oncogenesis shifted research and development focus toward the use of lentivirus, as detailed in section "Lentiviruses."

While these concerning outcomes slowed the progress of RVV-based therapies, efforts to modify the vectors to reduce risk of transcriptional dysregulation of proto-oncogenes as well as the progress of CAR-T and CAR-NK therapies have preserved interest in the use of gammaretroviral vectors. As of 2024, six CAR-T therapies have been approved by the FDA, of which two utilize gammaretroviruses for gene modification and the remainder use lentiviral vectors (Table 5). In 2017, the FDA approved the first CAR-T therapy, Yescarta® (Axicabtagene ciloleucel), for the treatment of certain types of large B-cell lymphoma in adults [121]. Within a few years, Tecartus® (brexucabtagene autoleucel) was also approved for treatment of B-cell acute lymphoblastic lymphoma [122]. Both Yescarta® and Tecartus® are generated by isolation of patient T-cells which are transduced ex vivo with gammaretroviral particles to insert a chimeric antigen receptor (CAR) which targets the CD19 receptor on the cancer cells. While the use of retroviral vectors for modification of T-cells appears to be less prone to causing malignancies in patients than the use of modified stem cells, there is still concern, and recently, the FDA has investigated the potential of T-cell lymphomas in CAR-T patients prompting boxed warnings on all CAR-T therapies, as of 2024 [123, 124].

4.4 Lentiviruses

Although the use of gammaretroviral vectors has historically been more common, the number of clinical trials and approved products using lentiviral vectors (LVV) for gene therapy is increasing. Lentiviruses offer several advantages over gammaretroviruses, including their ability to infect and integrate into both dividing and non-dividing cells, allowing them to target various tissue and cell types. This broad tropism makes them desirable for gene therapy applications. Additionally, LVV integration into the chromosome poses less concern than gammaretroviruses. Unlike gammaretroviruses that preferentially integrate near transcriptional start sites, LVV preferentially integrates within transcriptional units [114, 125, 126], reducing concerns regarding potential oncogenesis. Finally, lentiviruses transduce both dividing and non-dividing cells as compared to gammaretroviruses and carry a much larger gene cassette compared to AAV and gammaretroviruses (Fig. 15).

Table 5 Approved CAR-T therapies

Product	Retrovirus type	Target antigen	Target disease(s)	Marketing approval
Kymriah® (tisagenlecleucel)	Lentivirus	CD19	B-cell acute lymphoblastic leukemia (ALL) B-cell non-Hodgkin lymphoma (NHL)	FDA—2017 https://www.fda.gov/vaccines-blood-biologics/cellular-gene-therapy-products/kymriah-tisagenlecleucel EMA—2022 https://www.ema.europa.eu/en/medicines/human/EPAR/kymriah
Yescarta® (axicabtagene ciloleucel)	Gamma retrovirus	CD19	B-cell non-Hodgkin lymphoma (NHL) Follicular lymphoma	FDA—2017 https://www.fda.gov/vaccines-blood-biologics/cellular-gene-therapy-products/yescarta EMA—2022 https://www.ema.europa.eu/en/medicines/human/EPAR/yescarta
Tecartus® (brexucabtagene autoleucel)	Gamma retrovirus	CD19	Mantle cell lymphoma (MCL) B-cell acute lymphoblastic leukemia (ALL)	FDA—2021 https://www.fda.gov/vaccines-blood-biologics/cellular-gene-therapy-products/tecartus-brexucabtagene-autoleucel EMA—2022 https://www.ema.europa.eu/en/medicines/human/EPAR/tecartus
Breyanzi® (lisocabtagene maraleucel)	Lentivirus	CD19	B-cell non-Hodgkin lymphoma (NHL)	FDA—2022 https://www.fda.gov/vaccines-blood-biologics/cellular-gene-therapy-products/breyanzi-lisocabtagene-maraleucel EMA—2022 https://www.ema.europa.eu/en/medicines/human/EPAR/breyanzi
Abecma® (idecabtagene vicleucel)	Lentivirus	BCMA	Multiple myeloma	FDA—2021 https://www.fda.gov/vaccines-blood-biologics/abecma-idecabtagene-vicleucel EMA—2021 https://www.ema.europa.eu/en/medicines/human/EPAR/abecma
Carvykti® (ciltacabtagene autoleucel)	Lentivirus	BCMA	Multiple myeloma	FDA—2022 https://www.fda.gov/vaccines-blood-biologics/cellular-gene-therapy-products/carvykti EMA—2022 https://www.ema.europa.eu/en/medicines/human/EPAR/carvykti

4.5 Production of the Lentiviral Vectors

Despite the more complex genome of lentiviral vectors, production is very similar to gammaretroviral vectors. Transient transfection and stable producer cell line processes have been developed which provide the essential genes for vector production. Lentiviral vector production systems have been refined over time to improve safety and performance, reducing the risk of replication-competent viruses. The first-generation LVV systems contained a significant portion of the HIV genome, including the genes necessary for viral growth and replication (e.g., *gag*, *pol*, *tat*, and *rev*), as well as accessory genes *viv*, *vpr*, *vpu*, and *nef*. While the accessory genes were advantageous for survival and fitness of LVV in vivo, they are not necessary for growth in vitro, and therefore removed in the second-generation system [105]. LVV vectors are pseudotyped with other viral glycoproteins, commonly the glycoprotein of vesicular stomatitis virus (VSV-G), to allow for better tropism since VSV-G interacts with ubiquitous cellular receptor on diverse cells. Additionally, VSV-G increases LVV particle stability during purification steps [127, 128]. Furthermore, in an attempt to improve safety, third-generation systems currently in use split the viral genome into separate plasmids reducing the likelihood of replication-competent lentivirus (RCL). The LVV is made using three separate plasmids encoding the *gag/pol*, *rev*, and envelope protein, and a fourth (transfer) plasmid contains the TOI flanked with the LTRs [105, 129]. With the addition of a constitutively active promoter upstream of the LTRs in the transfer plasmid, the *tat* gene was made unnecessary [129]. Other modifications to the system to enhance safety include deletion in the 3' LTR to disrupt the promoter/enhancer activity creating a self-inactivating LVV. The deletion in the 3' LTR is transferred to the 5' LTR after one round of reverse transcription, resulting in loss of transcriptional activity of the provirus in the target cells, minimizing the risk of recombination to generate RCL as well as reducing the likelihood of aberrant expression of neighboring coding sequences adjacent to the vector integration site [105, 130, 131].

Production of LVV typically relies on co-transfection of the LVV production plasmids along with the genome plasmid containing the TOI into a cell line such as HEK293 or HEK293T. HEK293T cells express the SV40 T-antigen which renders them more efficient for vector production [132, 133]. Co-transfection of the four plasmids used in third-generation process (Figs. 17 and 18) provides all the components necessary to produce functional LVV particles which can be harvested from the culture medium. For large-scale production of LVV, the transient transfection process can be very costly, and scalability is unreliable. Development of stable producer cell lines presents a more affordable, scalable, and consistent solution to overcome these limitations. By integrating the genes necessary for LVV production including the transgene into the packaging cell line, the need for plasmid transfection is eliminated. The Cytegrity™ cell line technology, developed at St. Jude's Children's Hospital, is an example of cell line technology with the packaging genes under a tetracycline-inducible system to allow for generation of PCL for LVV production [134, 135]. The resulting upstream process using a PCL is streamlined, reproducible, and readily scalable.

Downstream processing of the LVV particles should result in a product with the desired concentration and purity among other desired quality attributes. Once harvested from the culture medium, processing the LVV particles typically starts with flow filtration and clarification of the bulk harvest to remove cell debris, followed by various purification steps for capture, polishing, concentration, and buffer exchange which may include chromatography steps (e.g., anion exchange, size exclusion, etc.), tangential flow filtration, and/or gradient ultracentrifugation. Following purification, the eluted fractions undergo filtration steps to remove remaining debris and for sterilization purposes (Fig. 18) [137, 138].

One caveat to the use of LVV as therapeutics is that the antigens are recognized by the immune system as foreign, limiting their applications for in vivo therapies. However, they continue to remain the popular choice for ex vivo applications such as chimeric antigen receptor T-cell (CAR-T) therapies and for modifying hematopoietic stem cells (HSCs) [138]. Modifications to the production system to reduce impurities can reduce immunogenicity, opening the door for in vivo applications of LVV. For example, using stable PCL instead of transfection methods eliminates residual plasmid DNA in the product. Additionally, modifying the PCL to disrupt genes needed for major histocompatibility complex or increasing CD47 expression have been shown to impact the proteins that are incorporated into the LVV envelope and reduce the immune response [139, 140].

4.6 Clinical Use of LVV

As of 2024, six CAR-T therapies have been approved by the FDA, and most of these are generated via transduction of the T cell with LVV to integrate the CAR (Table 5: Approved CAR-T). In 2016, Kymriah became the first approved CAR-T therapy to use lentivirus to transduce the T-cells ex vivo. Similar to the approved gammaretrovirus-mediated CAR-T products, Yescarta® and Tecartus®, the LVV-mediated CAR-T therapies, Kymriah® and Breyanzi®, modify the patient's T cells to recognize the CD19 target on the B cells to treat the B cell malignancies such as B cell acute lymphoblastic leukemia, B cell non-Hodgkins lymphoma, and follicular lymphoma. Other approved CAR-T therapies, Abecma® and Carvykti®, use LVV to modify the T cells to recognize the target antigen B cell maturation antigen (BCMA) for the treatment of multiple myeloma. Other immunotherapies utilizing gammaretroviruses and lentiviruses are being explored including CAR-NK therapies, which aim to enhance the cancer-fighting abilities of natural killer (NK) cells. CAR-NK therapies are similar to CAR-T therapies: isolated NK cells are genetically modified, often by transduction with retrovirus or lentivirus, to express the CAR targeting a specific antigen in the cancer cells. Currently, there are no approved CAR-NK therapies.

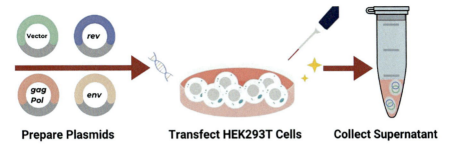

Prepare Plasmids   **Transfect HEK293T Cells**   **Collect Supernatant**

Fig. 17 Transient transfection of plasmids for lentiviral vector production. Four packaging plasmids are transfected into cells (e.g., HEK293T) for vector production. Vector particles are released into the media, which is collected and clarified of cell debris before further purification and/or concentration. (Adapted from [136])

LVVs have proven successful at stem cell therapies; however, as seen with the gammaretroviruses, oncogenic risks are still present and monitoring of patients is recommended. In the last 2 years, the FDA approved four therapies using hematopoietic stem cells (HSCs) transduced with LVV (Table 6). The first was Zynteglo™ in 2022. Zynteglo™ is a treatment for beta thalassemia by transducing a patient's HSCs ex vivo with a self-inactivating LVV carrying a globin gene. The modified cells are introduced back into the patient, where they replicate, and the stable integration of the functional globin gene is passed to the progeny [142]. The next one, Skysona™ to treat cerebral adrenoleukodystrophy (CALD), uses a LVV encoding the ATP-binding cassette, subfamily D, member 1 (ABCD1) to transduce ex vivo the patients' own HSC cells. The modified stem cells are provided back to the patient and expression of the functional ABCD1 gene to preserve neurological function in these patients [143]. In 2023, the Lyfgenia® was approved for the treatment of sickle cell disease (SCD) in patients 12 years and older (REF FDA site). Lyfgenia® uses a lentiviral vector for genetic modification of patient's HSCs to produce HbA^{T87Q} which functions similarly to a functional hemoglobin A, lowering the risk of sickling and occlusion that is present in SCD patients. While patients saw an improvement in symptoms, there was occurrence of hematologic malignancy (blood cancer) in a small number of patients treated with Lyfgenia® triggering inclusion of a black box warning on the label and requiring life-long monitoring for malignancies in these patients.

Most recently, the FDA approved Lenmeldy® (Libmeldy® in EU) in which HSCs genetically modified ex vivo using a lentiviral vector carrying a functional copy of the ARSA gene for treatment of metachromatic leukodystrophy (MLD). One major drawback for these and other cell and gene therapy products is the high price per treatment. These are among the priciest drugs on the market [144, 145]. The high costs and market access led to the withdrawal of Skysona™ and Zynteglo™ from the European market, although they remain on the market in the United States [146]. Despite the high price and risks associated with malignancies, these therapies and many more still in clinical studies are proving that the use of LVV for cell therapy

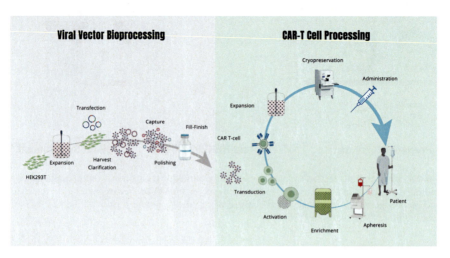

Fig. 18 Viral vector processing and CAR-T cell processing (Adapted from [141]). HEK293T cells are the most common producer cells used for lentivirus production. Cells are expanded and then transfected with plasmid DNA (an alternative way to make lentivirus is to use stably transfected producer cells like Cytegrity™ that abrogates the use of costly plasmids at scale). After transfection, vector production occurs and the vector produced buds into the spent media. This cell supernatant is usually treated with endonuclease to remove host cell DNA and plasmid DNA impurities and clarified to separate from cells and cell debris. The vector in the clarified harvest is then captured usually via anion exchange chromatography. The vector may be processed through an additional polishing step before concentration, buffer exchange into an appropriate formulation buffer, and filling into vials/bags. The filled vector product can then be used for CAR T cell processing. The cell process starts with isolating peripheral blood mononuclear cells (PBMCs) either from a patient or healthy donor. PBMCs are further processed to enrich T lymphocytes (usually using antibody conjugated magnetic beads). The enriched T cells are then activated using anti-CD3/CD28 antibody-conjugated beads followed by transduction with the lentiviral vector product. Post-transduction, the CAR therapeutic gene is integrated into the target T cell genome to produce fully functional CAR T cells. These cells are expanded in culture to reach an appropriate cell mass for therapy, harvested, conditioned, and cryopreserved. The cryopreserved cells are then shipped back to the clinic for administration to a patient

applications is effective and that promising treatments for many genetic diseases and cancers.

Research aimed at improving gene therapy using LVV and identifying novel uses for the vectors in clinical applications is ongoing. One mechanism for overcoming risk of insertional mutagenesis is development of non-integrating lentiviral vectors, where the viral vector maintains the ability to transduce dividing and non-dividing cells, yet the proviral genome does not integrate [147]. The proviral genome remains as a non-replicating episome, and transgene expression may be short-lived compared to integrated transgenes with episome loss during cell division [147]. While the lack of stable, long-term expression may limit their clinical applications, these may still be useful for applications for slower dividing cells such as mesenchymal stem cells (MSCs) or where long-term expression may not be necessary, such as CAR-T. Other applications for NILVs, such as vaccine development, are also being studied. Other applications for LVV being explored include the use to deliver gene editing components (e.g., guide RNAs for Crispr-Cas system) and as cancer vaccines through expressing tumor antigens or co-stimulatory signals on dendritic cells [105].

Overall, LVVs have proven useful for several therapeutic applications, such as CAR-T and HSC-based therapies, and the production systems have evolved to increase the safety profile and reduce the risk malignancies. Continued research into vector improvements, as well as novel applications holds the promise that LVVs will continue to be an important therapeutic tool.

5   Herpes Simplex Virus (HSV)

Considerable progress in the development of herpes simplex virus (HSV) vectors in gene therapy, oncolytics, and immunotherapy has proven their potential in safely and efficiently delivering these therapies [43, 148]. HSV infects mucosal epithelial cells leading to cell lysis, progeny virions then access sensory nerve terminals and are carried via retrograde axonal transport to the peripheral nerve cell nucleus, where the virus can undergo a lytic or a latent life cycle, and the genome can stay as a circular episome for life [149, 150]. Several biological features make it an attractive viral vector for gene therapy, particularly in nervous system disease indications. Specifically, HSVs can infect both dividing and non-dividing cells and show a broad host range with virus entry receptors ubiquitous in diverse cell types [150, 151]. Almost half of the viral genes can be deleted from HSV vectors since they are non-essential for virus growth in cultured cells providing a high cargo capacity of foreign DNA [149]. The latent viral genome does not integrate into the host DNA and can persist as episomes in the nucleus of infected cells, avoiding the risk of insertional mutagenesis and can become transcriptionally silenced over time in neurons [151, 152]. HSV is also known for its ability to evade the host immune system [153, 154].

HSV-1 is the most characterized member of the herpesviridae family; it contains a linear dsDNA genome of over 150 kb in length and encodes approximately 90 genes. The genome is composed of two segments, the unique long (UL) and unique short (US), each flanked by inverted repeats containing important viral regulatory elements (Fig. 19). HSV is an enveloped virus with an icosahedral capsid surrounded by a tegument layer, and its DNA replication occurs in the nucleus of infected cells. Viral gene expression is controlled as a transcriptional cascade of immediate-early, IE (ICP4, ICP27, ICP0, ICP22, ICP47), early, E, and late, L genes, resulting in viral DNA replication, formation of new particles, and lysis of infected cells (Fig. 20). During a latent cycle, only latency-associated transcripts (LATs) are accumulated [148, 149, 155].

HSV has proven to be a suitable and powerful tool as a gene therapy vector, with approved therapeutic products in different countries. While antitumor strategies using replication-defective vectors to deliver therapeutic transgenes have yielded

Table 6 Approved HSC therapies

Drug name	Cell type	Virus	Modified gene	Targeted disease(s)	Approval
Strimvelis (autologous CD34+ enriched cell fraction that contains CD34+ cells transduced with retroviral vector that encodes for the human ADA cDNA sequence)	CD34+ cells	RV	Adenosine deaminase (ADA)	Severe combined immunodeficiency due to adenosine deaminase deficiency (ADA-SCID)	EMA—2016 https://www.ema.europa.eu/en/medicines/human/EPAR/strimvelis
Skysona (elivaldogene autotemcel)	Autologous HSC	LV	ABCD1	Cerebral adrenoleukodystrophy (CALD)	FDA—2022 https://www.fda.gov/vaccines-blood-biologics/skysona EMA—2021 Authorization withdrawn
Zynteglo (betibeglogene autotemcel)	Autologous HSC	LV	Beta globin	Beta thalassemia	FDA—2022 https://www.fda.gov/vaccines-blood-biologics/zynteglo EMA—2019 Authorization withdrawn
Lyfgenia (lovotibeglogene autotemcel)	Autologous HSCs	LV	HbA^{T87Q}	Sickle cell disease	FDA—2023 https://www.fda.gov/vaccines-blood-biologics/lyfgenia
Lenmeldy (libmeldy—EU)atidarsagene autotemce	CD34+ cells	LV	ARSA (Arylsulfatase A)	Metachromatic leukodystrophy	FDA—2024 https://www.fda.gov/vaccines-blood-biologics/cellular-gene-therapy-products/lenmeldy EMA—2020 https://www.ema.europa.eu/en/medicines/human/EPAR/libmeldy

encouraging results, the fastest growing area in HSV vector-mediated cancer treatment lies with oncolytic viruses. However, HSV vector delivery to tumor cells is plagued by three major obstacles: inadequate penetration in tumor interstitial space; significant uptake of HSV vectors into normal organs; and host immune response to the HSV vector [157, 158]. New strategies are under development to overcome some of these limitations, like re-targeting vectors, combination therapies, armed vectors, and a carrier cell delivery approach [159]. The different types of HSV vectors and their clinical relevance are elucidated below including amplicon HSV, replication defective HSV, and replication-competent HSV (Oncolytic or oHSV).

5.1   Amplicon HSV

HSV amplicon vectors are plasmids or bacterial chromosomes that contain only two viral sequences: HSV-1 origin (ori) of replication and HSV-1 packaging signal (pac), an antibiotic resistance gene (AmpR), an *E. coli* origin of replication (ColE1), and a transgene expression cassette of up to 150 kb in size [160]. These vectors have the same properties as HSV-1 virions but contain a genome-length amplicon plasmid concatemer instead of the viral genome. Amplicon vectors remain extrachromosomal, and their production relies on a full complement of helper virus functions, typically a replication-defective HSV-1 to minimize amplicon stock contamination with replication-competent HSV; complementing cells are used to enable helper virus replication and amplicon replication and packaging [151, 161]. Amplicon vectors show no toxicity for target cells, due to the absence of viral protein expression. However, dependence on a helper virus for vector production and packaging makes this platform challenging to manufacture and validate.

5.2   Replication Defective Vectors

Different types of defective recombinant vectors have been developed. These vectors eliminate genes responsible for lytic viral gene expression and toxicity, while achieving durable transgene expression. To date, several replication-defective vectors have been constructed in which several IE genes are deleted in various combinations [151, 157] (Fig. 21). Replication-defective vectors have been tested as a gene delivery platform in many different preclinical and clinical studies of gene therapies in neuropathies, cancer gene therapy, and monogenic diseases [148, 162, 163].

The recently FDA-approved beremagene geperpavec (B-Vec, VYJUVEK) platform by Krystal Biotech is a defective-replication HSV vector designed as a topical therapy for the treatment of dystrophic epidermolysis bullosa (DEB) [164]. DEB is a serious, ultra-rare genetic blistering disease caused by mutations in the COL7A1 gene that encodes type VII collagen (COL7), assembled into anchoring fibrils, the

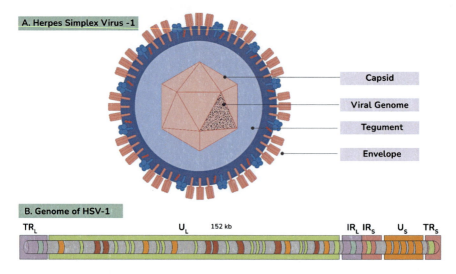

Fig. 19 Herpes simplex virus (HSV-1) structure. (**a**) The HSV virion is composed of an icosahedral nucleocapsid, surrounded by an amorphous tegument, and an outer lipid bilayer envelope containing viral glycoproteins essential for the attachment and penetration of the virus into the cells. (**b**) The viral genome is a linear dsDNA of over 150 kb in length, and is composed of two segments, the unique long (U_L) and unique short (U_s), each flanked by inverted repeats; terminal repeats (TR) and internal repeats (IR). (Adapted from [156])

basic membrane that anchors epidermis and dermis together. B-Vec vector is the first approved topical and redosable gene therapy product derived from wt HSV-1. It is deleted for ICP4 and ICP22 and contains two copies of COL7A1 gene and regulatory elements inserted in the ICP4 locus [154, 162, 165]. A Phase 3 study (GEM-3; NCT04491604) evaluating the efficacy and safety of B-VEC in patients with DEB showed complete wound healing at 3 and 6 months in patients with DEB when compared with placebo [162].

5.3 Oncolytic Viruses (OV)

Oncolytic HSV (oHSV) are attenuated replication-competent vectors engineered to reduce neurovirulence and enhance specific tumor lytic activity and immunogenicity [167] (Fig. 22). Several HSV-1-based oncolytic viruses are currently under investigation in various clinical trials, both as single agents and in combination with various immunomodulatory drugs [159]. The first-generation oHSVs were designed with a mutation in a single gene that restricted their replication in dividing cells [157, 168]. However, the deletion of the thymidine kinase (tk) in these vectors rendered the virus insensitive to antiherpetic drugs, an important safety feature [169]. Second generation of HSVs was created with multigenic mutations to increase safety and remove the risk of replication-competent virus production in addition to

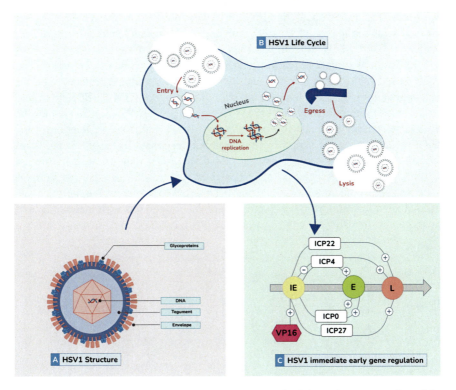

Fig. 20 HSV replication cycle and gene expression cascade. Schematic representation of the HSV-1 lytic cycle. (**a**) HSV-1 virion structure. (**b**) Viral glycoproteins mediate entry of HSV-1 by endocytosis or envelope fusion at the plasma membrane. Released capsid inserts DNA into nucleus, where DNA replication and packaging into new capsids take place. Complex envelopment steps occur at nuclear and Golgi membranes, resulting in cell lysis and virion release. (**c**) Cascade regulation of HSV gene expression and their expression dependence on prior viral gene products. The viral tegument protein VP16 activates the expression of immediate early genes (IE), which in turn activates promoters of early genes (E). Early gene products lead to viral DNA replication, releasing the expression of late genes (L), resulting in the packaging of viral genomes and formation of new virus particles. (Adapted from [151])

other safety advantages [170]. These vectors retained an intact tk gene making them susceptible to standard anti-HSV therapies [157, 169]. Currently, HSV-G207 is in clinic for glioma treatment and proven to be safe with potential treatment response: pediatric recurrent high-grade glioblastoma HGG (Phase 2; NCT04482933) and recurrent cerebellar brain tumors (Phase 1; NCT03911388) [171]. Third-generation triple-mutated oHSVs are further attenuated in normal cells yet enhance the stimulation of antitumor immune response) [172, 173]. Phase 1/2 clinical trials in patients with glioblastoma demonstrated safety and showed an increase in a 1-year survival rate, which led to its conditional approval in Japan for the treatment of glioma [173]. The last-generation vectors have been further modified to improve their antitumor efficacy, by incorporating expression cassettes for the delivery of various transgenes. The armed oncolytic virus can express genes to increase the anti-tumor effect of virus replication [157, 169]. Imlygic® (Talimogene laherparepvec, T-Vec, OncoVEX GM-CSF by Amgen) was the first oncolytic virus approved by the FDA in 2015 for the treatment of local recurrent unresectable melanoma. Imlygic® is an HSV1-based vector, with ICP34.5 and ICP47 genes deleted and replaced with the human granulocyte-macrophage colony-stimulating factor (GM-CSF) gene to render the virus more immunogenic [174, 175]. Deletion of ICP34.5 gene enables the virus to selectively replicate in tumors but not in normal tissues. Deletion of the ICP47 gene revokes the suppression of immune responses to the virus and helps activate the immune system. Expressed GM-CSF attracts dendritic cell infiltration for tumor antigen presentation and subsequently leads to an adaptive immune response against the tumor. Imlygic® has been tested in numerous other clinical trials, including in combination with other therapies for patients with melanoma, as a monotherapy in patients with a variety of other cancers, and in real-world (or

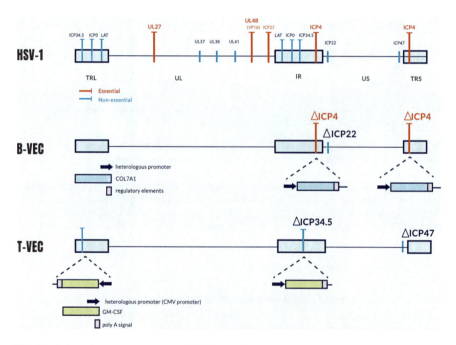

Fig. 21 Schematic genome structure of HSV-1 and two vector derivatives. The HSV-1 genome is composed of unique long (UL) and unique short regions separated by two internal regions (IRL and IRS) repeated in inverted orientation at the genome termini (TRL and TRS). The location of some non-essential (blue) and essential (red) genes is indicated. B-VEC vector was derived by deleting two copies of the immediate early gene ICP4; in addition, ICP22 was also deleted to reduce the cytotoxic effect. Two full-length copies of the human COL7A1 gene each with their own expression control elements were inserted into the ICP4 locus. T-VEC vector is deleted of genes encoding ICP34.5 and ICP47. The ICP34.5-encoding sequences have been replaced with a cassette consisting of the cytomegalovirus promoter, the gene encoding human GM-CSF, and the bovine growth hormone pA sequence. (Adapted from [151, 154, 166])

clinical practice) studies [166]. Presently, many other oncolytic viruses are undergoing clinical trials for applications in single therapy or combination therapy, and most of them are safe and show almost no dose-limiting toxicities (Table 7).

5.4 Manufacturing of HSV Vectors

Production of high-titer, infectious HSV stocks can be challenging because of its biology features: production efficiency and product safety profile are usually inversely correlated, in that rendering HSV vectors replication incompetent by genetic deletions also typically reduces HSV yield; and HSV particles are highly sensitive to production and processing conditions and can easily be inactivated during manipulation. Large-scale production of HSV vectors can be performed by infecting production cells (usually Vero) in cell factories or bioreactors [87, 176]. The infected culture is harvested, salt-treated, and clarified by centrifugation (for small scale only) and/or depth filtration. Residual DNA is digested with

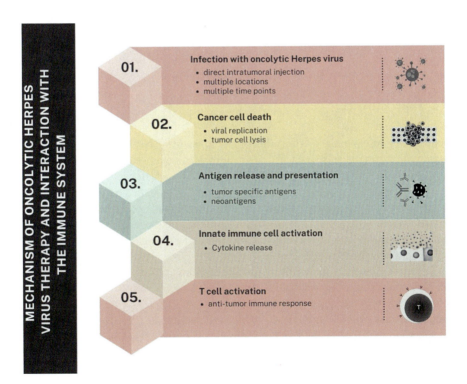

Fig. 22 Oncolytic virus therapy mechanism. Infection with oncolytic herpes virus results in tumor cell lysis leading to the release of tumor-specific antigens. Antigen presentation and virus-induced activation of innate immune cells in turn trigger activation of anti-tumor immune response. (Adapted from [167])

endonuclease and the vector purified by ion exchange chromatography. The product is further purified and concentrated by tangential-flow filtration (ultrafiltration/diafiltration), and the recovered vector is sterile filtered, formulated, filled into cryovials, and stored below −70 °C until use [163, 176, 177] (Fig. 23). Optionally, if sterile filtration is not an option (significant loss of virus during this step), the entire process will need sterile validation, an expensive endeavor. The final product is tested for safety (including absence of replication-competent vector), identity, purity, titer, and expression of transgene (see Table 12 in the analytics portion of this chapter).

HSV vectors present several properties that make them suitable gene therapy vectors such as high cargo capacity, no risk of insertional mutagenesis, and the ability to infect both quiescent and dividing cells. Over the years, significant efforts have been devoted to the development of safer vectors that could deliver efficient transgene expression, and numerous recombinant HSVs have been designed and tested in clinical studies. The approval of Imlygic® (T-VEC, OncoVex GM-CSF by Amgen) for the treatment of melanoma by the FDA in 2015 led to an increase in the application of HSV vectors in oncolytic strategies. Currently, oHSVs account for most of the studies in gene therapy clinical trials involving HSV vectors. There are still challenges for the use of oHSV in cancer therapy, and several studies are under development to improve specific delivery, and efficacy as a monotherapy or in combination with other therapies.

6 Gene Editing

Gene editing refers to the modification of an organism's DNA through deletion/correction of existing gene sequences or the addition of novel sequences. Human cells have evolved to preserve their genetic sequences. Two sets of every gene in somatic cells are maintained providing redundancy against DNA damage; further, there are many molecular pathways dedicated to the maintenance of DNA integrity and sequence fidelity (Fig. 24) [178]. Scientists have taken advantage of some of these innate mechanisms present in cells to effect precision changes to the genome using various molecular tools (Fig. 25).

The most prominent gene editing tools used today include zinc finger nucleases (ZFNs), transcription activator-like effector nucleases (TALENs), and clustered regularly interspaced short palindromic repeats (CRISPR)—CRISPR associated protein 9 (Cas9): all of these create double-stranded breaks (DSB) with high precision in genomic DNA (Fig. 26). In response, cells correct this DNA damage using either random insertions and deletions (indels) during the non-homologous end joining (NHEJ) repair of DSBs or gene corrections via homology-directed repair (HDR). The basic mechanistic requirements for these gene editing tools to function are broken down into a series of functions: first, the gene editing molecule will identify and bind to the target DNA sequence with very high specificity; second, the gene editing molecule will cause a double-stranded break in the targeted genomic

Table 7 HSV vectors in clinical trials (phases 2 and 3)

Product	Conditions	Phases	NCT number
Beremagene geperpavec[a]	Dystrophic epidermolysis bullosa	1|2	NCT03536143
	Dystrophic epidermolysis bullosa	3	NCT04491604
	Dystrophic epidermolysis bullosa	3	NCT04917874
G207	Glioma|astrocytoma|glioblastoma	1|2	NCT00028158
	High-grade glioma	2	NCT04482933
HSV1716	Malignant pleural mesothelioma	1|2	NCT01721018
KB105[a]	Autosomal recessive ichthyosis	2	NCT05735158
	Autosomal recessive congenital ichthyosis	1|2	NCT04047732
OH2	Solid tumor|gastrointestinal cancer	1|2	NCT03866525
	Melanoma	1|2	NCT04616443
	Central nervous system tumors	1|2	NCT05235074
	Non-muscle-invasive bladder cancer	1|2	NCT05232136
	Advanced bladder carcinoma	2	NCT05248789
	Advanced colorectal carcinoma	2	NCT05648006
	Melanoma	3	NCT05868707
	Solid tumor|melanoma	1|2	NCT04386967
RP1	Cutaneous squamous cell carcinoma	2	NCT04050436
	Advanced cutaneous malignancies	1|2	NCT04349436
	Triple negative breast neoplasms	1|2	NCT06067061
	Squamous cell carcinoma	1|2	NCT05858229
	Melanoma|non-melanoma skin cancer	2	NCT03767348
RP2| RP3	Refractory metastatic colorectal cancer	2	NCT05733611
RP3	Hepatocellular carcinoma	2	NCT05733598
T3011	Advanced solid tumor	1|2	NCT05598268
	Advanced solid tumor	1|2	NCT05602792
	Advanced solid tumors	1|2	NCT06214143
Talimogene laherparepvec	Breast cancer	1|2	NCT02779855
	Melanoma	2	NCT00289016
	Melanoma	2	NCT02965716
	Lymphomas or non-melanoma skin cancers	2	NCT02978625
	Soft tissue sarcoma	2	NCT02923778
	Melanoma	3	NCT00769704
TBI-1401(HF10)	Melanoma	2	NCT03153085
Vusolimogene oderparepvec (RP1)	Melanoma	3	NCT06264180

[a]Beremagene geperpavec and KB105 are replication-defective vectors

sequence; finally, the cell's DNA repair machinery will repair the DSB either through NHEJ or HDR if there is a suitable template DNA available [180].

ZFNs and TALENs proteins share similar functional domain composition in that they both have a specific DNA sequence binding domain (modular zinc fingers that

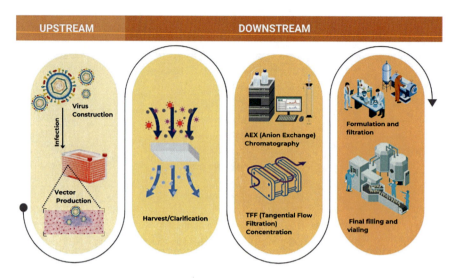

Fig. 23 HSV manufacturing. Large-scale HSV vector production and purification processes. Schematic representation of infection production method and purification processes. Production cells are grown in cell factories or fixed-bed bioreactors and infected with HSV seed virus. After a suitable incubation period, infected cells are lysed or treated with a virus-releasing agent, followed by endonuclease treatment to degrade host cell DNA, and depth filtration to remove cellular debris. The clarified filtered harvest is applied to anion exchange chromatography, concentrated by tangential flow filtration, and purified viral particles are suspended in an appropriate formulation buffer for final filling and vialing

bind to nucleotide triplets in ZFNs and modular repeat variable diresidues that can bind single nucleotides in TALENs) fused to an endonuclease domain (Fok1) which causes the DSB [182]. The modularity of the DNA-binding domains allows ZFNs and TALENs to be programmed to target very specific sequences in the genome. Although these technologies are powerful and programmable tools to edit the genome, the design, construction, and screening of different variants of the entire ZFN/TALEN protein for each target locus can be quite challenging and time-consuming [183].

ZFNs and TALENs were quickly overshadowed by the discovery and the utilization of bacterially derived CRISPRs in the CRISPR-Cas9 system. While CRISPR-Cas9 still works on the same principle of having a DNA targeting and binding function coupled to endonuclease function to create the DSB at the target site, the molecular mechanism is significantly different from ZFNs/TALENs. Specifically, the CRISPR-Cas9 proteins are brought to a target genomic sequence by a single co-delivered short strand of RNA known as a guide RNA (gRNA). Unlike ZFNs and TALENs, the same gene editing proteins can be used but novel gRNA sequences are synthesized for each target site. This ease of targeting and screening many DNA sites in a short time before identifying the ideal site for therapeutic gene editing has sparked the current era of high throughput precision gene editing [184].

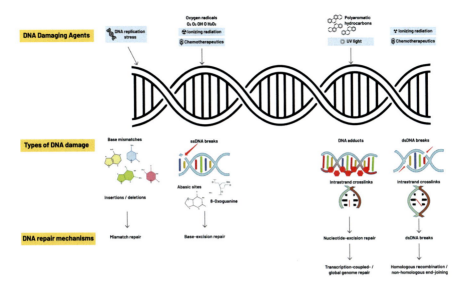

Fig. 24 Mechanisms to preserve DNA fidelity in eukaryotic cells. Several DNA-damaging agents may result in a range of DNA lesions. Each type of DNA damage can be corrected by a specific DNA repair mechanism, such as mismatch repair, base-excision repair, transcription-coupled/ global genome repair, homologous recombination (homology-directed repair, HDR), or non-homologous end-joining (NHEJ). (Adapted from [178])

Even with the rapid advances in gene targeting and gene editing capabilities in recent years, there are still a few roadblocks in the use of these targeted gene disruption tools. First among these is that all these tools target and bind short sequences of genomic DNA prior to creating DSBs in the DNA. Given the extremely large size of the human genome and the presence of many similar and repeating sequences spread out through the genome, there is a risk of off-target gene editing at a rate comparable to the target site itself with unwanted gene disruption occurring at homologous sites that are nearly identical to the target site throughout the genome [185]. These off-target edits may affect the viability of the cells being modified and, in some instances, give rise to oncogenic transformation of the target cells by the accidental activation of oncogenes. Because unwanted homologous site disruptions can be very rare events spread out through the entire population of cells modified, detection can be challenging. This makes it critical to reduce off-target gene disruption during the design stage itself.

With the availability of next-generation sequencing (NGS), it is now possible to obtain an exact sequence of the human genome being edited down to the single nucleotide level [186]. Combining the vast amount of genome sequence information with powerful bioinformatic analysis, it is now possible to screen the target genome sequence for potential homologous sites across the entire genome in silico. Following this, the target genome sequence may be slightly modified to significantly decrease or completely remove targeting to homologous sites that may be disrupted accidentally.

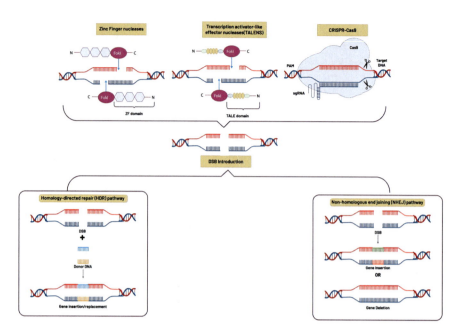

Fig. 25 Double-stranded break repair mechanisms. Genome editing nucleases (ZFNs, TALENs, and CRISPR/Cas) induce DNA double-strand breaks (DSBs) in targeted sites. DSBs can be repaired by nonhomologous end joining (NHEJ) or, in the presence of a donor template, by homology-directed repair (HDR). In the presence of donor DNA, recombination between homologous DNA sequences present on the donor template and a specific chromosomal site can facilitate targeted gene insertion/replacement. NHEJ introduces small base insertions or deletions that can result in gene disruption, and when two DSBs are induced simultaneously, the intervening genomic sequence can be deleted. (Adapted from [179])

Another cause of unwanted non-target genome sequence disruptions is the prolonged persistence of the gene editing molecule in the target cell. ZFN/TALEN proteins have a relatively long half-life within the target cells, and since they have both DNA binding and cutting functionality built into the same molecule, it could lead to accidental disruption of random DNA sites. To reduce off-target cutting and increase target sequence specificity, the Fok1 endonuclease domain was engineered to create mutants that function as an obligate heterodimer, i.e., both the mutant Fok1 domains must be bound to the target DNA site in close proximity for them to create the double-stranded break. Neither mutant Fok1 can create the DSB on its own. One of the mutant Fok1 domains was fused to a ZFN/TALEN that bound the sense strand and the other to the ZFN/TALEN binding the anti-sense strand. Only when both the sense and anti-sense ZFN/TALENs bind the genomic DNA on either side of the targeted cutting site, the obligate heterodimer mutant Fok1 domains are able to create a double-stranded break in the targeted DNA. In addition to reducing random off-target gene disruptions directly, this modification to the Fok1 domains forces the use of two ZFN/TALENs to disrupt a single DNA target. This doubles the recognition sequence for the ZFN/TALEN molecules and significantly increases the specificity of gene targeting [187] (Fig. 27).

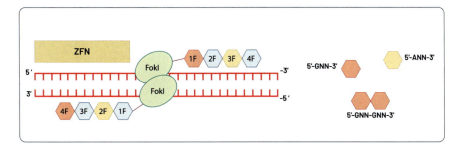

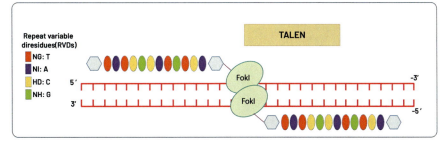

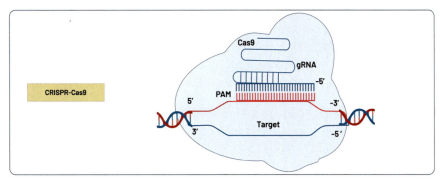

Fig. 26 ZFNs, TALENs, and CRISPR-Cas9—the molecular tools for gene editing. ZFNs recognize target sites that consist of two zinc-finger binding sites that flank a 5- to 7-bp spacer sequence recognized by the FokI cleavage domain. TALENs recognize TALE DNA-binding sites that flank a 12- to 20-bp spacer sequence recognized by FokI cleavage domain. The Cas9 nuclease is targeted to DNA sequences complementary to the targeting sequence within the guide RNA (gRNA) flanked by a specific protospacer adjacent motif (PAM) sequence. ZFNs (zinc-finger nucleases), TALENs (transcription activator-like effector nucleases), CRISPR-Cas9 (clustered regularly interspaced short palindromic repeats (CRISPR)-CRISPR-associated protein 9. (Adapted from [181])

Similarly, the CRISPR-Cas9 system is able to mitigate off-target cutting where the Cas9 molecule is mutated (D10A mutation) into a nickase rather than an endonuclease. This makes the use of dual targeting gRNA in combination with the CRISPR-Cas9 protein so that two nicks are made on opposite DNA strands in close proximity at the target genome site to disrupt the gene of interest. Importantly, the gRNAs that are co-delivered with the CRISPR-Cas9 protein have a much shorter

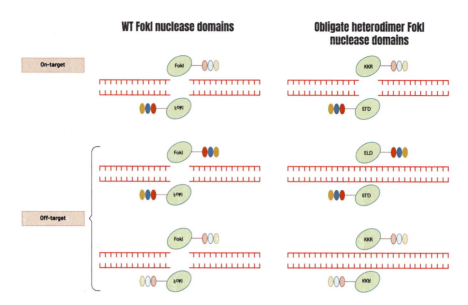

Fig. 27 Obligate heterodimer ZFNs and TALENS. To cleave a specific site in the genome, ZFNs and TALENs are designed as a pair that recognize sequences flanking the site on each DNA strand. Upon binding on either side of the site, the pair of FokI domains dimerize and cleave the DNA at the site, generating a double-strand break (DSB). To reduce off-targets, the FokI endonuclease domain was engineered to function as an obligate heterodimer to create a DSB, and only when both sense and anti-sense ZFN/TALEN are bound to the DNA on either side of the target site, the heterodimer FokI domains can create a DSB, increasing targeting specificity. (Adapted from [179])

half-life than ZFN/TALEN gene editing proteins causing the gRNA to be cleared out of the target cell in a very short span of time, limiting the duration of CRISPR-Cas9 activity and further reducing unwanted off-target gene disruption.

The clinical use of genome editing depends fundamentally on the ratio of on-versus off-target (inaccurate) genetic changes, and the fraction of on-target edits that produce the desired genetic outcome (typically disease gene knockouts or therapeutic gene insertions). Off-target editing occurs when CRISPR-induced DNA cleavage (DSBs) and repair happen at genomic locations not intended for modification, typically sites that have a sequence similar to the intended editing site [188]. DSBs created by nucleases such as Cas9 can result in indels, translocations, and rearrangements which are undesired byproducts [189]. The risks intrinsic to DNA cleavage-induced genome editing led to the development of CRISPR-Cas9-mediated genome regulation or editing methods that do not involve double-stranded DNA cutting. Recent advances in gene editing in mammalian cells include the development of programmable nucleases, base editors, and prime editors [183, 188].

Base editing is a gene editing method that generates precise point mutations in genomic DNA without directly generating DSBs, requiring a DNA donor template or relying on cellular HDR [189]. DNA base editors (BEs) are composed of a catalytically impaired Cas nuclease and a base-modification enzyme that operates on single-stranded DNA (ssDNA) but not double-stranded DNA (dsDNA). Upon binding to its target locus in DNA, base pairing between the gRNA and target DNA strand leads to the displacement of a small segment of single-stranded DNA in an "R-loop." The deaminase enzyme modifies the DNA bases within this single-stranded DNA bubble. To improve efficiency in eukaryotic cells, the catalytically disabled nuclease also generates a nick in the non-edited DNA strand, inducing cells to repair the non-edited strand using the edited strand as a template [189]. There are two classes of DNA base editors: cytosine base editors (CBEs) convert a C•G base pair into a T•A base pair, and adenine base editors (ABEs) convert an A•T base pair to a G•C base pair. Collectively, CBEs and ABEs can mediate all four possible transition mutations (C to T, A to G, T to C, and G to A) [189]. ABEs are an especially useful class of base editor because they reverse the most common type of pathogenic point mutation (C,G to T,A), which accounts for approximately half of known pathogenic single-nucleotide polymorphisms (SNPs) [189].

Additionally, CRISPR-Cas9 systems have been engineered to remove their DNA-cutting activity for applications that exploit their RNA-guided DNA-binding ability, generating a nuclease-dead version of the protein (dCas9). It was shown that targeting dCas9 to a coding sequence or its promoter resulted in transcription inhibition in *E. coli*. This approach was termed CRISPR interference (CRISPRi) and led to the development of different technologies for gene regulation and editing [190, 191]. Fusion of a repressor domain to dCas9 targeted to a specific genomic location resulted in silencing of gene expression where the fusion protein was bound in mammalian cells [192]. Since then, numerous fusions have been generated, including dCas9 fused to transcription activator domains to upregulate gene expression, a mechanism termed CRISPR activation (CRISPRa). This expanded the applications of the CRISPR system, involving gene expression and epigenetic modulation, which could provide the means for the development of new therapies based on the corrective activation of genes or pathways that fail to express during a disease [191, 193].

Since the initial discovery of Cas9, two key advances have enabled the expansion of the CRISPR technology, CRISPR systems that use other Cas proteins and engineering Cas molecules to enhance their functionality. All known DNA-targeting Cas nucleases require a targeted locus to be flanked by a specific sequence called a protospacer adjacent motif (PAM). In addition to their efficiency and specificity for DNA cutting, PAM sequences are important factors that can determine the genomic space that can be targeted by a particular Cas protein [191]. Several Cas proteins have been discovered that recognize different PAM sequences, which enabled a greater variety of targeting sequences [194]. Among these, Cas12 nucleases have also been used in DNA editing applications. Another group of proteins, known as Cas13, was shown to be able to bind and cleave single-stranded RNA and has been used for targeted RNA degradation, enabling changes to the transcriptome [191]. Several Cas9 and Cas12 variants have been engineered to broaden their range of targetable PAM sequences, and significant effort has been made to develop Cas variants that have lower off-target editing activity. High-fidelity Cas variants have been generated that can reduce off-target editing to a substantial extent; however, these variants generally showed lower on-target activity as well. Additionally, it was noted that the engineered Cas protein variants often were not as robust as their naturally occurring Cas counterparts [195, 196].

Through these and other molecular changes to ZFNs, TALENs, and CRISPR tools, increased gene modification specificity, efficiency, and reduction of unwanted side effects have been achieved. CRISPR-Cas9 system has emerged to be the preferred gene editing tool for the scientific community due to its low cost of reagent screening, higher specificity and efficiency of gene editing, suitability for multiplexed gene editing, and having a simple set of reagents that are relatively easier to deliver into target cells or organisms.

6.1 Clinical Applications

Advances in cell and gene therapy (CGT) have revolutionized medicine over the last decade. Erstwhile untreatable diseases are now treatable by using modified patient/donor cells as therapeutic effectors.

Zinc finger nucleases have been successfully used in several clinical trials—one of the most successful examples include the disruption of the HIV coreceptor CCR5 on CD4+ T cells and CD34+ HSCs to engineer HIV resistance [197, 198]. Additionally, clinical trials with zinc finger nucleases using AAV (or AdV) as a delivery vehicle have been shown to have some successful outcomes (Table 8) [181]. Similar promising clinical trials exist for CRISPR gene editing tools (Table 9). One of the most promising applications in this area is the use of chimeric antigen receptor T cells (CAR-T) to target and destroy cancer cells even in patients with advanced cancer that were deemed untreatable using conventional cancer therapies. The process of CAR-T cell therapy manufacturing involves the isolation of T cells from the cancer patient and genetically modifying them (typically through lentivirus or gamma retrovirus delivery of the CAR transgene—see retrovirus section of this chapter) to express a chimeric surface receptor that binds to specific biomarkers found on the patient cancer cells (such as CD19 in multiple myeloma cells). This modification activates the T cell initiating cell-killing activities to eliminate the targeted cancer cell. The manufacturing and treatment processes are very complex with the patient's T-cells being modified ex vivo to create the CAR-T cells (autologous cell therapy). While this therapy has been proven in multiple clinical trials across the world, the widespread implementation has been very challenging due to the high cost of the treatment (a significant portion of this cost arising from manufacturing). Efforts have been directed to produce allogeneic therapies, wherein a healthy donor's T-cells are modified ex vivo and expanded to multiple patients bringing the cost of therapy for each patient down to a much more affordable number. CRISPR gene editing technology for the precision gene knockout of the TCRab and MHC1 genes in the CAR-T cells to prevent any graft versus host or host versus graft reaction and rejection issues is one way to achieve allogeneic cell treatments. With CRISPR technology, other T cell-specific markers such as CD52 can be

Table 8 Zinc finger nucleases in the clinic

Clinical trial ID	Target disease	Phase	Mechanism: IV delivery of ZFN and transgene into albumin locus of hepatocytes	Vector	Company	Notes
NCT02695160	Hemophilia B	Phase 1	Transgene: Factor IX	AAV2/6	Sangamo (SB-FIX)	Terminated
NCT02702115	MPS I	Ph 1/2	Transgene: IDUA (α-L-iduronidase)	AAV2/6	Sangamo (SB-318)	Terminated
NCT03041324	MPS II	Ph 1/2	Transgene: IDS (Irudonate 2-sulfatase)	AAV2/6	Sangamo (SB-913)	Terminated
NCT04628871	Long-term follow-up for subjects who received SB-318, SB-913, and SB-FIX					
Clinical trial ID	**Target disease**	**Phase**	**Mechanism**	**Vector**	**Company**	**Notes**
NCT01252641	HIV	Ph 1/2	Ex vivo deletion of CCR5 on CD4+ T-cells using ZFN	AdV	Sangamo (SB-728-T)	Completed

knocked out and certain alloimmune defense receptor genes can be knocked in (Fig. 28) to further improve the functional performance as well as the persistence of the CAR-T cells in the patient blood. Precision gene editing using CRISPR technology can be a major factor to enable the broad commercialization of CAR-T therapies at affordable costs allowing maximum drug accessibility to patients worldwide.

Another example of a significant therapeutic breakthrough that leverages CRISPR technology is the recently approved gene therapy for hematopoietic diseases including sickle cell anemia (adult globin function insufficiency) and beta thalassemia (adult globin synthesis insufficiency). These are conditions that are debilitating and may require frequent blood transfusions to maintain quality of life. Stem cells from patients suffering from these conditions are harvested, and CRISPR-Cas9 technology is used to disrupt the BCL11 gene which is responsible for inhibiting the production of fetal hemoglobin protein beyond the infant stage (Fig. 29). The modified stem cells are engrafted back into the patients, and the red blood cells arising from these stem cells are able to maintain a functional and sufficiently high level of fetal hemoglobin in them which enables efficient oxygen transport by the RBCs. Patients treated with this gene therapy have significantly reduced to no dependence on blood transfusions to maintain their quality of life. The approval of Casgevy® marks the first gene-editing technology approval for any clinical indication [10].

Gene editing tools can be delivered to a patient in one of several ways. The CRISPR/Cas9 complex can be delivered to patient or healthy donor cells ex vivo in the form of plasmid DNA, mRNA, or ribonucleoprotein (RNP) [201]. RNP delivery is a robust and transient way to deliver these complexes and minimizes off-target events, insertional mutagenesis, and immune responses. RNPs can be delivered into the cell physically using microinjections and electroporation, among other

Table 9 CRISPR gene editing tools currently in clinical trials

Clinical trial ID	In vivo/ Ex vivo	Target disease	Phase	Type of gene editor	Company
NCT04853576	Ex vivo	Sickle cell disease (SCD)	Ph 1/2	CRISPR/ Cas12	Editas (EDIT-301)
NCT05444894	Ex vivo	Beta thalassemia (TDT)	Ph 1/2	CRISPR/ Cas12	Editas (EDIT-301)
NCT05456880	Ex vivo	Sickle cell disease (SCD)	Ph 1/2	Base Editor	Beam Tx (BEAM-101)
NCT05488340	In vivo (phage)	UTI by MDR *E. coli*	Ph 2/3	CRISPR/Cas3	Locus Biosciences (LBP-EC01)
NCT04601051	In vivo (LNP)	Hereditary transthyretin amyloidosis (hATTR)	Ph 1	CRISPR/Cas9	Intellia (NTLA-2001)
NCT05120830	In vivo (LNP)	Hereditary angioedema (HAE)	Ph 1/2	CRISPR/Cas9	Intellia (NTLA-2002)
NCT05643742	Ex vivo (allo. CAR T)	CD19+ malignancies/ autoimmune	Ph 1/2	CRISPR/Cas9	Crispr Tx (CTX-112)
NCT05795595	Ex vivo (allo. CAR T)	Solid tumors/ hematological malignancies	Ph 1/2	CRISPR/Cas9	Crispr Tx (CTX-131)
NCT04637763	Ex vivo (allo. CAR T)	B-cell non-Hodgkin lymphoma	Ph 1	CRISPR/Cas	Caribou Biosciences (CB010)
NCT05722418	Ex vivo (allo. CAR T)	Multiple lymphoma	Ph 1	CRISPR/Cas9	Caribou Biosciences (CB011)
NCT06128044	Ex vivo (allo. CAR T)	Acute myeloid lymphoma	Ph1	CRISPR/Cas9	Caribou Biosciences (CB012)
NCT05885464	Ex vivo (allo. CAR T)	T-cell lymphoblastic leukemia (T-LL)	Ph 1/2	Base Editor	Beam Tx (BEAM-201)
NCT05398029	In vivo (LNP)	Familial hypercholesteremia	Ph 1	Base Editor	Verve Tx (VERVE-101)
ACTRN12623000809639[a]	In vivo (LNP)	Cardiovascular disease	Ph 1	CRISPR/Cas9	Crispr Tx (CTX-310)
ACTRN12623001095651[a]	In vivo (LNP)	Cardiovascular disease	Ph 1	CRISPR/Cas9	Crispr Tx (CTX-320)
NCT05144386	In vivo (AAV9)	HIV	Ph 1	CRISPR/Cas9 dual gRNA	Excision bio (EBT-101)

[a]Records identified in WHO Trial Registration Data Set

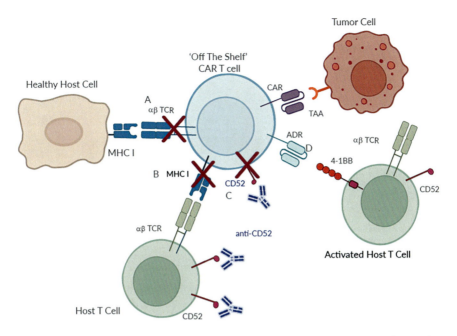

Fig. 28 Gene editing strategies to create allogeneic CAR-T cells. A. Disruption of TCRαβ to prevent CAR-T cells from recognizing healthy host cells, decreasing the risk of graft-versus-host-disease. B. MHC I gene disruption in CAR-T cells to avoid their detection and elimination by host T cells. C. Elimination of CD52 (T cell marker) from CAR-T cells to enable the use of CD52 antibody for the lymphodepletion of host T cells without affecting CAR-T cells, increasing engraftment success. D. Expression of alloimmune defense receptor (ADR) on CAR-T cells that selectively recognize 4-1BB marker expressed on host's activated T cells, leading to their destruction and decreasing the risk of rejection. (Adapted from [199])

techniques. Alternatively, they can be delivered using synthetic carriers such as lipid nanoparticles (LNPs) and exosomes among other techniques. If a donor sequence is needed to repair the edited area, the most common approach utilizes delivery of this sequence packaged in viral vectors such as AAV. Alternatively, gene editing tools can be transduced into target cells using these viral vectors (Table 10). In case of AAV, the size of the widely used SpCas9 (from *Streptococcus pyogenes*) causes limitations in packaging into AAV vectors. This can be overcome using smaller sized Cas9 proteins from different bacteria including *Staphylococcus aureus* Cas9 (SaCas9) and *Streptococcus thermophiles* Cas9 (St 1 Cas9) or different classes of Cas proteins like Cas3 or Cas13 that are much smaller in size [184, 202]. Another way to overcome the limited size of AAV is to deliver gRNA into cells that have been modified to stably express Cas9 protein. Yet another strategy is to transduce target cells with two AAVs to deliver these gene editing tools: one for the gRNA and the other for Cas9. However, the dual vector transduction method requires twice as much virus dosed resulting in a potential safety risk that in turn can reduce the therapeutic editing potential [203, 204]. Other alternatives for viral delivery of gene

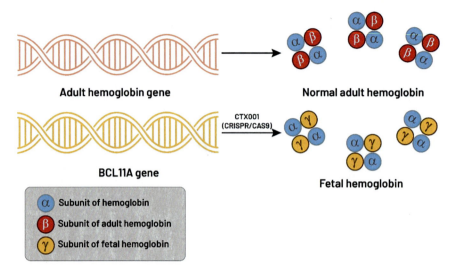

Fig. 29 Use of CRISPR in the first approved gene therapy for sickle cell disease and beta thalassemia. Normal adult hemoglobin comprises two α subunits and two β subunits. Fetal hemoglobin is formed by two α subunits and two γ subunits, is naturally present during fetal development, and is switched to the adult form after birth. BCL11A is a transcription factor that represses γ-globin expression and fetal hemoglobin in erythroid cells. CTX001 (CRISPR/Cas9) targets the erythroid-specific enhancer region of the *BCL11A* gene, and this edit leads to an increase of fetal hemoglobin production in red blood cells, thus providing functioning hemoglobin in sickle-cell disease and transfusion-dependent β-thalassemia patients. (Adapted from [200])

editing tools in vivo include LVV and AdV vectors, which allow for larger packaging capacity over AAV, however still have disadvantages as indicated in Table 10 [147, 205].

The development and innovations in gene editing tools continue to provide specificity and efficiency in gene targeting and gene editing capabilities in the clinic. While in vivo clinical trials with ZFNs have shown promise, most of these trials have currently been terminated and the current subjects moved to a long-term follow-up trial given the lack of investment in the current market. The CRISPR-Cas9 system has emerged to be the preferred gene editing tool for the scientific community due to its low cost of reagent screening, higher specificity and efficiency of gene editing, suitability for multiplexed gene editing, and having a simple set of reagents that are relatively easier to deliver into target cells or organisms. The approval of Casgevy® marks the first gene-editing technology approval—a huge milestone given the discovery of the CRISPR system as a gene editing tool only 11 years prior to this approval. With the continuous innovations in gene editing technologies to improve safety and substantially decrease off-target editing, this field of study continues to show promise.

Table 10 Advantages and disadvantages of several viral vectors for the delivery of CRISPR

Viral vector	Advantage	Disadvantage	Mitigation strategies
AAV	Limited host genome integration Multiple serotypes to improve cell tropism	Limited packaging capacity (<5 kb) Capsid immunogenicity	Dual AAV delivery Smaller Cas effector proteins
LVV	High packaging capacity Less immunogenic	Random integration into host genome Insertional mutagenesis	Non-integrating lentivirus vectors (NILV)
AdV	Low integration into host genome High packaging capacity	Highly immunogenic virus	Use of non-human AdV vectors Engineer AdVs through copolymer encapsulation

Fig. 30 Flowchart depicting a typical quality by design (QbD) approach

7 Gene Therapy Analytics

To expedite the development of manufacturing processes and quality control strategies, the quality by design (QbD) approach proves equally beneficial for gene therapy products as it has been for recombinant proteins. However, there is limited knowledge needed for implementing the QbD approach in the gene therapy field. QbD is defined as "a methodical approach to development that starts with predefined goals and prioritizes comprehension of product and process, alongside detailed process control" [206]. The QbD approach (Fig. 30) encompasses the establishment of a quality target product profile (QTPP) as exemplified in Table 11, the identification of critical quality attributes (CQAs), and the development of processes, including the definition of design spaceand determination of control strategies.

Within this framework, the pivotal step of identifying CQAs, which are described as "physical, chemical, biological, or microbiological properties or characteristics that must fall within specified limits, ranges, or distributions to guarantee the desired product quality" by the International Conference on Harmonization (ICH Guideline, Pharmaceutical Development, Q8(R2) [206], is crucial. This is because ensuring the quality, safety, and efficacy of a product relies on well-defined CQAs, underpinned by a robust manufacturing process and control strategy. Limits and ranges for the

Table 11 Example of quality target product profile [207]

Item	Target
Mechanism of action	Unique to the product being developed
Route of administration (RoA)	Intravenous
Dosage form	Suspension for injection
Storage condition	Under—60 °C
Shelf life	2 years
Empty capsid	Less than 20%
Replication-competent AAV	Under detection limit

CQAs may be broader during early development and refined and tightened as the product progresses through development and clinical phases, and knowledge of the product increases. Recombinant viral vectors are used for gene introduction and/or regulation in most gene therapy-based modalities currently in use. As detailed throughout this chapter, viral vectors are complex biological systems comprised of different structural proteins and a genome consisting of either DNA or RNA. Even seemingly simple viral vectors, such as rAAV, boast structures more complex than many recombinant proteins [3]. The inherently complex nature of viral vectors poses distinctive challenges for characterization and quality control testing. Gene therapy products must adhere to stringent safety standards, underscoring the criticality of developing appropriate analytical tools for comprehensive characterization.

This section explores the key analytical testing methods employed to ensure the safety, efficacy, and quality of diverse types of viral vectors utilized in gene therapy.

Throughout the process development phase from research scale to large-scale platforms, understanding the analytical needs within this relatively new therapeutic area is key to success. It is essential that developers and manufacturers can establish that the product meets safety and quality standards. Effective analytical methods are essential for verifying product identity, potency, and safety assessing the impact of product and process-related impurities comprehensively. These methods encompass a wide range of CQAs and can be evaluated using various techniques [208].

Viral vector developers are required to follow specific guidelines outlined by the FDA (United States) or EMA (EU) and other regulatory agencies as needed to ensure adherence to overall product characterization standards according to cGMP (current Good Manufacturing Practice) principles, following the SISPQ framework (Safety, Identity, Strength/Potency, Purity, and Quality) [55, 209]. Evaluation of product characteristics during early clinical development helps to identify CQAs used to specify key characteristics of the DS and DP and their specifications for later clinical studies. CQAs should be monitored through appropriate analytical testing throughout the manufacturing process to characterize product purity, heterogeneity, safety, and efficacy. Figure 31 outlines a representative example of interplay between viral vector manufacturing and analytics.

Key CQAs during viral vector manufacturing encompass viral potency, identity, purity, and safety. Several common methods are used by manufacturers to monitor

Fig. 31 Representative workflow for viral vector manufacturing and analytical testing

key attributes during the process development phase and throughout clinical development to ensure that required specifications are met. Table 12 below lists commonly used methods for monitoring viral vector quality attributes [210].

For an in-depth understanding of the viral vector properties, product developers should consider the following list of quality attributes along with analytical methods that are commonly utilized and their purpose.

1. *Virus Titer*: Virus titer or concentration refers either to infectious, genome, or capsid/particle titer.

   - *Physical Titer*: Physical titer quantifies intact virus and is expressed as viral particles per mL (VP/mL) or vector genome copies per mL (GC/mL or vg/mL). The genome titer represents the number of copies of the vector genome (i.e., virions) present, estimates the potency of a sample, and is primarily used for clinical dosing. Quantitative PCR (qPCR)-based assays are commonly used for genome titer determination; however, qPCR has been predominantly replaced by digital droplet PCR (ddPCR) for its higher sensitivity and reproducibility [211]. Other methods for physical titer determination include ELISA to quantify the capsid proteins present in the sample (e.g., capsid ELISA for AAV or p24 ELISA for LVV) or optical density assays (A260/280) for particle titer.
   - *Infectious or Functional Titer*: Functional titer, or infectious titer, measures the ability of virus particles to transduce and/or infect target cells (i.e., infectivity). This titer is lower than physical titer as not all virus particles are infectious. Quantification methods include in vitro assays in which a susceptible cell line is infected with serial dilutions of the viral sample and the concentration of virus particles is determined by calculation of the infectious particles that produce cytopathic effects (CPE) or visible plaques (e.g., plaque forming assay). Other methods, such as the TCID50 or endpoint dilution assay, quantify the amount of virus required to kill or cause CPE in 50% of the susceptible cells, and transgene expression assays [212, 213]. Furthermore, infectious lentiviral vector titer can be quantitated by flow cytometry in adherent HEK293T cells with an appropriate antibody to the expressed transgene. Another approach for infectious titer of lentivirus products is to determine the vector copy number (VCN) of lentivirus genomes integrated into the

Table 12 Examples of analytical assays used for viral vector drug substance and drug products

Attribute	Category	Testing material	Standard method(s)	AdV	AAV	RVV/LVV	HSV
Identity	Nucleic acid	DS, DP	Sequencing (e.g., NGS), PCR	✓	✓		✓
	Viral capsid	DS	Western blot, RP-HPLC, LC-MS, ELISA	✓	✓	X (WB/HPLC/ LC-MS) ✓ (ELISA)	✓
Quantity/strength/ potency	Viral genome titer	IP, DS, DP	qPCR, ddPCR, dPCR	✓	✓	✓	✓
	Capsid titer	IP, DS, DP	ELISA, HPLC	✓	✓	✓	✓
	Transgene expression	DS, DP	Transduction/ELISA	✓	✓	✓	✓
		DS, DP	Transduction/flow cytometry	✓	✓	✓	✓
	Antigen binding	DS, DP	Transduction/ELISA/flow cytometry	✓	✓	✓	✓
	Infectious Titer	DS, DP	Infectivity assay (flow cytometry)	✓	✓	✓	✓
		DS, DP	Infectious Titer Assay—Plaque assay, $TCID_{50}$	✓	✓	✓	✓
Purity	Aggregates	DP	DLS	✓	✓	✓	✓
	% Full capsid (full/empty capsids)	DS, DP	AEX/AUC/TEM	✓	✓	X	✓
	Sub-visible particles	DS, DP	USP<787,788, 789> (SEC, light obscuration, light microscopy)	✓	✓	✓	✓

Category	Parameter	Applicability	Method					
Process-related impurities	Residual host cell DNA	DS/DP	qPCR, ddPCR, CE	✓	✓	✓	✓	
	Residual host cell protein	DS/DP	ELISA	✓	✓	✓	✓	
	Residual plasmid DNA/residual antibiotic resistance gene	DS/DP	qPCR, ddPCR	✓	✓	✓	✓	
	Residual endonuclease	DS/DP	ELISA	✓	✓	✓	✓	
	Residual transfection reagent	DS/DP	HPLC, ELISA	✓	✓	✓	✓	
	Residual lysis buffers (detergents)	DS/DP	HPLC	✓	✓	✓	✗	
	Residual E1a, E2a, E4 and/or VA	DS	qPCR, ddPCR	✓	✓	✓	✓	
	Residual BSA	DS/DP	ELISA	✓	✓	✓	✓	
	Leachables	DS/DP	LC-MS, GC MS	✓	✓	✓	✓	
Safety	Sterility	DP	Culture method	✓	✓	✓	✓	
	Bioburden	IP, DS	EP 2.6.12 [48], USP <61> [51]	✓	✓	✓	✓	
	Mycoplasma	IP, DS	EP 2.6.7; USP <63> (culture); alternative method (PCR)	✓	✓	✓	✓	
	Adventitious viruses	IP, DS	ICH Q5A	✓	✓	✓	✓	
	Replication-competent virus	IP, DS	Co-culture with permissive cell line method	✓	✓	✓	✓	
	Endotoxin	IP, DS, DP	USP <85>, LAL	✓	✓	✓	✓	
General quality	Appearance	DS, DP	USP <630>	✓	✓	✓	✓	
	pH	DS, DP	USP <791>	✓	✓	✓	✓	
	Osmolality	DS, DP	<USP 785>	✓	✓	✓	✓	
	Extractable volume	DP	EP 2.9.17	✓	✓	✓	✓	
Other characterization assays	Deamidation	DS	SEC, DLS, MS	✓	✓	✓	✓	
	Glycosylation	DS	SEC, DLC, MS	✓	✓	✓	✓	
	Peptide mapping	DS	MS	✓	✓	✓	✓	

genome of the target cells using a qPCR/ddPCR output. In such a case, the lentiviral vector is used to transduce target cells in culture, the infected cell lysate is harvested and processed, and one set of primers/probe are used to target the gene of interest and another used to target a host cell gene (usually albumin).

2. *Identity*: Product identity distinguishes the product from any other product and hence the test used is specific to the product of interest.

- *Genetic Identity*: Molecular techniques such as PCR and high-throughput genome sequencing, such as next-generation sequencing (NGS), are utilized to confirm the genetic identity of the vector genome in the viral vector product [214]. For lentiviral vectors that are integrated into the host cell genome in ex vivo-modified cell therapies, additional tests may be required to confirm the identity of the modified cell. For example, the vector copy number (VCN) quantifies the number of vector genomes integrated in the host cell using qPCR or ddPCR techniques. Additionally, insertional site analysis using NGS is used to ensure that the sequence has not been inserted in undesirable sites in the genome that could impact patient safety caused by integration-induced toxicity [215].
- *Protein Identity*: Thorough characterization of capsid proteins is crucial for viral infectivity and vector potency. Thus, ensuring the identity, capsid protein stoichiometry, and integrity are indicators of product and process consistency. ELISA and Western blotting are preferred methods due to their cost-effectiveness, simplicity, and ease of execution [216]. Antibodies and ELISA kits are readily available to detect common capsid or viral proteins (e.g., natural capsids for AAV, p24 for LVV, etc.). Other, more specialized methods, such as mass spectrometry (MS) methods, coupled with chromatographic (LC-MS) and electrophoretic-separation (CE-MS) techniques, are utilized for peptide mapping. These methods separate peptide moieties differently and serve as orthogonal techniques to provide a comprehensive analysis of capsid proteins.

- Finally, capillary electrophoresis-sodium dodecyl sulfate (CE-SDS) is an automated, reproducible, high-throughput, and precise method, rapidly replacing the relatively more variable SDS-PAGE method for analyzing viral vector stoichiometry.

3. *Strength/Potency*

Potency denotes the ability or capacity of a product to elicit the desired effect, ultimately influencing the potential dose required and hence reflects the product's mechanism of action, i.e., relevant therapeutic activity or intended biological effect. Potency assays are conducted as part of product conformance testing, comparability studies, and stability testing. Tests for potency are either in vitro or in vivo tests or both and predominantly specific for each product (though it is desirable to use platform assays to trend multiple products in the clinic). Potency assays should be designed to represent the mechanism of action and be validated during the product development process. Regulatory guidelines recommend that whenever possible, a potency assay should measure the biological activity of the expressed gene product, not merely its presence [217]. However, it is important to note that a single test is unlikely to be enough to measure product attributes that predict clinical efficacy, further bolstering the need for multiple matrices to evaluate the efficacy of the product. This is true especially when the product is complex without a fully characterized mechanism of action, and/or when there are multiple active biological activities and/or a single bioassay is not sufficiently quantitative. There are challenges, however, to developing robust potency assays for cell and gene therapy products, including inherent variability of starting materials, lack of appropriate reference standards for novel gene therapy vectors, limited material for testing, complex mechanisms of action, and the in vivo fate of the product through due to the viral vector infection and transgene expression process [218]. Given these challenges, the FDA recommends an incremental approach to the implementation of potency assays. Early during product development, expression assays and/or functional assays with limited quantitative information may be exploited. As product development expands into later phases, with increased product knowledge more robust and quantitative assays should be developed to gather meaningful data about product efficacy.

4. *Purity/Impurities*

Vector purity, as well as impurities from upstream and downstream operations, including production cell lines and plasmids, are monitored and quantified to gauge both viral purity and potential immunogenicity risks.

- *Process Impurities*: Examples include cell culture media components, supplements, additives, and chemicals introduced during downstream purification processes. Quantification is achieved using MS and high-performance liquid chromatography (HPLC) methods.
- *Production Impurities*: These encompass unwanted host cell proteins (HCP) and nucleic acids from production cell lines, helper viruses, and plasmids. Detection methods include PicoGreen, DNA Threshold assay, and qPCR. Orthogonal methods like CE, CE-MS, LC-MS, and LC-MS/MS are also used.
- *Vector-Related Impurities*: Empty capsids, capsids containing non-desired nucleic acid sequences, non-vector DNA, and partial vector genomes are inefficient at transducing target cells for therapeutic effect. These inefficiencies can necessitate higher total capsid doses to deliver sufficient therapeutic cargo, which can trigger undesired immune responses from a patient [219]. Detection and characterization of these impurities are challenging but critical for product safety [220].

5. *Safety*

Safety testing ensures absence of contaminating agents posing risks to patients. Tests include sterility, endotoxin, and mycoplasma that are the same for any vector product. Detection of replication-competent viruses and other adventitious agents is a specific requirement for viral vectors.

- *Replication-Competent Virus (RCV)*: Recombinant viral vectors are engineered to be replication defective and hence the risk of generating replication-competent virus is low. While the incidence is low, these events can occur during any stage of vector manufacturing by means of stochastic recombination events during virus vector production and can increase with each successive amplification. Therefore, RCV testing is mandated by regulatory bodies to ensure the biosafety of the final product. Cell lines permissive to retro-, lenti-, or AAV infection are used to test for RCV in vitro. Detection methods include Southern blotting, qPCR, and cell-based assays.
- Retroviral integration into the genome is associated with severe adverse events; therefore, regulators have developed guidelines to test for replication-competent retrovirus or lentivirus (RCR/RCL) testing during manufacturing of retroviral-based products and continued patient monitoring [221]. These guidelines detail the stages during which RCR/RCL must be evaluated (e.g., cell banks, end-of-production cells, drug substance supernatant, continued patient monitoring), amount of sample to be tested, and appropriate testing. The assays should include culture of supernatant on permissive cell line allowing amplification of any potential RCR present, followed by detection using an appropriate indicator cell line, RVV-specific PCR, ELISA, or product-enhanced reverse transcriptase assay (PERT). Patient monitoring schedule recommendations include analysis prior to treatment, and then 3, 6, and 12 months after treatment, and annually for up to 15 years. Analysis of RCR in patients may include serologic detection of RCR-specific antibodies and analysis of patient PBMCs by PCR for RCR-specific DNA sequences.
- Adventitious viruses present a risk within any biomanufacturing process during both cell and viral production. Sources of viral contamination may be attributed to the use of raw materials of biological origin (e.g., serum), human and animal-derived cell lines, or the viral amplification within the bioreactor during production. The key steps to reduce the risk of viral contaminants are prevention of contamination, removal during downstream purification via viral clearance steps (if no negative impact on product), and detection. Even with prevention and removal steps, rigorous testing is needed to ensure no detectable adventitious viruses are present. Regulators have published guidelines with recommendations for stages at which screening for viral contamination should occur and types of testing at each stage [55, 222]. It is recommended to perform non-specific screen for adventitious virus for all cell banks and viral vector products. Testing includes using a selection of susceptible cell lines (in vitro screening) and in vivo testing (using suckling and adult mice, embryonated hen's eggs). Species-specific viral screening is recommended based on the potential source of contamination (e.g., raw materials used in the process; origin of the cell line). For example, PCR-based screening for a panel of human viruses should be performed on all cell banks of human origin, whereas the use of murine cell lines would require screening via mouse antibody production (MAP) test. Bovine and porcine viral testing should be performed for all cells exposed to animal-derived raw materials

(e.g., bovine serum or porcine trypsin) and methods are detailed in the Code of Federal Regulations, 9 CFR.

- Acceptance of molecular-based assays to replace animal assays is emerging. ICH Q5A(R2) has been updated to include molecular methods such as nucleic amplification techniques (e.g., PCR) or targeted or non-targeted next-generation sequencing (NGS) as a replacement for both in vitro and in vivo animal assays and is now encouraged [223]. The use of NGS can provide defined sensitivity and broader virus detection, while reducing the use of animals, sample material, and testing time.

6. *Stability*

Stability studies should be performed to provide evidence on how the quality of the drug substance and/or drug product varies over time when stored under various conditions (e.g., temperature, humidity, etc.). The stability study should demonstrate that the product is within acceptable limits for the duration of the clinical study. Per the Code of Federal Regulation, 21CFR211.166, the stability program will determine the appropriate product storage conditions and expiration date. When planning stability studies, the specific storage conditions and container, formulation, and product CQAs should be evaluated to determine the stability conditions, timepoints, and testing panel. Stability analysis may include measurement of container closure integrity, identity, purity, quality, and potency. Stability testing should cover the planned storage conditions, as well as stressed storage conditions. Stress testing is primarily a way to identify potential degradation products and delineate the intrinsic stability of the analyte as well as identify key stability indicating assays.

In addition to monitoring the stability of the drug product in the final container configuration, a well-planned stability program should also include testing of manufacturing process intermediates and bulk drug substance, if the intermediate and drug substance material is held or stored at any point during manufacturing process. For instance, the bulk drug substance of AAV products may be stored for some period of time prior to final formulation and filling into the drug product and therefore the stability of the AAV at this step should be evaluated. One of the key attributes to be evaluated during stability studies of any biotechnological product, including viral vectors for cell and gene therapy, is potency to ensure the product maintains the ability to achieve its intended effect. Another essential characteristic is to evaluate purity, in particular amount of degradation or aggregation of the product. One of the challenges of extended storage of viral vectors is viral vector aggregation, particularly for AAV capsids. Aggregation should be evaluated at the time of viral vector product manufacture and at specific timepoints during stability studies. Techniques such as DLS, MADLS, SEC-MALS, negative stain electron microscopy, and AUC aid in accurate evaluation of aggregation.

Stability studies should also include evaluation of other characteristics that are not specific to viral vectors, including appearance of the product throughout storage, sterility of the product at the end of storage (at a minimum), and container closure integrity to ensure the integrity of the storage container has remained intact under the storage conditions.

Apart from the commonly used viral vectors, different types of gene-editing products are currently under development [183], as described earlier in this chapter. Like other gene therapy products, control and characterization of the gene-editing products should focus on the definition of starting materials, active substance, identity, purity, and potency. Both product and process-related impurities need to be well controlled. For example, any variation in the sequence of guide RNA (gRNA) used in CRISPR/Cas9-mediated gene editing may impact the sequence editing at undesirable locations thereby increasing the risk of off-target effects [223].

During pre-clinical and early clinical phases (I and II), few product characterization tests can initially be performed using qualified assays with broad specifications. For clinical trials, investigational products must meet pre-determined QC testing and specifications set as per extensive characterizations. For pivotal clinical trials and commercialization, methods need to be validated. Appropriate analytical methods also need to be available for clinical trial samples (e.g., replication-competent lentivirus/retrovirus testing). Thus, a phase-appropriate strategy toward fit-for-purpose analytical development is required to ensure the necessary robustness as per regulatory guidelines.

8 Conclusion

Cell and gene therapies and other advanced therapies hold the key to cure or prevent genetic diseases and some cancers by targeting the root cause in a precise manner. The field has come a very long way especially in the last decade that has seen significant cell and gene therapy approvals with at least nine approved in 2023 in the United States alone. The CGT market is expected to expand exponentially over the next five years from $5.3 billion in 2022 to $19.9 billion by 2027 [224]. However, these therapies come with very expensive prices—most priced between $400,000 and $4.25 million per dose making it challenging for payers (private and government) to provide reimbursement or payment models such that all patients have access to these life-altering treatments. Regardless, gene therapy can be considered an elegant concept crudely executed. To elaborate, the tools and technologies including the delivery systems are extremely complex and like all novel, innovative technologies not completely without drawbacks. We are still unraveling the safety of these therapies and evaluating potential serious adverse effects associated with their use. As more of these products become commercially available with extensive data from their use and a better understanding of the outcomes, better vectors and clinical protocols can be developed to ensure more robust, effective, and safer alternatives. Regardless of the pitfalls we currently face, viral vectors are definitely here to stay, and as the field continues to grow and develop, we will be better able to master their biology and improve on their manufacture.

References

1. American Society of Gene & Cell Therapy. Gene, cell, & RNA therapy landscape report - Q1 2024 quarterly data report. 2024. https://www.asgct.org/publications/landscape-report
2. Alliance for Regenerative Medicine. Cell and gene therapy sector data. 2024. https://alliancerm.org/data/. https://alliancerm.org/data/. Accessed 10 Aug 2024.
3. Wang J-H, Gessler DJ, Zhan W, Gallagher TL, Gao G. Adeno-associated virus as a delivery vector for gene therapy of human diseases. Signal Transduct Target Ther. 2024;9:78.
4. American Society of Gene & Cell Therapy Timeline of Gene and Cell Therapy. In: https://www.asgct.org/about/timeline-history
5. Cline MJ. Perspectives for gene therapy: inserting new genetic information into mammalian cells by physical techniques and viral vectors. Pharmacol Ther. 1985;29:69–92.
6. Anderson WF. Human gene therapy. Science. 1992;256:808–13.
7. Kohn DB, Mitsuya H, Ballow M, Selegue JE, Barankiewicz J, Cohen A, Gelfand E, Anderson WF, Blaese RM. Establishment and characterization of adenosine deaminase-deficient human T cell lines. J Immunol. 1989;142:3971–7.
8. Gostimskaya I. CRISPR-Cas9: a history of its discovery and ethical considerations of its use in genome editing. Biochemistry (Mosc). 2022;87:777–88.
9. Ishino Y, Krupovic M, Forterre P. History of CRISPR-Cas from encounter with a mysterious repeated sequence to genome editing technology. J Bacteriol. 2018; https://doi.org/10.1128/JB.00580-17.
10. Food and Drug Administration. FDA approves first gene therapies to treat patients with sickle cell disease. 2023. https://www.fda.gov/news-events/press-announcements/fda-approves-first-gene-therapies-treat-patients-sickle-cell-disease
11. Crystal RG. Adenovirus: the first effective in vivo gene delivery vector. Hum Gene Ther. 2014;25:3–11.
12. Crystal RG, McElvaney NG, Rosenfeld MA, Chu C-S, Mastrangeli A, Hay JG, Brody SL, Jaffe HA, Eissa NT, Danel C. Administration of an adenovirus containing the human CFTR cDNA to the respiratory tract of individuals with cystic fibrosis. Nat Genet. 1994;8:42–51.
13. Zabner J, Couture LA, Gregory RJ, Graham SM, Smith AE, Welsh MJ. Adenovirus-mediated gene transfer transiently corrects the chloride transport defect in nasal epithelia of patients with cystic fibrosis. Cell. 1993;75:207–16.
14. Sibbald B. Death but one unintended consequence of gene-therapy trial. CMAJ. 2001; 164:1612.
15. Raper SE, Yudkoff M, Chirmule N, et al. A pilot study of in vivo liver-directed gene transfer with an adenoviral vector in partial ornithine transcarbamylase deficiency. Hum Gene Ther. 2002;13:163–75.
16. Gorsky K. Ghosts of gene therapies past- lessons learned from Jesse Gelsinger. Health Science Inquiry. 2019. https://doi.org/10.29173/hsi200
17. The Journal of Gene Medicine. Gene Therapy Clinical Trials Worldwide. 2024. https://a873679.fmphost.com/fmi/webd/GTCT. Accessed 21 Mar 2024.
18. Food and Drug Administration. ADSTILADRIN. 2023. https://www.fda.gov/vaccines-blood-biologics/cellular-gene-therapy-products/adstiladrin
19. Zhang H, Wang H, An Y, Chen Z. Construction and application of adenoviral vectors. Mol Ther Nucleic Acids. 2023;34:102027. https://doi.org/10.1016/j.omtn.2023.09.004.
20. Centers for Disease Control and Prevention. Adenoviruses – symptoms. 2024. https://www.cdc.gov/adenovirus/symptoms.html#:~:text=Adenoviruses%20can%20cause%20a%20wide,pneumonia%20(infection%20of%20the%20lungs). Accessed 21 Mar 2024.
21. Ison MG, Hayden RT. Adenovirus. Microbiol Spectr. 2016;4 https://doi.org/10.1128/microbiolspec.dmih2-0020-2015.
22. Leikas AJ, Ylä-Herttuala S, Hartikainen JEK. Adenoviral gene therapy vectors in clinical use—basic aspects with a special reference to replication-competent adenovirus formation and its impact on clinical safety. Int J Mol Sci. 2023; https://doi.org/10.3390/ijms242216519.

23. Bulcha JT, Wang Y, Ma H, Tai PWL, Gao G. Viral vector platforms within the gene therapy landscape. Signal Transduct Target Ther. 2021;6:53.
24. Gingeras TR, Sciaky D, Gelinas RE, Bing-Dong J, Yen CE, Kelly MM, Bullock PA, Parsons BL, O'Neill KE, Roberts RJ. Nucleotide sequences from the adenovirus-2 genome. J Biol Chem. 1982;257:13475–91.
25. Kulanayake S, Tikoo SK. Adenovirus core proteins: structure and function. Viruses. 2021; https://doi.org/10.3390/v13030388.
26. Schwartze J, Havenga M, Bakker W, Bradshaw A, Nicklin S. Adenoviral vectors for cardiovascular gene therapy applications: a clinical and industry perspective. J Mol Med. 2022;100:875–901.
27. Danthinne X, Imperiale MJ. Production of first generation adenovirus vectors: a review. Gene Ther. 2000;7:1707–14.
28. Graham FL, Smiley J, Russell WC, Nairn R. Characteristics of a human cell line transformed by DNA from human adenovirus type 5. J Gen Virol. 1977;36:59–72.
29. Louis N, Evelegh C, Graham FL. Cloning and sequencing of the cellular–viral junctions from the human adenovirus type 5 transformed 293 cell line. Virology. 1997;233:423–9.
30. Sharon D, Kamen A. Advancements in the design and scalable production of viral gene transfer vectors. Biotechnol Bioeng. 2018;115:25–40.
31. Gilgenkrantz H, Duboc D, Juillard V, Couton D, Pavirani A, Guillet JG, Briand P, Kahn A. Transient expression of genes transferred in vivo into heart using first-generation adenoviral vectors: role of the immune response. Hum Gene Ther. 1995;6:1265–74.
32. Yang Y, Su Q, Wilson JM. Role of viral antigens in destructive cellular immune responses to adenovirus vector-transduced cells in mouse lungs. J Virol. 1996;70:7209–12.
33. Trivedi PD, Byrne BJ, Corti M. Evolving horizons: adenovirus vectors' timeless influence on cancer. Gene Therapy Vaccines Viruses. 2023;15:2378.
34. Alba R, Bosch A, Chillon M. Gutless adenovirus: last-generation adenovirus for gene therapy. Gene Ther. 2005;12:S18–27.
35. Parks RJ, Chen L, Anton M, Sankar U, Rudnicki MA, Graham FL. A helper-dependent adenovirus vector system: removal of helper virus by Cre-mediated excision of the viral packaging signal. Proc Natl Acad Sci. 1996;93:13565–70.
36. Palmer D, Ng P. Improved system for helper-dependent adenoviral vector production. Mol Ther. 2003;8:846–52.
37. Liu J, Seol D-W. Helper virus-free gutless adenovirus (HF-GLAd): a new platform for gene therapy. BMB Rep. 2020;53:565–75.
38. Morral N, Parks RJ, Zhou H, Langston C, Schiedner G, Quinones J, Graham FL, Kochanek S, Beaudet AL. High doses of a helper-dependent adenoviral vector yield supraphysiological levels of α1-antitrypsin with negligible toxicity. Hum Gene Ther. 1998;9:2709–16.
39. Schiedner G, Morral N, Parks RJ, Wu Y, Koopmans SC, Langston C, Graham FL, Beaudet AL, Kochanek S. Genomic DNA transfer with a high-capacity adenovirus vector results in improved in vivo gene expression and decreased toxicity. Nat Genet. 1998;18:180–3.
40. Liu Q, Muruve DA. Molecular basis of the inflammatory response to adenovirus vectors. Gene Ther. 2003;10:935–40.
41. Muruve DA, Barnes MJ, Stillman IE, Libermann TA. Adenoviral gene therapy leads to rapid induction of multiple chemokines and acute neutrophil-dependent hepatic injury in vivo. Hum Gene Ther. 1999;10:965–76.
42. Lin D, Shen Y, Liang T. Oncolytic virotherapy: basic principles, recent advances and future directions. Signal Transduct Target Ther. 2023;8:156.
43. Zhao Y, Liu Z, Li L, Wu J, Zhang H, Zhang H, Lei T, Xu B. Oncolytic adenovirus: prospects for cancer immunotherapy. Front Microbiol. 2021; https://doi.org/10.3389/fmicb.2021.707290.
44. Mantwill K, Klein FG, Wang D, Hindupur SV, Ehrenfeld M, Holm PS, Nawroth R. Concepts in oncolytic adenovirus therapy. Int J Mol Sci. 2021; https://doi.org/10.3390/ijms221910522.
45. Packiam VT, Lamm DL, Barocas DA, et al. An open label, single-arm, phase II multicenter study of the safety and efficacy of CG0070 oncolytic vector regimen in patients with

BCG-unresponsive non–muscle-invasive bladder cancer: interim results. Urologic Oncol Semin Original Investig. 2018;36:440–7.

46. Wang P, Li X, Wang J, et al. Re-designing Interleukin-12 to enhance its safety and potential as an anti-tumor immunotherapeutic agent. Nat Commun. 2017;8:1395.

47. Mendonça SA, Lorincz R, Boucher P, Curiel DT. Adenoviral vector vaccine platforms in the SARS-CoV-2 pandemic. NPJ Vaccines. 2021;6:97.

48. He T-C, Zhou S, da Costa LT, Yu J, Kinzler KW, Vogelstein B. A simplified system for generating recombinant adenoviruses. Proc Natl Acad Sci. 1998;95:2509–14.

49. Hardy S, Kitamura M, Harris-Stansil T, Dai Y, Phipps ML. Construction of adenovirus vectors through Cre-lox recombination. J Virol. 1997;71:1842–9.

50. Aoki K, Barker C, Danthinne X, Imperiale MJ, Nabel GJ. Efficient generation of recombinant adenoviral vectors by Cre-lox recombination in vitro. Mol Med. 1999;5:224–31.

51. Zhou D, Zhou X, Bian A, Li H, Chen H, Small JC, Li Y, Giles-Davis W, Xiang Z, Ertl HCJ. An efficient method of directly cloning chimpanzee adenovirus as a vaccine vector. Nat Protoc. 2010;5:1775–85.

52. Lee CS, Bishop ES, Zhang R, et al. Adenovirus-mediated gene delivery: potential applications for gene and cell-based therapies in the new era of personalized medicine. Genes Dis. 2017;4:43–63.

53. Cheshenko N, Krougliak N, Eisensmith RC, Krougliak VA. A novel system for the production of fully deleted adenovirus vectors that does not require helper adenovirus. Gene Ther. 2001;8:846–54.

54. Lee D, Liu J, Junn HJ, Lee E-J, Jeong K-S, Seol D-W. No more helper adenovirus: production of gutless adenovirus (GLAd) free of adenovirus and replication-competent adenovirus (RCA) contaminants. Exp Mol Med. 2019;51:1–18.

55. Food and Drug Administration. Chemistry, manufacturing, and control (CMC) information for human gene therapy investigational new drug applications (INDs) guidance for industry. Silver Spring; 2020.

56. Hickey JM, Jacob SI, Tait AS, Vahid FD, Barritt J, Rouse S, Douglas A, Joshi SB, Volkin DB, Bracewell DG. Measurement of adenovirus-based vector heterogeneity. J Pharm Sci. 2023;112:974–84.

57. Krutzke L, Rösler R, Allmendinger E, Engler T, Wiese S, Kochanek S. Process- and product-related impurities in the ChAdOx1 nCov-19 vaccine. elife. 2022;11:e78513.

58. Chen G-X, Zhang S, He X, Liu S, Ma C, Zou X-P. Clinical utility of recombinant adenoviral human p53 gene therapy: current perspectives. Onco Targets Ther. 2014;7:1901.

59. Sakurai F, Tachibana M, Mizuguchi H. Adenovirus vector-based vaccine for infectious diseases. Drug Metab Pharmacokinet. 2022;42:100432.

60. European Medicines Agency COVID-19 medicines. https://www.ema.europa.eu/en/human-regulatory-overview/public-health-threats/coronavirus-disease-covid-19/covid-19-medicines

61. Gross M. New hopes for gene therapy. Curr Biol. 2014;24:R983–6.

62. Cooney AL, McCray PB, Sinn PL. Cystic fibrosis gene therapy: looking back, looking forward. Genes (Basel). 2018; https://doi.org/10.3390/genes9110538.

63. Collins DE, Reuter JD, Rush HG, Villano JS. Viral vector biosafety in laboratory animal research. Comp Med. 2017;67:215–21.

64. U.S. Food and Drug Administration. Briefing document – Cellular, Tissue, aand Gene Therapies Advisory Committee (CTGTAC) meeting #70. Silver Spring; 2021.

65. Kuzmin DA, Shutova MV, Johnston NR, Smith OP, Fedorin VV, Kukushkin YS, van der Loo JCM, Johnstone EC. The clinical landscape for AAV gene therapies. Nat Rev Drug Discov. 2021;20:173–4.

66. Au HKE, Isalan M, Mielcarek M. Gene therapy advances: a meta-analysis of AAV usage in clinical settings. Front Med (Lausanne). 2022;8

67. Zhao Z, Anselmo AC, Mitragotri S. Viral vector-based gene therapies in the clinic. Bioeng Transl Med. 2022;7:e10258.

68. Piemonti L, Scholz H, de Jongh D, et al. The relevance of advanced therapy medicinal products in the field of transplantation and the need for academic research access: overcoming bottlenecks and claiming a new time. Transpl Int. 2023;36 https://doi.org/10.3389/ti.2023.11633.

69. Food and Drug Administration. Cellular, Tissue, and Gene Therapies Advisory Committee September 2–3, 2021 Meeting Briefing Document- FDA. 2021.

70. Atchison RW, Casto BC. Hammon WMcD (1965) adenovirus-associated defective virus particles. Science. 1979;149:754–6.

71. Samulski RJ, Muzyczka N. AAV-mediated gene therapy for research and therapeutic purposes. Annu Rev Virol. 2014;1:427–51.

72. Bower JJ, Song L, Bastola P, Hirsch ML. Harnessing the natural biology of Adeno-associated virus to enhance the efficacy of cancer gene therapy. Viruses. 2021;13 https://doi.org/10.3390/v13071205.

73. Salganik M, Hirsch ML, Samulski RJ. Adeno-associated virus as a mammalian DNA vector. Microbiol Spectr. 2015; https://doi.org/10.1128/microbiolspec.MDNA3.

74. Asaad W, Volos P, Maksimov D, Khavina E, Deviatkin A, Mityaeva O, Volchkov P. AAV genome modification for efficient AAV production. Heliyon. 2023;9:e15071.

75. Gao G, Vandenberghe LH, Alvira MR, Lu Y, Calcedo R, Zhou X, Wilson JM. Clades of adeno-associated viruses are widely disseminated in human tissues. J Virol. 2004;78:6381–8.

76. Pupo A, Fernández A, Low SH, François A, Suárez-Amarán L, Samulski RJ. AAV vectors: the Rubiks cube of human gene therapy. Mol Ther. 2022;30:3515–41.

77. Pillay S, Carette JE. Host determinants of adeno-associated viral vector entry. Curr Opin Virol. 2017;24:124–31.

78. Gurda BL, DiMattia MA, Miller EB, et al. Capsid antibodies to different adeno-associated virus serotypes bind common regions. J Virol. 2013;87:9111–24.

79. Daya S, Berns KI. Gene therapy using adeno-associated virus vectors. Clin Microbiol Rev. 2008;21:583–93.

80. Li C, Samulski RJ. Engineering adeno-associated virus vectors for gene therapy. Nat Rev Genet. 2020;21:255–72.

81. Kozarsky K. ReGenX AAV vector technology: a tool for in vivo screening. Nat Methods. 2010;7:iii–iv.

82. Colella P, Ronzitti G, Mingozzi F. Emerging issues in AAV-mediated in vivo gene therapy. Mol Ther Methods Clin Dev. 2018;8:87–104.

83. Ayuso E, Mingozzi F, Bosch F. Production, purification and characterization of adeno-associated vectors. Curr Gene Ther. 2010;10:423–36.

84. Young P. Treatment to cure: advancing AAV gene therapy manufacture. Drug Discov Today. 2023;28:103610.

85. Alazard-Dany N, Nicolas A, Ploquin A, Strasser R, Greco A, Epstein AL, Fraefel C, Salvetti A. Definition of herpes simplex virus type 1 helper activities for adeno-associated virus early replication events. PLoS Pathog. 2009;5:e1000340.

86. Clément N, Grieger JC. Manufacturing of recombinant adeno-associated viral vectors for clinical trials. Mol Ther Methods Clin Dev. 2016;3:16002.

87. Clément N, Knop DR, Byrne BJ. Large-scale adeno-associated viral vector production using a herpesvirus-based system enables manufacturing for clinical studies. Hum Gene Ther. 2009;20:796–806.

88. Hakim CH, Clément N, Wasala LP, et al. Micro-dystrophin AAV vectors made by transient transfection and herpesvirus system are equally potent in treating mdx mouse muscle disease. Mol Ther Methods Clin Dev. 2020;18:664–78.

89. Wu Y, Mei T, Jiang L, Han Z, Dong R, Yang T, Xu F. Development of versatile and flexible Sf9 packaging cell line-dependent OneBac system for large-scale recombinant adeno-associated virus production. Hum Gene Ther Methods. 2019;30:172–83.

90. Srivastava A, Mallela KMG, Deorkar N, Brophy G. Manufacturing challenges and rational formulation development for AAV viral vectors. J Pharm Sci. 2021;110:2609–24.

91. Jiang Z, Dalby PA. Challenges in scaling up AAV-based gene therapy manufacturing. Trends Biotechnol. 2023;41:1268–81.
92. Trivedi PD, Yu C, Chaudhuri P, Johnson EJ, Caton T, Adamson L, Byrne BJ, Paulk NK, Clément N. Comparison of highly pure rAAV9 vector stocks produced in suspension by PEI transfection or HSV infection reveals striking quantitative and qualitative differences. Mol Ther Methods Clin Dev. 2022;24:154–70.
93. Bucher K, Rodríguez-Bocanegra E, Wissinger B, Strasser T, Clark SJ, Birkenfeld AL, Siegel-Axel D, Fischer MD. Extra-viral DNA in adeno-associated viral vector preparations induces TLR9-dependent innate immune responses in human plasmacytoid dendritic cells. Sci Rep. 2023;13:1890.
94. Wright JF. Codon modification and PAMPs in clinical AAV vectors: the tortoise or the hare? Mol Ther. 2020;28:701–3.
95. Penaud-Budloo M, François A, Clément N, Ayuso E. Pharmacology of recombinant adeno-associated virus production. Mol Ther Methods Clin Dev. 2018;8:166–80.
96. Hösel M, Broxtermann M, Janicki H, et al. Toll-like receptor 2-mediated innate immune response in human nonparenchymal liver cells toward adeno-associated viral vectors. Hepatology. 2012;55:287–97.
97. Muhuri M, Maeda Y, Ma H, Ram S, Fitzgerald KA, Tai PW, Gao G. Overcoming innate immune barriers that impede AAV gene therapy vectors. J Clin Invest. 2021;131 https://doi.org/10.1172/JCI143780.
98. Parker AND, Vargas J, Anand V, Qu G, Sommer J, Wright F, Couto L. 1013. In vivo performance of AAV2 vectors purified by CsCl gradient centrifugation or column chromatography. Mol Ther. 2003;7:S390–1.
99. Ferreira V, Petry H, Salmon F. Immune responses to AAV-vectors, the Glybera example from bench to bedside. Front Immunol. 2014;5 https://doi.org/10.3389/fimmu.2014.00082.
100. Yang O, Tao Y, Qadan M, Ierapetritou M. Process design and comparison for batch and continuous manufacturing of recombinant adeno-associated virus. J Pharm Innov. 2023;18:275–86.
101. Philippidis A. Novartis confirms deaths of two patients treated with gene therapy Zolgensma. Hum Gene Ther. 2022;33:842–4.
102. Mullard A. Gene therapy community grapples with toxicity issues, as pipeline matures. Nat Rev Drug Discov. 2021;20:804–5.
103. de Jong YP, Herzog RW. Liver gene therapy and hepatocellular carcinoma: a complex web. Mol Ther. 2021;29:1353–4.
104. Shen W, Liu S, Ou L. rAAV immunogenicity, toxicity, and durability in 255 clinical trials: a meta-analysis. Front Immunol. 2022;13 https://doi.org/10.3389/fimmu.2022.1001263.
105. Milone MC, O'Doherty U. Clinical use of lentiviral vectors. Leukemia. 2018;32:1529–41.
106. Coffin JM, Hughes SH, Varmus HE, editors. Retroviruses. Cold Spring Harbor: Cold Spring Harbor Laboratory Press; 1997.
107. Balvay L, Lopez Lastra M, Sargueil B, Darlix J-L, Ohlmann T. Translational control of retroviruses. Nat Rev Microbiol. 2007;5:128–40.
108. Chameettachal A, Mustafa F, Rizvi TA. Understanding retroviral life cycle and its genomic RNA packaging. J Mol Biol. 2023;435:167924.
109. Petropoulos C. Retroviral taxonomy, protein structures, sequences, and genetic maps. Retroviruses. 1997;
110. Feinberg MB, Baltimore D, Frankel AD. The role of Tat in the human immunodeficiency virus life cycle indicates a primary effect on transcriptional elongation. Proc Natl Acad Sci USA. 1991;88:4045–9.
111. Garoff H, Hewson R, Opstelten DJ. Virus maturation by budding. Microbiol Mol Biol Rev. 1998;62:1171–90.
112. Villanueva RA, Rouillé Y, Dubuisson J. Interactions between virus proteins and host cell membranes during the viral life cycle. Int Rev Cytol. 2005;245:171–244.

113. Cherukuri P Overcoming challenges in gammaretroviral vector development and manufacturing. https://www.genezen.com/insights/overcoming-challenges-in-gammaretroviral-vector-development-and-manufacturing-2/
114. Blaese RM, Culver KW, Miller AD, et al. T lymphocyte-directed gene therapy for ADA-SCID: initial trial results after 4 years. Science. 1995;270:475–80.
115. Morgan MA, Galla M, Grez M, Fehse B, Schambach A. Retroviral gene therapy in Germany with a view on previous experience and future perspectives. Gene Ther. 2021;28:494–512.
116. Hacein-Bey-Abina S, Von Kalle C, Schmidt M, et al. LMO2-associated clonal T cell proliferation in two patients after gene therapy for SCID-X1. Science. 2003;302:415–9.
117. Hacein-Bey-Abina S, Garrigue A, Wang GP, et al. Insertional oncogenesis in 4 patients after retrovirus-mediated gene therapy of SCID-X1. J Clin Invest. 2008;118:3132–42.
118. Braun CJ, Boztug K, Paruzynski A, et al. Gene therapy for Wiskott-Aldrich syndrome—long-term efficacy and Genotoxicity. Sci Transl Med. 2014;6:227ra33. https://doi.org/10.1126/scitranslmed.3007280.
119. Pai S-Y. Built to last: gene therapy for ADA SCID. Blood. 2021;138:1287–8.
120. Valsecchi MC. Rescue of an orphan drug points to a new model for therapies for rare diseases. Nat Italy. 2023; https://doi.org/10.1038/d43978-023-00145-1.
121. Food and Drug Administration. FDA approves CAR-T cell therapy to treat adults with certain types of large B-cell lymphoma. 2017. https://www.fda.gov/news-events/press-announcements/fda-approves-car-t-cell-therapy-treat-adults-certain-types-large-b-cell-lymphoma
122. Food and Drug Administration. TECARTUS. 2024. https://www.fda.gov/vaccines-blood-biologics/cellular-gene-therapy-products/tecartus
123. Food and Drug Administration. FDA requests class-wide boxed warning for CAR T-cell agents regarding secondary T-Cell malignancy risk. 2024. https://www.onclive.com/view/fda-requests-class-wide-boxed-warning-for-car-t-cell-agents-regarding-secondary-t-cell-malignancy-risk
124. Nelson R. FDA investigating safety risks in CAR T-cell recipients. Lancet. 2023;402:2181.
125. Tsuruyama T, Hiratsuka T, Yamada N. Hotspots of MLV integration in the hematopoietic tumor genome. Oncogene. 2017;36:1169–75.
126. Schröder ARW, Shinn P, Chen H, Berry C, Ecker JR, Bushman F. HIV-1 integration in the human genome favors active genes and local hotspots. Cell. 2002;110:521–9.
127. Burns JC, Friedmann T, Driever W, Burrascano M, Yee JK. Vesicular stomatitis virus G glycoprotein pseudotyped retroviral vectors: concentration to very high titer and efficient gene transfer into mammalian and nonmammalian cells. Proc Natl Acad Sci. 1993;90:8033–7.
128. Yee J-K, Friedmann T, Burns JC. Chapter 5 Generation of high-titer pseudotyped retroviral vectors with very broad host range. 1994. pp 99–112.
129. Dull T, Zufferey R, Kelly M, Mandel RJ, Nguyen M, Trono D, Naldini L. A third-generation lentivirus vector with a conditional packaging system. J Virol. 1998;72:8463–71.
130. Miyoshi H, Blömer U, Takahashi M, Gage FH, Verma IM. Development of a self-inactivating lentivirus vector. J Virol. 1998;72:8150–7.
131. Zufferey R, Dull T, Mandel RJ, Bukovsky A, Quiroz D, Naldini L, Trono D. Self-inactivating lentivirus vector for safe and efficient in vivo gene delivery. J Virol. 1998;72:9873–80.
132. Gama-Norton L, Botezatu L, Herrmann S, Schweizer M, Alves PM, Hauser H, Wirth D. Lentivirus production is influenced by SV40 large T-antigen and chromosomal integration of the vector in HEK293 cells. Hum Gene Ther. 2011;22:1269–79.
133. Ausubel LJ, Hall C, Sharma A, et al. Production of CGMP-grade lentiviral vectors. Bioprocess Int. 2012;10:32–43.
134. Lee C-L, Ringpis G-E, Yan M, Camba-Colon J, Zhang J, Bartlett JS. 668. Detailed comparison of self-inactivating lentiviral vectors produced by transient transfection and vectors produced by the Cytegrity™ stable cell line system. Mol Ther. 2015;23:S266.
135. Bonner M, Ma Z, Zhou S, Ren A, Chandrasekaran A, Gray JT, Sorrentino BP, Throm RE. 81 Development of a second generation stable Lentiviral packaging cell line in support of clinical gene transfer protocols. Mol Ther. 2015;23:S35.

136. Shaw A, Cornetta K. Design and potential of non-integrating lentiviral vectors. Biomedicines. 2014;2:14–35.
137. Merten O-W, Hebben M, Bovolenta C. Production of lentiviral vectors. Mol Ther Methods Clin Dev. 2016;3:16017.
138. Cherukuri P. Overcoming development and manufacturing challenges with lentiviral vectors. https://www.genezen.com/insights/overcoming-development-and-manufacturing-challenges-with-lentiviral-vectors/
139. Milani M, Annoni A, Bartolaccini S, et al. Genome editing for scalable production of alloantigen-free lentiviral vectors for in vivo gene therapy. EMBO Mol Med. 2017;9:1558–73.
140. Milani M, Annoni A, Moalli F, et al. Phagocytosis-shielded lentiviral vectors improve liver gene therapy in nonhuman primates. Sci Transl Med. 2019;11 https://doi.org/10.1126/scitranslmed.aav7325.
141. Labbé RP, Vessillier S, Rafiq QA. Lentiviral vectors for T cell engineering: clinical applications, bioprocessing and future perspectives. Viruses. 2021;13 https://doi.org/10.3390/v13081528.
142. Asghar AA, Khabir Y, Hashmi MR. Zynteglo: Betibeglogene autotemcel – an innovative therapy for β- thalassemia patients. Ann Med Surg (Lond). 2022;82:104624.
143. Food and Drug Administration. SKYSONA. 2024. https://www.fda.gov/vaccines-blood-biologics/skysona
144. Kansteiner F. Orchard sets new gene therapy price tag at $4.25M—the steepest of any drug. 2024. https://www.fiercepharma.com/pharma/kyowa-kirins-orchard-sets-new-gene-therapy-price-tag-425m-steepest-any-drug-us
145. Kansteiner F, Becker Z, Liu A, Sagonowsky E, Dunleavy K. Most expensive drugs in the US in 2023. 2023. https://www.fiercepharma.com/special-reports/priciest-drugs-2023
146. Dunleavy K. With the pricing situation "untenable" in Europe, bluebird will wind down its operations in the "broken" market. 2021. https://www.fiercepharma.com/pharma/situation-untenable-bluebird-will-wind-down-its-operations-broken-europe
147. Gurumoorthy N, Nordin F, Tye GJ, Wan Kamarul Zaman WS, Ng MH. Non-integrating lentiviral vectors in clinical applications: a glance through. Biomedicines. 2022;10 https://doi.org/10.3390/biomedicines10010107.
148. Marconi P, Argnani R, Epstein AL, Manservigi R. HSV as a vector in vaccine development and gene therapy. Adv Exp Med Biol. 2009;655:118–44.
149. Goins WF, Huang S, Cohen JB, Glorioso JC. Engineering HSV-1 vectors for gene therapy. Methods Mol Biol. 2014;1144:63–79.
150. Glorioso JC. Herpes simplex viral vectors: late bloomers with big potential. Hum Gene Ther. 2014;25:83–91.
151. Artusi S, Miyagawa Y, Goins WF, Cohen JB, Glorioso JC. Herpes simplex virus vectors for gene transfer to the central nervous system. Diseases. 2018;6 https://doi.org/10.3390/diseases6030074.
152. Morissette G, Flamand L. Herpesviruses and chromosomal integration. J Virol. 2010;84:12100–9.
153. Tognarelli EI, Palomino TF, Corrales N, Bueno SM, Kalergis AM, González PA. Herpes simplex virus evasion of early host antiviral responses. Front Cell Infect Microbiol. 2019;9:127.
154. Gurevich I, Agarwal P, Zhang P, et al. In vivo topical gene therapy for recessive dystrophic epidermolysis bullosa: a phase 1 and 2 trial. Nat Med. 2022;28:780–8.
155. Roizman B, Knipe D. Herpes simplex viruses and their replication, chapter 72. In: Fields virology. 4th ed. Philadelphia: Lippincott, Williams & Wilkins; 2001. p. 2399–459.
156. Baez MV, Aguirre AI, Epstein AL, Jerusalinsky DA. Chapter 19 - Using herpes simplex virus type 1-based amplicon vectors for neuroscience research and gene therapy of neurologic diseases. In: Gerlai RT, editor. Molecular-genetic and statistical techniques for behavioral and neural research. San Diego: Academic; 2018. p. 445–77.
157. Manservigi R, Argnani R, Marconi P. HSV recombinant vectors for gene therapy. Open Virol J. 2010;4:123–56.

158. Mody PH, Pathak S, Hanson LK, Spencer JV. Herpes simplex virus: a versatile tool for insights into evolution, gene delivery, and tumor immunotherapy. Virology (Auckl). 2020;11:1178122X20913274.

159. Nguyen H-M, Saha D. The current state of oncolytic herpes simplex virus for glioblastoma treatment. Oncolytic Virother. 2021;10:1–27.

160. Oehmig A, Fraefel C, Breakefield XO. Update on herpesvirus amplicon vectors. Mol Ther. 2004;10:630–43.

161. de Silva S, Bowers WJ. Herpes virus amplicon vectors. Viruses. 2009;1:594–629.

162. Guide SV, Gonzalez ME, Bağcı IS, et al. Trial of Beremagene Geperpavec (B-VEC) for dystrophic epidermolysis bullosa. N Engl J Med. 2022;387:2211–9.

163. Fink DJ, Wechuck J, Mata M, Glorioso JC, Goss J, Krisky D, Wolfe D. Gene therapy for pain: results of a phase I clinical trial. Ann Neurol. 2011;70:207–12.

164. Krystal Biotech Inc. Krystal Biotech receives FDA approval for the first-ever redosable gene therapy, VYJUVEK™ (beremagene geperpavec-svdt) for the treatment of dystrophic epidermolysis bullosa. 2023. https://ir.krystalbio.com/news-releases/news-release-details/krystal-biotech-receives-fda-approval-first-ever-redosable-gene

165. Parry T, Krishnan S, Prosdocimo DA. A new era of in vivo gene therapy: the applicability of a differentiated HSV-1 based vector platform for redosable medicines. Cell Gene Ther Insights. 2022;08:641–51.

166. Khushalani NI, Harrington KJ, Melcher A, Bommareddy PK, Zamarin D. Breaking the barriers in cancer care: the next generation of herpes simplex virus-based oncolytic immunotherapies for cancer treatment. Mol Ther Oncolytics. 2023;31:100729.

167. Koch MS, Lawler SE, Chiocca EA. HSV-1 oncolytic viruses from bench to bedside: An overview of current clinical trials. Cancers (Basel). 2020;12 https://doi.org/10.3390/cancers12123514.

168. Martuza RL, Malick A, Markert JM, Ruffner KL, Coen DM. Experimental therapy of human glioma by means of a genetically engineered virus mutant. Science. 1991;252:854–6.

169. Kardani K, Sanchez Gil J, Rabkin SD. Oncolytic herpes simplex viruses for the treatment of glioma and targeting glioblastoma stem-like cells. Front Cell Infect Microbiol. 2023;13:1206111.

170. Mineta T, Rabkin SD, Yazaki T, Hunter WD, Martuza RL. Attenuated multi-mutated herpes simplex virus-1 for the treatment of malignant gliomas. Nat Med. 1995;1:938–43.

171. Varela ML, Comba A, Faisal SM, Argento A, Franson A, Barissi MN, Sachdev S, Castro MG, Lowenstein PR. Gene therapy for high grade glioma: the clinical experience. Expert Opin Biol Ther. 2023;23:145–61.

172. Todo T, Martuza RL, Rabkin SD, Johnson PA. Oncolytic herpes simplex virus vector with enhanced MHC class I presentation and tumor cell killing. Proc Natl Acad Sci USA. 2001;98:6396–401.

173. Todo T, Ito H, Ino Y, Ohtsu H, Ota Y, Shibahara J, Tanaka M. Intratumoral oncolytic herpes virus G47Δ for residual or recurrent glioblastoma: a phase 2 trial. Nat Med. 2022;28:1630–9.

174. Andtbacka RHI, Kaufman HL, Collichio F, et al. Talimogene Laherparepvec improves durable response rate in patients with advanced melanoma. J Clin Oncol. 2015;33:2780–8.

175. Amgen. DA approves IMLYGIC™ (Talimogene Laherparepvec) as first oncolytic viral therapy in the US. press release. 2015. https://www.amgen.com/newsroom/press-releases/2015/10/fda-approves-imlygic-talimogene-laherparepvec-as-first-oncolytic-viral-therapy-in-the-us

176. Knop DR, Harrell H. Bioreactor production of recombinant herpes simplex virus vectors. Biotechnol Prog. 2007;23:715–21.

177. Ungerechts G, Bossow S, Leuchs B, Holm PS, Rommelaere J, Coffey M, Coffin R, Bell J, Nettelbeck DM. Moving oncolytic viruses into the clinic: clinical-grade production, purification, and characterization of diverse oncolytic viruses. Mol Ther Methods Clin Dev. 2016;3:16018.

178. Helena JM, Joubert AM, Grobbelaar S, Nolte EM, Nel M, Pepper MS, Coetzee M, Mercier AE. Deoxyribonucleic acid damage and repair: capitalizing on our understanding of the

mechanisms of maintaining genomic integrity for therapeutic purposes. Int J Mol Sci. 2018;19 https://doi.org/10.3390/ijms19041148.

179. Ochiai H, Yamamoto T. Genome editing using zinc-finger nucleases (ZFNs) and transcription activator-like effector nucleases (TALENs). In: Yamamoto T, editor. Targeted genome editing using site-specific nucleases: ZFNs, TALENs, and the CRISPR/Cas9 system. Tokyo: Springer Japan; 2015. p. 3–24.

180. Asmamaw M, Zawdie B. Mechanism and applications of CRISPR/Cas-9-mediated genome editing. Biologics. 2021;15:353–61.

181. Gaj T, Sirk SJ, Shui S-L, Liu J. Genome-editing technologies: principles and applications. Cold Spring Harb Perspect Biol. 2016;8 https://doi.org/10.1101/cshperspect.a023754.

182. Gaj T, Gersbach CA, Barbas CF. ZFN, TALEN, and CRISPR/Cas-based methods for genome engineering. Trends Biotechnol. 2013;31:397–405.

183. Raguram A, Banskota S, Liu DR. Therapeutic in vivo delivery of gene editing agents. Cell. 2022;185:2806–27.

184. Uddin F, Rudin CM, Sen T. CRISPR gene therapy: applications, limitations, and implications for the future. Front Oncol. 2020;10:1387.

185. Guo C, Ma X, Gao F, Guo Y. Off-target effects in CRISPR/Cas9 gene editing. Front Bioeng Biotechnol. 2023;11:1143157.

186. Satam H, Joshi K, Mangrolia U, et al. Next-generation sequencing technology: current trends and advancements. Biology (Basel). 2023;12 https://doi.org/10.3390/biology12070997.

187. Naeem M, Majeed S, Hoque MZ, Ahmad I. Latest developed strategies to minimize the off-target effects in CRISPR-Cas-mediated genome editing. Cells. 2020;9 https://doi.org/10.3390/cells9071608.

188. Doudna JA. The promise and challenge of therapeutic genome editing. Nature. 2020;578:229–36.

189. Rees HA, Liu DR. Base editing: precision chemistry on the genome and transcriptome of living cells. Nat Rev Genet. 2018;19:770–88.

190. Qi LS, Larson MH, Gilbert LA, Doudna JA, Weissman JS, Arkin AP, Lim WA. Repurposing CRISPR as an RNA-guided platform for sequence-specific control of gene expression. Cell. 2013;152:1173–83.

191. Chavez M, Chen X, Finn PB, Qi LS. Advances in CRISPR therapeutics. Nat Rev Nephrol. 2023;19:9–22.

192. Gilbert LA, Larson MH, Morsut L, et al. CRISPR-mediated modular RNA-guided regulation of transcription in eukaryotes. Cell. 2013;154:442–51.

193. Casas-Mollano JA, Zinselmeier MH, Erickson SE, Smanski MJ. CRISPR-Cas activators for engineering gene expression in higher eukaryotes. CRISPR J. 2020;3:350–64.

194. Cebrian-Serrano A, Davies B. CRISPR-Cas orthologues and variants: optimizing the repertoire, specificity and delivery of genome engineering tools. Mamm Genome. 2017;28:247–61.

195. Kim N, Kim HK, Lee S, Seo JH, Choi JW, Park J, Min S, Yoon S, Cho S-R, Kim HH. Prediction of the sequence-specific cleavage activity of Cas9 variants. Nat Biotechnol. 2020;38:1328–36.

196. Anzalone AV, Koblan LW, Liu DR. Genome editing with CRISPR-Cas nucleases, base editors, transposases and prime editors. Nat Biotechnol. 2020;38:824–44.

197. Perez EE, Wang J, Miller JC, et al. Establishment of HIV-1 resistance in CD4+ T cells by genome editing using zinc-finger nucleases. Nat Biotechnol. 2008;26:808–16.

198. Holt N, Wang J, Kim K, et al. Human hematopoietic stem/progenitor cells modified by zinc-finger nucleases targeted to CCR5 control HIV-1 in vivo. Nat Biotechnol. 2010;28:839–47.

199. Caldwell KJ, Gottschalk S, Talleur AC. Allogeneic CAR cell therapy—more than a pipe dream. Front Immunol. 2021;11

200. Cross R. CRISPR gene editing in humans appears safe, and potentially effective. 2019. https://cen.acs.org/business/CRISPR-gene-editing-humans-appears/97/i46. https://cen.acs.org/business/CRISPR-gene-editing-humans-appears/97/i46. Accessed 10 Aug 2024.

201. Zhang S, Shen J, Li D, Cheng Y. Strategies in the delivery of Cas9 ribonucleoprotein for CRISPR/Cas9 genome editing. Theranostics. 2021;11:614–48.
202. Wang D, Zhang F, Gao G. CRISPR-based therapeutic genome editing: strategies and in vivo delivery by AAV vectors. Cell. 2020;181:136–50.
203. Hayashi H, Kubo Y, Izumida M, Matsuyama T. Efficient viral delivery of Cas9 into human safe harbor. Sci Rep. 2020;10:21474.
204. Fang H, Bygrave AM, Roth RH, Johnson RC, Huganir RL. An optimized CRISPR/Cas9 approach for precise genome editing in neurons. elife. 2021;10 https://doi.org/10.7554/eLife.65202.
205. Stephens CJ, Kashentseva E, Everett W, Kaliberova L, Curiel DT. Targeted in vivo knock-in of human alpha-1-antitrypsin cDNA using adenoviral delivery of CRISPR/Cas9. Gene Ther. 2018;25:139–56.
206. International Council for Harmonisation. International Council for Harmonisation of technical requirements for pharmaceuticals for human use. 2009
207. Tanaka T, Hanaoka H, Sakurai S. Optimization of the quality by design approach for gene therapy products: a case study for adeno-associated viral vectors. Eur J Pharm Biopharm. 2020;155:88–102.
208. Alliance Regenerative Medicine Project A-Gene – a case study-based approach to integrating QbD principles in Gene Therapy CMC programs. https://alliancerm.org/manufacturing/a-gene-2021/
209. European Medicines Agency (EMA). Guideline on the quality, non-clinical, and clinical aspects of gene therapy medicinal products, 22 March 2018, EMA/CAT/80183/2014
210. King D, Schwartz C, Pincus S, Forsberg N. Viral vector characterization: a look at analytical tools. 2018. https://cellculturedish.com/viral-vector-characterization-analytical-tools/
211. Lock M, Alvira MR, Chen S-J, Wilson JM. Absolute determination of single-stranded and self-complementary adeno-associated viral vector genome titers by droplet digital PCR. Hum Gene Ther Methods. 2014;25:115–25.
212. Wright JF, Zelenaia O. Vector characterization methods for quality control testing of recombinant adeno-associated viruses. Methods Mol Biol. 2011;737:247–78.
213. Snyder RO, Audit M, Francis JD. rAAV vector product characterization and stability studies. Methods Mol Biol. 2012;807:405–28.
214. Sanmiguel J, Gao G, Vandenberghe LH. Quantitative and digital droplet-based AAV genome titration. Methods Mol Biol. 2019;1950:51–83.
215. Bushman FD. Retroviral insertional mutagenesis in humans: evidence for four genetic mechanisms promoting expansion of cell clones. Mol Ther. 2020;28:352–6.
216. Kuck D, Kern A, Kleinschmidt JA. Development of AAV serotype-specific ELISAs using novel monoclonal antibodies. J Virol Methods. 2007;140:17–24.
217. Ramsey JP, Khatwani SL, Lin M, Boregowda R, Surosky R, Andrew Ramelmeier R. Overview of analytics needed to support a robust gene therapy manufacturing process. Curr Opin Biomed Eng. 2021;20:100339.
218. Food and Drug Administration. Guidance for industry – potency tests for cellular and gene therapy products. 2011. https://www.fda.gov/files/vaccines,%20blood%20%26%20biologics/published/Final-Guidance-for-Industry%2D%2DPotency-Tests-for-Cellular-and-Gene-Therapy-Products.pdf
219. Wright JF. Quality control testing, characterization and critical quality attributes of adeno-associated virus vectors used for human gene therapy. Biotechnol J. 2021;16:e2000022. https://doi.org/10.1002/biot.202000022.
220. Gimpel AL, Katsikis G, Sha S, et al. Analytical methods for process and product characterization of recombinant adeno-associated virus-based gene therapies. Mol Ther Methods Clin Dev. 2021;20:740–54.
221. Food and Drug Administration. Testing of retroviral vector-based human gene therapy products for replication competent retrovirus during product manufacture and patient follow-up – guidance for industry. Silver Spring; 2020.

222. International Council for Harmonisation of Technical Requirements for Pharmaceuticals for Human Use. ICH Q5A(R2) Guideline on viral safety evaluation of biotechnology products derived from cell lines of human or animal origin. 2023. https://www.ema.europa.eu/en/documents/scientific-guideline/ich-q5ar2-guideline-viral-safety-evaluation-biotechnology-products-derived-cell-lines-human-or-animal-origin-step-5_en.pdf

223. Tavridou A, Rogers D, Farinelli G, Gravanis I, Jekerle V. Genome-editing medicinal products: the EMA perspective. Nat Rev Drug Discov. 2024;23:242–3.

224. Rabin J. The evolution and future of cell & gene therapy. 2023. https://www2.deloitte.com/us/en/blog/health-care-blog/2023/the-evolution-and-future-of-cell-and-gene-therapy.html

225. Ginn SL, Mandwie M, Alexander IE, Edelstein M, Abedi MR (2024) Gene therapy clinical trials worldwide to 2023—an update Abstract The J Gene Med. 26(8). https://doi.org/10.1002/jgm.v26.810.1002/jgm.3721

Process Development for Biologics Therapeutics

Ankur Bhatnagar, Partha Hazra, Harish V. Pai, Navratna Vajpai, Karthik Ramani, and Nagaraj Govindappa

Abstract This chapter, titled Process Development for Biologics Therapeutics, provides a comprehensive overview of the advancements and critical elements in the development of biologics production processes. It explores the evolution of process development, highlighting key innovations in cell line development, upstream and downstream processing, and drug product formulation. The chapter delves into the importance of a comprehensive set of analytical methods and process characterization studies to ensure the quality, safety, and efficacy of biologics. It also examines the regulatory framework outlined in ICH guidelines and the Quality by Design (QbD) approach, which guides the systematic design and development of robust and reliable processes. The discussion includes considerations for technology transfer, scale-up challenges, and the role of data science in optimizing process parameters and improving yields. Furthermore, the chapter addresses process intensification strategies, which aim to enhance production efficiency and sustainability. By bridging scientific principles with industrial practices, this comprehensive overview aims to provide a thorough understanding of the various facets of biologics process development, from early-stage research to large-scale manufacturing.

Keywords Biologics · Process · Engineering · Cell-lines · Upstream · Downstream · Formulation · Analytical

1 Introduction

The advent of recombinant DNA technology in the 1970s has revolutionized the global landscape of biologic drug development to address the unmet needs of patients. Since then, biologics have been produced in biological systems such as microbial fermentation or cell cultures. This is very different from the conventional

A. Bhatnagar (✉) · P. Hazra · H. V. Pai · N. Vajpai · K. Ramani · N. Govindappa
Biocon Biologics Limited, Biocon Research Center, Bangalore, Karnataka, India
e-mail: ankur.bhatnagar@biocon.com

small molecule drugs which are produced through chemical synthesis and are less complex. Continuous advancement in technology has been the mainstay of increased production and consistency in the quality of biologic therapeutic products [1]. Nonetheless, despite significant evolution and improved understanding, the manufacturing of biologic drugs is highly complex and is achieved through robust process development that involves structured design and optimization of processes to manufacture and produce safe, effective, and high-quality biologic drugs.

Process development for biologic therapeutic products is an intricate and iterative process that requires a multidisciplinary skill set and knowledge encompassing biology, engineering, pharmaceutical sciences, analytical chemistry, and regulatory affairs. A robust and well-optimized manufacturing process is imperative for producing biologic drugs that meet the highest standards of safety, efficacy, and quality. In recent years, there has been a growing trend of applying Quality by Design (QbD) principles during process development to scientifically design and optimize manufacturing processes leveraging an in-depth understanding of product and process characteristics [2–4]. This involves starting with an initial target product profile (TPP) of the product. The TPP of the product typically evolves as the product moves through different stages of development, and hence, it becomes an iterative cycle.

To ensure a robust and consistent process, the US Food and Drug Administration (FDA) outlines a structured approach comprising three stages for process validation: Process Design, Process Qualification, and Continued Process Verification. In the Process Design stage, manufacturers establish a robust manufacturing process based on scientific principles and risk management strategies. This involves defining critical process parameters (CPPs) and critical quality attributes (CQAs) to ensure consistent product quality. In the Process Qualification stage, manufacturers demonstrate that the process is capable of consistently producing products meeting predetermined specifications. This is achieved through a series of qualification runs under controlled conditions, verifying the process's reproducibility and performance. Finally, in Continued Process Verification, ongoing monitoring and analysis of process data ensure that the process remains in a state of control during routine manufacturing. This iterative approach to process validation ensures product quality and patient safety throughout the product lifecycle.

The key process steps involved in the manufacturing of biologic therapeutics are as follows:

1. *Cloning and cell line development*, where the gene coding for the protein of interest is inserted into the chosen host cells using recombinant DNA technology.
2. *Upstream*, which typically uses a fermenter or a bioreactor to create a controlled contamination-free environment and provide required nutrients for the cells producing the product of interest.
3. *Downstream, or purification* step which uses a series of unit operations based on protein separation techniques like chromatography, filtration, crystallization, etc. to purify and obtain the protein of interest into a highly purified state.

4. *Formulation and drug product*, which involves designing formulations that ensure the stability, efficacy, and safety of biological drugs as well as determining the most suitable drug delivery system.

Analytical methods are indispensable tools throughout the process development of biologic protein drugs, enabling scientists to monitor, optimize, and ensure the quality and performance of biologic drugs from early development through commercialization. They play a crucial role in providing essential insights into the characteristics, quality, and performance of the drug substance and drug product.

Additionally, functional bioassays are employed to evaluate the biological activity of biologic drugs by measuring their ability to elicit a specific pharmacological response in a pertinent biological system. These assays offer vital insights into the mechanism of action, potency, and efficacy of biologic drugs and are essential for demonstrating their therapeutic activity.

The process for manufacturing a biologic drug therapeutic is typically divided into two categories: drug substance and drug product. In the context of biologic therapeutic manufacturing processes, a drug substance refers to the active pharmaceutical ingredient (API), which is the biologically active component derived from living organisms. This substance serves as the foundation of the therapeutic product and is typically obtained after downstream. The drug substance undergoes extensive characterization and stability studies to ensure its purity, potency, and consistency.

The drug product encompasses the final formulated dosage form, incorporating the drug substance along with excipients, stabilizers, and other components necessary to optimize its stability, administration, and efficacy. With biologics, drug products can take various forms such as injectable solutions, lyophilized powders, or suspensions, each tailored to the specific therapeutic needs and patient preferences. Development and manufacturing processes for both the drug substance and drug product adhere to stringent regulatory standards, emphasizing safety, quality, and efficacy throughout the product lifecycle.

Apart from the country or region-specific regulatory guidelines, ICH provides internationally recognized standards and recommendations for ensuring the quality, safety, and efficacy of these products (Table 1). These guidelines offer a framework for biopharmaceutical companies to follow throughout the various stages of development, from preclinical research to commercialization. By adhering to ICH guidelines, companies can establish consistent and robust processes for product development, manufacturing, and regulatory submissions, which enhances regulatory compliance and facilitates global market access.

2 Evolution

Biologic therapeutic products have a rich and fascinating history that spans centuries. The development of these products has been influenced by significant scientific breakthroughs, technological advancements, and regulatory milestones. The early

Table 1 Relevant ICH guidelines for Biologic drug development

ICH Q5A(R1)	Quality of biotechnological products: stability testing of biotechnological/biological products	Provides recommendations on stability testing for biotechnological/biological products, including how to design stability studies, select batches for testing, and establish shelf-life and storage conditions
ICH Q5B	Quality of biotechnological products: Analysis of the expression construct in cells used for production of r-DNA derived protein products	Outlines considerations for evaluating the expression construct used in the production of recombinant DNA-derived protein products, including gene sequence, expression vector, and control elements
ICH Q6B	Test procedures and acceptance criteria for biotechnological/ biological products	Provides recommendations on test procedures and acceptance criteria for biotechnological/ biological products, covering aspects such as purity, identity, potency, and impurities
ICH Q8(R2)	Pharmaceutical development	Outlines principles and approaches for pharmaceutical development, including defining the quality target product profile (QTPP), identifying critical quality attributes (CQAs), and conducting risk assessments
ICH Q9	Quality risk management	Describes principles and approaches for quality risk management in pharmaceutical development and manufacturing, including risk assessment, risk control, and risk communication
ICH Q10	Pharmaceutical quality system	Provides guidance on establishing and implementing a pharmaceutical quality system to ensure product quality, including quality management, quality control, and continual improvement
ICH Q11	Development and manufacture of drug substances (chemical entities and biotechnological/biological entities)	Provides recommendations on the development and manufacture of drug substances, covering topics such as manufacturing process design, process validation, and technology transfer
ICH Q14	Analytical procedure development	Provides guidance on risk-based approaches for developing and maintaining analytical procedures suitable for the assessment of the quality of drug substances and drug products
ICH Q2 (R2)	Validation of analytical procedures	Provides guidance on the method validation of the analytical procedures for its intended purpose primarily linked to lot release and stability
USP <1225>	Validation of compendial methods	Provides guidance on the method validation of analytical procedures for its intended purpose
USP <1032>	Design and development of biological assays	Provides guidance on in vitro bioassays design and development for mechanism of action

years of biologic therapeutics were marked using serum therapy, where blood serum containing antibodies was administered to treat infectious diseases such as diphtheria and tetanus. Scientists such as Emil von Behring and Shibasaburo Kitasato pioneered the use of blood serum-containing antibodies to treat infectious diseases like diphtheria and tetanus. Serum therapy laid the groundwork for the use of biological molecules as therapeutic agents and showcased the possibility of utilizing the body's immune system to fight disease. The discovery of insulin by Frederick Banting, Charles Best, and John Macleod in the 1920s marked a significant milestone in biologic drug development. Insulin is a hormone naturally produced from animal sources.

However, it wasn't until the latter half of the twentieth century that strides in biotechnology revolutionized the field, paving the way for the development of complex protein-based therapeutics. The emergence of recombinant DNA technology in the 1970s proved to be a game-changer in biologic drug development. It was in 1978 and 1979 with the successful expression of two mammalian hormones, first somatostatin and then human insulin in *E. coli*. The successful production of human insulin in bacteria provided, for the first time, a practical, scalable source of human insulin resulting in the approval in 1982, for the treatment of diabetics [5]. This breakthrough enabled scientists to manipulate and engineer the DNA of living organisms, facilitating the production of recombinant proteins in bacterial, yeast, and mammalian cells. This technology paved the way for the development of a wide range of biologic therapeutics, including hormones, enzymes, growth factors, and monoclonal and bispecific antibodies [6].

The 1980s witnessed the emergence of monoclonal antibodies (mAbs) as a powerful class of biologic therapeutics. The development of hybridoma technology by Köhler and Milstein in 1975 paved the way to produce highly specific mAbs, leading to breakthrough therapies for cancer, autoimmune diseases, and other conditions.

During the 1980s and 1990s, substantial progress was achieved in bioprocessing techniques facilitating large-scale production of biologic drugs. This included improvements in cell line development, cell culture systems, fermentation technology, downstream purification methods, analytical instrumentation, and techniques.

Initially, cell line development involved laborious methods like random integration of gene constructs into host cell genomes. With the introduction of site-specific integration techniques such as homologous recombination, scientists gained better control over transgene integration sites and expression levels, resulting in more stable and predictable cell lines [7]. Technologies like CRISPR-Cas9 revolutionized cell line engineering, allowing precise genome editing and the introduction of desired traits such as increased protein expression and improved product quality [8]. Additionally, high-throughput screening methods and omics technologies such as genomics, transcriptomics, and proteomics enabled rapid identification and characterization of desirable cell line attributes [9].

Upstream processing has advanced significantly, optimizing cell culture conditions, media formulations, and bioreactor designs [10]. These improvements have led to higher cell densities, enhanced protein expression, and improved scalability. Media formulations have evolved to be entirely chemically defined and animal component-free, incorporating specialized supplements for optimal protein expression [11, 12]. Bioreactor design enhancements, such as single-use systems [13] and online monitoring sensors, have improved flexibility, scalability, and control. High-throughput small-scale bioreactor screening platforms have also streamlined process development and optimization.

Downstream processing has seen remarkable advancements in purification technologies, chromatography resins, and filtration methods, resulting in more efficient and cost-effective purification of biologic proteins with enhanced purity and yield. Traditional chromatography techniques, such as ion exchange, size exclusion, and affinity chromatography, have been refined with improved resins, ligands, and column designs. Introduction of novel chromatography modalities, including mixed-mode chromatography, hydrophobic interaction chromatography (HIC), and multicolumn chromatography systems, has enabled greater selectivity, throughput, and purification efficiency. Significant advancements have been made in filtration technologies for removing impurities and contaminants from biologic protein solutions. Another significant evolution in downstream process development is the adaptation of process intensification strategies to enhance productivity and reduce manufacturing costs. This includes the implementation of continuous processing techniques, such as continuous chromatography, simulated moving bed (SMB) chromatography, and integrated continuous bioprocessing (ICB), offering higher throughput, shorter cycle times, and increased process efficiency compared to traditional batch processes.

A similar evolution was seen in *drug product development*. Initially, the focus was on basic formulation strategies and drug delivery systems to produce protein therapeutics. However, with the growing complexity of biologic molecules and the increasing demand for safer and more effective treatments, drug product development has evolved to incorporate sophisticated analytical methods, innovative delivery systems, and advanced manufacturing processes. This evolution has led to the development of novel formulations [14], such as sustained-release formulations, high conc. Antibodies (reduced injection volumes and frequencies) [15], targeted delivery systems, and combination therapies, all aimed at enhancing patient experience and therapeutic efficacy. Drug product scientists have been continuously working toward developing novel delivery systems that improve drug bioavailability, reduce dosing frequency, and enhance patient convenience by shifting the necessity of hospitalization to medication at home settings.

One of the most significant advances is observed in *Analytical Sciences,* enabling improvements in the Process streams. Initially, analytical methods for biologic protein drugs focused on basic physicochemical characterization, such as protein concentration, molecular weight, and purity. The emergence of advanced analytical platforms, such as ultra-high-performance liquid chromatography (HPLC), capillary electrophoresis (CE), and multi-dimensional chromatography, has facilitated the separation, quantification, and characterization of complex mixtures of biologic proteins with improved resolution, sensitivity, and throughput. Additionally, the integration of automation, robotics, and high-throughput screening technologies has enabled the rapid and efficient analysis of large numbers of samples, accelerating drug development and manufacturing processes. As in separation sciences, the evolution of analytical techniques for biologic protein drugs has expanded to encompass a wide range of physicochemical characterization methods to assess structural integrity, stability, and product-related impurities. These include spectroscopic techniques such as UV-visible spectroscopy, fluorescence spectroscopy, circular dichroism spectroscopy, and NMR spectroscopy, which provide insights into protein structure and conformational changes. Additionally, mass spectrometry (MS)-based techniques, such as peptide mapping (multi-attribute monitoring (MAM)), intact mass analysis, and glycan analysis, enable identification and characterization of protein variants, post-translational modifications, and product-related impurities.

Functional bioassays have evolved from simple cell-based assays to specialized kit-based assays tailored to specific therapeutic modalities. For example, receptor binding assays, enzyme activity assays, and cell-based signaling assays are frequently used to assess the activity of biologic drugs targeting receptors, enzymes, or intracellular signaling pathways. Additionally, the advent of reporter gene assays, bioimaging techniques, and high-content screening platforms has facilitated the quantification and visualization of complex cellular responses to biologic drugs with greater sensitivity and specificity. Some of these highly sensitive assays have replaced animal testing studies, thus having a big ethical benefit too.

In the twenty-first century, the introduction of biosimilars—follow-on versions of biologic drugs with similar efficacy, safety, and quality profiles—has further expanded the realm of biologic therapeutics. This has fostered increased competition, reduced costs, and improved access to biologic therapies. Unlike generic drugs, which are chemically identical to their brand-name counterparts, biosimilars are not exact replicas of the reference biologic drug due to the complexity of biologic molecules and the intricacies of the manufacturing processes. Developing a biosimilar entails a significant challenge in process development, aiming to achieve a manufacturing process that yields a product with highly similar critical quality attributes (CQAs) to the reference product.

The future of biologic drug development is shaped by trends toward continuous manufacturing [16] and the integration of Industry 4.0 technologies such as automation, artificial intelligence, and data analytics [17]. These advancements hold the promise of further enhancing efficiency, flexibility, and quality control in biologic manufacturing.

In summary, the evolution of biologic therapeutic products has been characterized by continuous innovation, driven by advancements in science, technology, and regulatory frameworks. As the field continues to progress, process development remains a cornerstone of biologic drug development, ensuring the production of safe, effective, affordable, and high-quality therapies for patients worldwide. The historical evolution of biologic drug development is a testament to human ingenuity and scientific progress. From humble beginnings in serum therapy to the present era of precision medicine, biologic therapeutics have revolutionized disease treatment and significantly improved the quality of life for millions globally.

3 Cell Line Development

3.1 Selection of Host Expression Systems

The need for producing recombinant proteins for pharmaceutical applications continues to rise rapidly. After identifying the type of biologic molecule to be produced, the next critical decision is selecting the host cell line for expression. *These host cells are the **living machines** used to make these complex products.* The choice of the host significantly impacts product quality, yield, and manufacturing efficiency [7, 18, 19].

Factors influencing the selection include:

Expression Level and Productivity: Vital to meet the commercial demand for the quantity and Cost of Goods (COGs).

Post-translational Modifications (PTMs): Some biologic drugs require specific PTMs, such as glycosylation, phosphorylation, or disulfide bond formation, for proper folding, stability, and biological activity. The host cell system should possess the necessary machinery to perform these PTMs accurately.

Biological Safety: Host cells must be free from contaminants and pathogens, ensuring product safety. They must be characterized for the absence of these agents before using for recombinant protein production.

Genetic Stability and Manipulability: Essential for consistent product quality and genetic optimization.

Manufacturability and Regulatory Considerations: Scalability and regulatory approval compatibility are crucial. Host cell lines should have well-characterized history and traceability.

Both prokaryotic (*E. coli*, *Bacillus*, etc.) and eukaryotic (yeasts, CHO cells, etc.) expression systems are available, each with distinct advantages and limitations too (Table 2) [20].

Table 2 Benefits and drawbacks of different expression systems

Expression systems	Bacterial	Yeast	Mammalian
Benefits	Rapid growth (short batch durations), simple economical media, easy to genetically manipulate, high yields	Suitable for large-scale production, typically expresses proteins in extracellular medium, ease of genetic modifications, performs simple post-translational modifications	Perform human-like post-translational modifications. Good for expressing complex biologics like mAbs, bispecific etc.
Drawbacks	Protein usually forms inclusion bodies, negligible post-translation modifications, endotoxins contamination	Post-translational modifications differ from humans	Expensive media, shear sensitive, challenges to run very large scale

This section below will describe few recombinant protein expression systems which are being used extensively in the therapeutic industry.

Different organisms have unique codon preferences, impacting protein production. Customizing the gene sequence to match the host's codon preferences is crucial for maximizing protein expression. Bioinformatics tools with codon optimization algorithms facilitate the design of genes tailored for optimal expression in specific cellular environments [12].

3.2   Bacterial Expression System

Bacterial expression is the most common expression system employed to produce recombinant proteins, with *Escherichia coli* BL21, *Escherichia coli* K12, *Bacillus brevis*, *Bacillus megaterium*, and *Bacillus subtilis*, to list a few [21]. *E. coli* remains the favored choice due to its well-characterized nature, rapid growth to high cell densities, cost-effective culture media requirements, and ease of scalability [22]. The entire fermentation process typically gets over in 1–2 days. *E. coli* systems can express proteins ranging from 15 to 70 kDa, either in the inclusion bodies [23] cytoplasm or periplasmic [24].

Some of the commonly used promoters are T7 and TRP, which are inducible by IPTG, and Arabinose, which can be induced by arabinose in a controlled manner. There are a few other promoters also reported and used [18]. The *E. coli* BL21 strain engineered with T7 RNA polymerase is a commonly employed system for protein expression due to its high selectivity and efficiency. This strain facilitates a high frequency of transcription initiation and effective elongation, resulting in faster RNA elongation compared to *E. coli* RNA polymerase. The T7 promoter, utilized in this system, is notably stronger than *E. coli* RNA polymerases, enhancing the transcription process. Developed by William Studier at Brookhaven National Laboratory, this technology is widely utilized for the production of various therapeutic proteins [6].

E. coli is widely utilized in the production of numerous biologics, benefiting millions of patients globally. Despite its prevalence, *E. coli* lacks the machinery for complex post-translational modifications necessary for certain proteins. It also presents challenges like inclusion body formation and endotoxin contamination. However, its cost-effective nature makes *E. coli* a valuable tool for recombinant protein production.

3.3   Yeast and Fungi as Eukaryotic Microbial Expression Systems

Unlike bacterial systems, these eukaryotic hosts offer the advantage of post-translational modifications and efficient protein secretion to the extracellular medium [25, 26]. Additionally, they do not have endotoxins or oncogenes, which makes the purification stages easier compared to bacterial processes [27].

Yeast offers several other advantages too, including rapid growth, ease of genetic manipulation, and well-established fermentation protocols. Many species of yeasts and fungi are given GRAS (Generally Regarded as Safe) status, meeting the requirement of regulatory bodies. Most commonly used yeast hosts are *Saccharomyces cerevisiae* [28], *Pichia pastoris, Aspergillus niger, Aspergillus oryzae,* and various *Trichoderma* species.

3.4 Mammalian Expression System for the Complex Biotherapeutic Protein Production

The evolution of mammalian expression host systems for biologics production has been a transformative journey marked by significant advancements in molecular biology, cell culture technology, and bioprocess engineering. Some of the common mammalian expression platforms are the following.

Chinese Hamster Ovary (CHO) Cells

CHO cells are the leading mammalian expression platform for biologic production due to their robust growth, genetic stability, and ability to perform human-like post-translational modifications, particularly glycosylation. They are widely employed for manufacturing complex biologics like monoclonal antibodies and therapeutic proteins, ensuring high expression levels and proper folding. Adaptation to suspension culture enables scalable production in bioreactors, while stable cell lines allow for long-term protein manufacturing. Various CHO cell variants, including CHO-K1, CHO-S, and CHO-DG44, offer specific advantages for different applications. The sequencing of the CHO cell genome in 2011 provided critical insights for optimizing culture conditions and developing high-producing cell lines. CHO cells have a proven track record of safety and regulatory acceptance by agencies like the FDA and EMA in therapeutic protein manufacturing.

Human Embryonic Kidney (HEK) 293 Cells

HEK 293 cells are derived from human embryonic kidney cells and offer high transfection efficiency, making them suitable for transient protein expression studies and rapid protein production. They are commonly used to produce vaccines, and proteins requiring human-like PTMs. HEK 293 cells are particularly useful to produce proteins that are difficult to express in other mammalian cell lines.

Human Cell Lines (e.g., PER.C6, CAP-T, and Expi293F)

Various human cell lines have been developed for recombinant protein expression, offering advantages such as enhanced protein quality and human-like PTMs. PER.C6 cells are derived from human retinal cells and are used to produce viral vectors and vaccines. CAP-T cells are derived from human T-lymphocytes and are engineered to support high-level protein expression. Expi293F cells are a suspension-adapted derivative of HEK 293 cells, offering high transfection efficiency and rapid protein expression kinetics.

Mouse Myeloma Cells (NS0 and Sp2/0)

Mouse myeloma cell lines like NS0 and Sp2/0 originate from tumors in mouse bone marrow, known as myelomas. Derived from B-lymphocytes, these cells have a unique ability to secrete large amounts of antibodies and other proteins, making them valuable for commercial protein production. They are extensively used for recombinant protein expression due to their high productivity and compatibility with human proteins.

Insect Cell Lines (e.g., Sf9 and High Five)

Insect cell lines like *Spodoptera frugiperda* (Sf9) and *Trichoplusia ni* (High Five) offer a robust platform for producing recombinant proteins, including challenging ones like membrane proteins and viral antigens. The baculovirus expression vector system (BEVS) is commonly used to infect these cells, enabling high-level protein expression under a strong viral promoter [29]. While insect cells perform glycosylation, they differ from other mammalian cells, primarily involving high-mannose-type glycans and lacking complex-type glycans. The resulting glycan structures in insect cells are simpler due to the absence of terminal sialic acid and galactose residues commonly found in mammalian cells.

3.5 Regulatory Expectations for Clones, Cell Lines, and Expression Systems for Biotherapeutic Protein Production

Regulatory agencies like the FDA and EMA have stringent expectations for clones, cell lines, and expression systems used in producing biotherapeutic proteins. Monoclonality evidence is required for cell lines (Joel T. Welch∗, N. Sarah Arden, Biologicals 62 (2019) 16–21. Considering "clonality": A regulatory perspective on the importance of the clonal derivation of mammalian cell banks in biopharmaceutical development). Genetic characterization involves karyotyping, DNA sequencing, and PCR analysis to detect abnormalities and viral contaminants. Authentication compares the genetic profile to reference databases like International Cell Line Authentication Committee (ICLAC). Long-term stability studies assess genetic and phenotypic stability over passages. Once the lead clone is identified, cells are banked for long-term storage and use. A two-tier cell banking system is typically prepared, including the Master Cell Bank (MCB) and Working Cell Bank (WCB). MCB serves as the primary source, extensively characterized for genetic stability. WCB undergoes rigorous testing before routine production. Both are stored in liquid nitrogen vapor-phase storage tanks. Regulatory agencies require comprehensive data demonstrating safety, purity, and consistency, with guidelines such as ICH

Table 3 Emerging trends in cell line development

Genome editing and engineering	Techniques like CRISPR/Cas9 allow researchers to precisely edit the host cell genome to enhance protein expression, improve glycosylation patterns, or introduce specific modifications to optimize protein production [31, 32]
Omics technologies	Combining genomics, transcriptomics, proteomics [33] and metabolomics data offers a comprehensive understanding of CHO cell biology, aiding in optimizing culture conditions and developing cell lines for improved recombinant protein production
High-throughput screening (HTS) platforms	HTS platforms automate the screening of numerous cell clones and culture conditions, expediting the identification and optimization of high-producing cell lines
Cell engineering tools	Advanced tools, including viral vectors, non-viral delivery systems (e.g., Maxcyte, electroporation systems like 4D-Nucleofector® and Neon NxT) enable precise control over gene expression and regulation in CHO cells, generating high transcript levels for Gene of Interest (GOI), facilitated by incorporating enhancer elements, non-coding RNAs/introns, and strong promoters. Transposon-based genomic integration enables for insertion of multiple copies of GOI, yielding high expression levels [34]
Transforming bacterial and yeast hosts	Genetic modifications to mimic mammalian cell post-translational modifications (PTMs). Strategies involve introducing eukaryotic glycosyltransferases, chaperones, and proteases into hosts like *E coli* and *Saccharomyces cerevisiae*. These modifications enable production of proteins with more complex PTMs, akin to mammalian cells, bolstering therapeutic potential [35]

Q5A(R1), ICH Q5D, FDA Guidance, EMA Guideline, and USP General Chapter <1043>.

3.6   Looking Ahead

The evolution of expression host systems for biologics production continues to be driven by advancements in cell engineering, omics technologies, and bioprocess optimization [30].

Some of the emerging technologies are listed in Table 3.

4   Upstream Process Development

In the upstream processing stage, various types of microorganisms and mammalian cells are utilized to produce biologic products such as proteins, antibodies, enzymes, and vaccines using a bioreactor. Each host system has its advantages and limitations, and selection is based on the specific needs of the bioprocess and the desired characteristics of the final biologic product.

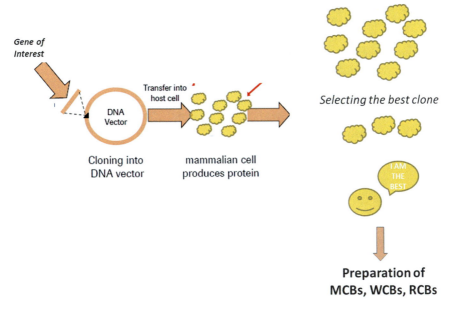

Fig. 1 Process of cloning and selection of the top/lead clone based on multiple attributes as shown. RCB Research Cell Bank, MCB Master Cell Bank, WCB Working Cell Bank

After the cloning is completed, multiple criteria are used to select the top/lead clone from the generated clones (Fig. 1). The top clone is expanded, banked, and preserved for long-term storage under cryo conditions. These banks can be classified as Research, Master, Working cell banks depending on the conditions used during their preparation and testing. It may be a better option to also keep a back-up clone available in case any challenge is found with the lead clone during the course of development of the program.

Once the top clone from the cell line development is identified, the optimization of the upstream process starts. The upstream process optimization generally relies on two main aspects: media and feeds providing nutrients for the living cells and the environment conditions inside the bioreactors [36].

4.1 Optimization of Media and Feeds

Optimizing media and feeds is vital for maximizing cell growth, productivity, and product quality in bioprocessing. Basal media formulations provide essential nutrients like carbon and nitrogen sources, minerals, vitamins, and salts necessary for robust cell growth and product synthesis. Buffered basal media help maintain pH, while supplements like yeast extract or peptones enhance cell metabolism and protein synthesis.

Concentrated nutrient feeds are added during cultivation to sustain growth and production. Fed-batch feeding strategies maintain optimal nutrient concentrations, supporting high cell densities and protein yields while minimizing by-products. The nutritional environment significantly influences product quality by affecting protein folding, modifications, and by-product formation. Optimizing media and feeds thus promotes higher product purity and potency.

4.1.1 Chemically Defined Media

Chemically defined media for cell culture are formulations that contain precisely defined components, including salts, amino acids, vitamins, trace elements, and other essential nutrients required for cell growth and proliferation. These media are specifically designed to provide a controlled and reproducible culture environment, free from variability associated with complex supplements such as serum or hydrolysates [37] (Fig. 2).

4.1.2 Media and Feed Optimization Approaches

Various approaches are used to optimize media and feeds for cell culture. Empirical optimization involves systematically varying components one factor at a time (OFAT) to find optimal formulations, albeit being time-consuming. Statistical Design of Experiments (DoE) allows simultaneous variation of multiple factors, identifying optimal conditions more efficiently [38–40]. Metabolic flux analysis (MFA) quantitatively analyzes cellular metabolism to pinpoint inefficiencies. High-throughput screening (HTS) rapidly screens large libraries of components. Omics

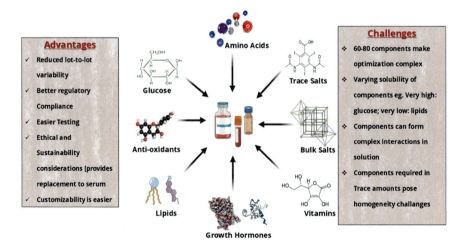

Fig. 2 Components of chemically defined cell culture media

technologies provide insights into cellular responses. However, optimizing media with numerous components presents challenges due to complex interactions and varied solubilities, ranging from grams to micrograms per liter. Achieving homogeneity amidst such diversity is challenging.

Optimizing media and feed for cell culture involves a delicate balance of art and science, especially for mammalian cells. With compositions typically comprising 60–100 components, achieving optimal conditions for different cell types is complex. The interactions among these components are intricate and nonlinear, with varying solubilities ranging from highly soluble compounds like glucose and sodium chloride to insoluble ones like lipids. Moreover, the quantities required span from grams per liter to micrograms per liter for trace elements, posing challenges for achieving perfect homogeneity.

4.2 Optimization of Bioreactor Conditions

The following are the important aspects of bioreactor optimization.

4.2.1 Choice of Bioreactor

The initial step in optimizing conditions involves selecting the appropriate bioreactor type tailored to the cell line's requirements, production process, and operational scale. Options include stirred-tank stainless steel bioreactors, airlift bioreactors, wave bioreactors, and disposable single-use bioreactors. Each type offers distinct advantages and limitations concerning mixing efficiency, oxygen transfer rates, scalability, and operational simplicity. The choice of bioreactor is typically determined by factors such as availability in existing facilities and the specific needs of the manufacturing process. Stirred-tank stainless steel bioreactors are preferred for large-scale biopharmaceutical manufacturing due to their robustness, scalability, and reliability. They ensure efficient mixing and uniform distribution of nutrients and oxygen, making them suitable for high-demand products like insulin. Wave bioreactors utilize rocking or wave-induced motion for gentle mixing and oxygenation, offering scalability and ease of operation for research and smaller production-scale applications [41]. Single-use bioreactors provide scalability, flexibility, and rapid setup, making them ideal for research, process development, and medium- to large-scale manufacturing with short turnaround times.

4.2.2 Mode of Bioreactor Operation

Fed-batch and perfusion cultures are key strategies for maximizing cell productivity and viability in bioreactor systems. In fed-batch processes, nutrients are added incrementally to sustain cell growth and protein production, leading to high cell densities and product titers. However, this method may face challenges due to nutrient depletion or waste accumulation. Conversely, perfusion involves continuous removal of spent media and addition of fresh media, maintaining constant cell density and offering steady-state conditions for prolonged culture and higher productivity. While perfusion systems require specialized equipment and may be more complex to operate, the choice between the two methods depends on product requirements, process economics, and facility capabilities.

4.2.3 Optimization of Operating Parameters

Optimizing bioreactor conditions involves controlling parameters such as temperature, pH, dissolved oxygen, agitation, and aeration, which directly influence cell physiology and protein production. Other factors like redox potential and dissolved carbon dioxide levels also impact cell performance. Sensitivity to shear varies depending on the cell type, with mammalian cells being particularly sensitive. Quality by Design principles advocate for establishing a "Design Space," a parameter range ensuring consistent production meeting quality standards, determined through systematic experimentation.

4.2.4 Process Monitoring and Control

Various tools are available for monitoring the upstream process, including standard probes for temperature, pH, dissolved oxygen, and biomass and glucose levels. Additional analyzers like online sensors for cell growth estimation and spectroscopic techniques such as Raman and near-infrared (NIR) provide valuable data for multiple attributes [42]. These serve as process analytical technology (PAT) tools, facilitating early detection and control of process deviations [43]. Advanced biochemical analyzers and mass spectrometers can now measure up to 50–100 metabolites, offering vital insights during process optimization.

4.3 Emerging Trends in Upstream Processing

Emerging trends in upstream processing for biopharmaceutical manufacturing focus on improving productivity, quality, and efficiency, while ensuring scalability and regulatory compliance. Some of them are listed in Table 4.

Table 4 Emerging trends in upstream processing

Cell-line engineering	Advancements in genome editing enable development of high-producing cell lines with enhanced stability, product quality, metabolic activity
Single-use bioreactors	Offer flexibility, scalability, enhanced sensor technology, better aeration, and mixing
Continuous bioprocessing	Provides higher productivity and smaller footprint supported by improved perfusion devices and culture media
Microfluidic platforms and DoE approach	Accelerate process development through high-throughput screening and optimization techniques [44]
Digitalization and data analytics	Enable predictive modeling, virtual process simulation, and advanced control strategies for optimization
Cell-free bioprocessing	Offers a rapid alternative for protein synthesis without traditional cell culture methods

5   Downstream Process Development

5.1   Overview

Downstream purification is a crucial stage in the process development of biological products, and ensures that the final product meets quality, safety, and regulatory standards. The overall approach toward downstream process development for a biologic therapeutic product involves a systematic and iterative process aimed at achieving high product purity, potency, and safety while optimizing process efficiency and scalability. The downstream process steps are designed to reduce or remove process-related, host cell-related, and product-related impurities (Table 5). This phase usually starts with the harvest from cell culture bioreactors, containing expressed active pharmaceutical ingredients (API) in an impure form, and ends with a highly purified and appropriately concentrated product, ready for final formulation and packaging [45, 46].

Purification process development encompasses four main stages. Throughout each stage, meticulous planning, evaluation, and adjustments are crucial to ensure the success and efficiency of the development process.

1. Define specific goals for the downstream process based on the QTPP.
2. Set up essential analytical tools to assess the target molecule and impurities.
3. Develop an iterative approach for optimization and refinement of the purification process across different scales, from small laboratory setups to larger pilot facilities.
4. Adopts a linear scale-up approach for technology transfer to larger production settings.

Downstream purification strategies are developed through collaborative discussions, leveraging existing literature and in-house expertise. These strategies undergo thorough assessment for factors like yield, purity, scalability, and feasibility.

Table 5 Example of potential impurities and contaminants typically removed during downstream processing

Origin	Contaminants
Host related	Viruses (usually for mammalian cells), host cell protein, host cell DNA, bacteria, mycoplasma, endotoxin (usually for bacterial host), etc.
Process related	Bacteria, adventitious viruses, cell culture media components, purification reagents, metals, column leachable, leachable from single-use systems, etc.
Product related	Oxidized variants, aggregates, dimers, deamidated species, protein fragments, succinimide formation, disulfide pairing variants, etc.

Compatibility with existing manufacturing facilities is ensured for seamless integration during scale-up [47, 48].

A platform approach is often preferred for downstream development, providing a standardized framework for unit operation development and characterization [49]. This streamlined approach accelerates the scale-up of bio-separation operations like centrifugation, chromatography, and filtration, ensuring consistency and reliability across projects and manufacturing scales. However certain exceptions could be when a molecule has a unique technical challenge or a new technology needs to be introduced.

5.2 Major Activities of Downstream Process Development

To perform the downstream process development activities methodically, the whole development activities can be divided into different stages.

5.2.1 Understanding the Characteristics of Upstream Product Output

- Assess the outputs of the upstream process to determine the starting material's quality, quantity, and impurity profile.
- Compare with the desired target product profile to identify the downstream purification process development targets.

5.2.2 Outline the Unit Operations for the Downstream Process

The downstream process flow varies depending on the source of the product (mammalian, bacteria, or yeast). Each expression system produces distinct types of process and product-related impurities. Consequently, the strategy for downstream processing is customized and largely influenced by the chosen expression system [50].

Some of the key unit operations commonly employed in downstream processing are given in Table 6.

Table 6 Commonly employed unit operations in downstream processing

Harvesting/ clarification	Separating cells, cell debris, solid particles from culture medium to improve the clarity and purity of the product stream before downstream processing Can be achieved through centrifugation, microfiltration, or depth filtration
Capture	Target biomolecule is selectively captured from clarified supernatant. Chromatography techniques such as affinity, ion exchange, or size exclusion are commonly used
Inclusion bodies (IB) solubilization/ refolding	Dissolution of insoluble protein aggregates (IBs) formed during protein expression in bacterial hosts. Helps release the target protein from IBs to make it accessible for downstream purification Refolding—Renaturation of the solubilized protein to its native, biologically active conformation
Intermediate purification	Purification steps to purify the target biomolecule and remove remaining impurities; involves chromatography steps, precipitation/crystallization, or filtration
Polishing	Additional purification step(s) involve additional appropriate chromatography steps to further purify the biomolecules. This is the crucial step and considered to be the final step of purification to meet the target product and process-related variants
Concentration/buffer exchange	Concentration, volume reduction for processing, and/or buffer exchange of the purified protein solution. Also used for adjusting the final concentration and pH of the product and exchanging the buffer for a formulation suitable for storage and administration Ultrafiltration or diafiltration is typically used for this application
Bioburden reduction	To ensure intermediates and final product are free from microbial contamination

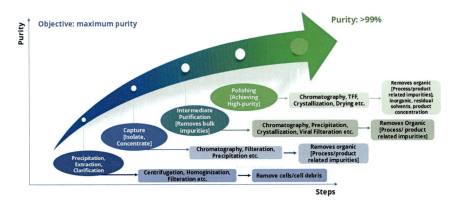

Fig. 3 A summary of the four major stages of the downstream purification process showing the technique used and purpose served for the stages

These unit operations are often performed sequentially, with each step contributing to the overall purification and isolation of the target biomolecule [51]. The purpose of each major stage of downstream process and the unit operation involved in the downstream process is also captured in Fig. 3. The specific unit operations used can vary depending on the characteristics of the biologic product and the desired purity and potency requirements.

In ADC (antibody drug conjugate) development, downstream purification faces added complexity due to the presence of conjugated cytotoxic drug payloads and linker molecules. Specialized chromatographic methods and/or any other separation methodology are developed to separate and purify the ADC from impurities like unconjugated antibody, free drug, linker species, and residual reagents. Handling cytotoxic compounds presents challenges, necessitating specialized containment measures and equipment to ensure operator safety and minimize exposure during purification and formulation.

The downstream purification processes for bi-specific mAbs typically involve more complex chromatographic schemes to separate various product isoforms, such as monospecific and bispecific species, as well as potential product-related impurities.

5.2.3 Optimization of Selected Unit Operations

The key aspects of process development for some common downstream unit operations used in biologic product manufacturing are given in Table 7.

DOE is crucial in downstream process development for biologic therapeutics, streamlining optimization, and characterization of unit operations. By exploring multiple parameters concurrently, it identifies critical factors influencing performance and quality. Efficient experimental plans, data analysis, and mathematical modeling predict process behavior, enhancing efficiency and reliability.

Effective removal of process-related impurities, such as host cell proteins, DNA, and viruses, is crucial to meet regulatory requirements and ensure product safety.

Table 7 Key aspects and parameters of process optimization for downstream unit operations

Chromatography	Resin and column selection, buffer composition, flow rates, and gradient profiles. Aims to achieve high purity and yield of target molecule while minimizing impurities, ensuring robustness and reproducibility
Filtration	Selecting appropriate filter media; optimizing process parameters such as pressure, temperature, and flow rates; validating the filtration step to ensure efficient removal of impurities while retaining the product
Centrifugation	Optimizing parameters such as rotor speed, time, and temperature to achieve efficient separation of components based on density differences
Precipitation	Optimizing conditions such as pH, temperature, and addition of precipitating agents to selectively precipitate target molecule, leaving impurities in solution. Process development focuses on achieving high yield and purity of the precipitated product
Crystallization	Controlling parameters such as temperature, cooling rate, agitation, and seeding to achieve the desired crystal size, shape, and purity
Ultrafiltration and diafiltration (UF/DF)	Membrane selection, pore size, transmembrane pressure, and buffer composition to achieve efficient separation and concentration of product while removing impurities

5.2.4 Process Integration and Optimization

- Streamline individual purification steps to create an integrated process flow.
- Optimize the entire downstream process to maximize yield and purity while minimizing processing time and resource consumption.
- Implement strategies like continuous processing, single-use systems, and automation, wherever possible, to improve efficiency and productivity.

5.2.5 Develop Appropriate Checks and Controls

Downstream processing is pivotal for controlling both process- and product-related impurities in biologic therapeutics, with analytics serving as a cornerstone for quality assessment and process monitoring (Fig. 4). Certain proteins are vulnerable to degradation during downstream processing, with factors like pH, temperature, light exposure, hold times, exacerbating degradation risks. Analytical techniques like high-performance liquid chromatography (HPLC), capillary electrophoresis (CE), mass spectrometry (MS), and gel electrophoresis are commonly employed for characterizing the quality of the product under purification. Analytical methods are also crucial for detecting and quantifying process-related impurities like host cell proteins, DNA, and antifoam.

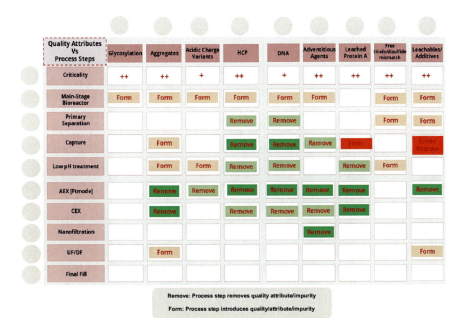

Quality Attributes Vs Process Steps	Glycosylation	Aggregates	Acidic Charge Variants	HCP	DNA	Adventitious Agents	Leached Protein A	Free thiols/disulfide mismatch	Leachables/Additives
Criticality	++	++	+	++	+	++	++	++	++
Main-Stage Bioreactor	Form	Form	Form	Form	Form	Form		Form	Form
Primary Separation				Remove	Remove			Form	Form
Capture		Form		Remove	Remove	Remove	Form		Form/Remove
Low pH treatment		Form	Form	Remove	Remove		Remove	Form	
AEX [Ft mode]		Remove	Remove	Remove	Remove	Remove	Remove		Remove
CEX		Remove		Remove	Remove	Remove	Remove		
Nanofiltration						Remove			
UF/DF		Form							Form
Final Fill									

Remove: Process step removes quality attribute/impurity

Form: Process step introduces quality/attribute/impurity

Fig. 4 Formation and removal of various product quality attributes in different process steps [52, 53]

In-process tests and controls are vital for maintaining product quality and optimizing process efficiency. These tests occur throughout purification, monitoring key parameters like protein concentration, impurity levels, and product quality. Detecting deviations early allows for timely adjustments, ensuring process robustness and product consistency. Real-time feedback from in-process controls empowers operators to make informed decisions, optimizing process conditions for optimal performance.

The end of downstream processing is the drug substance. This may be taken directly for drug product preparation or stored. The general expectation from the drug substance specifications and release are: Identity (confirming the correct molecular identity of the drug substance), Purity (ensuring the substance is free from impurities and contaminants), Potency (determining the biological activity or effectiveness of the substance), Quantity (establishing the concentration or content of the active ingredient), Stability (assessing the stability of the substance under specified conditions), Characterization (providing detailed information on physicochemical properties and structure), Microbial limits (setting limits for microbial contamination), Residual solvents (ensuring compliance with acceptable levels of residual solvents), Endotoxin levels (establishing limits for endotoxin contamination), and Host cell protein and DNA (setting limits for host cell-derived impurities). Additionally, stability studies need to be performed to establish the storage conditions and shelf life.

5.3 Emerging Trend and Future Perspective

Emerging trends in downstream processing for biologics focus on integrating real-time monitoring systems, adopting single-use technologies for flexibility and contamination risk reduction, and emphasizing sustainability [54–56]. There's also a growing emphasis on sustainability and green processing methods. Collaboration among stakeholders, academia, and regulatory agencies is crucial for innovation and addressing challenges, while evolving regulatory frameworks ensure product safety and quality amidst technological progress.

Here are some key trends and perspectives shaping the future of downstream processing in biologics:

- Continuous processing: Adoption of continuous chromatography and membrane-based separations for higher efficiency.
- Advanced chromatography resins: Development of improved resins for enhanced selectivity and capacity.
- Filtration technologies: Advancements in filtration for better impurity removal.
- Process intensification: Strategies to boost productivity and cut costs.
- Sustainability: Focus on eco-friendly processes and reduced waste.
- Automation and digitalization: Integration for better control and monitoring.
- Personalized medicine: Tailoring processes for customized therapeutics.

6   Drug Product Development

The downstream process ends with the preparation of a drug substance, which is the biologically active pharmaceutical ingredient. However, the full realization of therapeutic potential hinges not only on the drug substance's potency but also on its formulation, delivery, and stability. This is where drug product development comes into play, serving as the bridge between the drug substance and its intended clinical application. From formulation optimization to packaging design, drug product development encompasses a myriad of disciplines, each contributing to the final product's safety, efficacy, and patient experience.

Creating a well-defined target product profile (TPP) or quality target product profile (QTPP) is a crucial strategic step before commencing development activities for the drug product. The TPP outlines key aspects such as the patient population, administration route, dosage form, strength, storage requirements, and container-closure systems (e.g., vial, pre-filled syringes-PFS), including specific delivery platforms like pens or autoinjectors, as well as desired drug product quality attributes. This comprehensive document aligns development, manufacturing, and commercialization approaches and integrates patient-centric needs, enhancing acceptance of the final product. It is important to note that the TPP is dynamic, especially for a novel molecule, and should be regularly updated as additional knowledge about the product is gained through various stages of development, from discovery to pre-clinical to clinical to market.

Following lead candidate identification in the discovery stage, product development progresses through early and late-stage activities. Early-stage studies focus on adapting platform formulations and manufacturing processes to advance the product through toxicology and initial clinical studies (Phase 1 and/or Phase 2). Despite a platform approach, this phase involves multiple parallel activities, including drug substance (DS) process development, pre-formulation and formulation studies for the drug product (DP), analytical development supporting DS and DP process development, and initial manufacturability assessments to scale up DS/DP for clinical supply manufacturing.

6.1   Early-Stage Development

In early-stage studies, challenges often stem from limited material availability for robust formulation studies. Changes in upstream and downstream purification processes can affect drug product (DP) quality and stability, necessitating collaboration between formulation development scientists and colleagues in upstream and purification activities. Depending on time and budget constraints, the choice of dosage form for evaluation may evolve. For instance, storing and shipping the drug substance (DS) in a frozen state allows flexibility in preparing dosing regimens for safety and efficacy evaluation without engaging in costly DP manufacturing processes. This approach also eliminates the need for separate stability studies for DS and DP. Alternatively, lyophilizing the DS using a standard platform cycle and excipient selection facilitates easier shipment to clinical trial centers and provides flexibility in reconstitution volumes and dosing regimen preparation. Lyophilized products typically offer greater stability than their solution counterparts and simpler handling. In some cases, a liquid formulation filled in vials may be developed based on molecule nature and stability, providing valuable scientific data and experience for specification and shelf-life determination throughout development stages.

In the early stages of product development, a variety of factors affecting drug product quality and stability are assessed, as outlined in Table 8. Physico-chemical and biophysical analytical tools, along with bioassays if necessary, are employed in both traditional and high-throughput formats to gather a comprehensive dataset. This allows development scientists to understand product characteristics and recommend appropriate processing, storage, and handling instructions. Formulation is more than just mixing ingredients; choice of excipients Table 9, quality, sequence, and addition method all impact product quality and stability, making attention to these aspects crucial from both development and process perspectives [57]. Collaboration with process engineering and technical operations teams is recommended to understand unit operations the product will undergo, identifying risk factors and mitigation approaches early in the development program. Critical quality attributes (CQA) and potential critical manufacturing process parameters (CPP) are defined at this stage, correlating to develop potential control strategies explored further in late-stage studies.

6.2 Late-Stage Development

Activities and assessments undertaken in the early-stage development continue in this stage and focus on finalizing the formulation (strength, composition), analytical methods, process and manufacturing facility to support the pivotal clinical studies, final container closure system (and administration device), process validation, and commercial launch of DP.

In this stage, the TPP/QTPP is modified as appropriate and applicable. This evolved profile will then serve as the basis for manufacturing DS and DP for pivotal clinical studies and typically should be representative of the commercial launch.

In the early stages of development, data collected from lab, pilot scale, and scale-up activities are used for detailed process risk assessment. Various tools, such as Failure Mode and Effect Analysis (FMEA), are employed to map out potential failures across the intended commercial process and facility. This comprehensive assessment covers all aspects from raw materials and equipment to environmental conditions and packaging components. By identifying potential failure modes, cross-functional teams can prioritize and manage risks effectively, refining critical process parameters (CPPs) and their impact on critical quality attributes (CQAs). This facilitates the development of control strategies to minimize any adverse effects on the final product.

Table 8 List of factors typically affecting drug product quality and stability

Focus area	Typical activities	Impact area	Analytical tools
Pre-formulation/formulation	pH, ionic strength, excipients, metal ions, viscosity (for high-concentration products)	Conformational stability, aggregation, charge variants, impurity profiles, fragmentation, colloidal stability	Differential scanning calorimetry (DSC) / fluorimetry (DSF), size exclusion chromatography (SEC), ion-exchange chromatography (IEX), reverse phase chromatography (RP-HPLC), hydrophobic interaction chromatography (HIC), capillary electrophoresis (CE), dynamic light scattering (DLS), microflow imaging (MFI), UV spectroscopy, Raman spectroscopy, Fourier transform infrared spectroscopy (FT-IT), circular dichroism (CD), appropriate potency and or biological assays if available and applicable
Process related	Freezing and thawing, agitation, lyophilization, filtration, product contact parts such as tubing, pumping system, mixing system	Aggregation, fragmentation, conformational stability, particulate formation, colloidal stability, protein adsorption	
Environmental factors	Temperature and light	Conformational stability, impurities	
Container closure	Freezing and thawing, agitation	Conformation and colloidal stability	
In-use stability (based on route of administration)	Compatibility studies with appropriate IV diluents and bags, infusion systems, closed system transfer devices, (CSTD), syringe compatibility	Conformation and colloidal stability, particle formation, protein adsorption, storage and handling of product in the clinic	

Table 9 Commonly used excipients in protein formulations

Excipients class	Function	Examples
Stabilizers	Prevent degradation during storage and transport. Sugars form hydrogen bonds with proteins and prevent denaturation and aggregation. Surfactants serve as stabilizers by coating protein surfaces, preventing their aggregation and adsorption onto container surfaces	Sugars: Sucrose, trehalose. Surfactants: Polysorbates (PS) (e.g., PS 20, PS 80), Poloxamer 188
Buffers	Maintain formulation pH, ensuring optimal stability and activity	Phosphate, acetate, citrate, histidine, succinate, Tris
Tonicity adjusters	To adjust the osmolarity of the formulation to match physiological conditions	Sodium chloride, mannitol, glycerol, sucrose, trehalose
Anti-oxidants	Prevent oxidation	Methionine, EDTA
Viscosity-reduction	Lower viscosity	Arginine, sodium chloride, glycine, proline
Lyo/cryoprotectants	Protect biologic molecules during lyophilization (freeze-drying) processes.	Sucrose, trehalose
Preservatives	Added to multidose formulations to prevent microbial growth and maintain product sterility	Benzyl alcohol, phenol, m-cresol

Quality and stability data collected from pre-clinical to PQ lots are used to establish specifications and assign shelf life for the product, utilizing different statistical approaches to assess quality trends. For biologics, real-time data gathering is emphasized to minimize extrapolation on shelf-life claims due to the product's complexity and varied degradation pathways. Stability of critical excipients is monitored to prevent negative impacts on the quality of the active biologic. In multi-dose products with preservatives, the stability program includes product withdrawal to assess preservative efficacy against microbial contamination during actual administration.

Other important focus areas at late stage include: (a) conducting appropriate risk assessments on extractables and leachables and experimental studies of potential leachates that can enter the product by virtue of the product encountering various parts in the manufacturing process; (b) risk assessment on nitrosamine and demonstrating that species of interest are absent or within acceptable levels.

Transport validation studies are crucial for ensuring the integrity of biologic therapeutics during shipment. They begin with simulation tests to assess the effects of shock, vibration, and temperature variations during transportation by road and air. These simulations inform the design of appropriate packaging and shipment strategies to mitigate risks to product quality. Actual transportation studies then evaluate air and sea shipments from manufacturing sites to distribution centers worldwide. Both packaging and product integrity are rigorously assessed to ensure compliance with predefined quality criteria. Some of the stress factors therapeutic proteins may encounter are listed in Table 10.

In summary, DP manufacturing is highly complicated involving several steps, and requires coordinated efforts between different functional teams. With the goal being ensuring a high quality efficacious and safe product to the end user (patients), the development should be grounded in solid and established scientific concepts. It should involve a thorough understanding of the biologic, its properties, structure-function relationship, and stability and ably supported by high-end analytical tools. It is important to consciously acknowledge the fact that the product being manufactured for the patients is the "lead actor" and it is necessary for the manufacturer to ensure that the requirements of the "lead actor" are never compromised. The key here is to ensure that the manufacturer implements practices which are consistent and sustainable.

7   General Concepts in Process Development

7.1   Choice of Raw Materials

Raw materials play a fundamental role in the manufacturing of biologic therapeutics, as they serve as the building blocks to produce the final product. Therefore, their selection is crucial, considering factors such as their source, purity, stability,

Table 10 Stress conditions that a therapeutic protein drug may encounter

Stress conditions	Stages where protein may encounter these conditions
Temperature excursions, fluctuations	Storage, transportation, and manufacturing steps such as purification, formulation, and fill-finish operations
Oxidative stress	Manufacturing processes, e.g., purification, during storage and administration. Exposure to air, light, and metal ions can promote oxidation
Shear stress	Manufacturing processes, e.g., filtration, mixing, and pumping, during transport, administration via injection or infusion
pH	Manufacturing steps like purification, formulation, dilution, freezing, and storage. Dilution in infusion liquid, administration
Light exposure	Manufacturing, storage, transportation, and handling
Chemical interactions	Formulation step, where proteins may interact with buffers, excipients, or surfaces of containers and equipment
Container interactions	Storage and transportation, where proteins may adsorb to the surface of glass or plastic containers, or during administration when using delivery devices such as syringes or infusion pumps

compatibility with the manufacturing process, and the desired product attributes. Raw materials can include cell culture media components, growth factors, buffers, purification resins, excipients, and packaging materials, among others. Each raw material must be thoroughly evaluated for quality, consistency, and potential risks, such as contamination with microorganisms, viruses, or adventitious agents.

In the pharmaceutical industry, regulatory agencies such as the FDA (Food and Drug Administration) in the United States and the EMA (European Medicines Agency) in Europe provide guidelines and requirements for the management of raw materials in biologic therapeutics manufacturing. These regulations emphasize the importance of establishing and implementing robust quality management systems to ensure the safety, efficacy, and consistency of the final product. Some of the key considerations are given in Table 11.

Effective management of raw materials is essential throughout the entire manufacturing process, from cell culture and purification to formulation and packaging. By adhering to stringent quality standards and regulatory requirements, manufacturers can ensure the integrity and consistency of biologic therapeutics, ultimately safeguarding patient health and well-being.

7.2 Scale Down Model (SDM) and Process Characterization (PC)

Scale-down model development involves creating a smaller, representative version of a larger-scale process or system. This smaller model is designed to mimic the key characteristics and behaviors of the full-scale process, allowing researchers to study

Table 11 Some key considerations during raw material (RM) selection and management

Risk management	Conducted to identify potential hazards associated with RMs, such as allergens, toxins, or variability in composition. Mitigation strategies, such as supplier qualification, testing, and monitoring, should be implemented to minimize these risks [58, 59]
Supplier qualification	Thorough evaluation of suppliers to ensure they meet and adhere to GMP. Conducting audits, inspections, and assessments of supplier's facilities, processes, and quality system
Material testing	Comprehensive testing of RMs for identity, purity, potency, and absence of contaminants
Change control	Any changes to RM specifications, suppliers, or sourcing regions should be carefully evaluated and managed through a formal change control process. These changes may require requalification, retesting, or validation activities to ensure they do not impact product quality or regulatory compliance.

and optimize process parameters, test different scenarios, and predict performance without the need for large-scale production.

The development of scale-down models typically involves several steps:

Identification of Critical Parameters: Identify the critical parameters and variables that influence the performance of the full-scale process. Following are some examples for:

> Upstream—cell density, nutrient concentrations, pH levels, dissolved oxygen levels, agitation speed, aeration rates, etc.
> Downstream—resin type, column dimensions, flow rates, buffer compositions, operating conditions, filtration rates, etc.
> Formulation and Drug Product—raw material characteristics, excipient concentrations, mixing conditions, processing temperatures, storage conditions, etc.

Design of Experimental Studies: Based on critical parameters, a small-scale experimental setup is designed to replicate key features of the large scale. This involves selecting suitable small-scale bioreactors for upstream, chromatography columns, and filtration units for downstream, as well as mixers, lyophilizers, and filling setups for formulation and drug product stages. Scaling criteria are determined to ensure the scale-down model accurately reflects the full-scale process behavior.

Validation and Calibration: Once set up, the scale-down model (SDM) undergoes validation by comparing its performance across various conditions with the full-scale process. This includes assessing cell growth kinetics, metabolic profiles, and product quality in upstream processes, and purification yields, impurity removal efficiencies, and product purity in downstream processes. Formulation steps focus on comparing product characteristics and stability profiles. Once validated, the SDM becomes useful for testing and optimization trials.

Process Characterization is a systematic and comprehensive approach to under-standing and defining the behavior of a manufacturing process [60]. It involves identifying and evaluating critical process parameters (CPPs) and critical quality attributes (CQAs) that impact the quality, safety, and efficacy of the final prod-uct. The goal of process characterization is to establish a thorough understanding of the process and its variability, enabling robust control and optimization to ensure consistent product quality. The key steps involved in process characteriza-tion studies are as follows:

Identification of Potential Critical Process Parameters: Based on risk assessments like Failure Mode and Effect Analysis (FMEA) and historical data, potential Critical Quality Attributes (CQAs) are identified. In upstream processing, critical parameters include cell density, viability, nutrient concentrations, pH, dissolved oxygen levels, and agitation rates. Downstream processing may involve chroma-tography conditions (e.g., pH, conductivity), filtration parameters (e.g., pressure, flow rate), and process timing. Formulation and Drug Product stages may include mixing time, temperature, pH, and excipient concentrations as critical parameters.

Design of Experiments (DOE): DOE approach is utilized to perform experiments, using validated scale-down models, to evaluate the effect of various potential critical process parameters on the CQAs. The parameters affecting the CQAs are classified as CPPs.

Establishment of Process Ranges and Limits: Based on the data analysis of the DOE experiments, the acceptable ranges and the limits for each CPP are established to ensure product quality and consistency. These ranges become the Proven Acceptable Ranges (PAR) for various input parameters.

The outcome of the process characterization studies is vital for developing an appropriate control strategy and process validation approach. Scale down models and at Scale experiences are combined to define the design space for various unit operations and establish the normal operating ranges (NOR) within which routine manufacturing operations should occur. The NOR is typically within the PAR for various input parameters ensuring that control strategies are in place to maintain product consistency and meet product quality criteria.

7.3   Process Intensification Strategies

Process intensification in biologic therapeutic process development involves enhancing productivity, efficiency, and cost-effectiveness through various strategies in both upstream and downstream processes. In upstream processes, which involve cell culture and protein expression, intensification aims to maximize cell growth and protein production within a smaller footprint and shorter time frame. This can be achieved through increased cell densities in the production or seed bioreactors [61], higher specific productivity, and improved culture conditions.

One approach to upstream process intensification is the use of high-cell-density cultures, where cells are cultivated at much higher concentrations than traditional batch cultures [62]. This is typically accomplished through fed-batch or perfusion systems, allowing for sustained cell growth and prolonged production phases. Perfusion systems, in particular, enable continuous removal of spent media and addition of fresh nutrients, maintaining optimal conditions for cell growth over extended periods.

In downstream processes, which involve purification and isolation of the target protein, intensification focuses on improving yield, purity, and process efficiency. This can be achieved through the implementation of continuous chromatography, membrane-based separations, and integrated purification platforms [63]. Continuous chromatography systems enable uninterrupted protein purification, reducing downtime and increasing overall productivity. Membrane-based separations offer advantages such as higher throughput, reduced solvent consumption, and improved scalability compared to traditional column chromatography.

The benefits of process intensification in biologic therapeutic process development are numerous. By increasing productivity and efficiency, intensification strategies can lead to higher yields of therapeutic proteins, reduced manufacturing costs, and faster time-to-market for biopharmaceutical products. Additionally, intensification allows for the use of smaller production facilities and equipment, resulting in reduced capital investment and operating expenses [64].

However, process intensification also presents challenges, particularly in terms of process control, scalability, and regulatory compliance. Continuous processes require robust monitoring and control systems to maintain consistent product quality and ensure compliance with regulatory requirements. Scale-up from bench-scale to commercial production can be complex and may require extensive process validation and optimization.

From a regulatory perspective, process intensification in biologic therapeutic process development must adhere to established guidelines and standards set forth by regulatory agencies such as the FDA and EMA. Manufacturers must demonstrate the comparability of intensified processes to traditional batch processes in terms of product quality, safety, and efficacy. Additionally, continuous manufacturing processes may require novel regulatory approaches and acceptance criteria, necessitating close collaboration between industry stakeholders and regulatory authorities.

In summary, process intensification offers significant potential for advancing biologic therapeutic process development by increasing productivity, efficiency, and cost-effectiveness. While challenges exist, overcoming these obstacles through innovation and collaboration will be essential for realizing the full benefits of intensified biopharmaceutical manufacturing.

7.4   Applications of Data Sciences in Process Development

Data science and analytics play a crucial role in process development for biologic products, as they offer valuable insights and drive optimization efforts across various stages of the development lifecycle. Some of the areas where data science can significantly contribute are as follows:

Design of Experiments (DOE): DOE enables systematic experimentation to identify optimal process conditions by analyzing data statistically. This approach accelerates the optimization of biologic processes, enhancing product yield, purity, and performance. In upstream processes, DOE optimizes media and feed strategies, while in drug product development, it streamlines formulation optimization.

Predictive Modeling and Simulation Techniques: They use data-driven approaches to develop mathematical models predicting process behavior and outcomes. These models, including statistical, AI/ML-based, and mechanistic ones, elucidate complex relationships between process parameters and product attributes, aiding in optimization and decision-making. Statistical and AI/ML models leverage historical process data to simulate performance under diverse scenarios, optimizing parameters and predicting outcomes. Mechanistic modeling, particularly in chromatography optimization, enhances understanding and optimization efforts.

Process Monitoring and Control: This involves real-time data capture through sensors for parameters like pH, temperature, dissolved oxygen, and more. Advanced data tools analyze large datasets, removing noise and identifying patterns. Chemometric models correlate spectra with sample attributes, while machine learning classifies samples based on criteria. Statistical control techniques detect deviations, triggering corrective actions for process optimization. Artificial intelligence has recently emerged as a data-driven approach with multiple applications in bioprocess [65].

Troubleshooting and Root Cause Analysis: This leverages data analytics techniques to identify underlying causes of process deviations or product failures. Tools like multivariate analysis and machine learning algorithms analyze historical process data to correlate process parameters with product attributes, aiding in issue identification. Data-driven root cause analysis enhances process understanding and supports continuous improvement initiatives.

Process Scale-Up and Technology Transfer: Data analytics supports process scale-up and technology transfer activities by analyzing data from pilot-scale and commercial-scale manufacturing runs. By leveraging historical data and predictive models, researchers can assess the scalability of processes, identify scale-dependent parameters, and optimize process conditions for successful scale-up.

Quality by Design (QbD) Implementation: Data analytics supports the principles of Quality by Design (QbD) by enabling a systematic and data-driven approach to process development and optimization. Through the analysis of process data and identification of critical process parameters (CPPs) and critical quality attributes (CQAs), QbD methodologies ensure that process variability is minimized, and product quality is consistently maintained.

Data Integration and Management: Biologic process development generates vast amounts of data from various sources such as experiments, sensors, and analytical instruments. Data science techniques facilitate the integration and management of these diverse datasets, allowing for centralized storage, retrieval, and organization of data for analysis.

Statistical software packages like R, SAS, and SPSS are commonly used for data analysis, hypothesis testing, and experimental design, offering a wide array of statistical functions. Dedicated DOE software such as JMP, Design-Expert, and Minitab are tailored for planning and analyzing designed experiments, featuring user-friendly interfaces. Data visualization tools like Tableau, Spotfire, and Power BI enable interactive visualizations for exploring relationships in process data. MVDA software such as SIMCA, Unscrambler, and PLS Toolbox analyze multivariate datasets using methods like PCA and PLS. Process data historians like OSIsoft PI and Wonderware Historian collect and analyze real-time data for monitoring parameters and equipment performance. Machine learning platforms like TensorFlow and scikit-learn are applied for predictive modeling and anomaly detection in biologics process development, utilizing algorithms such as neural networks and random forests. Additionally, custom data analysis scripts developed using Python and MATLAB are integrated into workflows for advanced data processing. Leveraging data science techniques accelerates process development, enhances product quality, and enables data-driven decision-making throughout the product lifecycle.

8 Analytical Methods in Biologics Development

Drug development is a complex process involving research, discovery, testing, and regulatory approval. Recent advancements, including high-throughput screening and computational modeling, have expedited this process. Analytical methods are integral, serving as the "five senses" guiding drug development by monitoring Critical Quality Attributes (CQAs) such as molecular structure, impurities, and potency. These methods ensure regulatory compliance and product quality at every stage, from discovery to manufacturing to clinical trials. Apart from the methods which are employed to evaluate CQAs, multiple orthogonal techniques are used as part of extensive characterization of the biologic drug during different stages of development (Fig. 5).

For a novel drug, the clinical efficacy of the biologics drug is mainly established through clinical trials, while for a biosimilar drug, rigorous analytical characterization is crucial due to reduced clinical sample sizes (Fig. 6). During the process development, constant feedback between analytical and process scientists is required for optimal results.

While there is a slight difference in the two approaches, it does not reduce the rigor of characterization of biologic drugs. These drugs are complex due to their large molecular shape and size, hence require evaluation using multiple orthogonal methods as per the need of the development stage. Table 12 captures different methods used for the assessment of biologic drug products across different stages of product development (Fig. 7).

Analytical methods used to evaluate the quality of a particular drug attribute(s) should be validated and/or qualified prior to the intended purpose. Multiple

Fig. 5 Analytical sciences
cater to every stage of the
drug development process

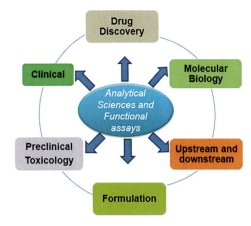

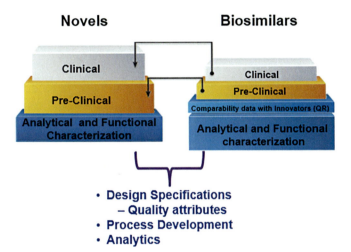

Fig. 6 Pictorial representation of the different approaches used during different stages of novel
versus a biosimilar product

compendial guidelines are available such as USP <1225>, ICH Q2(R2), and ICH
Q14, which outline various aspects and requirements for different categories of the
analytical methods. Table 13 captures requirements for method validation based on
its category.

During the development and validation of these methods, system suitability testing (SST) parameters are established for each method type to ensure adequate control over routine usage. These parameters address minor day-to-day variations in
instruments, electronics, analytical operations, and sample preparations, enhancing
the method's ruggedness. They align with regulatory requirements outlined in phar-
macopeial standards and guidelines. Balancing regulatory compliance with the need
for efficiency and innovation poses a key challenge in drug development.

Table 12 Different methods used for assessment of biologics drug product across stages of the product development

Stage of process development	Information obtained on	Attributes	General description
Early discovery and late stage (drug substance and drug product)	Primary structure	Intact mass and reduced mass	Mass spectrometry (MS): Identifies and quantifies molecules based on their mass-to-charge ratio, providing information on molecular weight and structure
		Reduced and non-reduced peptide mass fingerprinting	
		Multi-attribute Monitoring (MAM)	
		Edman sequencing	
		Extinction coefficient	
Early development and then late-stage lot release (drug substance or drug product)	Glycoforms	Glycan profiling	Assessing glycoforms is crucial to understand how different cell lines used in upstream processes affect drug candidates' post-translational effects. These glycoform levels can significantly impact the drug's safety, efficacy, and immunogenicity based on its mechanism of action.
		Exoglycosidase array	
		Monosaccharide	
		Afucosylation	
		Sialic acid	
Early to late stage (all stages of development and stability)	Purity	Protein content by affinity chromatography	Separation techniques such as chromatography (such as HPLC and GC) and charge electrophoresis (CE): Separate and quantify components of a mixture based on their size as well as interactions with a stationary phase and a mobile phase
		Size exclusion—SE	
		Charge variant by IEX	
		Reverse phase—RP	
		Hydrophobic interaction—HIC	
		CE-SDS non-reduced	
		CE-SDS reduced	
		PTMs (glycation, oxidation, deamidation)	
		Sequence variants	
		Free cysteine estimation	
		pI variant by cIEF	

Late stage (drug substance and drug product)	Higher order structure	Far and near UV CD spectroscopy	Spectroscopic techniques are used for identity as well as characterization of the molecule's higher-order structure. Multiple techniques are used as these techniques provide orthogonal information and are needed to ensure product's desired three-dimensional structure to ensure safety and efficacy
		Fourier transform infrared spectroscopy	
		Intrinsic and extrinsic fluorescence	
		Differential scanning calorimetry	
		X-ray crystallography	
		NMR spectroscopy	
		Multi-angle light scattering	
		Dynamic light scattering	
		Native-MS	
		HDX-MS	
		IM-MS	
		SAXS	
Early to late stage (all stages of development and stability)	Biological function	Binding to receptor or ligand	To evaluate the functional capability, based on drug candidates' mechanism of action or disease area. These will be utilized as functional testing for lot release assays and monitoring the stability of batches too
		Cell-based functional assay	
Late stage (drug substance and drug product)		Fc effector binding (FcγRIa, FcγRIIa, FcγRIIb, FcγRIIIa, FcγRIIIb, FcRn, C1q)	As a part of product characterization and mechanism of action, additional testing is essential to be performed
		Fc effector function (ADCC, CDC, ADCP)	
Late stage (drug product)	Product attributes	Protein content	Determines the size distribution of particles in a sample, which is important for drug formulation and delivery
		pH	
		Excipients	
		Visible and sub-visible particulates	

Reference: [66]

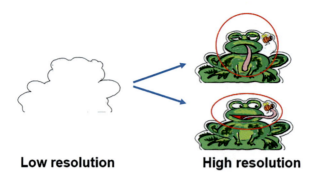

Low resolution   **High resolution**

Fig. 7 Orthogonal techniques play a key role in evaluating structure and function. While low-resolution techniques ensure the overall shape of the molecule, high-resolution techniques are needed to ensure local details. The cartoon depicts how a combination of the two orthogonal tools (low resolution and high resolution) ensures the correct configuration of the tongue to catch the fly (biological function)

Table 13 Description of method validation parameters

Parameter	Identity	Qualitative	Assay	Quantitative	Description of the attribute
Specificity	Yes	Yes	Yes	Yes	No interference due to sample matrix or any closely related compound
Precision	No	Yes	Yes	Yes	Precision refers to the consistency and reproducibility of results obtained from repeated measurements. It includes system and method precision, which assesses repeatability under the same conditions, intermediate precision, which evaluates repeatability across different conditions or instruments, and reproducibility, which measures consistency across different laboratories
Accuracy	No	No	Yes	Yes	Accuracy for the method describes the closeness of the results obtained from the method to its true value
Linearity	No	No	Yes	Yes	Linearity of the method establishes proportionality of the results to the concentration of the analyte
Range	No	No	Yes	Yes	Range for the method indicates the method performance
Robustness	Yes	Yes	Yes	Yes	Method robustness is performed to assess method's performance under normal usage conditions by deliberately altering the method parameters
Detection limit	No	No	No	Yes	Limit at which the related compounds or impurity can be detected above the noise level
Quantitation limit	No	No	Yes	Yes	Limit at which the related compounds or impurity can be quantitated accurately and precisely

Overcoming these challenges necessitates a multidisciplinary approach, fostering collaboration among analytical chemists, pharmacologists, regulatory experts, and other stakeholders. Continuous optimization and refinement of analytical methods throughout the drug development process are crucial for ensuring the success of novel drug candidates (Table 14).

9 Technology Transfer and Scale-Up

Technology transfer is crucial for scaling up biologic products, involving the transfer of knowledge, processes, and methodologies between different scales or manufacturing sites. Thorough planning is essential, encompassing project objectives, timelines, and resource needs. Key stakeholders are identified, cross-functional teams formed, and risk assessments conducted to address potential challenges.

Some of the key strategic decisions to be taken before technology transfer are as follows:

1. Scale of manufacturing based on the short- and long-term projections of the drug
2. Manufacture internally or outsourced and the investment plan
3. Target markets and regulatory requirements for those market
4. Target cost of goods for the product

Some of the key considerations for a technology transfer are:

Robust process understanding and characterization: Understanding the process intricately, including critical parameters, controls, and variability, is key for seamless technology transfer. Process characterization studies, using validated small-scale models, identify and quantify variabilities to establish robustness. Additionally, conducting due diligence to evaluate intellectual property rights, licensing agreements, and potential liabilities is imperative before scaling up to commercial production.

Facility fit assessment: It evaluates the compatibility of an existing manufacturing facility with the requirements of a new process. It includes assessing equipment size, capacity, functionality, and compatibility with process parameters. This assessment identifies any needed facility modifications or regulatory compliance issues.

Training and knowledge transfer: Activities are essential to equip personnel at the receiving organization with the necessary skills and understanding to execute the transferred process effectively. Adequate training ensures that personnel are familiar with the standard operating procedures (SOPs), protocols, and methodologies involved, reducing the risk of errors, deviations, or non-compliance issues.

Risk assessment and mitigation: These are crucial for successful technology transfer. Thorough assessment identifies potential challenges, enabling the development of strategies to ensure project success. Supply chain risks, such as dependence on external suppliers for critical materials, can impact reliability.

Table 14 Typical analytical methods used during different stages of biologic manufacturing

Category	Testing	Specifications/limits	In-process	DS Release	DP Stability	DS/DP Release	Stability	Extended characterization
Product characteristics	Color	Compendial		X	X	X	X	
	Clarity	Compendial		X	X	X	X	
	pH	Compendial		X		X		
	Osmolality	Compendial		X		X		
	Particulate matter (SVP)	Compendial		X	X	X	X	
	Visual appearance (visible particles)	Compendial		X	X	X	X	
Identity	Peptide mapping—Sequence confirmation	In-house		X		X		
	Identity by SDS-PAGE (R and NR)	In-house		X		X		
Product heterogeneity	Purity by SEC	In-house	X	X	X	X	X	X
	Purity by CE (NR)	In-house	X	X	X	X	X	X
	CE-SDS (R)	In-house	X	X	X	X	X	X
	cIEF	In-house	X	X	X	X	X	X
	Desialylated cIEF	In-house	X	X	X	X	X	X
	PTMs—Glycosylation variants	In-house	X	X	X			X
	PTMs—Product variants (oxidation, deamidation, charge variants)	In-house	X	X	X	X	X	X
	IEX	In-house	X	X	X	X	X	X
Potency	Protein concentration	In-house	X	X	X			X
	Relative potency—In vitro bioassay	In-house	X	X	X		X	X

(continued)

Drug delivery and device	Delivered volume	Compendial				X	X	X
	Extractable volume	Compendial			X	X	X	X
	Container closure integrity test (CCIT)	Compendial			X	X	X	X
Safety	Break loose and glide force	Compendial			X	X	X	X
	Bacterial endotoxin	Compendial		X	X	X	X	X
	Microbial contaminants	Compendial		X	X	X	X	X
	Sterility testing	Compendial		X	X	X		X
Process-related impurities	Host cell protein	Compendial	X		X	X		
	Host cell DNA	Compendial	X		X	X		
Excipient	Excipients	In-house	X			X	X	X
Structural characterization—physicochemical	Intact and reduced mass—LC-MS	In-house						X
	Multi-attribute monitoring (MAM)—LC-MS	In-house						X
	Far, near UV—CD	Qualitative						X
	FTIR	Qualitative						X
	Intrinsic fluorescence (IF)	Qualitative						X
	DSC	Qualitative	X					X
	SEC-MALS	Qualitative						X
	DLS	Qualitative	X					X
	AUC	Qualitative						X
	HDX-MS	Qualitative						X
	IM-MS	Qualitative						X
	NMR	Qualitative						X

(continued)

Table. 14 (continued)

| Category | Testing | Specifications/limits | DS | | DP | DS/DP | | Extended |
			In-process	Release	Stability	Release	Stability	characterization
Functional characterization—	Effector functions	In-house						X
In vitro bioassays	Fc receptor function assays	In-house						X
	FcγRIa	In-house						X
	FcγRIIa	In-house						X
	FcγRIIb	In-house						X
	FcγRIIIa	In-house						X
	FcγRIIIb	In-house						X
	FcRn	In-house						X
	C1q	In-house						X
	ADCC	In-house						X
	CDC	In-house						X

SEC Size Exclusion Chromatography, *IEX* Ion Exchange Chromatography, *CE* Capillary Electrophoresis, *PTM* Post-Translational Modifications, *CD* Circular Dichroism, *NMR* Nuclear Magnetic Resonance, *DLS* Dynamic Light Scattering, *FTIR* Fourier Transform Infrared Spectroscopy, *DSC* Differential Scanning Calorimetry, *AUC* Analytical Ultracentrifugation, *HDX-MS* Hydrogen deuterium Exchange, *cIEF* Capillary Isoelectric Focusing, *MALS* Multi-Angle Light Scattering, *R* Reduced, *NR* Non-Reduced, *SVP* Sub-Visible Particulate, *IM-MS* Ion Mobility Mass Spectrometry

Derisking strategies like engaging alternate vendors or long-term agreements address these risks, ensuring continuity.

Technology transfer involves detailed documentation, including process descriptions, SOPs, batch records, and specifications, ensuring consistency and clarity. Collaboration among R&D, process development, manufacturing, QA/QC, regulatory affairs, engineering, and project management is essential. Validation and qualification studies follow initial batches, demonstrating process robustness and compliance.

9.1 Scale-Up Strategies for Upstream

Scale-up strategies for upstream processing involve transferring a laboratory-scale process to larger production scales while maintaining product quality, yield, and process robustness. During lab-scale development, optimal cell culture environment conditions are derived from experiments that result in desired process outcomes like product titer and quality. The approach for scale-up remains to replicate, at large scale, these environmental conditions as close as possible to the lab scale. This will result in getting a similar product outcome as targeted during development (Fig. 8).

Maintaining geometric similarity across scales is crucial for consistent performance. It involves preserving similar geometrical aspects like aspect ratio and

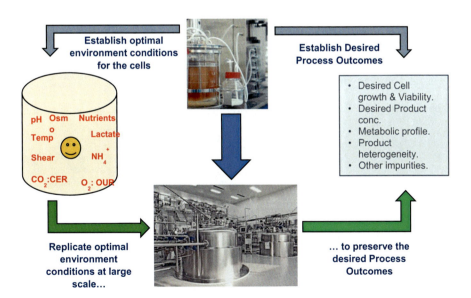

Fig. 8 Once the cell culture conditions are optimized at a small scale to get the desired process outcomes in terms of titer and product quality, the effort during scale-up is to replicate similar conditions at a large scale to preserve the desired process outcomes

impeller dimensions between laboratory and large-scale bioreactors. This approach ensures comparable fluid dynamics and mass transfer characteristics. In cases where geometric similarity isn't feasible, scale-down experiments become essential to simulate mixing and mass transfer from larger scales.

In upstream processes, parameters can be grouped into two categories: scale-independent and scale-dependent. Scale-independent parameters, like pH, temperature, dissolved oxygen levels, and nutrient concentrations, stay consistent across scales for uniform cell growth and product quality. Robust control systems are crucial for real-time monitoring and adjustment of these parameters. Scale-dependent parameters, such as bioreactor volume and agitation rates, vary with scale and require careful control for optimal process performance.

As the bioreactor scale increases, adjustments are made to maintain similar hydrodynamic conditions, such as mixing efficiency, gas-liquid mass transfer, and heat dissipation, as those observed at smaller scales. Agitation and aeration rates are adjusted to achieve similar mixing and gas-liquid mass transfer characteristics at larger scales. KLa (volumetric mass transfer coefficient), P/V (power per unit volume), and impeller tip speed are important parameters used to characterize and optimize gas-liquid mass transfer, particularly oxygen transfer, in bioreactor systems. These parameters are often used during scale-up calculations as explained in Table 15.

Computational fluid dynamics (CFD) is a useful tool to get insights into fluid flow behavior, mixing dynamics, and mass transfer in bioreactor. CFD models aid in optimizing bioreactor designs, operating conditions, and scale-up strategies. Engineers can predict crucial parameters like shear stress, oxygen concentration, and temperature gradients, facilitating informed decision-making.

For a scenario where development has been done using glass or steel bioreactors and large-scale uses disposable bioreactors, ensure that the materials used in the disposable bioreactor are compatible with the cell culture system like culture media, feeds, and process conditions.

Table 15 Parameters like Kla, P/V, and tip speed are often used for scale-up calculations in upstream

Kla (volumetric mass transfer coefficient)	Measures gas-liquid mass transfer efficiency in bioreactors. Quantifies oxygen transfer from gas to liquid per unit liquid volume and is usually determined experimentally. It is crucial for aerobic cell metabolism. Engineers try to maintain similar Kla levels during scale-up by adjusting aeration and agitation
P//V (power per unit volume)	Illustrates the energy needed for agitation and mixing in bioreactors. It is determined by directly measuring agitator motor power or indirectly via torque sensors/electrical current. Effective scale-up involves adjusting agitation speeds and impeller designs (if permissible) to ensure efficient mixing while avoiding cell damage due to stress
Tip speed	Is the linear velocity at the outer edge of impeller blades. It directly influences energy dissipation rate, turbulence, gas-liquid mass transfer, and oxygen transfer rates. For shear-sensitive cultures or high-viscosity cultures, precise control of tip speed is vital to prevent cell damage and ensure proper mixing and oxygenation

Scaling up upstream processes presents various challenges that must be addressed:

- Ensuring uniform mixing, nutrient distribution, and gas-liquid mass transfer is more difficult in larger bioreactors due to the formation of stratified zones and gradients.
- Oxygen transfer limitations arise in larger-scale bioreactors, leading to oxygen gradients and limitation, particularly problematic for fast-growing microbial cultures like bacteria and yeast.
- Managing heat dissipation and temperature control becomes more complex with increased metabolic heat generation and reduced surface area-to-volume ratios at larger scales.
- Maintaining sterility and preventing contamination becomes increasingly critical with larger bioreactor volumes.

9.2 Scale-Up Strategies for Downstream

Once the scale of operations for the upstream is finalized, the downstream needs to be scaled up to match the inputs from the upstream bioreactors. The most widely used unit operations in downstream are liquid/solid handling, chromatography, and filtration.

Liquid/Solid Handling

The liquid/solid handling covers a wide range of operations including centrifugation (removing biomass, crystals, etc.), buffer and cleaning solutions preparations, in-process holds steps, and storage.

For liquid handling, determining vessel size at a large scale typically involves considering the minimum concentration of the product in each unit operation. Additional volume should be allocated for steps requiring mixing, pH adjustments, or the addition of other solutions. Temperature control requirements are also considered. Scale-down experiments help ensure homogeneity in mixing steps while minimizing product stress. Compatibility assessments between the product and bag film should be conducted for disposable bags, considering scale differences and duration.

Centrifugation, especially using disc stack centrifuges, is prevalent in industrial processes for solid handling, such as biomass from cell culture broth. Its efficiency hinges on factors like solids volume fraction, effective clarifying surface (V/D), and acceleration factor ($\omega^2 r/g$). Scale-up relies on the Sigma Factor (Σ), representing the centrifuge's equivalent area, which varies with type and speed. Challenges include submicron particle generation from cells and debris due to centrifugal shear, potentially affecting product stability.

Due to the challenges and high capital costs associated with centrifugation, depth filtration is gaining popularity, especially for lower biomass culture harvests. Scaling up depth filtration often involves maintaining a constant filtration flux.

However, this approach can lead to a linear increase in required filtration area with increasing process volume, potentially becoming impractical at larger scales. In such cases, alternatives like centrifugation or specialized filters, such as those with charged membranes, may be necessary.

Membrane Filtration

Membrane filtration is crucial in biopharmaceutical manufacturing, serving roles like depth filtration, ultrafiltration, and sterilization. Scaling up involves not just adjusting filter area but also considering format and operation mode changes. Due to scale discrepancies, "safety factors" are used to estimate filter area needs at larger scales, ensuring process effectiveness across different scales.

Normal flow or dead-end filtration is typically used for bioburden reduction, depth filtration, and virus filtration. This can be either flux-limited or capacity-limited.

Flux limitation arises when the fluid flow rate through a filter membrane surpasses the maximum sustainable level for the system. It's influenced by membrane characteristics (like pore size), applied pressure, and solution properties. Exceeding this limit can result in issues such as membrane fouling, reduced permeability, or membrane damage.

Capacity limitation arises when a filter membrane becomes saturated with retained particles or solutes, which become barrier that hinders further filtration, reaching its maximum holding capacity. This is influenced by factors like particle concentration, membrane surface area, and filtration duration. Strategies such as membrane cleaning, pre-filtration, or using higher-capacity membranes can address this limitation and extend the filtration system's lifespan.

Irrespective of the method used to determine filter capacity, scale-up is achieved by assuming a safety factor (typically $>=1.5$) to accommodate feed and membrane variability.

In tangential flow filtration (TFF), the product stream's sweeping action mitigates concentration polarization effects, such as solute build-up on the membrane surface and osmotic pressure effects. Like dead-end filtration, the size of a TFF system depends on filter capacity, defined as the volume of feed processed per unit membrane area. Depending on whether the filtration operates at constant flux or constant pressure, this volume corresponds to reaching a maximum pressure drop or an unacceptable permeate flow rate.

Scale-up of a TFF step can be generally considered simple because membrane cartridges (cassettes or hollow fibers) are linearly scalable. Overall filtration time should be a main consideration for the UF/DF step. One of the important aspects of UF/DF or microfiltration systems should be the minimization of yield losses and reduction of dead volume, which reduces wash volumes.

Chromatography

Sizing of a chromatography column can be performed based on both chromatographic and non-chromatographic factors, as well as by considering all the facility and product quality-related constraints. Typically, the column size chosen for chromatography is based on the following set of equations:

$$CV = \frac{Mass_{batch}}{Load_{cycle} N_{cycles}}$$

$$N_{cycles} = \frac{Time_{Purif}}{Time_{cycle}}$$

where CV is the column/resin volume [L], $Load_{cycle}$ is the load per cycle [kg/L_{resin}], including a safety factor to account for feed and resin variability, $Mass_{batch}$ is the mass delivered from a bioreactor [kg], N_{cycles} is the number of cycles per batch, $Time_{Purif}$ is the total allocated purification time for a single batch [h], and $Time_{cycle}$ is the time for a single chromatography cycle [h].

The equation derives the packed resin volume (CV) needed to purify a specified mass within a given time frame. CV serves as a normalization factor for different solution volumes in chromatography steps. Chromatography methods using column volumes (CV) offer scale-invariant applicability across different scales, maintaining parameters like bed height and sample concentration relative to resin volume.

In constant bed height scale-up, increasing process volume involves enlarging column diameter. Vigilant monitoring of parameters such as flow rate, pressure, and pH is essential. Conversely, constant residence time scale-up adjusts bed height and column diameter to maintain steady productivity, particularly effective for bind-and-elute chromatography modes.

For a polishing step based on flow-through mode where the remaining impurities are adsorbed on the column and the antibody flows through, the linear scale-up criterion can be recommended.

When scaling up a chromatography step, not only the column but also a chromatography system needs to be considered. Typically, the chromatography systems are chosen on the flow requirement and pressure specifications to match desired production rates.

9.3 Scale-Up Strategies for Drug Product

Scaling up the production of drug products for biologic therapeutics involves several distinct considerations tailored to the complexities of biologics. The formulation process must be adapted to accommodate larger batch sizes while maintaining the stability, efficacy, and safety of the product. One critical aspect of scaling up drug product manufacturing for biologics is ensuring the compatibility of the formulation with larger-scale manufacturing equipment and processes. This may involve conducting compatibility studies to assess the impact of increased mixing times, shear forces, and temperature fluctuations on product stability and integrity.

Issues such as protein aggregation, degradation, or particle formation can arise from prolonged mixing times or increased shear forces in larger-scale equipment.

Additionally, the scalability of filling processes, including vial or syringe filling, lyophilization (freeze-drying), and packaging, must be carefully evaluated and optimized. This may involve conducting feasibility studies and pilot-scale trials to assess the performance of equipment and processes under scaled-up conditions. Process parameters such as filling speed, fill volume accuracy, lyophilization cycle times, and container closure integrity must be validated to ensure compliance with regulatory requirements and product specifications.

10 Conclusion

In conclusion, effective process development for biologics therapeutics relies on a comprehensive approach that blends scientific innovation with regulatory rigor and operational efficiency. This chapter reviewed the foundational steps, starting with cell line development, which is pivotal to establishing productive and stable biologics manufacturing. Key elements of upstream and downstream processing were explored, focusing on maximizing yields and ensuring high purity levels, while drug product formulation secures the stability and efficacy needed for clinical success. Analytical methods and process characterization are essential components, meeting rigorous quality requirements and regulatory standards such as ICH guidelines. Quality by Design (QbD) was emphasized as a proactive framework that embeds quality into every process step, minimizing risk and improving consistency. Addressing scale-up and technology transfer challenges also requires strategic planning and a data-driven approach to maintain product performance across scales. Additionally, the application of data science and process intensification strategies offers promising avenues to boost efficiency and reduce production costs. This integrated approach to process development ultimately supports the reliable and scalable production of biologics, bringing advanced therapies to patients and meeting the evolving needs of the healthcare industry.

References

1. Kelley B. Industrialization of mAb production technology: the bioprocessing industry at a crossroads. MAbs. 2009;1:443–52.
2. Pramod K, Tahir MA, Charoo NA, Ansari SH, Ali J. Pharmaceutical product development: a quality by design approach. Int J Pharm Investig. 2016;6:129–38.
3. Rathore AS, Winkle H. Quality by design for biopharmaceuticals. Nat Biotechnol. 2009;27:26–34.
4. Meitz A, Sagmeister P, Langemann T, Herwig C. An integrated downstream process development strategy along QbD principles. Bioengineering. 2014;1:213–30.

5. Riggs AD. Making, cloning, and the expression of human insulin genes in bacteria: the path to humulin. Endocr Rev. 2021;42:374–80.

6. Studier FW, Moffatt BA. Use of bacteriophage T7 RNA polymerase to direct selective high-level expression of cloned genes. J Mol Biol. 1986;189:113–30.

7. Tripathi NK, Shrivastava A. Recent developments in bioprocessing of recombinant proteins: expression hosts and process development. Front Bioeng Biotechnol. 2019;7:420.

8. Grav LM, la Cour Karottki KJ, Lee JS, Kildegaard HF. Application of CRISPR/Cas9 genome editing to improve recombinant protein production in CHO cells. Methods Mol Biol. 2017;1603:101–18.

9. Becker J, Hackl M, Rupp O, Jakobi T, Schneider J, Szczepanowski R, Bekel T, Borth N, Goesmann A, Grillari J, Kaltschmidt C, Noll T, Pühler A, Tauch A, Brinkrolf K. Unraveling the Chinese hamster ovary cell line transcriptome by next-generation sequencing. J Biotechnol. 2011;156:227–35.

10. Wurm FM. Production of recombinant protein therapeutics in cultivated mammalian cells. Nat Biotechnol. 2004;22:1393–8.

11. Yao T, Asayama Y. Animal-cell culture media: history, characteristics, and current issues. Reprod Med Biol. 2017;16:99–117.

12. Sharp PM, Li W-H. The codon adaptation index-a measure of directional synonymous codon usage bias, and its potential applications. Nucleic Acids Res. 1987;15:1281–95.

13. Eibl R, Kaiser S, Lombriser R, Eibl D. Disposable bioreactors: the current state-of-the-art and recommended applications in biotechnology. Appl Microbiol Biotechnol. 2010;86:41–9.

14. Mieczkowski CA. The evolution of commercial antibody formulations. J Pharm Sci. 2023;112:1801–10.

15. Desai M, Kundu A, Hageman M, Lou H, Boisvert D. Monoclonal antibody and protein therapeutic formulations for subcutaneous delivery: high-concentration, low-volume vs. low-concentration, high-volume. MAbs. 2023;15:2285277.

16. Fisher AC, Kamga MH, Agarabi C, Brorson K, Lee SL, Yoon S. The current scientific and regulatory landscape in advancing integrated continuous biopharmaceutical manufacturing. Trends Biotechnol. 2019;37:253–67.

17. Chen Y, Yang O, Sampat C, Bhalode P, Ramachandran R, Ierapetritou M. Digital twins in pharmaceutical and biopharmaceutical manufacturing: a literature review. Processes. 2020;8:1088.

18. Sodoyer R. Expression systems for the production of recombinant pharmaceuticals. BioDrugs. 2004;18:51–62.

19. Schütz A, Bernhard F, Berrow N, Buyel JF, Ferreira-da-Silva F, Haustraete J, van den Heuvel J, Hoffmann JE, de Marco A, Peleg Y, Suppmann S, Unger T, Vanhoucke M, Witt S, Remans K. A concise guide to choosing suitable gene expression systems for recombinant protein production. STAR Protoc. 2023;4:102572.

20. Yin J, Li G, Ren X, Herrler G. Select what you need: a comparative evaluation of the advantages and limitations of frequently used expression systems for foreign genes. J Biotechnol. 2007;127:335–47.

21. Terpe K. Overview of bacterial expression systems for heterologous protein production: from molecular and biochemical fundamentals to commercial systems. Appl Microbiol Biotechnol. 2006;72:211–22.

22. Choi JH, Keum KC, Lee SY. Production of recombinant proteins by high cell density culture of Escherichia coli. Chem Eng Sci. 2006;61:876–85.

23. Hartley DL, Kane JF. Properties of inclusion bodies from recombinant Escherichia coli. Biochem Soc Trans. 1988;16:101–2.

24. Chang JY-H, Pai R-C, Bennett WF, Bochner BR. Periplasmic secretion of human growth hormone by Escherichia coli. Portland Press Ltd.; 1989.

25. Govindappa N, Hanumanthappa M, Venkatarangaiah K, Periyasamy S, Sreenivas S, Soni R, Sastry K. A new signal sequence for recombinant protein secretion in Pichia pastoris. J Microbiol Biotechnol. 2014;24:337–45.

26. Salomon D, Orth K. What pathogens have taught us about posttranslational modifications. Cell Host Microbe. 2013;14:269–79.
27. Carabetta VJ, Hardouin J. Bacterial post-translational modifications. Front Microbiol. 2022;13:874602.
28. Hou J, Tyo KE, Liu Z, Petranovic D, Nielsen J. Metabolic engineering of recombinant protein secretion by Saccharomyces cerevisiae. FEMS Yeast Res. 2012;12:491–510.
29. Kost TA, Condreay JP, Jarvis DL. Baculovirus as versatile vectors for protein expression in insect and mammalian cells. Nat Biotechnol. 2005;23:567–75.
30. Kuo CC, Chiang AW, Shamie I, Samoudi M, Gutierrez JM, Lewis NE. The emerging role of systems biology for engineering protein production in CHO cells. Curr Opin Biotechnol. 2018;51:64–9.
31. Cho SW, Kim S, Kim JM, Kim JS. Targeted genome engineering in human cells with the Cas9 RNA-guided endonuclease. Nat Biotechnol. 2013;31:230–2.
32. Cong L, Ran FA, Cox D, Lin S, Barretto R, Habib N, Hsu PD, Wu X, Jiang W, Marraffini LA, Zhang F. Multiplex genome engineering using CRISPR/Cas systems. Science. 2013;339:819–23.
33. Baycin-Hizal D, Tabb DL, Chaerkady R, Chen L, Lewis NE, Nagarajan H, Sarkaria V, Kumar A, Wolozny D, Colao J, Jacobson E, Tian Y, O'Meally RN, Krag SS, Cole RN, Palsson BO, Zhang H, Betenbaugh M. Proteomic analysis of Chinese hamster ovary cells. J Proteome Res. 2012;11:5265–76.
34. Schmieder V, Fieder J, Drerup R, Gutierrez EA, Guelch C, Stolzenberger J, Stumbaum M, Mueller VS, Higel F, Bergbauer M. Towards maximum acceleration of monoclonal antibody development: leveraging transposase-mediated cell line generation to enable GMP manufacturing within 3 months using a stable pool. J Biotechnol. 2022;349:53–64.
35. Choi BK, Bobrowicz P, Davidson RC, Hamilton SR, Kung DH, Li H, Miele RG, Nett JH, Wildt S, Gerngross TU. Use of combinatorial genetic libraries to humanize N-linked glycosylation in the yeast Pichia pastoris. Proc Natl Acad Sci USA. 2003;100:5022–7.
36. McNeil B, Harvey L. Practical fermentation technology. Wiley; 2008.
37. Zhang J, Greasham R. Chemically defined media for commercial fermentations. Appl Microbiol Biotechnol. 1999;51:407–21.
38. Singh V, Haque S, Niwas R, Srivastava A, Pasupuleti M, Tripathi CK. Strategies for fermentation medium optimization: an in-depth review. Front Microbiol. 2016;7:2087.
39. Mandenius CF, Brundin A. Bioprocess optimization using design-of-experiments methodology. Biotechnol Prog. 2008;24:1191–203.
40. Möller J, Kuchemüller KB, Steinmetz T, Koopmann KS, Pörtner R. Model-assisted design of experiments as a concept for knowledge-based bioprocess development. Bioprocess Biosyst Eng. 2019;42:867–82.
41. Singh V. Disposable bioreactor for cell culture using wave-induced agitation. Cytotechnology. 1999;30:149–58.
42. Claßen J, Aupert F, Reardon KF, Solle D, Scheper T. Spectroscopic sensors for in-line bioprocess monitoring in research and pharmaceutical industrial application. Anal Bioanal Chem. 2017;409:651–66.
43. Maruthamuthu MK, Rudge SR, Ardekani AM, Ladisch MR, Verma MS. Process analytical technologies and data analytics for the manufacture of monoclonal antibodies. Trends Biotechnol. 2020;38:1169–86.
44. Legmann R, Schreyer HB, Combs RG, McCormick EL, Russo AP, Rodgers ST. A predictive high-throughput scale-down model of monoclonal antibody production in CHO cells. Biotechnol Bioeng. 2009;104:1107–20.
45. Tang S, Tao J, Li Y. Challenges and solutions for the downstream purification of therapeutic proteins. Antib Ther. 2024;7:1–12.
46. DiPaolo B, Pennetti A, Nugent L, Venkat K. Monitoring impurities in biopharmaceuticals produced by recombinant technology. Pharm Sci Technol Today. 1999;2:70–82.
47. Clarke K. Downstream processing; 2013. p. 209–34.

48. Baumann P, Hubbuch J. Downstream process development strategies for effective bioprocesses: trends, progress, and combinatorial approaches. Eng Life Sci. 2017;17:1142–58.
49. Shukla AA, Hubbard B, Tressel T, Guhan S, Low D. Downstream processing of monoclonal antibodies—application of platform approaches. J Chromatogr B Analyt Technol Biomed Life Sci. 2007;848:28–39.
50. Najafpour GD. Chapter 7 – Downstream processing. In: Najafpour GD, editor. Biochemical engineering and biotechnology. Amsterdam: Elsevier; 2007. p. 170–98.
51. M.C. College, Downstream processing, 2016.
52. Chan C. Critical quality attributes assessment and testing strategy for biotherapeutics development. Am Pharm Rev. 2019;22:29–33.
53. T. Stangler, and S. Biopharmaceuticals, What to control? CQAs and CPPs, 2011 BWP workshop on setting specifications, London, UK. 2011.
54. Matte A. Recent advances and future directions in downstream processing of therapeutic antibodies. LID – https://doi.org/10.3390/ijms23158663 [doi] LID – 8663.
55. Cramer SM, Holstein MA. Downstream bioprocessing: recent advances and future promise. Curr Opin Chem Eng. 2011;1:27–37.
56. Rathore AS, Zydney AL, Anupa A, Nikita S, Gangwar N. Enablers of continuous processing of biotherapeutic products. Trends Biotechnol. 2022;40:804–15.
57. Goswami S, Wang W, Arakawa T, Ohtake S. Developments and challenges for mAb-based therapeutics. Antibodies. 2013;2:452–500.
58. Low D, Rathore A. Managing raw materials in the QbD paradigm, part 1: understanding risks. BioPharm Int. 2010;23
59. Low D, Rathore A. Managing raw materials in the QbD paradigm, part 2: risk assessment and communication. BioPharm Int. 2010;23
60. Hakemeyer C, McKnight N, John RS, Meier S, Trexler-Schmidt M, Kelley B, Zettl F, Puskeiler R, Kleinjans A, Lim F. Process characterization and design space definition. Biologicals. 2016;44:306–18.
61. Hecht V, Duvar S, Ziehr H, Burg J, Jockwer A. Efficiency improvement of an antibody production process by increasing the inoculum density. Biotechnol Prog. 2014;30:607–15.
62. Yang WC, Lu J, Kwiatkowski C, Yuan H, Kshirsagar R, Ryll T, Huang YM. Perfusion seed cultures improve biopharmaceutical fed-batch production capacity and product quality. Biotechnol Prog. 2014;30:616–25.
63. Somasundaram B, Pleitt K, Shave E, Baker K, Lua LH. Progression of continuous downstream processing of monoclonal antibodies: current trends and challenges. Biotechnol Bioeng. 2018;115:2893–907.
64. Fisher AC, Kamga M-H, Agarabi C, Brorson K, Lee SL, Yoon S. The current scientific and regulatory landscape in advancing integrated continuous biopharmaceutical manufacturing. Trends Biotechnol. 2019;37:253–67.
65. Cheng Y, Bi X, Xu Y, Liu Y, Li J, Du G, Lv X, Liu L. Artificial intelligence technologies in bioprocess: opportunities and challenges. Bioresour Technol. 2023;369:128451.
66. Berkowitz SA, Engen JR, Mazzeo JR, Jones GB. Analytical tools for characterizing biopharmaceuticals and the implications for biosimilars. Nat Rev Drug Discov. 2012;11:527–40.

Manufacturing of Biologic Therapeutics

Dhananjay Patankar

Abstract Biologic manufacturing ensures a consistent supply of high-quality drugs throughout their lifecycle. The production process includes drug substance manufacturing, divided into upstream and downstream stages, and drug product manufacturing, which covers formulation, filling, labeling, and packaging. Preventing microbial contamination is critical and achieved by using closed systems and operating in clean rooms as per regulatory guidelines.

Good Manufacturing Practices (GMP) form the foundation of biologics production, requiring predefined specifications for equipment and materials, standardized procedures, and comprehensive documentation to ensure product quality and compliance. Before commercial use, the manufacturing process undergoes validation to demonstrate its robustness and consistency. Once validated, any changes to the process or facility require regulatory approval.

Key commercial considerations include scheduling, capacity planning, inventory management, and supply chain optimization to meet market demands while controlling costs. Emerging trends in biologics manufacturing, such as continuous production, personalized therapies, automation, and digital technologies, are driving efficiency and innovation in the field. These advancements promise to improve scalability, reduce costs, and enhance product accessibility while maintaining stringent quality standards. Together, these practices and innovations ensure the safe and reliable delivery of biologic drugs to patients worldwide.

Keywords Drug Substance · Drug Product · Clean room · GMP · Validation · Specifications · Unit Operations

D. Patankar (✉)
Independent Consultant, Bengaluru, Karnataka, India

1 Introduction

In earlier chapters, we have looked at the development of manufacturing processes. The objective of development is to come up with a process that can deliver the desired product quality reliably and cost-effectively. During process development, the output of the development effort consists of data, knowledge, and understanding. The material produced in the course of the experiments is used for analytical characterization and studying other properties like stability. The material is not directly used as a drug.

Manufacturing is that activity where the objective is to deliver a quantity of the product for an intended use. For a biologic drug, the intended use can be for a study such as toxicity study or clinical study, or the intended use can be distribution in the market for use by patients in a commercial setting.

Traditionally, manufacturing is carried out in a batch mode. The process starts with a vial of cell bank containing the engineered cells that produce the desired product. The process goes through multiple unit operations, as described in earlier chapters, and results in a certain quantity of drug substance (or drug product) that constitutes one batch or lot. The lot is identified by a unique lot number. A small sample of the product is withdrawn and undergoes testing for quality specifications. If the sample meets all of the quality specifications of the product as approved by the licensing authority (FDA, EMA, or other national regulatory authority), the entire lot is released for use (in the case of drug substance, it is released for conversion to drug product, while in the case of drug product, it is released for supply to clinic or the market).

Implementing a manufacturing operation has the following elements:

1. Design of the manufacturing facility and the physical arrangement of the different operating sections, movement of operating people, raw materials and intermediate products from one stage to another.
2. Translating the developed process, received from the development division via technology transfer, into a routine operation that faithfully reproduces the process conditions as described, including appropriate process controls.
3. Mechanisms to minimize the risk of any errors and build assurance that the process runs as expected every time, that the final product meets the intended quality, and that every dose supplied to the patient is of the right quality.
4. Managing the scheduling of batches and associated logistics and supply chain in an operationally efficient and cost-effective way.

Drug manufacturing is governed by regulations known as Good Manufacturing Practices (GMP). Compliance to GMP is an integral and overarching element of any manufacturing operation and is internally monitored by the Quality Assurance division.

2   Drug Substance Manufacturing

The biologic drug substance manufacturing process consists of multiple unit operations grouped into upstream process and downstream process. In the upstream process, the cells are grown in vessels of increasing size (called the inoculum train) to
obtain a large number of cells, which are then inoculated into the production reactor.
The production step lasts from about 1 day for bacterial processes to 4–6 days for
yeast-based processes and to 12–15 days for mammalian processes, during which
the cells grow in number and also produce the desired product. The last step in the
upstream process is the harvest, when the cells are separated from the product
(which is secreted out into the medium) by filtration or centrifugation. In cases
where the product is expressed inside the cells, the upstream process may end with
the separation of the cells from the fermentation media, or may continue with lysis
of the cells, and separation of the crude protein from the cell debris. The harvested
product may in some cases be stored, usually frozen, before it is taken to the downstream process. The downstream process is a series of steps involving purification,
as well as conversion of the product to its final molecular form (such as when the
product is expressed as an inclusion body that needs to be refolded to its final form,
or the protein does not have its final sequence and has to be modified with an enzymatic reaction, or if the protein molecule is to be conjugated to another molecule in
products such as pegylated proteins or antibody-drug conjugates). Purification steps
typically involve the use of chromatography and filtration.

2.1   Manufacturing Facility Layout

A manufacturing plant therefore is usually grouped into the following sections:
upstream process (USP) section, downstream process (DSP) section, and a section
for preparing the media, buffers and other solutions required in the process (see
Fig. 1 for an illustrative layout of a manufacturing plant). The three sections are run
independently. Operators enter and exit each section separately without moving
from one section directly to the other; only the product passes from one section to
the other. The product is transferred through material transfer hatches (passboxes)
for transfer of small containers, or material transfer airlocks for large containers.
Such passboxes or airlocks are designed such that one operator places a container in
it from one side and another operator pulls it from the other side. Sometimes the
product may also be piped from one section to the other directly and taken into a
holding tank on the other side.

The three sections described above are usually divided into several individual
rooms as below:

Upstream section: (1) Inoculum or seed preparation room. This is where the cell
  bank vial is thawed and taken through multiple rounds of cell expansion in flasks
  or small reactors. (2) Bioreactor or production room. This contains the production

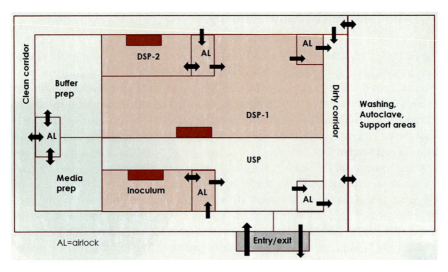

Fig. 1 Illustrative layout of a biologics drug substance manufacturing plant
Different sections are shown as blocks. Arrow are the directions of personnel movement across the different boundaries. Darker colored areas represent higher grades of clean air. The layout is intended to conceptually describe the different sections and should not be taken as representing or recommending any particular facility. Different layouts are possible based on the scales, operational requirements, and preferences of the organization.

reactor, as well as often the last one or two steps of the seed train. (3) Harvest room. This is for cell separation from the medium and for cell lysis. In some facilities the harvest activity may be done in the bioreactor room itself. Normally the intermediate product going from the upstream section to the downstream section has to be cell-free. Although the recombinant organisms used in biologic production are safe, it is desirable to contain the live organisms in the upstream section and inactivate the cells by adding disinfectant or by steam sterilization before they are discarded.

Downstream section: The downstream section may consist of multiple rooms, usually at least two. In products based on mammalian expression, all purification steps until the last viral clearance step, usually the viral filter, are done in one room (DSP-1), whereas all the steps after the viral filter are done in another room (DSP-2), which is especially controlled to protect it from any potential entry of adventitious virus contamination. In the case of microbial products, the rooms may separate different purification steps. The final drug substance is sterile-filtered, sampled for quality testing, and stored frozen.

Support section: The support section contains areas for media preparation and buffer preparation. The support section also includes a washing area for cleaning of small containers and an autoclave for sterilization of washed items. Large containers are cleaned and sterilized or sanitized in place, most commonly by circulation of sodium hydroxide solution.

A common philosophy in manufacturing facility design is unidirectional movement of personnel and material. Under this concept, personnel enter the facility from one point, pass through a corridor, enter a process room for running the operations, and exit the process room from the other end. The personnel have to return to the starting point before they can change gowning and go into a different process room or section (such as upstream or downstream). Similarly, raw materials or clean items such as holding bags and tubings enter a process room from a clean side, while the used materials exit the process room on a "dirty" side from where the material may be taken for washing or for discarding.

2.2   Clean Room Design and Operations

One of the key risks in biologic manufacturing is microbial contamination, as the media, buffers, and the product itself are rich potential nutrients for microbial growth. Therefore, prevention of microbial contamination is a critical requirement during manufacturing. This is accomplished by two approaches: use of closed operations and manufacturing in clean rooms.

Clean room operations   The process steps are carried out in clean rooms or in clean air cabinets. Clean rooms and cabinets are classified into different grades based on the number of non-viable particles (i.e., inert particles) of certain size that can be found per unit volume of air in the room or space as well as the number of viable particles (i.e., microorganisms) seen when measured in different ways. EU authorities classify clean rooms as Grade A, B, C, D (see Tables 1 and 2) [Ref. 1], whereas the USFDA follows a different classification system described by ISO (*ISO 14644-1:2015*) [Ref. 2]. The maintenance of clean room conditions requires air to be circulated through the room at a certain rate, higher for the higher grades, and filtered through high-efficiency particulate air (HEPA) filters. Further, it requires stringent cleaning procedures in the rooms and the operators need to wear full body gowns that prevent shedding of particles, as shedding from skin and hair from operators is a large potential source of contamination.

Critical operations in DS manufacturing are carried out in Grade C rooms, and open manipulations or transfers of product are carried out in Grade A cabinets placed in Grade C rooms. Less critical operations or closed operations can be carried out in Grade D rooms.

Table 1 Clean room classification as per EU-GMP [See Ref. 1]

| Grade | Maximum limits for total particles $\geq 0.5\ \mu m/m^3$ | | Maximum limits for total particles $\geq 5\ \mu m/m^3$ | |
	At rest	In operation	At rest	In operation
A	3520	3520	Not defined	Not defined
B	3520	352,000	Not defined	2930
C	352,000	3,520,000	2930	29,300
D	3,520,000	Not defined	29,300	Not defined

Table 2 Clean room viable particles limits as per EU-GMP [See Ref. 1]

Grade	Air sample CFU/m^3	Settle plates (diameter 90 mm) CFU/4 hours	Contact plates (diameter 55 mm) CFU/plate
A	No growth		
B	10	5	5
C	100	50	25
D	200	100	50

To maintain the air quality in compliance with the designated classification, clean rooms are protected by airlocks which operators must pass through to go from a corridor to a process room, or a process room with one classification to one with a different classification. These airlocks create a barrier preventing air from a lower classification area entering the higher classification area, even when doors are opened for personnel movement.

Gowning Personnel entering the facility usually change into clean factory garments before entering the main corridor of the facility (See Entry/Exit point in Fig. 1). Prior to entering the process rooms where manufacturing takes place, they usually wear an overgown with a second layer, which helps prevent any risk of contamination from the person's body or clothing to be transferred to the product. In Grade C rooms, the secondary gowns are generally full body suits (see Fig. 2) to prevent particle shedding and contamination of the product. The secondary gown is removed while exiting the specific process room, and the person changes back into their street clothes after returning to the entrance change area.

Closed operations Microbial contamination can be prevented by running the process in a closed system, i.e., where the equipment as well as the transfer lines to take the product from one operation to the next are closed and the product is not exposed to the external environment. Such vessels and the transfer lines are cleaned and sterilized before use, and once in operation, they prevent any external contamination of the product. When manufacturing has open operations where the product is exposed to the room environment (such as mixing or collection of eluate from a chromatography column), the open steps are done in a Grade A cabinet in a Grade C room. Closed equipment are more robust in preventing microbial contamination, even when placed in a Grade D room. It is also practically more convenient and cheaper to build and maintain a Grade D environment than a Grade C environment with Grade A cabinets; thus, closed operations are preferred whenever feasible.

2.3 Automation

Manufacturing equipment can be controlled using various degrees of automation. Process conditions such as temperature, pH, stirring, or flow rates, once set, are maintained using controllers on the individual equipment, but in highly manual

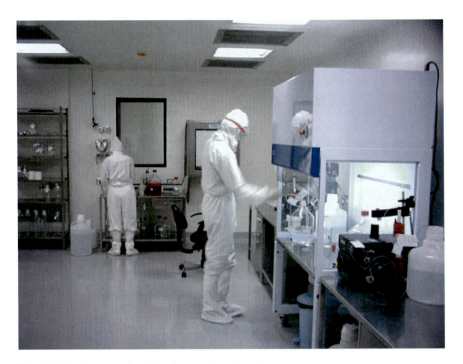

Fig. 2 Typical gowning in a biologic manufacturing plant
The room is a Grade C clean room, with a Grade A clean cabinet, and small-scale manual operations.

operations, the parameters are set and changed manually. The start and stop of an operation, start and stop of collection of eluate in a chromatography step, or transfer of material from one equipment or holding vessel to another after the completion of a step is also carried out manually. Highly manual operations are only common at small scales and in early clinical manufacturing. At an intermediate degree of automation, referred to as islands of automation, all of the steps in a given unit operation are automated, with recipes pre-loaded—for example, changing set points, adding feeds based on a time schedule or in response to certain online measurements, or collection of chromatography eluates; however, each unit operation runs independently and has to be started and stopped manually. In highly automated systems, multiple unit operations, including the transfer of product from one equipment to another, may be programmed together and executed automatically.

In order to control the process performance, it is necessary to monitor and control various process parameters. Additional parameters are monitored to verify whether they are in the expected range. Such in-process testing can be done offline, online, or inline. Inline testing is where the sensor or probe is directly in contact with the process itself and does not require removal of a sample. Online testing is where a sample is drawn from the process and diverted to an analytical instrument which is connected to the process equipment. Offline refers to a sample being drawn from the process and taken to a separate analytical instrument for testing.

2.4   Capacity and Throughput Planning

Optimum utilization of the facility requires equipment to be in operation most of the time with minimal idle time. Also, the batches should follow one another with minimum time lost in between for switchover.

Optimum equipment utilization is achieved by duplicating equipment that have long occupancy. In a typical mammalian process, such as monoclonal antibody production, the upstream production reactor is a 12–15 day process, whereas the entire downstream process happens in 4–5 days, with each equipment being occupied for 1–2 days. The most optimal design therefore has three to four bioreactors with a single downstream train, such that one reactor is inoculated at a 3–4 day interval in turns, giving a capacity of six to seven batches a month, with high utilization of all equipment. Most monoclonal antibody manufacturing plants have such a configuration, with the production bioreactors ranging from 2000 to 20,000 L capacity each.

3   Drug Product Manufacturing

Drug product manufacturing is operationally simpler than drug substance manufacturing but more critical as the product directly goes to the patient. Biologic drugs are generally administered as injections, so sterility is the most important quality attribute that the drug product (DP) has to meet, as any contamination can have serious adverse effects on the patient. Since microbial contamination is a chance event, sterility cannot be assured by testing of a sample of the final product, but rather has to be assured by a variety of control measures including design and maintaining of clean room conditions, operational and procedural controls during manufacturing, and periodic process simulation runs. In a process simulation run, the actual manufacturing process is carried out with microbial nutrient medium instead of the product, and the vials are incubated to confirm absence of any microbial growth. This confirms that the process as established is robust enough to deliver sterile product.

3.1   Manufacturing Process

DP manufacturing starts with the preparation of the bulk formulation, which involves mixing the drug substance with any excipients and adjusting the concentration if required. The product is filter sterilized by passing through sterilizing filters placed inline and filled aseptically into vials or syringes. The filling machine has a clean air of Grade A standard and is placed in a room of Grade B. In modern facilities, an even higher standard of clean air is provided by conducting the entire filling operation in a Grade A isolator, which is sealed from the outside environment and provides a very high level of assurance of sterility. In the case of lyophilized products, the vials after being filled are moved to a lyophilizer, where the water is sublimated, leaving a cake containing the product with other formulation ingredients. As the vials have to be open during this entire process, maintenance of sterility is critical when the vials are moved from the filling station to the lyophilizer and within the lyophilizer. Overall, regulatory agencies such as the FDA have very strict standards on maintenance of sterility assurance in the manufacturing of drug products; this is often the subject of FDA observations during inspections.

3.2   *Manufacturing, Packaging, and Labeling*

Commercial batch sizes of biologic drug products range from a few thousand to several hundred thousand doses, with filling speeds of the order of 100 or more vials (or syringes) per minute. After filling (and lyophilization where required), the drug product containers undergo visual inspection to ensure absence of any contaminating particles. Containers where any particles are present (usually the result of dust from the air, or sometimes improper washing of the vials, or interactions with the product) are rejected. The visual inspection may be carried out manually or by an automated sensor, which can automatically reject containers having any particulates. After visual inspection, the containers are labeled with the drug name, strength, lot number, and other details.

Some drug products are supplied with devices for convenient administration. The product may be filled in cartridges that are then inserted into metered administration devices such as pens. More complex devices such as infusion pumps, inhalers, sprays, etc., are not common in biologic drugs, except for insulin pumps. When the product is to be sold in such devices, the final step in DP manufacturing will involve assembly of the device with the drug cartridge.

The containers within which the product is filled, such as vials, syringes, or cartridges, are called the primary packing material. Subsequently the primary containers are placed inside cartons called secondary packaging material. These are the cartons in which the product is usually available commercially.

4   Good Manufacturing Practices

Regulatory authorities approve for marketing a product of a defined quality produced with a given manufacturing process. It is the manufacturer's responsibility to ensure that every dose of the drug that reaches the patient is manufactured only according to the approved process and matches the quality that was approved by the regulatory authority. To enable building such assurance, there are regulations called Good Manufacturing Practice or GMP. The standards of what constitutes good manufacturing practice evolve with time, with introduction of new drug modalities, increasing knowledge about likely risks, innovation in manufacturing processes, availability of new technologies for analysis and control, and changes in the business environment in which manufacturing of drugs is carried out. For these reasons, GMP is often referred to as cGMP or "current good manufacturing practice," and regulatory agencies expect manufacturers to remain compliant to the "current" standards. The FDA guideline is described in the Code of Federal Regulations Title 21 (21CFR) Parts 210, 211, 212, and 600, whereas the EU guideline is described in EudraLex Volume 4, which has multiple parts and annexes. The regulatory guidelines are described in broad terms whereas more detailed guidances on specific topics are put out by other public organizations and industry bodies such as ICH (International Conference on Harmonization), USP (United States Pharmacopeia), and PDA (Parenteral Drugs Association). They are also inferred from actual FDA interactions and inspections.

Good manufacturing practices are designed to ensure consistency of product quality and reduce the risk that a batch of product will fail to meet the expected quality standards. There are three kinds of risks to product quality:

1. Risk of contamination and cross contamination from external sources
2. Variability of output quality due to variability of input materials, process conditions, and operating procedures
3. Risk of errors and mixups during the operations

These risks can be minimized through appropriate facility and equipment design, minimizing variability of input materials and procedures and minimizing opportunities for errors and mixups. All of these aspects, therefore, form part of GMP guidelines and practices.

4.1 Facility Design and Equipment

Control of the risk of microbial contamination has been dealt with in the section on facility design. Proper facility design and ongoing maintenance of cleanroom conditions are parts of GMP compliance. Cross contamination means contamination of a product batch with material from another batch. This can occur due to carryover of residues resulting from inadequate cleaning. Cleaning of multi-use equipment is therefore a critical point in any GMP compliance inspection. In recent years, biologics manufacturing is increasingly shifting to use of single-use disposables including bioreactors, holding vessels, transfer tubings, etc. A major driver for this shift has been that it avoids any risks due to improper cleaning.

Consistency of production can be achieved by reducing the variability of input materials, equipment functioning, and operating procedures. All equipment must be designed according to predetermined specifications, and any replacement of parts must use the same specifications. Equipment must undergo preventive maintenance and calibration of all gauges, sensors, and measuring devices to ensure that equipment performance, and especially any measurements of parameters and conditions, remains consistent.

4.2   Raw Materials and Consumables

Raw materials quality can have a significant impact on the process performance and product quality, either due to the presence of impurities, or in their ability to affect cell growth, protein production, or protein quality. Therefore, raw materials must meet predetermined specifications and must be purchased from vendors who have been qualified not only to supply a consistent quality of material but also to inform the drug manufacturer if they make changes to the manufacturing process of the raw material which might in turn impact the quality of the material and impact the drug manufacturing process. Specifications are also applicable for all containers and flow paths which are in contact with the product or with the media and buffers throughout the process to ensure that they are compatible with the chemical composition and pH of the process streams and do not adsorb the product, react with it, or leach out any impurities into the product.

Water is the most used raw material and there are different grades of water quality defined by the regulations. Water For Injection (WFI) is the highest grade and is used for biologics manufacturing processes.

Raw materials and consumables used in the process are stored in the warehouse. In line with GMP principles, all incoming materials have to be evaluated or tested to confirm that they meet the specifications set for them. The materials are then transferred to the manufacturing area where they are needed.

4.3   Procedural Controls

Manual operations are susceptible to person-to-person and batch-to-batch variations in how a certain operation is executed. It is therefore desirable to automate processes. Where manual processes are to be used, the procedures to be followed are written down in detail and operators are trained on the procedures, so that there is no person-to-person variability in how the various steps are executed. It is a necessary part of GMP that every activity, no matter how trivial, have a detailed written procedure, also known as an SOP or standard operating procedure.

Manufacturing under GMP also involves a variety of mechanisms to minimize the risk of inadvertent errors and mixups. Intermediate checks are put in place to ensure that the equipment and materials meet the prescribed requirements, the clean rooms meet the required standards, equipment are calibrated and cleaned, buffers and intermediate products are tested as specified before the next operation begins, and all steps are executed according to the written instructions. Mixups can be prevented by physical segregation and appropriate labeling (such as between approved and rejected material, or cleaned versus to-be-cleaned equipment).

4.4 Documentation

Generation of written evidence that all the requirements and instructions have been followed is an important part of building assurance that the production is carried out as intended, meeting all process conditions, fulfilling all prerequisites, and with procedures executed as written. Documentary records must be generated for every activity conducted. These include equipment installation and qualification records, instrument calibration records, facility cleaning records, personnel training records, raw material weighing and dispending records, batch manufacturing records, analytical testing records, and so on. Absence of such written records imply that the quality of a batch of manufactured product cannot be assured, and in this case, the product may be withheld from being released for use. This has led to a pithy maxim of GMP: "document what you do, do what you document; if it is not documented, it never happened."

4.5 Quality Assurance and Compliance

GMP regulations require an independent quality assurance division whose function is to create the enabling SOPs and the monitoring mechanisms described above, carry out the intermediate checks during the batch, take decisions on whether or not to release a batch based on the results of the quality control testing, and review the manufacturing and testing records.

Compliance to GMP is ensured by the quality assurance division. It is expected that the QA division also conducts periodic internal audits to verify ongoing compliance with the GMP regulations. Further, the regulatory agencies (FDA and other national authorities) conduct periodic inspections at all manufacturing sites to verify compliance. Any observations of non-compliance have to be rectified in a time-bound manner; failure to do so can result in a manufacturer not being allowed to continue to sell a drug in the market.

5 Product Manufacturing Operations

The topics described above cover the manufacturing environment: manufacturing facility design and operations, GMP practices, etc. Now we look at implementing the manufacturing of a given product in such a facility.

5.1   Production Planning, Batch Manufacturing, Testing and Release

Production schedules are planned based on market forecasts. Raw material procurement is planned to align with the production schedules. For the initiation of a production batch, the quality assurance division has to verify that all prerequisites are met and issue a batch number or lot number.

At the end of each batch, a sample is collected and sent to the quality control division for testing against the specifications. The sample must be representative of the whole batch and the process of sampling itself must not cause any contamination in the product. The specifications are a list of attributes such as strength or concentration, purity or level of specific impurities, potency or biological activity, and distribution of various glycosylated or chemically modified species of the product. The written specifications define the testing procedure and the limits of each attribute that the batch must fall within. GMP regulations also govern the testing procedure analogous to manufacturing: all instruments and reagents must meet predefined criteria, all gauges and measuring devices must be calibrated, the testing procedure must be written down in detail, analysts must be trained, there must be intermediate checkpoints to ensure the testing is done as per procedure, and the test meets its own performance criteria. All test results are compiled in a Certificate of Analysis (CoA).

In an ongoing and complex operation, it is inevitable that some manual errors happen, process conditions deviate from the set acceptable ranges, or a unit operation has to be paused for a while due to an equipment breakdown. The checks and documentation process described earlier ensure that such deviations are recorded. An important component of GMP is to assess the likely impact of such deviations on the product quality and to release the batch only if it is concluded that there is no actual or likely impact on the product quality, else to reject the batch. It is also expected that the cause of the deviation is investigated so that corrective actions can be taken to prevent recurrence of the deviation.

It is the responsibility of QA to review the CoA, verify the testing itself was carried out according to the requirements, review the manufacturing batch records to verify that the manufacturing process was carried out according to the requirements, and review any deviations for likely impact. Upon satisfactory review, the batch is released for further use, else it is rejected. This final decision is referred to as batch disposition.

5.2   Supply Chain and Distribution

Raw materials and consumables used for manufacturing are stored in the warehouse. The warehousing system should have clear demarcation between material at different stages:

- Material under quarantine, i.e., material that is received but currently under testing
- Approved material, i.e., material that has been approved post-quality testing and can be used for manufacturing
- Rejected material, i.e., material that does not meet the set specifications and must be returned or destroyed

The product post-manufacturing is moved into a finished products storage warehouse, where the product is stored at its designated storage condition (usually frozen for drug substance and refrigerated for drug product). Similar to raw material storage, the finished product storage should also have designated spaces for product that is under quarantine, released for use, or rejected.

At the time of distribution, the product is packed in larger boxes, called tertiary packaging, which are transported in cold or frozen conditions, as needed.

5.3 Validation

The FDA defines validation as "establishing documented evidence that provides a high degree of assurance that a specific process will consistently produce a product meeting its predetermined specifications and quality attributes." Traditionally this meant running multiple (usually three) consecutive batches of a process and demonstrating that the performance is consistent in terms of yield and quality at different intermediate stages as well as at the final product stage. However, in 2011, FDA issued a new guideline that expanded the concept of validation to a comprehensive approach covering process development and characterization (called process design as per the FDA guidance) and full-scale process performance qualification (PPQ). The first stage, process design, is where you build robustness into the process and verify it through small-scale experiments (called process characterization) wherein you establish tolerance limits of operating parameters within which the process is known to give the desired performance. The second stage, PPQ, is equivalent to the old concept of process validation, with multiple consecutive full-scale batches showing consistent performance. Once a process is validated, it should not be changed. Changes to a validated manufacturing process require following a strict procedure known as change control, wherein the implications of the change on product quality are studied, and fresh validation is carried out to show that the changed process also produces material of the same quality. Process changes require approval from regulatory authorities before being implemented in commercial production.

Once a validated process is in routine operation, FDA expects continuous monitoring and trending of process performance, called continuous process verification, which is considered as the third stage of process validation.

Validation as a concept applies not only to a manufacturing process but to any other activity, for example:

- Analytical method validation, which is a demonstration that the analytical procedure performs consistently,
- Cleaning validation, which is a demonstration that a given equipment, when cleaned using a specific cleaning procedure, gives consistent cleaning performance as defined by measuring residual carryover in a rinse or swab sample,
- Shipping validation, which is a demonstration that a packing and shipping procedure, especially for controlled temperature shipment (such as cold shipment, which is common for biologic drug products), maintains temperature within the specified acceptable range.

5.4   Monitoring, Trending, and Continuous Process Verification

When a manufacturing process is in routine commercial use, the ongoing performance is trended to see if there are any drifts or shifts in quality, or any other indications that the process is not running consistently. Graphical plots of these trends are called control charts. The control charts can be analyzed using statistical and other approaches. The trends can also be analyzed to study the impact of specific changes such as introduction of a new raw material source or the use of a particular equipment within the plant.

Such trending is not only a regulatory requirement but also helps the manufacturer understand better any factors that impact process performance so that they can take actions to further improve the consistency and quality of the product. It can also provide pointers to potential process improvements that can lead to higher yield and lower costs. Of course, any such changes can only be introduced following the process of change control described previously.

6   Post-Approval Changes

Changes to a biologic drug manufacturing process, ranging from improvements in the process to changes to the formulation, labeling, packaging, etc., may be necessary post-initial approval of the product by regulatory authorities. These changes may be necessary to improve product quality, enhance manufacturing efficiency, address safety concerns, or respond to changes in market demand. Some of the key drivers for the changes include:

- Manufacturing process optimization to improve efficiency and reduce costs. These changes may involve changes to process parameters, equipment upgrades, or implementation of new technologies to enhance product quality or yield, and reduce costs.

- Supplier or raw material changes to ensure continuity of supply and process consistency or to reduce costs.
- Changes in manufacturing scale and/or sites to adjust to changes in the market demand of the product.
- Enhance process robustness and control which may involve refining control strategies, implementing real-time monitoring technologies, etc.

The strategies and requirements for post-approval changes can vary between different regulatory agencies, leading to differences in how manufacturers must approach these changes. These requirements are covered through guidelines and regulations from the regulatory agencies.

A robust change control to capture and assess the impact of proposed changes on product quality, safety, and efficacy is typically required. Extensive comparability studies to compare the modified product with the originally approved product need to be performed. The comparability protocols outline the study design, acceptance criteria, and analytical methods used to assess product comparability before and after the change.

Depending on the nature and magnitude of the change, manufacturers need to submit different types of regulatory applications to obtain approval for post-approval process changes. These applications are called variations, supplements, or notifications, depending on the jurisdiction and regulatory requirements.

Regulatory authorities review submissions for post-approval process changes to evaluate the scientific rationale for the change, supporting data, and potential impact on product quality, safety, and efficacy. Regulatory approval is granted based on the assessment of the change's impact and compliance with regulatory requirements.

7   Commercial Considerations

While maintenance of product quality and compliance with regulatory requirements including GMP are mandatory for any product, commercial considerations are not far when manufacturing any product for sale. Apart from profitability considerations, drug pricing also has a social dimension in terms of being affordable and accessible to patients. Hence, control of manufacturing cost is important.

Cost of goods, or manufacturing cost, has the following major components (excluding indirect costs such as corporate overheads):

(a) Raw material and consumables cost (as actually used in the manufacturing process)
(b) Fixed cost of investment in plant construction
(c) Operating cost of running the plant, including utilities (electrical power, water, steam and gases), cleaning and disinfection, repair and maintenance, and expenses required to keep the facility in GMP compliance (periodic qualification, calibration and monitoring of equipment and facilities)
(d) Cost of manpower in manufacturing and support services

The relative contribution of each factor depends on the geography, the plant design, the manufacturing process, the nature and quantities of materials used, and the yield from the manufacturing process. Costs can be controlled by efficient facility design, cost-efficient sourcing of materials, high process yields, and controlling other operating costs. Other than the materials cost, the other items are largely fixed costs, so often the most important factor influencing cost of manufacturing is the occupancy of the facility: a high occupancy rate results in a lower per-unit manufacturing cost, whereas a facility with fewer batches manufactured through the year results in the entire fixed cost being loaded on to a smaller quantity of manufactured product.

8 Future Directions

8.1 Alternative Manufacturing Formats

Throughout this chapter, we have discussed the conventional batch manufacturing of a drug which is then distributed commercially. Some new trends coming up are continuous manufacturing and personalized manufacturing.

Continuous Manufacturing Unlike batch manufacturing, here the product coming out of one step is continuously fed to the next step. For example, bioreactor harvest can be continuously fed to the first downstream chromatography column and its eluate can be continuously fed to the next operation, and so on. This has the benefit of requiring smaller equipment, lower capital investment, and, therefore, lower cost of goods. However, it is operationally more complex, as it requires fine-tuning of these linked operations, and more precise on-line monitoring. Several companies are now pursuing these approaches [See Ref. 3, 4].

Personalized Manufacturing New treatment modalities are emerging, such as cell and gene therapies, which include autologous therapies. In this, cells from individual patients are collected and sent to the manufacturing unit for certain kind of processing, and then the cells are infused back into the patient. Although still produced in batch mode, every batch is unique in that it is meant for an individual patient. This demands a completely different approach to the concept of GMP, batch consistency, manufacturing scale, automation, etc.

8.2 Analytical Tools, Data Analysis, and AI

As with all areas of business and industry, data analysis is becoming increasingly important in biologics manufacturing, often referred to as "Pharma 4.0." It would be obvious from the earlier discussion that process monitoring, trending, and analyzing the effects of multiple variables on product quality are areas highly amenable for multivariate data analysis or AI-based predictive analysis.

Other trends are evident in analytical technologies, with more in-line measurements (referred to as process analytical technologies or PAT) which, combined with the availability of databases, algorithms, and data analysis tools, can provide real-time information about the state of a manufacturing process, enabling real-time decision-making for process control [see Ref. 5]. This has profound implications for the concept of process control during a batch, batch release testing, deviations assessment, and change control. However, it also throws up new challenges in terms of how to demonstrate consistency and reliability of such algorithms.

References

1. The rules governing medicinal products in the European Union–Volume 4 EU guidelines to good manufacturing practice - medicinal products for human and veterinary use - Annex 1 - Manufacture of sterile medicinal products. Brussels: European Commission; 2022.
2. ISO 14644-1: 2015 Cleanrooms and associated controlled environments - Part 1: classification of air cleanliness. Geneva: International Organization for Standardization; 2015.
3. Dream R, et al. Opportunities in continuous manufacturing of large molecules. Pharmaceutical Engineering. 2021 from: https://ispe.org/pharmaceutical-engineering/july-august-2021/opportunities-continuous-manufacturing-large-molecules.
4. Chen R, et al. Where do we stand on adopting continuous manufacturing for biologics?. Bioprocess Online. 2022 from: https://www.bioprocessonline.com/doc/where-do-we-stand-on-adopting-continuous-manufacturing-for-biologics-0001.
5. Gerson G, et al. Process analytical technologies—advances in bioprocess integration and future perspectives. J Pharma Biomed Anal. 2022;207 https://doi.org/10.1016/j.jpba.2021.114379.

Pharmacology of Protein-Based Therapies

Afsana Trini, Glareh Azadi, Laxmikant Vashishta, Kaushik Datta, and Vibha Jawa

Abstract The development of biologics necessitates a comprehensive understanding of pharmacokinetics (PK) and pharmacodynamics (PD) due to their complexity. This chapter highlights the importance of quantitative pharmacology in integrating PKPD data to guide dosing, efficacy, and safety decisions. It covers essential aspects such as in vitro and in vivo pharmacology, bioanalytical methods, and drug exposure-based modeling for toxicology and first-in-human (FIH) dose estimation. Key considerations include drug metabolism and pharmacokinetics (DMPK), ADME processes, and immunogenicity risk assessment. Advances in bioanalytical assay technologies, including ligand binding assays (LBA), radio-immunoassays (RIA), and liquid chromatography-mass spectrometry (LC-MS/MS), are discussed for their role in measuring drug concentrations and ensuring therapeutic efficacy. The chapter also addresses the development and validation of bioanalytical methods, emphasizing the importance of accurate quantification for dose-response studies and the pharmacokinetics-pharmacodynamics (PK-PD) relationship.

Keywords Biologics · Protein-therapeutics · Pharmacokinetics · Pharmacodynamics · Analytics · Immunogenicity · Prediction

A. Trini · V. Jawa (✉)
Translational Medicine and Bioanalysis, Bristol Myers Squibb, Lawrenceville, NJ, USA
e-mail: Vibha.jawa@bms.com

G. Azadi
Preclinical Optimization, DMPK, Bristol Myers Squibb, Brisbane, CA, USA

L. Vashishta
Alvotech Biosciences India PVT. Ltd., Bangalore, India

K. Datta
Nonclinical Safety (Toxicology), Summit/New Brunswick, NJ, USA

1   Introduction

1.1   Pharmacokinetics (PK) and Pharmacodynamics (PD) of Biologics

Protein and large molecule-based therapeutics require a thorough understanding of pharmacokinetics (PK) and pharmacodynamics (PD) due to their complexity compared to small molecules. Quantitative pharmacology plays a vital role in integrating PKPD data, enabling researchers to make informed decisions about dosing, efficacy, and safety. By considering these key components and implementing PKPD strategies, researchers can enhance the development of biologics and improve patient outcomes. This chapter highlights the importance of PKPD in biologics development, the key in vitro and in vivo pharmacology activities, and bioanalytical methods used to support drug exposure-based modeling for estimating toxicology and FIH doses, and the timing of implementation of these models. It sets the stage for further exploration of bioanalytical strategies, absorption, distribution, metabolism, and elimination (ADME) considerations for large molecules, and the role of quantitative pharmacology in optimizing drug development.

Drug metabolism and pharmacokinetics (DMPK) and ADME considerations are integral parts of early pharmacology activities for biologics. These studies help in understanding how the drug is absorbed, distributed, metabolized, and eliminated from the body, which is crucial for determining the drug's safety and efficacy. Early activities related to pharmacology for biologics include both in vitro and in vivo studies. While in vitro studies are crucial for understanding the mechanism of action of biologics, in vivo studies support further understanding of on- and off-target safety and efficacy.

1.2   Immunogenicity Risk Assessment of Biologics

Immunogenicity is another important consideration in the early stages of development of biologics that can impact exposure and safety and hence proactive mitigation strategies to address immunogenicity need to be implemented. Some key considerations to prepare for immunogenicity in clinic include identifying and mitigating risks during preclinical stage of development [1]. Other approaches include developing a clinical strategy by leveraging information from preclinical in vivo pharmacology and IND enabling toxicology studies and use of quantitative systems-based pharmacology (QSP) models of immunogenicity [2]. Mechanistic understanding of processes driving immunogenicity has driven the development of phenotypic cell marker-based and functional predictive assays following clinical administration of biologics [3–5].

1.3   Key Components to Support DMPK Activities

1.3.1   Bioanalytical Assays and Strategies

Bioanalysis generally refers to the quantitative measurement of the drug and its metabolites in biological matrices, such as blood, serum, plasma, urine, cerebrospinal fluids, saliva, tissue biopsies, etc. Bioanalysis is an integral part of the drug development process. Bioanalytical methods are used to characterize the pharmacokinetics, pharmacodynamics, and toxicity of the drug, which is crucial for the safety and efficacy assessment of the potential drug candidate. It supports regulatory filings, such as investigational new drug application (IND) as well as clinical trial–related bioanalysis leading to the submission of the Biologics license application and commercial approval process.

Bioanalytical Assay Formats

Bioanalytical assays are broadly discussed in three categories: ligand binding assays, radioimmunoassay, and chromatography-based assay. The type of bioanalytical assay is chosen based on the structure/design of the drug compound (such as single or multi-domain biologics, small molecule, or peptides) and specific information on analytes (active or free drug, total drug, binding antibodies, neutralizing antibodies (Nab), target or receptor levels, etc.) that are required to evaluate exposure- and toxicology-related levels of drug to interpret PK or TK studies [6–9].

Ligand Binding Assays (LBA) or Immunoassay
LBA assays are mostly used for biotherapeutics bioanalysis, primarily for high molecular weight compounds. These assays are crucial for monitoring therapeutic drug levels and anti-drug antibodies (ADA) in both preclinical and clinical stages. LBA assay relies on the antibody or antigen affinity to the analyte of interest. The success of the LBA assay depends on the availability of quality critical reagents (antigen or antibody, soluble receptors, or ligands, used for capture and detection). In addition, as often the clinical trials run for extended periods of time up to 2–3 years, the lot-to-lot consistency and performance of the critical reagents is paramount. The reagents in the LBA assays are developed to be highly specific so that they can bind to the target analyte with high affinity without the prior need of complex sample pre-treatment [10–12]. LBA technology has significantly evolved from radioimmunoassay to highly diverse methodologies in terms of signal read-out platform, assay sensitivity, specificity, and multiplexing. Some widely used applications of LBA assays are PK assays, receptor occupancy assays, immunogenicity, and biomarker assessment of biotherapeutics. Different platforms are used for the quantification, which includes classical enzyme-linked immunosorbent assay (ELISA), hybridization ELISA, and electrochemiluminescent (ECLIA) assays with diverse types (ligation, dual hybridization).

Radioimmunoassay (RIA)

The classic RIA method is based on the competitive binding principle. This assay uses a radiolabeled antigen to compete with a sample antigen (unlabeled antigen) for binding to an antibody with the appropriate specificity. The amount of radioactivity is inversely proportional to the antigen concentration in the sample [13]. RIA use is limited for sensitive measurement of hormones and small peptides due to the safety concern with the use and handling of radiolabeled material [14].

Liquid Chromatography Mass Spectrometry (LC-MS/MS)

Liquid chromatography-mass spectrometry (LC-MS/MS) is sensitive and specific, making it ideal for preclinical bioanalysis of small molecules [15]. It measures target analytes based on their mass/charge ratio and can quantify drugs and metabolites at exceptionally low concentrations (often pg/mL). LC-MS/MS requires sample preparation to remove proteins, typically through protein precipitation, liquid-liquid extraction, or solid-phase extraction [16].

Recent advancements have improved LC-MS/MS sensitivity and data analysis, enhancing its use in large-molecule bioanalysis, particularly for antibody-drug conjugates (ADCs). For ADCs, three main assays are used: total antibody (tAb), total conjugated antibody to payload (tADC), and unconjugated payload. LC-MS/MS is commonly used for the unconjugated payload, while tAb and tADC can be measured by both LC-MS/MS and ligand-binding assays (LBA) [17]. Often, a hybrid LBA–LC-MS/MS approach is employed for ADCs with cleavable linkers, combining affinity capture with LC-MS detection. LC-MS/MS offers superior selectivity and accuracy compared to LBA, particularly for differentiating between drugs, metabolites, and isoforms [17, 18].

Advances in High-Sensitivity Immunoassays for Low Drug Concentrations

The development of more potent therapeutics at lower doses has increased the need for sensitive platforms to measure low drug concentrations in various matrices (e.g., serum, plasma, CSF). Advances in bead-based methods and single-molecule detection, which enhance assay sensitivity and signal amplification, have significantly improved immunoassay performance. Notable platforms include Single Molecule Counting (SMC™) by Singulex, Single Molecule Arrays (Simoa™) by Quanterix, and Immuno-PCR (Imperacer®) by Chimera Biotec. These technologies address the bioanalytical challenges in biopharmaceutical drug development by boosting assay sensitivity [7–9, 19–22].

Bioanalytical Assay Development and Validation

The bioanalytical method aims to establish its design, operating conditions, limitations, and suitability for validation.

Method Development

Method development involves defining procedures for quantifying the analyte and characterizing key elements such as reference standards, critical reagents, calibration curves, quality control samples, selectivity, specificity, sensitivity, accuracy, precision, recovery, analyte stability, and minimum required dilution (MRD).

In preclinical toxicity studies, the goal is to evaluate drug toxicity, safety, and systemic exposure. Preclinical PK assays typically use generalized reagents to detect total drugs and often employ target capture with anti-human IgG detection. While preclinical ADA assays do not predict clinical immunogenicity, they can reveal ADA incidence, indicating potential safety issues such as hypersensitivity or exposure loss [23]. Preclinical ADA assays generally focus on cut point assessment, QC precision, sensitivity, stability, and drug tolerance, often using only a screening tier instead of the full three-tier approach used in clinical assays. GLP-supported assays are usually validated, but a qualified assay—similar to a validated one but without a quality assurance component—may be used if the drug has low immunogenicity risk and shows minimal issues in early studies.

Method Validation

Method validation requirements vary by drug development phase. In early phases, multiple drug candidates are screened, often with limited high-quality reagents and short development times. Despite these challenges, the bioanalytical team must provide reliable data for go/no-go decisions. Early-phase assays may be validated to a fit-for-purpose standard, showing sufficient accuracy and precision for initial biologic characterization. Preclinical in vivo assays must demonstrate suitability for their intended purpose.

Validation Criteria Differences

1. *Selectivity and Specificity*

   - *FIH Studies*: Assays must detect biologics amid endogenous substances and assess cross-reactivity.
   - *Clinical Studies*: Greater specificity is needed, including testing against a wider range of potential interferences.

2. *Sensitivity*

   - *FIH Studies*: Sensitivity should detect biologics at expected concentrations, with established LOD and LLOQ.
   - *Clinical Studies*: Higher sensitivity is required for detailed PK profiles, with reassessed and potentially lower LOD and LLOQ.

3. *Accuracy and Precision*

   - *FIH Studies*: Assays must be accurate and precise within acceptable ranges but with more flexible criteria.
   - *Clinical Studies*: Stricter accuracy and precision are required, following ICH M10 guidelines for consistency [24].

4. *Calibration Curve*

   - *FIH Studies*: Calibration curves should cover the concentration range, with QC samples for performance monitoring.

- *Clinical Studies*: More calibration points and stringent QC criteria are needed to ensure high performance across batches and sites.

5. *Stability*

   - *FIH Studies*: Basic stability studies for biologics and reagents under expected conditions.
   - *Clinical Studies*: Extensive stability studies, including long-term and multiple storage conditions.

6. *Reproducibility*

   - *FIH Studies*: Initial reproducibility assessments for short-term consistency.
   - *Clinical Studies*: Comprehensive reproducibility studies across multiple analysts, equipment, and sites.

7. Immunogenicity Assessment

   - *FIH Studies*: Preliminary assays for detecting ADAs and NAbs, focusing on initial safety.
   - *Clinical Studies*: Rigorous validation with detailed characterization of ADAs and NAbs for understanding immune response and its impact.

Pivotal clinical studies should adhere to ICH M10 guidelines for method validation. For non-regulatory studies, internal validation standards apply. Key validation parameters include accuracy, precision, selectivity, specificity, calibration curve, range, stability, and reproducibility for chromatographic methods, and specificity, selectivity, and accuracy for LBAs. If identical matrices are unavailable, surrogate matrices may be acceptable [25].

Relevance of Drug Level and Free Drug Quantitation

Drug Level Quantitation: Measuring systemic drug exposure (e.g., serum or plasma concentrations) is crucial in dose-response studies to understand the relationship between dose and response, especially when there's significant variability in pharmacokinetics or a nonlinear dose-concentration relationship. Key reasons for measuring drug levels include:

1. Both parent drug and metabolites are active.
2. Different exposure metrics (e.g., Cmax, AUC) show varying relationships with efficacy or safety.
3. Limited fixed doses in dose-response studies.
4. High variability in responses, necessitating exploration of underlying causes.

Free Drug Quantitation: Free drug refers to the unbound active form of a biotherapeutic, which can interact with target receptors or enzymes [9]. Since measuring unbound drug concentrations at action sites is often impractical, blood plasma concentrations are used to establish the pharmacokinetics-pharmacodynamics (PK-PD) relationship.

Case Study: Intravitreally Administered Aflibercept
For intravitreally administered VEGF inhibitor aflibercept (EYLEA) used in treating wet AMD or DME, free aflibercept concentrations in plasma are very low due to its binding with endogenous VEGF in the eye. This necessitates highly sensitive assays to measure both free and bound aflibercept. Total aflibercept concentration is calculated by combining free and bound forms, with a correction factor used to determine the concentration of aflibercept in the VEGF complex [26].

1.3.2 Use of Preclinical In Vitro and In Vivo Experimental Models

In Vitro Assays

In vitro models provide efficient screening assays that assess a wide range of ADME properties for drug candidates. These studies evaluate pharmacokinetic outputs such as solution properties, drug absorption and transport, metabolic stability, drug–drug interactions (DDI), and toxicity. Advances in cell culture technologies, including the development of 3D organoid systems, such as self-organizing 3D cell assemblies, enhance the prediction of drug metabolism and toxicity [27]. Additionally, using patient-derived cells in tissue-specific systems allows for personalized assessment of drug effects, improving specificity and speed in evaluating drug candidates [28].

Cellular Binding Assays: Surface plasmon resonance (SPR) is used to characterize the binding of the therapeutic drug to its target in real time. The equilibrium dissociation constant (K_D) or affinity of the drug to its target is measured by immobilizing the target on the chip surface followed by the addition of solution containing the potential binding partners over the surface to allow for the binding interactions [29, 30]. After selecting candidates based on SPR affinity (1:1 binding), binding potency is evaluated in cell-based assays. It is crucial to choose cell lines with target expression levels like those in humans, as bivalent interactions can enhance binding potency [29]. The SPR assay is used for the tissue cross-reactivity (TCR) study for the biologics. The TCR study can reveal the off-target binding of the drug in human or any other relevant animal species. In drug development, even with a known drug target, it is important to identify its binding to additional proteins. This off-target binding can negatively impact on the drug's safety and pharmacokinetics [31]. Cell microarray technology is used to screen for specific off-target binding interactions of the drug candidate [32]. If off-target hit is detected from this microarray, those off-target interactions can be further investigated using an optical label-free technique, biolayer interferometry (BLI), that measures molecular interaction in real time.

Functional Activity Assays: Lead candidates are assessed using in vitro and in vivo functional assays to generate dose-response curves and evaluate pharmacodynamic (PD) endpoints or biomarkers. Examples include cytokine release assay, cell activation/proliferation, or target cell cytotoxicity assay. Excessive cytokine release, also known "cytokine storm," is one of the adverse effects of therapeutic monoclonal antibodies. Therefore, during the drug candidate development, it is important to assess its cytokine modulation potential via cytokine release assay (CRA). CRA is performed in the whole blood and then in the isolated plasma cells with appropriate positive controls.

Cell lines used in the functional activity assay should have target expression and fitness comparable to humans, and various expression levels may be tested to account for target heterogeneity. These assays are particularly important if in vivo animal models are not cross-reactive or if PD cannot be assessed, aiding in human dose projection.

In Vivo Models

Understanding the additional ADME characteristics of a drug, such as oral bioavailability, clearance, distribution, and toxicity, is crucial for assessing its suitability for further development. These insights are essential for evaluating whether a lead drug candidate possesses the necessary ADME properties to justify filing an Investigational New Drug (IND) application and advancing clinical studies. Pharmacologically relevant species are chosen to establish mechanism of action (MoA), proof of biology, PK/PD relationships, and identify potential toxicities, including exaggerated pharmacology. If the lead molecule does not cross-react with disease models, a surrogate molecule with similar affinity and exposure is tested, usually in rodents. These models are essential for developing and validating mechanistic mathematical models based on in vivo exposure-response to aid in human dose projection. Target expression in animal models should closely match that in humans. Common in vivo models include xenograft, transgenic, and humanized mice, selected based on the mechanism of action.

1.4   Guidelines and Strategies for Selecting First-in-Human (FIH) Doses for New Molecular Entities (NMEs)

Selecting a First-in-Human (FIH) dose for a new molecular entity (NME) can be driven by either toxicology or pharmacology data. Toxicology-based approaches use metrics like the No Observed Adverse Effect Level (NOAEL), the Highest Non-Severely Toxic Dose (HNSTD) in non-rodents, or the Severely Toxic Dose to 10% of rodents (STD10). Pharmacology-driven methods focus on the Pharmacologically Active Dose (PAD) and the Minimum Anticipated Biologically Active Level (MABEL). While NOAEL, HNSTD, and STD10 are less conservative and often used in oncology or severe conditions, MABEL is a more conservative approach, emphasizing safety primarily for immunostimulatory biologics with agonistic MOA, especially when animal models do not represent heterogeneity of human

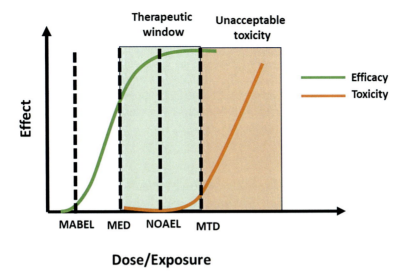

Fig. 1 Graphical representation of FIH dose as a function of effect and toxicity; MTD maximum tolerated dose and MED maximum effective dose

disease. The positions of NOAEL and MABEL on a dose-pharmacological response curve in relation to the therapeutic window and toxicity are illustrated in Fig. 1.

In 2005, the FDA recommended using NOAEL or PAD to calculate the Maximum Recommended Starting Dose (MRSD) for FIH trials. However, the TGN1412 incident highlighted the need for a more cautious approach. In response, the EMA recommended the MABEL approach, which uses the most sensitive functional assay to ensure a conservative dosing strategy, especially when animal models are not suitable.

After determining the FIH dose strategy, the Human Equivalent Dose (HED) is calculated based on body surface area or weight, with a safety factor applied to determine the MRSD for healthy volunteers. This safety factor varies based on the steepness of the dose-response curve, toxicity levels, and available preclinical data. For instance, a safety factor of 6 has been used in advanced cancer patients. However, HED calculation is not required for an NME with a molecular weight >100,000 Daltons, and such NME human doses should be normalized to mg/kg as obtained from toxicity studies with nonclinical species.

For PAD-based FIH doses, an integrated approach using both in vitro and in vivo data establishes the dose-response relationship. The MABEL approach is particularly conservative and commonly used for agonists and immune activators. The receptor occupancy (RO) can be estimated using the Hill equation, though this method is less suitable for bispecific constructs [33–37].

Mechanistic models that incorporate pharmacokinetics-pharmacodynamics (PK-PD) relationships are recommended for determining FIH doses. These models account for drug exposure, target characteristics, and target engagement duration.

As QSP models evolve, they offer refined approaches for determining FIH doses, balancing between traditional MABEL methods and newer quantitative systems pharmacology (QSP) models. For instance, the FIH dose of glofitamab, CD20 T cell engager, was estimated using a PK/PD approach while considering cytokine release (TCE MoA-based) data comparison between healthy animals (in vitro/in vivo) and B cell lymphoma patients (in vitro). This PK/PD approach-based FIH dose was approximately 35-fold higher than the MABEL-based starting dose calculated using EC_{20} for cytotoxicity/T cell activation [36, 37].

1.5 Optimizing First-In-Human Dose Selection for Biotherapeutics: Strategies and Considerations

Preclinical safety studies are essential for determining an appropriate starting dose and subsequent dose escalation for human trials. These studies must be conducted in pharmacologically relevant species, which, according to ICH S6 guidance [38], are species where the test material is active due to receptor expression, epitope recognition, and/or target-related pharmacological activity. However, traditional animal models may not always reflect the clinical candidate's binding affinity or functional activity, and some candidates might exhibit toxic effects in long-term studies due to high immunogenicity. In such cases, alternative approaches, as outlined by S6, include using transgenic or gene knockout models, disease-specific animal models, or surrogate molecules that closely resemble the clinical candidate in pharmacological activity, formulation, and production process. Comprehensive characterization of these surrogates through both in vitro and in vivo assessments is crucial for designing relevant toxicity studies and reducing clinical development risks.

When selecting an FIH dose for a biotherapeutic with no cross-reactivity in preclinical species, a conservative approach should be adopted. The FIH dose should be based on pharmacological activity (using the MABEL approach) rather than NOAEL, considering predicted pharmacokinetics (PK) and pharmacodynamics (PD) from surrogate studies. Quantitative risk assessment (i.e., standard safety margin calculation) is not suitable for surrogate molecules. This approach aligns with the ICH S6(R1) guideline, which notes that homologous protein studies are useful for hazard detection but not for quantitative risk assessment.

For example, when selecting an FIH dose for an immunomodulatory biotherapeutic, in vitro data such as cytokine release, phagocytosis, and protein binding selectivity should indicate no increased risks to human subjects. The starting dose should be projected based on Cmax and pharmacological effects modeled from human concentration-time curves. Doses can be escalated up to a predetermined level if no stopping criteria are met, enabling a comprehensive assessment of the exposure-response relationship. Additionally, minimal physiologically based pharmacokinetic (PBPK) models should be updated to cover variability and ensure adequate exposure for efficacy studies.

For multi-domain immunomodulatory biotherapeutics, such as T-cell and NK-cell engagers, the MABEL approach is recommended due to novel mechanisms of action and potential immune activation. MABEL-based FIH doses often are significantly lower than those considered efficacious in cancer patients. To avoid using significantly low FIH doses, original and expanded "non-MABEL approaches" were proposed respectively in a data- and/or risk-based manner for all biopharmaceuticals. These approaches provide a risk assessment-based decision tree to guide whether a MABEL approach should be considered [39].

1.6   Translating Animal Data to Human Doses: Key Considerations for Accurate Dose Prediction

When predicting human starting doses for new drugs, it is essential to account for differences in target expression, drug-target affinity, and the duration of target engagement between animals and humans. Minimizing these differences enhances the confidence in translating animal data to human dosing.

Accurate prediction of an efficacious dose in humans is crucial for designing successful clinical trials. This involves integrating both in vitro and in vivo data to forecast the appropriate dose. For drugs cross-reactive with non-human primates (NHP), pharmacokinetics (PK) and pharmacodynamics (PD) or biomarkers are assessed in these models. Rodent models are typically used to understand exposure-response kinetics, which helps in developing mathematical PK/PD models and correlating in vitro findings with in vivo results.

When human PK projection is needed, allometry based on body weight is applied using PK data from higher-order species, such as NHP. The following equation is used to predict human clearance:

$$Clearance(human) = clearance(monkey) * (BW_{monkey} / BW_{human})^w \; W \; is \; 0.85.$$

For drugs exhibiting target-mediated drug disposition (TMDD) and non-linear PK, often clearance from doses where TMDD is saturated is used for human projection [26]. This is because non-linear parameters like Michaelis-Menten constants (Km and Vmax) are not easily translatable. Mechanistic mathematical models with human parameters can also be utilized where target-specific kinetic is fixed based on NHP data.

1.7 Case Studies on the FIH Dose Prediction for Immunomodulatory Biotherapeutics: Monoclonal Antibody and Multi-domain Antibody Therapeutics

FIH dose selection for an immunomodulatory biotherapeutic has been illustrated via a thermometer plot in Fig. 2. The in vitro pharmacology data for the biotherapeutic does not indicate any specific or increased risks to human subjects. At the starting dose, the Cmax is projected, and the pharmacological effect is predicted based on modeling of the anticipated human concentration-time curve. A minimal physiologically based pharmacokinetic (PBPK) model should be updated to account for variability and uncertainty, ensuring adequate exposure coverage across a range of potentially efficacious doses to be studied in the diseased population.

In addition, the starting dose selection for a multi-domain immunomodulatory biotherapeutics (i.e., T-cell and NK-cell engager) is also based on the MABEL approach [40]. An example of a multi-domain immunomodulatory biotherapeutic is presented in Fig. 3. The reasons for choosing MABEL in this case are the novel mechanism of action and potential for activation of immune cells. Often, the MABEL-based approach led to an FIH dose at least 100-fold lower than the doses safely administered to cancer patients [40]. For a multi-specific monoclonal antibody, functional assays are more relevant for PA calculation than receptor occupancy based on high affinity targets. Weight of evidence from nonclinical

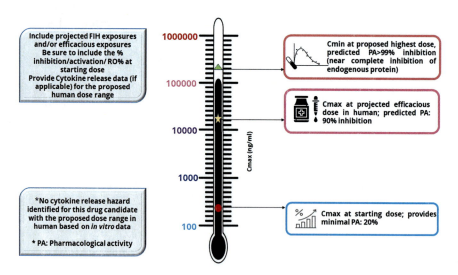

Fig. 2 FIH dose selection of an immunomodulatory biotherapeutic by MABEL approach: Dose range provides assessment over PA range, with minimal PA at starting dose. Here, the pharmacological activity (PA) is based on the target-dependent inhibition of a disease relevant peptide. In the nonclinical studies, a surrogate molecule was used due to lack of cross-reactivity of clinical candidate in any animal species. Therefore, no safety margin was calculated based on NOAEL

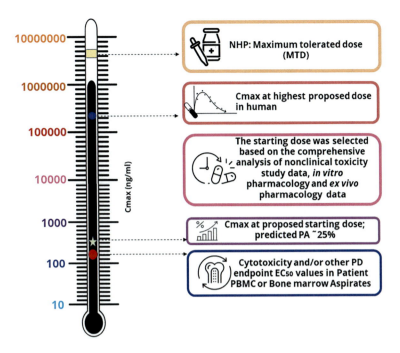

Fig. 3 FIH dose selection for a immunomodulatory multi-domain biotherapeutic based on MABEL approach due to its novel mechanism of action. Proposed dose with ~25% PA refers to only MABEL-based FIH dose selection. NOAEL was considered for FIH dose selection exercise, not affected FIH dose selection. No cytokine release and target organ-related toxicities were observed in the toxicity studies in relevant pharmacological species

pharmacology studies, toxicology, and available clinical experience with same target agents is most appropriate for FIH dose justification.

2 Conclusion

In conclusion, the successful development of biologics relies on a thorough understanding of pharmacokinetics (PK) and pharmacodynamics (PD), which are inherently more complex than those of small molecules. This chapter highlights the vital role of quantitative pharmacology in integrating PKPD data to inform dosing, efficacy, and safety decisions, ultimately improving patient outcomes. It emphasizes the importance of early pharmacology activities, including in vitro and in vivo studies, along with immunogenicity risk assessments and proactive strategies to mitigate potential risks.

Additionally, the chapter discusses key components supporting drug metabolism and pharmacokinetics (DMPK) activities, particularly the role of bioanalytical assays in measuring drug levels and metabolites. Various assay formats, including ligand binding assays and LC-MS/MS, are essential for accurate toxicity assessments and regulatory compliance. Method development and validation are critical, ensuring assays meet the required selectivity, sensitivity, and accuracy. By employing robust bioanalytical strategies, researchers can enhance the safety and efficacy of biologics, facilitating successful clinical outcomes.

Acknowledgments The authors would like to thank Dr. Rhea Bhattacharya for her help with figures and Dr. Yanshan Dai for his help with references.

References

1. Pineda C, Castaneda Hernandez G, Jacobs IA, Alvarez DF, Carini C. Assessing the immunogenicity of biopharmaceuticals. BioDrugs. 2016;30(3):195–206. https://doi.org/10.1007/s40259-016-0174-5. PubMed PMID: 27097915; PubMed Central PMCID: PMCPMC4875071.
2. Shakhnovich V, Meibohm B, Rosenberg A, Kierzek AM, Hasenkamp R, Funk RS, et al. Immunogenicity in clinical practice and drug development: when is it significant? Clin Transl Sci. 2020;13(2):219–23. https://doi.org/10.1111/cts.12717. PubMed PMID: 31762152; PubMed Central PMCID: PMCPMC7070797 for Clinical and Translational Science, Valentina Shakhnovich was not involved in the review or decision process for this paper
3. Cohen S, Myneni S, Batt A, Guerrero J, Brumm J, Chung S. Immunogenicity risk assessment for biotherapeutics through in vitro detection of CD134 and CD137 on T helper cells. MAbs. 2021;13(1):1898831. https://doi.org/10.1080/19420862.2021.1898831. PubMed PMID: 33729092; PubMed Central PMCID: PMCPMC7993230.
4. Jarvi NL, Balu-Iyer SV. A mechanistic marker-based screening tool to predict clinical immunogenicity of biologics. Commun Med (Lond). 2023;3(1):174. https://doi.org/10.1038/s43856-023-00413-7. PubMed PMID: 38066254; PubMed Central PMCID: PMCPMC10709359.
5. Chen X, Haddish-Berhane N, Moore P, Clark T, Yang Y, Li H, et al. Mechanistic projection of first-in-human dose for bispecific immunomodulatory P-cadherin LP-DART: an integrated PK/PD modeling approach. Clin Pharmacol Ther. 2016;100(3):232–41. https://doi.org/10.1002/cpt.393.
6. Dejager L, Banton S, Marques P, Rinikova G, Lory S, Hickford ES, et al. BiSim tool: a binding simulation tool to aid and simplify ligand-binding assay design and development. Bioanalysis. 2024;16(11):519–33. https://doi.org/10.4155/bio-2023-0242. PubMed PMID: 38629337; PubMed Central PMCID: PMCPMC11299794.
7. Lee JW, Kelley M, King LE, Yang J, Salimi-Moosavi H, Tang MT, et al. Bioanalytical approaches to quantify "total" and "free" therapeutic antibodies and their targets: technical challenges and PK/PD applications over the course of drug development. AAPS J. 2011;13(1):99–110. https://doi.org/10.1208/s12248-011-9251-3. PubMed PMID: 21240643; PubMed Central PMCID: PMCPMC3032085.
8. Staack RF, Jordan G, Heinrich J. Mathematical simulations for bioanalytical assay development: the (un-)necessity and (im-)possibility of free drug quantification. Bioanalysis. 2012;4(4):381–95. https://doi.org/10.4155/bio.11.321.
9. Talbot JJ, Calamba D, Pai M, Ma M, Thway TM. Measurement of free versus Total therapeutic monoclonal antibody in pharmacokinetic assessment is modulated by affinity, incubation time, and bioanalytical platform. AAPS J. 2015;17(6):1446–54. https://doi.org/10.1208/s12248-015-9807-8. PubMed PMID: 26265093; PubMed Central PMCID: PMCPMC4627453.

10. Rup B, O'Hara D. Critical ligand binding reagent preparation/selection: when specificity depends on reagents. AAPS J. 2007;9(2):E148–55. https://doi.org/10.1208/aapsj0902016. PubMed PMID: 17614357; PubMed Central PMCID: PMCPMC2751403.

11. Staack RF, Stracke JO, Stubenrauch K, Vogel R, Schleypen J, Papadimitriou A. Quality requirements for critical assay reagents used in bioanalysis of therapeutic proteins: what bioanalysts should know about their reagents. Bioanalysis. 2011;3(5):523–34. https://doi.org/10.4155/bio.11.16.

12. King LE, Farley E, Imazato M, Keefe J, Khan M, Ma M, et al. Ligand binding assay critical reagents and their stability: recommendations and best practices from the global bioanalysis consortium harmonization team. AAPS J. 2014;16(3):504–15. https://doi.org/10.1208/s12248-014-9583-x. PubMed PMID: 24687208; PubMed Central PMCID: PMCPMC4012044.

13. Goldsmith SJ. Radioimmunoassay: review of basic principles. Semin Nucl Med. 1975;5(2):125–52. https://doi.org/10.1016/s0001-2998(75)80028-6.

14. Huang PC, Tsai CH, Liang WY, Li SS, Huang HB, Kuo PL. Early phthalates exposure in pregnant women is associated with alteration of thyroid hormones. PLoS One. 2016;11(7):e0159398. https://doi.org/10.1371/journal.pone.0159398. PubMed PMID: 27455052; PubMed Central PMCID: PMCPMC4959782.

15. Jian W, Edom RW, Weng N. Important considerations for quantitation of small-molecule biomarkers using LC-MS. Bioanalysis. 2012;4(20):2431–4. https://doi.org/10.4155/bio.12.247.

16. Pitt JJ. Principles and applications of liquid chromatography-mass spectrometry in clinical biochemistry. Clin Biochem Rev. 2009;30(1):19–34. PubMed PMID: 19224008; PubMed Central PMCID: PMCPMC2643089

17. Wang J, Gu H, Liu A, Kozhich A, Rangan V, Myler H, et al. Antibody-drug conjugate bioanalysis using LB-LC-MS/MS hybrid assays: strategies, methodology and correlation to ligand-binding assays. Bioanalysis. 2016;8(13):1383–401. https://doi.org/10.4155/bio-2016-0017.

18. Myler H, Rangan VS, Wang J, Kozhich A, Cummings JA, Neely R, et al. An integrated multi-platform bioanalytical strategy for antibody-drug conjugates: a novel case study. Bioanalysis. 2015;7(13):1569–82. https://doi.org/10.4155/bio.15.80.

19. Chilewski SD, Dickerson WM, Mora JR, Saab A, Alderman EM. Evaluation of acoustic membrane microparticle (AMMP) technology for a sensitive ligand binding assay to support pharmacokinetic determinations of a biotherapeutic. AAPS J. 2014;16(6):1366–71. https://doi.org/10.1208/s12248-014-9659-7. PubMed PMID: 25245223; PubMed Central PMCID: PMCPMC4389748.

20. Fischer SK, Joyce A, Spengler M, Yang TY, Zhuang Y, Fjording MS, et al. Emerging technologies to increase ligand binding assay sensitivity. AAPS J. 2015;17(1):93–101. https://doi.org/10.1208/s12248-014-9682-8. PubMed PMID: 25331105; PubMed Central PMCID: PMCPMC4287293.

21. Mora J, Given Chunyk A, Dysinger M, Purushothama S, Ricks C, Osterlund K, et al. Next generation ligand binding assays-review of emerging technologies' capabilities to enhance throughput and multiplexing. AAPS J. 2014;16(6):1175–84. https://doi.org/10.1208/s12248-014-9660-1. PubMed PMID: 25193269; PubMed Central PMCID: PMCPMC4389750.

22. Purushothama S, Dysinger M, Chen Y, Osterlund K, Mora J, Chunyk AG, et al. Emerging technologies for biotherapeutic bioanalysis from a high-throughput and multiplexing perspective: insights from an AAPS emerging technology action program committee. Bioanalysis. 2018;10(3):181–94. https://doi.org/10.4155/bio-2017-0196.

23. Vandivort TC, Horton DB, Johnson SB. Regulatory and strategic considerations for addressing immunogenicity and related responses in biopharmaceutical development programs. J Clin Transl Sci. 2020;4(6):547–55. https://doi.org/10.1017/cts.2020.493. PubMed PMID: 33948231; PubMed Central PMCID: PMCPMC8057416.

24. ICH. M10 Bioanalytical Method Validation and Study Sample Analysis. https://www.ema.europa.eu/en/ich-m10-bioanalytical-method-validation-scientific-guideline2022.

25. FDA. Bioanalytical Method Validation Guidance for Industry. https://www.fda.gov/files/drugs/published/Bioanalytical-Method-Validation-Guidance-for-Industry.pdf2018.

26. Kaiser PK, Kodjikian L, Korobelnik JF, Winkler J, Torri A, Zeitz O, et al. Systemic pharmacokinetic/pharmacodynamic analysis of intravitreal aflibercept injection in patients with retinal diseases. BMJ Open Ophthalmol. 2019;4(1):e000185. https://doi.org/10.1136/bmjophth-2018-000185. PubMed PMID: 30997397; PubMed Central PMCID: PMCPMC6440611.

27. Wu X, Su J, Wei J, Jiang N, Ge X. Recent advances in three-dimensional stem cell culture systems and applications. Stem Cells Int. 2021;2021:9477332. https://doi.org/10.1155/2021/9477332. PubMed PMID: 34671401; PubMed Central PMCID: PMCPMC8523294.

28. Yang H, Sun L, Liu M, Mao Y. Patient-derived organoids: a promising model for personalized cancer treatment. Gastroenterol Rep (Oxf). 2018;6(4):243–5. https://doi.org/10.1093/gastro/goy040. PubMed PMID: 30430011; PubMed Central PMCID: PMCPMC6225812.

29. Schneider CS, Bhargav AG, Perez JG, Wadajkar AS, Winkles JA, Woodworth GF, et al. Surface plasmon resonance as a high throughput method to evaluate specific and non-specific binding of nanotherapeutics. J Control Release. 2015;219:331–44. https://doi.org/10.1016/j.jconrel.2015.09.048. PubMed PMID: 26415854; PubMed Central PMCID: PMCPMC4656072.

30. Wang X, Phan MM, Sun Y, Koerber JT, Ho H, Chen Y, et al. Development of an SPR-based binding assay for characterization of anti-CD20 antibodies to CD20 expressed on extracellular vesicles. Anal Biochem. 2022;646:114635. https://doi.org/10.1016/j.ab.2022.114635.

31. Fujii E, Kato A. Therapeutic antibodies: technical points to consider in tissue cross-reactivity studies. J Toxicol Pathol. 2024;37(3):101–7. https://doi.org/10.1293/tox.2024-0033. PubMed PMID: 38962261; PubMed Central PMCID: PMCPMC11219190.

32. Freeth J, Soden J. New advances in cell microarray technology to expand applications in target deconvolution and off-target screening. SLAS Discov. 2020;25(2):223–30. https://doi.org/10.1177/2472555219897567.

33. EMA. Guideline on strategies to identify and mitigate risks for first-in-human and early clinical trials with investigational medicinal products. https://www.ema.europa.eu/en/documents/scientific-guideline/guideline-strategies-identify-and-mitigate-risks-first-human-and-early-clinical-trials-investigational-medicinal-products-revision-1_en.pdf2017.

34. FDA. Guidance for Industry Estimating the Maximum Safe Starting Dose in Initial Clinical Trials for Therapeutics in Adult Healthy Volunteers https://www.fda.gov/media/72309/download2005.

35. McCarthy MW. Clinical pharmacokinetics and pharmacodynamics of Eravacycline. Clin Pharmacokinet. 2019;58(9):1149–53. https://doi.org/10.1007/s40262-019-00767-z.

36. Frances N, Bacac M, Bray-French K, Christen F, Hinton H, Husar E, et al. Novel in vivo and in vitro pharmacokinetic/Pharmacodynamic-based human starting dose selection for Glofitamab. J Pharm Sci. 2022;111(4):1208–18. https://doi.org/10.1016/j.xphs.2021.12.019.

37. Dudal S, Hinton H, Giusti AM, Bacac M, Muller M, Fauti T, et al. Application of a MABEL approach for a T-cell-bispecific monoclonal antibody: CEA TCB. J Immunother. 2016;39(7):279–89. https://doi.org/10.1097/CJI.0000000000000132.

38. EMA. ICH S6 (R1) Preclinical safety evaluation of biotechnology-derived pharmaceuticals – Scientific guideline. https://www.ema.europa.eu/en/ich-s6-r1-preclinical-safety-evaluation-biotechnology-derived-pharmaceuticals-scientific-guideline1998.

39. Matsumoto M, Polli JR, Swaminathan SK, Datta K, Kampershroer C, Fortin MC, et al. Beyond MABEL: an integrative approach to first in human dose selection of Immunomodulators by the Health and Environmental Sciences Institute (HESI) Immuno-safety technical committee (ITC). Clin Pharmacol Ther. 2024; https://doi.org/10.1002/cpt.3316.

40. Saber H, Gudi R, Manning M, Wearne E, Leighton JK. An FDA oncology analysis of immune activating products and first-in-human dose selection. Regul Toxicol Pharmacol. 2016;81:448–56. https://doi.org/10.1016/j.yrtph.2016.10.002.

Biomarkers

Pradip Nair, O. S. Bindhu, and Sayeeda Mussavira

Abstract Biomarkers have emerged as pivotal tools in the realm of clinical development, driving significant advancements in precision medicine. Their relevance lies in their ability to provide crucial insights into the molecular underpinnings of diseases, which facilitates the stratification of patients based on disease subtypes and predicted responses to therapies. This enables the design of more targeted and effective clinical trials, thereby accelerating the drug development process. Biomarkers also play a crucial role in early diagnosis, disease monitoring, and the assessment of treatment efficacy and safety, ensuring that therapeutic interventions are tailored to individual patient profiles. The integration of biomarkers into clinical development pipelines has not only enhanced our understanding of disease mechanisms but has also led to the identification of novel therapeutic targets. Challenges remain, including the need for rigorous validation and standardization of biomarkers, but the potential benefits underscore their critical importance. This chapter explores the multifaceted roles of biomarkers, providing a comprehensive overview of their contributions to clinical development and their impact on the future of personalized medicine. By elucidating the challenges and opportunities inherent in biomarker research, we aim to highlight their transformative potential in the pursuit of optimized healthcare outcomes.

Keywords Normal · Pathogenic · Pharmacology · Therapeutic · Diagnosis · Prognosis · Monitoring · Antecedent · Screening · Diagnostic · Staging · Discovery · Qualification · Validation · Regulatory · Systems biology · Machine learning · Computational omics · Surrogate markers · Companion diagnostics · Accredited laboratories

P. Nair (✉)
Syngene International (For Bicara Therapeutics), Bangalore, India

O. S. Bindhu
JAIN (Deemed-to-be University), Bangalore, India

S. Mussavira
Maharani Cluster University, Bangalore, India

1 Introduction

Biomarkers, which are biological molecules indicative of normal or pathological processes, or responses to therapeutic interventions, serve as critical tools in understanding disease mechanisms and enhancing drug development. The rise of biomarker-based clinical diagnosis and outcomes, in recent years, has played a transformative role in the evolution of precision medicine.

Traditionally, clinical trials followed a one-size-fits-all approach, often leading to variable outcomes and suboptimal therapeutic efficacy across diverse patient populations. Biomarkers address this challenge by facilitating a more personalized approach to medicine, allowing for the stratification of patients based on specific biological characteristics. This targeted stratification not only improves the selection of appropriate patient cohorts for clinical trials but also enhances the probability of success in drug development by aligning therapeutic strategies with the unique genetic and molecular profiles of patients.

In the realm of early diagnosis, biomarkers hold immense promise. They enable the detection of diseases at nascent stages, often before clinical symptoms manifest, thus paving way for timely intervention and improved prognosis. In disease monitoring and therapeutic assessment, biomarkers provide real-time insights into the biological responses to treatment, allowing clinicians to gauge the efficacy and safety of therapeutic regimens promptly. This dynamic monitoring capability is crucial for making informed decisions about treatment modifications and management strategies.

Moreover, the identification and validation of novel biomarkers open new avenues for therapeutic target discovery. By elucidating the molecular pathways involved in disease pathogenesis, biomarkers drive the development of innovative therapies that are precisely tailored to address specific disease mechanisms.

Despite the remarkable potential and achievements of biomarkers, several challenges persist. Rigorous validation, standardization, and integration into clinical practice require concerted efforts from researchers, clinicians, regulatory authorities, and industry stakeholders. This chapter delves into the multifaceted roles of biomarkers in clinical development, exploring their impact on personalized medicine, challenges, and future directions. Through a detailed examination of key concepts, methodologies, and case studies, we aim to provide a comprehensive understanding of the crucial role biomarkers play in shaping the future of healthcare.

2 Relevance of Biomarkers

2.1 History

If one were to consider the broad description of a biomarker as referring to a biological parameter that can be measured or quantified accurately and reproducibly, then it would cover diverse clinical measurements spanning physiological measurements (e.g., diastolic pressure) to molecular (e.g., liver enzymes, blood glucose), histological, and imaging characteristics (e.g., angiography). With this broad definition, it could be said that biomarkers have been used for centuries as indicators of human health or for the diagnosis of pathological conditions. One of the oldest biomarkers used to diagnose disease was the measurement of arterial pulse as documented in Indian, Chinese, Egyptian, and Greek medicine [1, 2]. However, the currently accepted description of a "biomarker" has evolved over the last five decades tracking the scientific research and clinical progress during this period [3]. It is believed that the word "biomarker" was first coined in 1973 by Rho et al. [4] to describe the presence or absence of specific biologic material. However, terms such as "biochemical markers" by Mundkur (1949) [5] and "biological markers" by Porter in 1957 [6] predate the first use of the word "biomarker" by Rho. Since the 1980s, another term "surrogate" usually described as "surrogate endpoint" or "surrogate marker" has been used in the context of biomarker related to improvement in a specific disease. Today, usage of "biomarkers" is significantly higher than the use of "biochemical or biological markers" described earlier [7–9].

2.2 Definitions, Categories, and Relevance

To reduce ambiguity, National Institute of Health (NIH) biomarker definition working group (2000) defined "biomarkers" as indicators of normal biologic and pathogenic processes and/or pharmacological responses to a therapeutic intervention that are frequently measured and evaluated. The FDA, while broadly accepting the above definition, described "biomarkers" as a measurable indicator that has the potential to be useful across the entire disease process; research and development of therapies; complicating disease diagnosis, prognosis, and monitoring; or disease progression or response to treatment [10, 11].

There is no well-established categorization system for biomarkers, and multiple classification approaches have been reported by diverse study groups [12]. "Disease-related biomarker" and "therapy-related biomarkers" are the two broad categories of biomarkers presented. The former with subtypes of antecedent, screening, diagnostic, and staging and prognostic is intended to provide information about an individual's risk toward a disease in the future or the natural course of an already existing disease. The latter (predictive and monitoring) describes the response of the marker to a therapeutic intervention [13] (Table 1).

3 Characteristics of a Biomarker

3.1 "Biomarkers, EndpointS, and Other Tools" (BEST) Resource

Due to differing impressions about the appropriate definitions of biomarkers based on contexts of use, a joint task force was formed to forge common definitions and to make them publicly available through a continuously updated online document by

Table 1 Biomarker classification

Biomarker category	Discussion	Examples	References
Disease-related biomarker			
Antecedent biomarkers	For identifying the risk of developing an illness in future for preventive interventions. Currently focused from the background of personal genomic testing and the challenge to interpret the results in terms of clinical consequences	Cancer susceptibility genes BRCAI and BRCA2 for breast cancer	[14]
Screening biomarkers	For detecting subclinical disease. Enabling intervention at an earlier and potentially more curable stage than under usual clinical diagnostic conditions	PSA for prostate cancer	[15, 16]
Diagnostic and staging biomarkers	To recognize overt disease and categorizing disease severity respectively	Troponin T for acute coronary syndrome, pBNP for cardiomyopathy	[17]
Prognostic biomarkers	For predicting future disease course and outcome of individual patients or groups of patients in terms of a clinical endpoint, including recurrence of disease	Her2 and cholesterol for breast cancer and cardiovascular diseases respectively	[18, 19]
Therapy-related biomarkers			
Predictive biomarkers	To predict therapeutic intervention efficacy before its administering. Link a specific treatment with a specific clinical endpoint. For identifying patients most likely responding/not responding to a certain drug	Bcr-Abl mutation in CML for imatinib treatment HER2 expression in breast cancer for trastuzumab Ras mutation for not responding to panitumumab in colorectal cancer	[18, 20, 21]
Monitoring biomarkers	For information about therapeutic response post therapy. Contribute to adjusting the level/dose of intervention on a dynamic and personal basis	Blood glucose or HbAlc level in diabetes Tumor markers in oncology	[22, 23]

name "Biomarkers, EndpointS, and Other Tools" (BEST) [24]. Established in 2015, with the primary focus on terms related to study endpoints and biomarkers, FDA jointly with NIH identified the harmonization of terms used in translational science and medical product development as a priority need [10].

According to BEST, there are seven categories of biomarkers based on their clinical utility such as susceptibility and risk, diagnostic, monitoring, prognostic, predictive, pharmacodynamic and treatment response, and safety [25]. In recent years, biomarkers have become increasingly relevant to identify disease state, to progression, and in pharmaceutical discovery. Biomarkers have been key to identify a drug's mechanism of action (in vitro and in vivo pharmacology), investigating toxicity and efficacy signals at an early stage of the development process. Biomarkers have been used for patient stratification and identification of patients who are likely to respond to therapy. Finally, biomarkers have also been used to evaluate a patient's response to therapy [24].

BEST concept could significantly contribute to an improved matching ability between a biomarker with its suitable application and to improve pace, efficiency, and precision in the development of useful diagnostic and therapeutic technologies and strategies along with development and implementation of public health policies [10].

3.2   Susceptibility/Risk, Diagnostic, Monitoring, Prognostic, Predictive, Pharmacodynamic/Response, and Safety Biomarkers

Susceptibility/risk biomarkers are those that have got huge potential in terms of conductance of epidemiological studies for delineating the risk of disease. This group can indicate the likelihood for an individual to develop (increase/decrease chance) a disease or medical condition who currently does not have any apparent disease. Such biomarkers may be detected years/decades before the clinical presentation of signs and symptoms and have no relationship to any specific treatment. Genetic markers indicating an individual's susceptibility for cancer development such as BRCA1/2 mutation in breast cancer and level of LDL and HDL cholesterol relating to the risk of coronary artery disease are a few examples in clinical practice that guide towards preventive strategies. Age, sex, smoking status, and family history are also routinely considered additional factors in models of risk to improve the precision of the predictions [10, 24, 26].

Diagnostic/disease biomarkers are the tools to detect or confirm the presence of a disease or condition of interest or to identify individuals with a subtype of the disease and thus can play a role as diagnostic, prognostic, or disease classification marker. These biomarkers give the information by screening, disease monitoring or by measuring the level of internal precursors that probably can be attained by disease [10, 27]. Cardiac troponin (cardiac muscle injury), 3-hydroxy-fatty acids (planctomycetes), selected glycans (cancer), glutamate (visceral obesity / altered metabolism), catestatin (psychological stress in heart patients), cystatin-C (glomerular filtration), and liver-type fatty acid-binding protein (renal injury or oxidative stress) are a few examples to quote [24, 28, 29].

Monitoring biomarkers hold a key role in disease management and treatment. They are measured repeatedly over time for assessing disease progression (occurrence of new disease effects, worsening of previously existing abnormalities, or change in disease severity or specific abnormalities) and treatment response during a disease or condition (favorable / unfavorable). Examples include hemoglobin A1c

(HbA1c) in diabetes to monitor long-term glycemic control (disease progression or the treatment effectiveness), brain natriuretic peptide (BNP) level in heart failure (assess severity and guide treatment decisions), and CEA in colorectal cancer [30–32]. This biomarker category can assist in many ways during the course of an intervention. They help determine how a drug is metabolized by a patient (monitoring drug concentration), detect therapeutic effect or disease progression while on or following treatment, or detect toxicity, for example, PT and partial thromboplastin time (PTT) values to maintain warfarin or heparin levels within therapeutic range for patients undergoing anticoagulation treatment and monitoring of HCV-RNA levels to assess both the presence of hepatitis C infection that would benefit from treatment and evidence of response or non-response to treatment. Patients undergoing leflunomide and methotrexate therapy for rheumatoid arthritis are routinely assessed for evidence of liver toxicity by periodic measurement of liver enzymes. Serial imaging studies are used routinely for monitoring disease status in patients with solid tumors to detect regression or progression during or after therapy or to detect recurrence after disease-free status is achieved with initial therapy. This biomarker type also found application throughout medical product development such as during therapeutic or prevention trials of new drugs, biologics, or devices [10]. Such biomarkers may also be used for individual or population level survelliance to understand presence of disease or the risk of developing a disease. Annual physical examinations of healthy adults (serum cholesterol, blood glucose, and urine creatinine to evaluate the risk for, and to detect the emergence of, medical conditions such as hypercholesterolemia, diabetes, and impaired kidney function), periodic examinations of individuals by the National Health and Nutrition Survey (NHANES) (to monitor health, including tobacco use, and diet), epidemiologic study designed by the Air Force Health Study (AFHS) to assess health effects of exposure to herbicides (dioxin contaminants used by Air Force personnel during the Vietnam conflict), etc., can be considered as other examples [10, 33].

A *prognostic biomarker* is used to identify the likelihood of a clinical event, disease recurrence, or disease progression (identifying "rapid vs. slow" progressing) in patients with a disease or medical condition [27, 34]. They are routinely used in clinical trials to set trial entry and exclusion criteria to identify higher-risk populations. They are additionally important for predicting the risk of an event or poor outcome in an individual which helps in decision-making regarding duration of hospital stay and/or in intensive care units. Yet another major use of prognostic biomarkers is for resource allocation in population health: by stratifying the risk for both negative clinical and financial outcomes, a healthcare organization can distinguish which patients could benefit from more intensive evaluation while allowing others to avoid unnecessary additional diagnostic tests or medical interventions [24]. Prognostic biomarkers are often used as eligibility criteria in clinical trials to identify patients who are more likely to have clinical events or disease progression. Thus, they are widely used as enrichment factors in drug development [35]. The following are a few examples used for assessing diverse likelihoods in different medical conditions—BRCA1/2 gene mutations in breast cancer (second breast cancer), chromosome 17p deletions and TP53 mutations in chronic lymphocytic leukemia (death), increasing PSA in patients with prostate cancer during follow-up (progression), plasma fibrinogen to select patients with chronic obstructive pulmonary disease at high risk (exacerbation and/or all-cause mortality) for inclusion in interventional clinical trials, CRP level to identify patients with unstable angina or a history of acute myocardial infarction (recurrent coronary artery disease events), Gleason score for evaluating patients with prostate cancer (progression), and total kidney volume to select patients with autosomal dominant polycystic kidney disease (progressive decline in renal function) for inclusion in interventional clinical trials [10].

Predictive biomarkers are those whose presence or change assist in predicting an individual or group of individual's likelihood to experience a favorable or unfavorable effect from the exposure to a medical product or environmental agent. These play a crucial role in the designing and execution of clinical studies enriching them for more targeted approach. They assist in predicting patients who are likely to respond to a particular treatment or experience adverse effects. Proving a biomarker to be an effective predictive tool therefore has been a challenge [24]. The utility of predictive biomarkers can also be handy tools to assist in informing patient care decisions (determining the patients who might benefit from a particular treatment / in selecting among multiple interventions). Beyond these, they also find application in studies of exposures to environmental toxins, tobacco smoke, nicotine, alcohol, food additives, environmental or occupational radiation, or infectious agents or to studies of the unintended ancillary effects of interventions. Squamous differentiation in non-small-cell lung cancer (to identify patients to avoid pemetrexed treatment), specific mutations in cystic fibrosis transmembrane conductance regulator (CFTR) gene (to select patients more likely to respond to particular treatments), BRCA1/2 mutations in ovarian cancer (to identify patients likely to respond to poly (ADP-ribose) polymerase inhibitors) etc., are a few examples to present under this section [36–38].

Pharmacodynamic/response biomarker indicates the biologic activity of a medical product or environmental agent without necessarily drawing conclusions about efficacy or disease outcome or necessarily linking this activity to an established mechanism of action. They are potentially applied during dose selection, in clinical proof-of-concept studies or while measuring a response to medical products and/or environmental agents. If validated, these markers can also used as secondary endpoints in clinical trials. Circulating B lymphocytes (in systemic lupus erythematosus to assess response to a B-lymphocyte stimulator inhibitor), sweat chloride (cystic fibrosis to assess response to cystic fibrosis transmembrane regulator (CFTR) potentiating agents), urinary glycosaminoglycan levels (response to enzyme replacement therapy in mucopolysaccharidosis type 1), etc., are a few examples in practice [39–41]. Surrogate endpoint biomarkers are a closely related category which is an endpoint used in clinical trials as a substitute for a direct measure of how a patient feels, functions, or survives. A surrogate endpoint does not measure the clinical benefit of primary interest in and of itself, but rather is expected to predict that clinical benefit or harm based on epidemiologic, therapeutic, pathophysiologic, or other scientific evidence. Reduction in HbA1c, HIV-RNA, and LDL cholesterol are validated surrogate endpoints for reduction of microvascular complications, human immunodeficiency virus (HIV) clinical disease control and cardiovascular events, respectively [42–44].

Biomarkers measured before or after an exposure to a medical product or an environmental agent to indicate the likelihood, presence, or extent of toxicity as an adverse effect are known as safety biomarkers [10]. They help in the monitoring of organ-specific toxicity/ adverse effects that are critical to ensure the safe sustenance of a given therapy. The toxicity signal generated via change in biomarker can facilitate suitable modification of the treatment and avoidance of severe toxicity states. Measuring granulocyte count during clozapine use and serum potassium during diuretic use represent typical examples of this biomarker group. They can also direct the needed treatment; for example, hypokalemia with a diuretic can indicate the need for potassium supplementation, and hyperkalemia with an aldosterone antagonist can indicate the need for dose adjustment or increase in loop diuretics. Such biomarkers would ideally signal developing toxicity. Creatinine phosphokinase for drugs that can cause muscle damage, serum creatinine for potentially nephrotoxic drugs, and transaminases for potentially hepatotoxic drugs are a few examples [45–47].

4 Types of Biomarkers

Based on their characteristics, biomarkers are categorized into three major types: molecular, cellular, and imaging.

4.1 Molecular Biomarkers

These markers exhibit biophysical characteristics and can be analyzed in diverse biological samples (serum, plasma, cerebrospinal fluid, bronchoalveolar lavage fluid, and biopsies) with the help of proteomic and genomic techniques. They essentially find their value in disease diagnosis applications across analytic epidemiology, randomized clinical trials, disease prevention, prognosis, and management [48, 49]. This group encompasses a wide range of molecules (small to large molecules) including peptides, proteins, lipids metabolites, nucleic acids (DNA and RNA), and other molecules [50, 51]. Further this group is subdivided into three subtypes: chemical, protein, and genetic.

The data on inborn errors produced from metabolism or genetic conditions associated with cancers, metabolic disorders, infectious diseases, dietary intake, drug, chemical, exposure to pollution, etc., constitute the chemical subtype of molecular biomarkers. MarkerDB, the online database of molecular biomarkers, reveals 1089 chemical biomarkers' association with 448 conditions/diseases and 106 exposures, most of which can be quantified with accuracy and precision [52–70]. Protein biomarkers are extremely beneficial and indicate evolution of inflammation and immunity-/stress-associated diseases such as cancer, diabetes, cardiovascular, neurological disorders, and other syndromes. Information pertaining to 142 protein biomarkers covering more than 160 diseases can be found in MarkerDB [48, 61, 71]. Genetic variations such as DNA mutation, DNA single nucleotide polymorphism, karyotypic features, etc., have been the most extensively used under the genetic biomarker category for the diagnosis of diverse disorders. They can be determined in the DNA of nucleated cells sourced from any biological sample such as tumor tissue whose cells tend to accumulate somatic mutations. There are 26,374 genetic biomarkers and 154 karyotype biomarkers in MarkerDB. The association of DNA biomarkers (biggest category) with more than 319 diseases or conditions has been specified [61, 72, 73].

4.2   Cellular Biomarkers

These are biological and measurable parameters used in clinical/laboratory tests that can be often accessed in diverse body fluids or soft tissue. They are used to assess prognosis or probability to respond to a specific treatment. This category of biomarkers allows morphological and physiological property-based isolation, sorting, quantification, and characterization of cells [28]. Cell morphology and live-cell biophysical parameters (cell motility, contractility, force generation, and cytoskeletal dynamics) are considered as cellular phenotypic biomarkers. Such biomarkers can provide an integrated readout of several underlying biochemical and biological processes when compared to molecular biomarkers that evaluate discrete components of multiple specific biochemical pathways. Selected examples of cellular biomarkers in specific pathologies are given in Table 2 [74, 75].

4.3   Imaging Biomarkers

This category has been the most commonly used due to their ease of accessibility, cost-effectiveness, and non-invasiveness predominantly to understand how a disease develops, behaves, and responds to treatment [76]. These biomarkers, with the clinical advantage of early disease detection, are objectively evaluated and measured as an indicator of normal pathogenic/biological processes, or pharmacologic responses to a therapeutic intervention. They are of three types—positron emission tomography (PET), computed tomography (CT), and magnetic resonance imaging (MRI) [28, 77–81]. Table 2 is a compilation of diverse biomarker types and their representative examples along with the associated pathologies.

Accumulated literature also mentions an additional way of sorting biomarkers (Table 3) by their characteristic features. These can be quantified with objectivity and assessed as indicators of normal bio-processes, pathological procedures, or pharmacological outcomes of various drug therapies [27, 96].

Table 2 Types of biomarkers and their representative examples

Sl. No.	Biomarker type		Selected examples and associated disorders	References
1	Molecular biomarkers	Chemical biomarkers	Isofuran/f2-Isoprostanes in oxidative stress (R)-3-Hydroxybutyric acid in aggression D-serine in kidney/Alzheimer's disease N-acetyl aspartate in psychiatric/neurological disorders/age-related changes Myeloperoxidase in oxidative stress/ inflammation, rheumatoid arthritis/ cardiovascular diseases/atherosclerosis/ obesity/neurodegenerative diseases/diabetes/ liver diseases/cystic fibrosis, and cancer Tyrosine in endurance exercise/metabolic syndrome risk Glutamate in neuropsychiatric disorders N-acetyl-beta-D-glucosamine in renal impairment in diabetic nephropathy/acute rheumatic fever Hyaluronic acid in liver fibrosis Pyridinoline in osteoarthritis Cystatin-C as an alternative to creatinine in GFR (kidney function) Cholesterol in cancer/cardiovascular/ metabolic disorders	[52–54, 56, 59–61, 64, 66, 68, 70, 82]
		Protein biomarkers	Glycosylated or glycated hemoglobin (HbA1c) to diagnose and monitor diabetes Brain natriuretic peptide (BNP) to monitor heart failure C-reactive protein (CRP) in rheumatoid arthritis, lupus, and cardiovascular diseases Prostate-specific antigen PSA in the diagnosis and monitoring of prostate cancer Ki-67 as a prognostic biomarker in breast and prostate cancer HER2/neu status as predictive biomarker in breast cancer α-chain of haptoglobin as serum biomarker for ovarian cancer	[57, 58, 65, 67]
		Genetic biomarker	BRCA1 and BRCA2 genes as susceptibility/ risk biomarker in breast and ovarian cancer BRAF mutations in melanoma EGFR mutation status in non-small-cell lung cancer KRAS mutations as prognostic biomarkers in pancreatic cancer	[62, 83–86]

(continued)

Table 2 (continued)

Sl. No.	Biomarker type	Selected examples and associated disorders	References
2	Cellular biomarkers	Live cell phenotypic biomarkers in tumorigenesis (cell tortuosity, nucleus area and perimeter dynamics, cell spreading dynamics, cell migration velocity, cell stiffness, cell area/perimeter dynamics, retrograde actin flow velocity) Circulating tumor cells, cancer-associated macrophage-like cells and immune cells with phagocytosed tumor material (utility in prognostication and treatment responsiveness assessment) Circulating hybrid cells (tumor–macrophage fusion) (role in the metastatic cascade) Immune cells (diagnosis/prediction of long COVID)	[74, 75, 87, 88]
3	Imaging biomarkers	PET tracks and measures glucose uptake in the body cells (to assess the efficacy of oncologic treatments) CT (a diagnostic modality using ionizing radiation for monitoring tumor status and to get cross and 3D images from the human body) MRI (with high spatial resolution and superior soft-tissue contrast image provision to understand neurodegenerative and cancer diseases) Prostate Imaging Reporting and Data System (diagnosis of prostate cancer) Breast Imaging Reporting and Data System (diagnosis of breast cancer) Liver Imaging Reporting and Data System (diagnosis of hepatocellular cancer) Other methods for diagnosis of various diseases (X-ray, endoscopy, optical coherence tomography, mammography, ultrasonography, near-infrared spectroscopy)	[89–95]

Table 3 Biomarker categorization considering basic characteristics

Gene based biomarkers	Protein baed biomarkers	Metabolism based biomarkers	Source based biomarkers	Disease based biomarkers	Other types
DNA (Nuclear, Mitochondrial) RNA (mRNA, MicroRNA)	Proteins Peptides Antibodies	Lipids Carbohydrates Enzymes Metabolites	Blood Serum Plasma Tissue	Prediction Detection Diagnosis Prognosis	Imaging In-silico Pathological

5 Discovery and Validation of Biomarkers

Biomarker development has been understood as a systematic and directed activity that starts with recognition of the need for a biomarker followed by discovery and validation of candidate biomarkers. This includes diverse studies that aim at verifying the accuracy and reliability of the detection method, formulating a hypothesis for the context of use, and testing the association between the biomarker and the clinical outcomes reflecting the presence of the disease/its prognosis, therapeutic target engagement, response to an intervention, and/or potential to respond to an intervention. There exists a parallel connect between the progress of an intended purpose of the biomarker from research to clinical trials/clinical practice and the degree of validation evidence supporting the use of the biomarker [97]. Following part of this chapter focuses on the key strategies and assisting tools adopted in the process of biomarker discovery.

5.1 Preclinical Tools and Strategy in Development of Biomarkers

In vitro studies are predominantly conducted in controlled laboratory environments outside of living organisms, using patient-derived and established or primary and established cell lines and tissues. These studies play a key role in biomarker discovery by allowing researchers to investigate the molecular mechanisms underlying disease progression and response to therapy enabling the identification of potential biomarkers. In vitro studies provide valuable insights into how specific genes, proteins, or other molecules behave in response to different stimuli or conditions. Researchers can use in vitro studies to screen for biomarkers by exposing these cells to various treatments or disease-related factors and observing changes in gene expression, protein levels, or other molecular characteristics in two-dimensional or three-dimensional cell culture models. For example, researchers might compare gene expression profiles between healthy and diseased cells to identify potential biomarkers that are upregulated or downregulated in disease states. Additionally, the alteration of the potential biomarkers when interrogated with a specific therapeutic or a combination of drugs can also be assessed. Correlation of potential biomarkers with functional changes in proliferation, migration, or response to treatment can also be assessed. Recently, organoid co-cultures studies have led to accelerated efforts in anti-cancer drug screening and enabled better understanding of the evolution of this disease. Further, these studies have enabled and advanced our fundamental understanding of the mechanisms of action using high-throughput platforms that interrogate various biomarkers of "clinical" efficacy [98].

While in vitro models enable high-throughput screening of drugs, they fail to consider the full complexity of a living organism. For this reason, researchers conduct in vivo studies to understand biomarkers associated with disease progression and therapy response within the complex environment of a living organism. In vivo studies can assess the diagnostic, prognostic, or predictive value of biomarkers by measuring their levels in biological samples such as blood, urine, or tissue. For example, researchers might analyze the correlation between a biomarker's expression levels and disease severity or treatment response in animal models or patient populations. In vivo studies can also provide important information on the stability, distribution, and clearance of biomarkers in the body, which are key considerations for their clinical use. In vivo models using patient-derived biopsies involve implanting biopsy material either subcutaneously (PDx models) or orthotopically, allowing it to expand in vivo. These models theoretically preserve some histology, gene expression, and somatic genetics of the patient tumor. They have been extensively used recently to understand disease progression and response to therapy. Transgenic mice, mutant mice, and humanized animal models (human immune system reconstituted in mice) have been used to further establish the mechanism of action and detect potential biomarkers [99, 100].

By integrating data from both in vitro and in vivo studies, researchers can gain a comprehensive understanding of potential biomarkers and their relevance for clinical applications (Fig. 1). In vitro studies help identify candidate biomarkers and elucidate their underlying mechanisms, while in vivo studies validate their diagnostic or prognostic value in a physiological context. Together, these studies contribute to the development of robust biomarkers that have the potential to improve patient care by enabling early detection, more accurate diagnosis, and personalized treatment strategies. In summary, in vitro and in vivo studies are integral components of biomarker discovery efforts, providing complementary insights into the molecular mechanisms of diseases and the clinical relevance of potential biomarkers. By

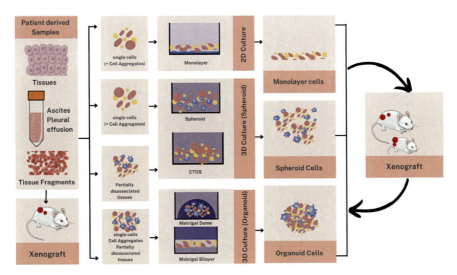

Fig. 1 In vivo *and* in vitro strategies adopted to identify potential biomarkers. (Figure adapted from Maru and Hippo [101])

combining data from both types of studies, researchers can advance our understanding of disease biology and develop biomarkers with diagnostic, prognostic, and therapeutic implications for improved patient outcomes [27, 28].

5.2   Systems Biology, Machine Learning, Computational Omics

Systems biology, machine learning, and computational omics are powerful tools that play a significant role in preclinical studies aimed at identifying potential clinical biomarkers. These approaches enable the integration of complex biological data, high-dimensional omics datasets, and network analyses to unravel disease mechanisms, identify novel biomarkers, and ultimately improve diagnostic and therapeutic strategies. Such data sets have evolved over the past decades with rapid advances in high throughput techniques such as genomic analysis including genome screening, whole exome screening, RNA analysis including sequencing and profiling, micro RNA analysis, and DNA methylation status at the cellular as well as tissue level [102]. Systems biology approaches allow the integration of genomics, transcriptomics, proteomics, metabolomics, and other omics data to comprehensively analyze molecular pathways and interactions associated with disease phenotypes. By combining multiple data types, researchers can uncover hidden patterns and identify potential biomarkers that may not be apparent through the analysis of individual omics datasets alone [103].

Machine learning algorithms, including random forest, support vector machines, and deep learning models, are extensively used in preclinical studies to analyze complex omics data and identify predictive biomarkers (Fig. 2). Machine learning

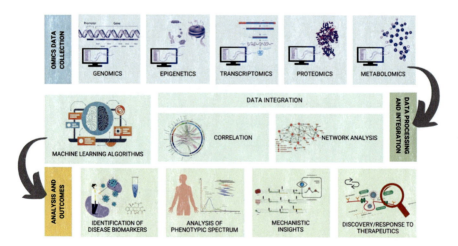

Fig. 2 Identification of predictive biomarkers using machine learning algorithms and other analytical tools. These can then be refined to develop artificial intelligence–based tools for biomarker detection and diagnosis. (Figure adapted form Chen et al. [105])

typically treats biomarker significance as a classification problem, such as determining the presence of a disease or the response to therapy. Before employing AI-based tools for biomarker detection and significance, machine learning should be able to assess accuracy, precision, recall, F1 score among others using a confusion matrix trained on the training data set and tested on the test dataset [104]. These algorithms can reveal subtle relationships between molecular features and clinical outcomes, aiding in the discovery of reliable biomarkers for disease diagnosis, prognosis, and treatment response prediction. Computational omics tools, such as network analysis and pathway modeling, assist researchers in understanding the interconnectedness of biological processes, identifying key regulatory nodes, and prioritizing potential biomarker candidates. By developing molecular interaction networks and pathway maps, researchers gain insights into disease mechanisms and identify biomarkers with diagnostic or therapeutic importance.

By leveraging the capabilities of systems biology, machine learning, and computational omics in preclinical studies, researchers can advance the identification and validation of potential clinical biomarkers with improved accuracy, efficiency, and translational value. These integrated approaches offer a holistic view of disease biology and facilitate the development of personalized medicine strategies based on molecular signatures and biomarker profiles [106–108]. As of 2023, FDA has now cleared 700 AI healthcare algorithms, more than 76% in radiology.

5.3 Challenges in Transition of Potential Biomarkers to Clinically Validated Targets

The journey from biomarker discovery to validation of a clinical biomarker is a critical process that determines the success and impact of the biomarker in healthcare. The success of this transition depends on several key factors and the following discussion focusses on those.

1. *Discovery Phase:* During the discovery phase, researchers identify potential biomarkers that show promise in association with a specific disease, condition, or treatment response. These biomarkers are often discovered through advanced technologies like genomics, proteomics, metabolomics, or imaging studies. Success in this phase is marked by the identification of novel biomarker candidates that exhibit significant associations with the intended clinical outcome.

2. *Preclinical Validation:* Before biomarkers can be tested in clinical settings, they need to undergo preclinical validation studies to demonstrate their biological relevance, specificity, and sensitivity. Success in this phase involves confirming the biomarker's performance in well-controlled experimental models, ensuring its potential clinical utility.

3. *Analytical Validation:* Once a biomarker candidate has been identified, it must undergo analytical validation to establish the technical parameters of the assay used for its measurement. Success in this phase is achieved by demonstrating the reproducibility, accuracy, precision, and reliability of the biomarker assay across different laboratories and platforms.

4. *Clinical Validation:* Clinical validation involves testing the biomarker in clinical samples to assess its diagnostic, prognostic, or predictive performance in relevant patient populations. Success in this phase is marked by robust evidence showing the biomarker's ability to differentiate disease states, predict outcomes, guide treatment decisions, or monitor therapeutic responses.

5. *Validation in Large Cohorts:* To further strengthen the validity of a biomarker, it is essential to validate its performance in large, diverse cohorts' representative of the target patient population. Success in this phase involves replicating the biomarker's performance across different populations, ensuring its applicability in real-world clinical settings.

6. *Validation in Prospective Studies:* Conducting prospective studies where biomarkers are evaluated in a longitudinal manner can provide stronger evidence of their clinical utility. Success in this phase involves demonstrating the biomarker's ability to impact patient outcomes and clinical decision-making when integrated into routine practice.

7. *Independent Validation:* Independent validation by multiple research groups or institutions is crucial to confirm the reliability and generalizability of a biomarker's performance. Success in this phase involves consistent replication of the biomarker's predictive or diagnostic value across independent validation studies.

8. *Regulatory Approval:* For a biomarker to be translated into clinical practice, it must meet regulatory standards set by agencies like the FDA. Success in this phase involves submitting comprehensive data on the biomarker's analytical and clinical performance, safety, and efficacy to obtain regulatory approval for its use in healthcare settings.

9. *Clinical Utility:* Ultimately, the success of a clinical biomarker lies in its demonstrated clinical utility and impact on patient outcomes. Biomarkers that effectively inform diagnosis, treatment selection, monitoring, or prognosis are more likely to be adopted in clinical practice and improve patient care.

Overall, the success of biomarker discovery to validation of a clinical biomarker hinges on rigorous scientific validation, robust evidence generation, regulatory compliance, and demonstration of clinical utility. Collaboration among researchers, clinicians, industry partners, and regulatory agencies is essential to navigate the complexities of biomarker validation and bring novel biomarkers from discovery to meaningful clinical applications. These complex steps and rigor could probably explain the relatively low translation of potential biomarkers to clinically validated biomarkers in reality (less than 5%) [27, 109].

6   Clinical Biomarkers—Validation and Case Studies

6.1   Concept of Companion Diagnostics

Companion diagnostics are tests or assays that are used in conjunction with a specific drug to identify which patients are most likely to benefit from the treatment and which patients may experience adverse effects. These tests help personalize treatment by providing information about a patient's genetic makeup or biomarkers that can influence their response to a particular therapy. In essence, companion diagnostics allow for a more targeted and precise approach to healthcare [110]. One of the key benefits of companion diagnostics is that they can help healthcare providers determine the most appropriate treatment for individual patients, optimizing the effectiveness of the therapy while minimizing potential risks. By identifying patients who are most likely to respond well to a particular drug, companion diagnostics can improve patient outcomes and reduce unnecessary exposure to medications that may not be effective. Moreover, companion diagnostics play a crucial role in the field of precision medicine, which aims to tailor medical treatment to the unique characteristics of each patient. By identifying genetic mutations or biomarkers that are relevant to a specific disease or drug response, companion diagnostics enable a more personalized and efficient approach to healthcare [110].

Overall, companion diagnostics represent a significant advancement in the field of personalized medicine, offering the potential to revolutionize how we treat a wide range of diseases, from cancer to infectious diseases. By helping healthcare providers identify the most appropriate treatment for individual patients based on their specific characteristics, companion diagnostics hold the promise of improving patient outcomes and enhancing the overall quality of care [110, 111].

Companion diagnostics play a crucial role in the field of oncology by helping identify patients who are most likely to benefit from a specific treatment. Here are some examples of approved companion diagnostics in oncology:

1. HercepTest: HercepTest is a companion diagnostic test approved by the FDA for use in identifying breast cancer patients who are candidates for treatment with Herceptin (trastuzumab). This test detects the overexpression of the HER2 protein in breast cancer tumor tissue, which makes patients eligible for targeted therapy with Herceptin [112].

2. PD-L1 Tests: Various companion diagnostic tests that detect PD-L1 expression levels in tumor tissue are used to identify patients with certain types of cancer, such as non-small-cell lung cancer and melanoma, who are likely to benefit from immune checkpoint inhibitors targeting the PD-1/PD-L1 pathway, such as Keytruda (pembrolizumab) and Opdivo (nivolumab) [113].

3. EGFR Mutation Test: Companion diagnostic tests that detect mutations in the EGFR gene are used to identify non-small-cell lung cancer patients who are suitable for treatment with EGFR tyrosine kinase inhibitors (TKIs) like Tarceva (erlotinib) or Iressa (gefitinib). These tests help personalize treatment by targeting specific genetic alterations in the tumor [114].

4. BRCA Testing: BRCA testing is used in patients with breast, ovarian, or pancreatic cancer to identify mutations in the BRCA1 and BRCA2 genes. Patients with BRCA mutations may be eligible for targeted therapies, such as PARP inhibitors like Lynparza (olaparib) or Talzenna (talazoparib), which are designed to exploit DNA repair deficiencies in cancer cells [115].

5. ALK Fusion Test: Companion diagnostics that detect ALK gene rearrangements are used to identify non-small-cell lung cancer patients who may benefit from treatment with ALK inhibitors like Xalkori (crizotinib), Alecensa (alectinib), or

Zykadia (ceritinib). These tests help guide treatment decisions for patients with this specific genetic alteration [116, 117].

6. KRAS Mutation Test: KRAS mutation testing is used in colorectal cancer patients to identify mutations in the KRAS gene, which can impact the effectiveness of certain targeted therapies, such as anti-EGFR antibodies like Erbitux (cetuximab) and Vectibix (panitumumab). Patients with wild-type KRAS tumors are more likely to respond to these treatments [118].

7. ROS1 Fusion Test: Companion diagnostics that detect ROS1 gene fusions are used to identify non-small-cell lung cancer patients who may benefit from targeted therapy with drugs like Xalkori (crizotinib) or Rozlytrek (entrectinib). These tests help pinpoint specific genetic alterations that can drive cancer growth and guide treatment selection [119, 120].

These are just a few examples of approved companion diagnostics in oncology that are essential for personalizing treatment and improving outcomes for cancer patients. By identifying specific biomarkers or genetic alterations in tumors, these tests help oncologists make more informed decisions regarding the most effective and appropriate therapies for individual patients. Companion diagnostics are not only limited to oncology but are also increasingly being developed for other diseases to personalize treatment based on individual patient characteristics.

Here are some examples of companion diagnostics for diseases other than cancer:

1. Cystic Fibrosis (CF):
CFTR Mutation Testing: Cystic fibrosis transmembrane conductance regulator (CFTR) mutation testing serves as a companion diagnostic for CFTR modulator therapies in patients with cystic fibrosis. Identifying specific CFTR mutations helps determine eligibility for targeted therapies that can improve lung function and quality of life in CF patients [121].

2. HIV/AIDS:
HIV Drug Resistance Testing: Drug resistance testing is used as a companion diagnostic for antiretroviral therapy in HIV/AIDS management. Identifying drug-resistant mutations in the HIV genome can guide the selection of effective antiretroviral regimens tailored to individual patients, optimizing treatment outcomes and preventing treatment failure [122].

3. Cardiovascular Disease:
CYP2C19 Genotype Testing: CYP2C19 genotype testing is a companion diagnostic for antiplatelet therapy in patients undergoing percutaneous coronary intervention (PCI). Variants in the CYP2C19 gene affect the metabolism of clopidogrel, a commonly prescribed antiplatelet drug, and can influence treatment response. Genotype-guided therapy helps identify patients at increased risk of adverse cardiovascular events and enables personalized antiplatelet treatment selection [123].

4. Alzheimer's Disease:
APOE Genotyping: Apolipoprotein E (APOE) genotyping is a companion diagnostic for Alzheimer's disease risk assessment. The APOE gene plays a role in the risk of developing late-onset Alzheimer's disease, with specific APOE

alleles associated with increased or decreased disease susceptibility. Genotyping for APOE variants can provide valuable information for assessing Alzheimer's disease risk and guiding personalized care and management strategies [124].

5. Infectious Diseases:

HLA-B*5701 Testing for HIV and HCV: HLA-B*5701 testing serves as a companion diagnostic for abacavir (HIV) and certain direct-acting antiviral drugs (HCV) to prevent severe hypersensitivity reactions. Screening for the HLA-B*5701 allele prior to initiating treatment helps identify individuals at risk of serious adverse drug reactions and enables the selection of safe and effective antiretroviral or antiviral therapies [125].

6. Celiac Disease:

HLA-DQ2/DQ8 Genotyping: HLA-DQ2/DQ8 genotyping is a companion diagnostic for celiac disease risk assessment. The presence of HLA-DQ2 and/or HLA-DQ8 haplotypes is strongly associated with susceptibility to celiac disease, a gluten-sensitive autoimmune condition. Genotyping for HLA-DQ2/DQ8 can aid in diagnosing celiac disease, especially in atypical or asymptomatic cases, and inform dietary management strategies [126].

7. Psychiatric Disorders:

Genetic Testing for Psychotropic Medications: Genetic testing can serve as a companion diagnostic to guide the selection and dosing of psychotropic medications used in the treatment of psychiatric disorders such as depression, anxiety, and bipolar disorder. Pharmacogenomic testing identifies genetic variants that influence drug metabolism and response, helping clinicians optimize medication choices, dosage adjustments, and minimize adverse effects in mental health treatment [127].

8. Hypercholesterolemia:

LDLR Gene Mutation Testing: LDL receptor (LDLR) gene mutation testing can act as a companion diagnostic for familial hypercholesterolemia, an inherited condition associated with high cholesterol levels and increased cardiovascular risk. Identifying pathogenic LDLR mutations can aid in diagnosing familial hypercholesterolemia, guiding treatment decisions, and implementing personalized lipid-lowering therapies to reduce cardiovascular morbidity and mortality [128].

By expanding the application of companion diagnostics beyond cancer, healthcare providers can harness the power of precision medicine to tailor treatments to individual patients' characteristics, improve therapeutic outcomes, minimize adverse effects, and advance personalized healthcare across a wide range of diseases and conditions.

6.2 Regulatory Pathway for Companion Diagnostics and Biomarkers

The regulatory pathway for the approval of biomarkers and companion diagnostics involves several key steps to ensure their safety, efficacy, and accuracy in guiding treatment decisions for patients. In the United States, biomarkers and companion diagnostics are regulated by the Food and Drug Administration (FDA) under the Centre for Devices and Radiological Health (CDRH) and the Center for Drug Evaluation and Research (CDER). The regulatory pathway for approval typically involves the following steps [129]:

1. Biomarker Discovery: The process begins with the identification of a biomarker that shows promise in predicting treatment response or disease progression. This may involve preclinical studies, epidemiological research, or other forms of scientific investigation [129].
2. Analytical Validation: Before a biomarker or companion diagnostic can be used in clinical practice, its analytical performance must be thoroughly evaluated. This includes assessing factors such as sensitivity, specificity, accuracy, precision, and reproducibility [129].
3. Clinical Validation: Once the biomarker or companion diagnostic has been analytically validated, it must undergo clinical validation to demonstrate its clinical utility in a real-world setting. This involves conducting well-designed clinical studies to evaluate the biomarker's ability to predict treatment outcomes or disease progression [129].
4. Regulatory Submission: The sponsor of the biomarker or companion diagnostic must submit a regulatory application to the FDA for review and approval. The type of submission will depend on the intended use of the product, its risk classification, and other factors [110].
5. FDA Review: The FDA will review the submission to determine whether the biomarker or companion diagnostic meets the necessary regulatory standards for safety and effectiveness. This review may involve consultation with an advisory committee of external experts [130, 131].
6. Post-Market Surveillance: After approval, the biomarker or companion diagnostic may be subject to post-market surveillance requirements to monitor its performance in real-world clinical practice. This helps ensure ongoing safety and effectiveness [132].

It is important to note that the regulatory pathway for biomarkers and companion diagnostics may vary depending on the specific intended use of the product, its risk classification, and other factors. For example, some biomarkers or companion diagnostics may be regulated as medical devices under the FDA's premarket notification (510(k)) or premarket approval (PMA) pathways, while others may be considered in vitro diagnostic tests or laboratory-developed tests.

Overall, the regulatory pathway for the approval of biomarkers and companion diagnostics aims to ensure that these products are safe, effective, and accurate in guiding treatment decisions for patients. By following these steps, developers can bring innovative biomarkers and companion diagnostics to market and improve patient outcomes in a responsible and evidence-based manner.

6.3 Examples of Validated Biomarkers and Case Studies

This section covers a few examples of biomarkers validated in the clinic across different diseases:

1. *Breast Cancer:* HER2 (human epidermal growth factor receptor 2, a receptor tyrosine kinase overexpressed in some breast tumours) is a biomarker that predicts response to targeted therapies like trastuzumab in breast cancer patients. Its overexpression is associated with aggressive tumor behavior. HER2 testing has become standard practice in determining treatment options for breast cancer patients. Multiple qualified test assay kits are commercially available now for accurate determination of HER2 status (at protein and gene levels) to evaluate tumors and to support optimal patient management. ELISA kits for Human Her-2 (ErbB2) facilitates quantification in various samples such as plasma, serum, supernatant (Invitrogen). A simple, sensitive, reproducible quantitative real-time PCR assay developed by Mylab Discovery Solutions supports the amplification of HER2 gene along with a reference gene in the same tube, analyze the copy-number ratio between them, and facilitate HER2 detection and quantitation [133]. FDA-approved dual-probe fluorescence in situ hybridization (FISH) assay kit, PathVysion pharmDx (dual-probe) and INFORM (mono-probe) FDA-approved FISH kits, the SPoT-Light CISH kits, immunohistochemistry (IHC) staining assay, HercepTest, Dako and Pathway, Ventana are a few options available currently [134, 135].

2. *Prostate Cancer:* Prostate-specific antigen (PSA) is a biomarker used for screening, diagnosis, and monitoring prostate cancer. Elevated levels of PSA can indicate prostate cancer or other prostate conditions. PSA testing is a highly sensitive serum biomarker that has revolutionized the early detection of prostate cancer over the past 20 years. In blood PSA occurs in two forms—complexed PSA which is protein-attached form and percent-free PSA that floats freely [15]. A standard PSA test typically measures total PSA—both attached and unattached forms. With an abnormal PSA test result, testing different types of PSA will be considered to decide the need of biopsy [16]. Human blood-/serum- or plasma-based lateral flow chromatographic immunoassay (On Site PSA Semi-quantitative Rapid Test) for the semi-quantitative detection of PSA is a preferred noninvasive screening method for prostate cancer [136]. Blood-/serum-/plasma-based diagnostic kit developed by AccuQuik™ (Prostate-Specific Antigen Test Kit) detects the level of total PSA using colloidal gold conjugate and anti-PSA antibodies. Ichroma Carcinoembryonic Antigen (CEA) kit, I-Chroma PSA Test Kit, and Wondfo Finecare PSA (Prostate Specific Antigen) Rapid Quantitative Test are few such examples [137].

3. *Alzheimer's Disease:* Loss of neurons, formation of extracellular senile plaques (amyloid β-protein as primary component), and intracellular neurofibrillary tangles (hyperphosphorylated form of Tau protein as primary component) are considered as the major hallmarks in the pathology of AD. During AD progression, tau is hyperphosphorylated and subsequently dissociated from microtu-

bule and polymerized into paired helical filaments. The degenerative process associated with AD well before its clinical onset and hence its early (presymptomatic) detection is crucial for its efficient management and treatment [138]. Efforts made to develop diagnostic indices of AD include the development of diagnostic kits like WO2002/088706 and WO2010/144634 which targets glutamine synthetase as an index and the degree of DNA methylation as an epigenetic marker. Considering the expression of miRNAs in AD pathology, studies have come out with new methods for early diagnosis such as fluorescent nanoparticle imaging-based kit with a probe molecule comprising an oligonucleotide capable of detecting one or more AD-specific microRNAs and biomarkers related to AD [139]. Cerebrospinal fluid biomarkers such as amyloid-beta and tau proteins are used in the diagnosis of Alzheimer's disease. These biomarkers even help in distinguishing Alzheimer's from other forms of dementia and tracking disease progression. There is also report on multiplex diagnostic kit, QPLEX™ Alz plus assay kit, capturing amyloid-β1-40, galectin-3 binding protein, angiotensin-converting enzyme, and periostin simultaneously. This kit utilizes microliters of peripheral blood and utilizes an optimized algorithm for early clinical diagnosis of AD by correlating with cerebral amyloid deposition [140]. Sysmex recently launched an assay kit to identify amyloid beta (Aβ) accumulation in brain [141]. The MILLIPLEX® MAP Human Amyloid Beta and Tau Panel is a 4-plex kit containing all reagents needed for simultaneous quantification of Aβ40, Aβ42, total tau proteins, and phosphorylated tau Thr181 (pTau181) in CSF samples [142].

4. *Colorectal Cancer:* Carcinoembryonic antigen (CEA) is a biomarker used for monitoring colorectal cancer. CEA levels in the blood are often elevated in patients with colorectal cancer and can be used to track treatment response or disease recurrence [15]. CEA rapid test kit (AccuQuik™ Diagnostic Detection) is an immunoassay test manufactured by AdvaCare Pharma. This kit contains antibodies that will react with antigens present in a patient's sample (serum or plasma). The rapid diagnostic test kits are produced either as a cassette containing a test strip or as an individual strip, both of which provide a quick and convenient method of diagnosis [143].

5. *Diabetes:* Hemoglobin A1c (HbA1c) is a biomarker used for monitoring long-term glucose control in patients with diabetes. HbA1c levels reflect average blood sugar levels over the past 2–3 months and are essential for adjusting diabetes treatment plans [144]. Numerous kits from diverse manufactures are available that measures HbA1C as a valuable indicator for long-term diabetic control. Atlas Direct Enzymatic Hemoglobin A1C (glycated hemoglobin A1C) reagents is one such example which is designed for the quantitative determination of stable HbA1C in the whole blood [145]. The Getein HbA1c Fast Test Kit and the hemoglobin A1c (HbA1c) Test Kit by Poclight are examples of tools used for monitoring glycemic control in diabetics [146, 147].

6. *Cardiovascular disease:* High-sensitivity C-reactive protein (hs-CRP) is a biomarker associated with inflammation and cardiovascular risk. Elevated levels of hs-CRP can indicate an increased risk of cardiovascular events, making it a valuable tool for risk stratification [148]. Human C-Reactive Protein ELISA Kit

for serum, plasma, cell culture supernatants, and urine by Sigma-Aldrich [149], hs-CRP+CRP Fast Test Kit (immunofluorescence assay) by Cosmic Scientific Technologies [150], and HS-CRP ELISA kit (Human High-Sensitivity C-Reactive Protein ELISA Kit) by MyBioSource [151] are a few representative examples.

7. *Lung Cancer:* Epidermal growth factor receptor (EGFR) mutations are bio-markers that predict response to EGFR inhibitors in non-small-cell lung cancer. Testing for EGFR mutations has become standard practice in selecting targeted therapies for lung cancer patients [152]. Cobas® EGFR Mutation Test v2 by Roche Diagnostics [153] is a real-time PCR-based approach to identify 42 mutations in the EGFR gene (in exons 18, 19, 20, and 21). This test follows a patient-friendly procedure which depends on liquid biopsy (plasma and not tis-sue biopsy/ Fine Needle Aspirate) from the patient, and it can be sampled fre-quently without putting patients at risk. This is designed to enable testing of both tissue and plasma specimens with a single kit. EntroGen is another manu-facturer of EGFR mutation analysis kits. Its *EGFR mutation analysis kit* detects the following mutations—Exon 18 (4 mutations), Exon 19 (44 mutations), Exon 20 (5 mutations), Exon 21 (2 mutations). Further, its *EGFR mutation analysis extension kit* detects mutations in Exon 18 (8 mutations) and Exon 20 (2 mutations) [154].

8. *HIV:* Viral load, measured as the amount of HIV RNA in the blood, is a crucial biomarker for monitoring HIV infection and response to antiretroviral therapy. Viral load testing helps in assessing treatment effectiveness and disease pro-gression [155]. Abbott RealTime HIV-1 assay is an in vitro reverse transcription-polymerase chain reaction assay for the quantitation of HIV-1 in whole blood spotted on cards as dried blood spots or human plasma from HIV-1 infected individuals [156]. AltoStar® HIV RT-PCR Kit 1.5 [157] and TRUPCR® HIV Viral Load Kit [158] are other such examples.

9. *Rheumatoid Arthritis:* Rheumatoid factor (RF) and anti-cyclic citrullinated peptide (anti-CCP) antibodies are biomarkers used in the diagnosis and progno-sis of rheumatoid arthritis. These biomarkers help differentiate rheumatoid arthritis from other forms of arthritis and predict disease severity [159, 160]. Getein Anti-CCP Fast Test Kit is used for the in vitro quantitative determination of anti-CCP in diverse samples (serum/plasma/whole blood) to aid in the diag-nosis of this pathology [161]. Anti-CCP ELISA test kit, an indirect solid-phase enzyme immunometric assay by Diagnostic Automation Inc., is designed for the quantitative determination of IgG class antibodies directed against CCP in serum or plasma [162]. Similarly there exists diverse fast performing diagnostic assay kits for the in vitro quantification of RF in serum/plasma/ whole blood [163].

10. *Ovarian Cancer:* Cancer antigen 125 (CA-125) is a protein antigen often found on the surface of ovarian cancer cells and is considered as a biomarker for the management of ovarian cancer. Elevated CA-125 levels can indicate ovarian cancer or other gynecological conditions, and monitoring CA-125 levels can help in assessing treatment response and disease progression [164]. CA-125 AccuBind ELISA Kits [165], Check-1 CA-125 Quantitative Rapid Test for

Easy Reader+® 15mins (74091) [166], CA125 Rapid Test Cassette (whole blood) [167], etc., are a few examples of approved diagnostic test kits.

These examples demonstrate the diverse applications of biomarkers in the clinic across a wide range of diseases. Validated biomarkers play a crucial role in personalized medicine, guiding treatment decisions, monitoring disease progression, and improving patient outcomes.

6.4 Requirements for Accredited Laboratories

Clinical Laboratory Improvement Amendments (CLIA) accreditation is a crucial certification for laboratories performing diagnostic testing on human samples. This federal mandate, overseen by the Centers for Medicare & Medicaid Services (CMS), ensures that laboratories meet specific quality standards, ensuring accuracy, reliability, and timeliness of test results regardless of where the test is performed. For laboratories focusing on biomarkers, which are biological molecules found in blood, other body fluids, or tissues indicating a normal or abnormal process, or a condition or disease, CLIA accreditation is particularly significant. Achieving CLIA accreditation involves several key components specific to biomarker testing:

1. *Proficiency Testing:* Laboratories must participate in proficiency testing, a program that compares a laboratory's test results against a standard or with results from other laboratories. This is critical for biomarkers to ensure that results are consistent and accurate across different testing platforms.
2. *Personnel Qualifications:* The regulations specify that staff members have the appropriate education, experience, and training to perform tests on biomarkers safely and effectively. This includes continuous education to keep up with advancements in biomarker testing technologies.
3. *Quality Control and Assurance:* Laboratories are required to implement rigorous quality control (QC) and quality assurance (QA) programs. These programs are crucial for biomarker testing, where the precision and accuracy of results can significantly impact patient diagnosis and treatment plans.
4. *Facility and Equipment Standards:* The physical laboratory environment, including the equipment used for testing biomarkers, must meet specific standards to ensure tests are performed safely and effectively.
5. *Inspections and Audits:* To maintain CLIA accreditation, laboratories undergo regular inspections and audits, ensuring ongoing compliance with CLIA standards. These inspections are critical for identifying any issues that could compromise the quality of biomarker testing. For researchers, clinicians, and patients, CLIA accreditation serves as an assurance that the laboratory adheres to stringent quality standards, making biomarker-based diagnostics and treatments more reliable and effective. It underlines the laboratory's commitment to excellence and safety in biomarker testing, which is fundamental in advancing personalized medicine and improving patient care outcomes [168].

References

1. Metwaly AM, et al. Traditional ancient Egyptian medicine: a review. Saudi J Biol Sci. 2021;28(10):5823–32.
2. Hajar R. The air of history: early medicine to galen (part I). Heart Views. 2012;13(3):120–8.
3. Jalalian SH, et al. Exosomes, new biomarkers in early cancer detection. Anal Biochem. 2019;571:1–13.
4. Rho JH, et al. A search for porphyrin biomarkers in Nonesuch Shale and extraterrestrial samples. Space Life Sci. 1973;4(1):69–77.
5. Mundkur BD. Evidence excluding mutations, polysomy, and polyploidy as possible causes of non-Mendelian segregations in Saccharomyces. Ann Mo Bot Gard. 1949;36(3):259–80.
6. Porter KA. Effect of homologous bone marrow injections in x-irradiated rabbits. Br J Exp Pathol. 1957;38(4):401–12.
7. Scarpelli DG, William B. "disease". Encyclopedia Britannica, 30 March 2024., https://www.britannica.com/science/disease. Accessed 4 May 2024.
8. Sturmberg JP, et al. Health and disease—emergent states resulting from adaptive social and biological network interactions. Front Med. 2019;6:59.
9. Landeck L, et al. Biomarkers and personalized medicine: current status and further perspectives with special focus on dermatology. Exp Dermatol. 2016;25(5):333–9.
10. FDA-NIH Biomarker Working Group. BEST (Biomarkers, EndpointS, and other Tools) Resource [Internet]. 2016. Available from: https://www.ncbi.nlm.nih.gov/books/NBK326791/.
11. Biomarkers Definitions Working Group. Biomarkers and surrogate endpoints: preferred definitions and conceptual framework. Clin Pharmacol Ther. 2001;69(3):89–95.
12. Detalle L, et al. 2.20 - Translational aspects in drug discovery. In: Chackalamannil S, Rotella D, Ward SE, editors. Comprehensive medicinal chemistry III. Oxford: Elsevier; 2017. p. 495–529.
13. Wacheck V. Biomarkers. In: Müller M, editor. Clinical pharmacology: current topics and case studies. Vienna: Springer Vienna; 2010. p. 225–39.
14. Brody LC, Biesecker BB. Breast cancer susceptibility genes. BRCA1 and BRCA2. Medicine (Baltimore). 1998;77(3):208–26.
15. Duffy MJ. Carcinoembryonic antigen as a marker for colorectal cancer: is it clinically useful? Clin Chem. 2001;47(4):624–30.
16. Tosoian J, Loeb S. PSA and beyond: the past, present, and future of investigative biomarkers for prostate cancer. ScientificWorldJournal. 2010;10:1919–31.
17. Rallidis LS, et al. NT-proBNP/cardiac troponin T ratio >7.5 on the second day of admission can differentiate Takotsubo from acute coronary syndrome with good accuracy. Hell J Cardiol. 2024;76:22–30.
18. Mitri Z, Constantine T, O'Regan R. The HER2 receptor in breast cancer: pathophysiology, clinical use, and new advances in therapy. Chemother Res Pract. 2012;2012:743193.
19. Jeong SM, et al. Effect of change in total cholesterol levels on cardiovascular disease among young adults. J Am Heart Assoc. 2018;7(12):e008819.
20. Branford S, et al. Detection of BCR-ABL mutations in patients with CML treated with imatinib is virtually always accompanied by clinical resistance, and mutations in the ATP phosphate-binding loop (P-loop) are associated with a poor prognosis. Blood. 2003;102(1):276–83.
21. Dean L, Kane M. Panitumumab therapy and RAS and BRAF genotype. In: Pratt VM, et al., editors. Medical genetics summaries [Internet]. Bethesda: National Center for Biotechnology Information (US); 2020.
22. Organization, G.W.H. Use of glycated haemoglobin (HbA1c) in the diagnosis of diabetes mellitus. Abbreviated Report of a WHO Consultation 2011. Available from: https://www.ncbi.nlm.nih.gov/books/NBK304271/.
23. Nagpal M, et al. Tumor markers: a diagnostic tool. Natl J Maxillofac Surg. 2016;7(1):17–20.

24. Califf RM. Biomarker definitions and their applications. Exp Biol Med (Maywood). 2018;243(3):213–21.
25. Shah A, Grimberg DC, Inman BA. Classification of molecular biomarkers. Société Internationale d'Urologie J. 2020;1(1):8–15.
26. FDA-NIH Biomarker Working Group. Susceptibility/risk biomarker. BEST (Biomarkers, EndpointS, and other Tools) Resource [Internet]. 2016 Dec 22 [Updated 2020 Aug 27]. Available from: https://www.ncbi.nlm.nih.gov/books/NBK402288/.
27. Ahmad A, Imran M, Ahsan H. Biomarkers as biomedical bioindicators: approaches and techniques for the detection, analysis, and validation of novel biomarkers of diseases. Pharmaceutics. 2023;15(6):1630.
28. Bodaghi A, Fattahi N, Ramazani A. Biomarkers: promising and valuable tools towards diagnosis, prognosis and treatment of Covid-19 and other diseases. Heliyon. 2023;9(2):e13323.
29. Group, F.-N.B.W. Diagnostic biomarker. BEST (Biomarkers, EndpointS, and other Tools) Resource [Internet]. Available from: https://www.ncbi.nlm.nih.gov/books/NBK402285/.
30. Gillery P. HbA(1c) and biomarkers of diabetes mellitus in Clinical Chemistry and Laboratory Medicine: ten years after. Clin Chem Lab Med. 2023;61(5):861–72.
31. Dahiya T, et al. Monitoring of BNP cardiac biomarker with major emphasis on biosensing methods: a review. Sensors Int. 2021;2:100103.
32. Kankanala VL, Mukkamalla SKR. Carcinoembryonic antigen. 2024 2023 Jan 23. Available from: https://www.ncbi.nlm.nih.gov/books/NBK578172.
33. FDA-NIH Biomarker Working Group. BEST (biomarkers, EndpointS, and other tools) resource [Internet]. Monitoring Biomarker. 2016. Available from: https://www.ncbi.nlm.nih.gov/books/NBK402282/.
34. Bischoff REH, Peter, editors. Comprehensive biomarker discovery and validation for clinical application. p.D.D.S.V. RSC Publishing.
35. U.S.F.a.D.A. 2012.
36. Liang J, et al. Mechanisms of resistance to pemetrexed in non-small cell lung cancer. Transl Lung Cancer Res. 2019;8(6):1107–18.
37. Brodlie M, et al. Targeted therapies to improve CFTR function in cystic fibrosis. Genome Med. 2015;7:101.
38. Faraoni I, Graziani G. Role of BRCA mutations in cancer treatment with poly(ADP-ribose) polymerase (PARP) inhibitors. Cancers (Basel). 2018;10(12).
39. Parodis I, Gatto M, Sjöwall C. B cells in systemic lupus erythematosus: targets of new therapies and surveillance tools. Front Med (Lausanne). 2022;9:487–507.
40. Sontag MK. Sweat chloride: the critical biomarker for cystic fibrosis trials. Am J Respir Crit Care Med. 2016;194(11):1311–3.
41. Kakkis E, Marsden D. Urinary glycosaminoglycans as a potential biomarker for evaluating treatment efficacy in subjects with mucopolysaccharidoses. Mol Genet Metab. 2020;130(1):7–15.
42. Ambrosi P, et al. Glycosylated hemoglobin as a surrogate for the prevention of cardiovascular events in cardiovascular outcome trials comparing new antidiabetic drugs to placebo. Cardiology. 2020;145(6):370–4.
43. Gilbert PB, et al. Virologic and regimen termination surrogate end points in AIDS clinical trials. JAMA. 2001;285(6):777–84.
44. Weintraub WS, Lüscher TF, Pocock S. The perils of surrogate endpoints. Eur Heart J. 2015;36(33):2212–8.
45. van Staa TP, et al. Predictors and outcomes of increases in creatine phosphokinase concentrations or rhabdomyolysis risk during statin treatment. Br J Clin Pharmacol. 2014;78(3):649–59.
46. Zeleke TK, et al. Nephrotoxic drug burden and predictors of exposure among patients with renal impairment in Ethiopia: a multi-center study. Heliyon. 2024;10(2):e24618.
47. Meunier L, Larrey D. Drug-induced liver injury: biomarkers, requirements, candidates, and validation. Front Pharmacol. 2019;10:1482.

48. Dhama K, et al. Biomarkers in stress related diseases/disorders: diagnostic, prognostic, and therapeutic values. Front Mol Biosci. 2019;6:91.

49. Picó C, et al. Biomarkers of nutrition and health: new tools for new approaches. Nutrients. 2019;11(5)

50. Rosenwald A, et al. The use of molecular profiling to predict survival after chemotherapy for diffuse large-B-cell lymphoma. N Engl J Med. 2002;346(25):1937–47.

51. Lukas A, Heinzel A, Mayer B. Biomarkers for capturing disease pathology as molecular process hyperstructure. bioRxiv. 2019:573402.

52. Ahn C, et al. Serum total cholesterol level as a potential predictive biomarker for neurological outcomes in cardiac arrest survivors who underwent target temperature management. Medicine (Baltimore). 2022;101(46):e31909.

53. Zmysłowski A, Szterk A. Oxysterols as a biomarker in diseases. Clin Chim Acta. 2019;491:103–13.

54. Brown AJ, Ikonen E, Olkkonen VM. Cholesterol precursors: more than mere markers of biosynthesis. Curr Opin Lipidol. 2014;25(2):133–9.

55. Ferguson TW, Komenda P, Tangri N. Cystatin C as a biomarker for estimating glomerular filtration rate. Curr Opin Nephrol Hypertens. 2015;24(3):295–300.

56. Whipp AM, et al. Ketone body 3-hydroxybutyrate as a biomarker of aggression. Sci Rep. 2021;11(1):5813.

57. Madeira C, et al. d-serine levels in Alzheimer's disease: implications for novel biomarker development. Transl Psychiatry. 2015;5(5):e561.

58. Baymeeva NV, Miroshnichenko II. [N-acetylaspartate is a biomarker of psychiatric and neurological disorders]. Zh Nevrol Psikhiatr Im S S Korsakova. 2015;115(8):94–8.

59. Makhlouf MMA, et al. Hyaluronic acid as a biomarker for liver fibrosis in chronic hepatitis "C" patients. Minia J Med Res. 2019;30(4):7–9.

60. Sheira G, et al. Urinary biomarker N-acetyl-β-D-glucosaminidase can predict severity of renal damage in diabetic nephropathy. J Diabetes Metab Disord. 2015;14:4.

61. Wishart DS, et al. MarkerDB: an online database of molecular biomarkers. Nucleic Acids Res. 2021;49(D1):D1259–67.

62. Patron J, et al. Assessing the performance of genome-wide association studies for predicting disease risk. PLoS One. 2019;14(12):e0220215.

63. Ware LB, et al. Plasma biomarkers of oxidant stress and development of organ failure in severe sepsis. Shock. 2011;36(1):12–7.

64. Milne GL, et al. Quantification of F2-isoprostanes as a biomarker of oxidative stress. Nat Protoc. 2007;2(1):221–6.

65. Mohorko N, et al. Elevated serum levels of cysteine and tyrosine: early biomarkers in asymptomatic adults at increased risk of developing metabolic syndrome. Biomed Res Int. 2015;2015:418681.

66. Khan AA, Alsahli MA, Rahmani AH. Myeloperoxidase as an active disease biomarker: recent biochemical and pathological perspectives. Med Sci (Basel). 2018;6(2):33.

67. Matsui T, et al. Tyrosine as a mechanistic-based biomarker for brain glycogen decrease and supercompensation with endurance exercise in rats: a metabolomics study of plasma. Front Neurosci. 2019;13:200.

68. Maltais-Payette I, et al. Circulating glutamate concentration as a biomarker of visceral obesity and associated metabolic alterations. Nutr Metab (Lond). 2018;15:78.

69. Lin C-H, Hashimoto K, Lane H-Y. Editorial: Glutamate-related biomarkers for neuropsychiatric disorders. Front Psychiatry. 2019;10:904.

70. Parveen U, et al. Pyridinoline: a narrative biomarker for osteoarthritis. Pharma Innov J. 2018;7(6):136–40.

71. Karlson EW, et al. Gene-environment interaction between HLA-DRB1 shared epitope and heavy cigarette smoking in predicting incident rheumatoid arthritis. Ann Rheum Dis. 2010;69(1):54–60.

72. Corella D, Ordovás JM. Biomarkers: background, classification and guidelines for applications in nutritional epidemiology. Nutr Hosp. 2015;31 Suppl 3:177–88.
73. Sharifi-Rad J, et al. Biological activities of essential oils: from plant chemoecology to traditional healing systems. Molecules. 2017;22(1):70.
74. Sant GR, Knopf KB, Albala DM. Live-single-cell phenotypic cancer biomarkers-future role in precision oncology? npj Precis Oncol. 2017;1(1):21.
75. Sutton TL, Walker BS, Wong MH. Circulating hybrid cells join the fray of circulating cellular biomarkers. Cell Mol Gastroenterol Hepatol. 2019;8(4):595–607.
76. Garner CE, et al. Volatile organic compounds from feces and their potential for diagnosis of gastrointestinal disease. FASEB J. 2017;21(8):1675–88.
77. Ziegler SI. Positron emission tomography: principles, technology, and recent developments. Nucl Phys A. 2005;752:679–87.
78. Garvey CJ, Hanlon R. Computed tomography in clinical practice. BMJ. 2002;324(7345):1077–80.
79. Ghantous CM, et al. Advances in cardiovascular biomarker discovery. Biomedicines. 2020;8(12):552.
80. Young PNE, et al. Imaging biomarkers in neurodegeneration: current and future practices. Alzheimers Res Ther. 2020;12(1):49.
81. Dregely I, et al. Imaging biomarkers in oncology: basics and application to MRI. J Magn Reson Imaging. 2018;48(1):13–26.
82. Lin CH, Hashimoto K, Lane HY. Editorial: glutamate-related biomarkers for neuropsychiatric disorders. Front Psych. 2019;10:904.
83. Dai M, et al. KRAS as a key oncogene in the clinical precision diagnosis and treatment of pancreatic cancer. J Cancer. 2022;13(11):3209–20.
84. Fu K, et al. Therapeutic strategies for EGFR-mutated non-small cell lung cancer patients with osimertinib resistance. J Hematol Oncol. 2022;15(1):173.
85. Ascierto PA, et al. The role of BRAF V600 mutation in melanoma. J Transl Med. 2012;10:85.
86. Jara L, et al. Mutations in BRCA1, BRCA2 and other breast and ovarian cancer susceptibility genes in Central and South American populations. Biol Res. 2017;50(1):35.
87. Klein J, et al. Distinguishing features of long COVID identified through immune profiling. Nature. 2023;623(7985):139–48.
88. Nitschke C, et al. Circulating cancer associated macrophage-like cells as a potential new prognostic marker in pancreatic ductal adenocarcinoma. Biomedicines. 2022;10(11):2955.
89. Nioka S, Chen Y. Optical tecnology developments in biomedicine: history, current and future. Transl Med UniSa. 2011;1:51–150.
90. Chernyak V, et al. Liver imaging reporting and data system (LI-RADS) version 2018: imaging of hepatocellular carcinoma in at-risk patients. Radiology. 2018;289(3):816–30.
91. Magny SJ, Shikhman R, Keppke AL. Breast imaging reporting and data system. 2023 [cited 2024]. Available from: https://www.ncbi.nlm.nih.gov/books/NBK459169/.
92. Park KJ, et al. Performance of prostate imaging reporting and data system version 2.1 for diagnosis of prostate cancer: a systematic review and meta-analysis. J Magn Reson Imaging. 2021;54(1):103–12.
93. Hajjo R, et al. Identification of tumor-specific MRI biomarkers using machine learning (ML). Diagnostics (Basel). 2021;11(5):742.
94. Fass L. Imaging and cancer: a review. Mol Oncol. 2008;2(2):115–52.
95. Croteau E, et al. PET metabolic biomarkers for cancer. Biomark Cancer. 2016;8(Suppl 2):61–9.
96. Das S, et al. Biomarkers in cancer detection, diagnosis, and prognosis. Sensors (Basel). 2023;24(1):37.
97. Davis KD, et al. Discovery and validation of biomarkers to aid the development of safe and effective pain therapeutics: challenges and opportunities. Nat Rev Neurol. 2020;16(7):381–400.

98. Dhandapani M, Goldman A. Preclinical cancer models and biomarkers for drug development: new technologies and emerging tools. J Mol Biomark Diagn. 2017;8(5):356.

99. Wilmes A, et al. Translational biomarkers, in vitro and in vivo. In: Bal-Price A, Jennings P, editors. In vitro toxicology systems. New York: Springer New York; 2014. p. 459–78.

100. Abdolahi S, et al. Patient-derived xenograft (PDX) models, applications and challenges in cancer research. J Transl Med. 2022;20(1):206.

101. Maru Y, Hippo Y. Current status of patient-derived ovarian cancer models. Cells. 2019;8(5):505.

102. Heo YJ, et al. Integrative multi-omics approaches in cancer research: from biological networks to clinical subtypes. Mol Cells. 2021;44(7):433–43.

103. Pinu FR, et al. Systems biology and multi-omics integration: viewpoints from the metabolomics research community. Metabolites. 2019;9(4):76.

104. Carracedo-Reboredo P, et al. A review on machine learning approaches and trends in drug discovery. Comput Struct Biotechnol J. 2021;19:4538–58.

105. Chen C, et al. Applications of multi-omics analysis in human diseases. MedComm (2020). 2023;4(4):e315.

106. Zhang B, Horvath S. A general framework for weighted gene co-expression network analysis. Stat Appl Genet Mol Biol. 2005;4:Article17.

107. Ma S, et al. Machine learning identifies transcriptional signatures of response to neoadjuvant chemoradiation in rectal cancer. Sci Rep. 2020;10(1):1–12.

108. Wang H, et al. Integrative omics analysis for the discovery of biomarkers and drug targets in triple-negative breast cancer. Int J Mol Sci. 2019;20(22):5610.

109. Wagner PD, Srivastava S. New paradigms in translational science research in cancer biomarkers. Transl Res. 2012;159(4):343–53.

110. Valla V, et al. Companion diagnostics: state of the art and new regulations. Biomark Insights. 2021;16:11772719211047763.

111. Jørgensen JT. Companion diagnostics: the key to personalized medicine. Expert Rev Mol Diagn. 2015;15(2):153–6.

112. Jørgensen JT, et al. A companion diagnostic with significant clinical impact in treatment of breast and gastric cancer. Front Oncol. 2021;11:676939.

113. Vranic S, Gatalica Z. PD-L1 testing by immunohistochemistry in immuno-oncology. Biomol Biomed. 2023;23(1):15–25.

114. Hammoudeh ZA, et al. Detecting EGFR mutations in patients with non-small cell lung cancer. Balkan J Med Genet. 2018;21(1):13–7.

115. Ragupathi A, et al. Targeting the BRCA1/2 deficient cancer with PARP inhibitors: clinical outcomes and mechanistic insights. Front Cell Dev Biol. 2023;11:1133472.

116. Thorne-Nuzzo T, et al. A sensitive ALK immunohistochemistry companion diagnostic test identifies patients eligible for treatment with Crizotinib. J Thorac Oncol. 2017;12(5):804–13.

117. Hoang T, et al. Efficacy of crizotinib, ceritinib, and alectinib in ALK-positive non-small cell lung cancer treatment: a meta-analysis of clinical trials. Cancers (Basel). 2020;12(3):526.

118. Soulières D, et al. KRAS mutation testing in the treatment of metastatic colorectal cancer with anti-EGFR therapies. Curr Oncol. 2010;17 Suppl 1(Suppl 1):S31–40.

119. Frampton JE. Entrectinib: a review in NTRK+ solid tumours and ROS1+ NSCLC. Drugs. 2021;81(6):697–708.

120. Davies KD, et al. Identifying and targeting ROS1 gene fusions in non-small cell lung cancer. Clin Cancer Res. 2012;18(17):4570–9.

121. Richards CS, et al. Standards and guidelines for CFTR mutation testing. Genet Med. 2002;4(5):379–91.

122. Clutter DS, et al. HIV-1 drug resistance and resistance testing. Infect Genet Evol. 2016;46:292–307.

123. Gower MN, et al. Clinical utility of CYP2C19 genotype-guided antiplatelet therapy in patients at risk of adverse cardiovascular and cerebrovascular events: a review of emerging evidence. Pharmgenomics Pers Med. 2020;13:239–52.

124. Kim J, Basak JM, Holtzman DM. The role of apolipoprotein E in Alzheimer's disease. Neuron. 2009;63(3):287–303.

125. Mallal S, et al. HLA-B*5701 screening for hypersensitivity to abacavir. N Engl J Med. 2008;358(6):568–79.

126. Szałowska-Woźniak DA, et al. Evaluation of HLA-DQ2/DQ8 genotype in patients with celiac disease hospitalised in 2012 at the Department of Paediatrics. Prz Gastroenterol. 2014;9(1):32–7.

127. Mrazek DA. Psychiatric pharmacogenomic testing in clinical practice. Dialogues Clin Neurosci. 2010;12(1):69–76.

128. Futema M, et al. Genetic testing for familial hypercholesterolemia—past, present, and future. J Lipid Res. 2021;62:100139.

129. Ou FS, et al. Biomarker discovery and validation: statistical considerations. J Thorac Oncol. 2021;16(4):537–45.

130. Predictive Tests. Available from: https://www.ncbi.nlm.nih.gov/books/NBK220030/, I.o.M. U.P.I.i.t.D.o.P.M.i.O.W.S.W.D.N.A.P.U.R.o.

131. Scherf U, et al. Approval of novel biomarkers: FDA's perspective and major requests. Scand J Clin Lab Invest. 2010;70(sup242):96–102.

132. Guidance for post-market surveillance and market surveillance of medical devices, i.i.v.d.G. and W.H.O.L.C.B.-N.-S. IGO.

133. Ltd, M.D.S.P. HER2 DNA amplification kit. Available from: https://mylabglobal.com/clinical/cancer/her2-dna-amplification-kit/.

134. Cayre A, et al. Comparison of different commercial kits for HER2 testing in breast cancer: looking for the accurate cutoff for amplification. Breast Cancer Res. 2007;9(5):R64.

135. Dandachi N, Hauser-Kronberger C. 5 - Detection of HER-2 oncogene in human breast carcinoma using chromogenic in situ hybridization. In: Hayat MA, editor. Handbook of immunohistochemistry and in situ hybridization of human carcinomas. Academic; 2002. p. 279–88.

136. CTK Biotech Inc (CTK). PSA semi-quantitative rapid test. Available from: https://protect.checkpoint.com/v2/___https://ctkbiotech.com/product/psa-semi-quantitative-rapid-test/___.YXBzMTpzeW5nZW5lOmM6bzoxYmJmNmZlNzdkMTIyYWJhMGE0 0NzI3M2U2NmQwNzExZDo2OjkwN2U6NDU1NGExMTdkNTM5MzFlMjg1NDY0Ym MyOTA0NmJhMWIwZTY0MjNkNzA0NDQxNmFjOTM4NWWJjODE1Mzg 4MDA0MzpwOlQ.

137. AccuQuik™. Prostate-specific antigen test kit. [cited 2 June 2024]. Available from: https://www.accuquiktestkits.com/products/prostate-specific-antigen-test-kit.

138. Anoop A, Singh PK, Jacob RS, Maji SK. CSF biomarkers for Alzheimer's disease diagnosis. Int J Alzheimers Dis. 2010;2010:606802. https://doi.org/10.4061/2010/606802. PMID: 20721349; PMCID: PMC2915796.

139. Park JS, et al. A novel kit for early diagnosis of Alzheimer's disease using a fluorescent nanoparticle imaging. Sci Rep. 2019;9(1):13184.

140. Na H, et al. The QPLEX™ plus assay kit for the early clinical diagnosis of Alzheimer's disease. Int J Mol Sci. 2023;24(13):11119.

141. Sysmex Corporation. Sysmex launches an assay kit to identify amyloid beta (Aβ) accumulation in the brain, a cause of Alzheimer's disease, using a small amount of blood. Kobe, Japan.

142. MILLIPLEX®. MILLIPLEX® Human amyloid beta and tau magnetic bead panel - multiplex assay. Available from: https://www.sigmaaldrich.com/IN/en/product/mm/hnabtmag68k?icid=sharepdp-clipboard-copy-productdetailpage.

143. Pharma, A., CEA Test Kit (Carcinoembryonic Antigen).

144. Weykamp C. HbA1c: a review of analytical and clinical aspects. Ann Lab Med. 2013;33(6):393–400.

145. Medical A. Atlas direct enzymatic hemoglobin A1C (glycated hemoglobin A1C). Available from: https://atlas-medical.com/product/208036/HbA1c.

146. Getein Biotech. HbA1c fast test kit (Immunofluorescence Assay). Available from: https://www.getein.com/hba1c-fast-test-kit-immunofluorescence-assay_p62.html.
147. Poclight. Hemoglobin A1c (HbA1c) test kit. Available from: https://www.poclightbio.com/.
148. Yousuf O, et al. High-sensitivity C-reactive protein and cardiovascular disease: a resolute belief or an elusive link? J Am Coll Cardiol. 2013;62(5):397–408.
149. Merck. Human C-reactive protein ELISA kit. Available from: https://www.sigmaaldrich.com/IN/en/product/sigma/rab0096?utm_source=google&utm_medium=cpc&utm_campaign=15000381723&utm_content=129438260635&gclid=CjwKCAjwrcKxBhBMEiwAIVF8rGDs0qmbBPe2IPsSifWz7SUNBKBMTVtJ_KDFQFyYMJxjvVHiRABSUBoChQAQAvD_BwE&icid=sharepdp-clipboard-copy-productdetailpage.
150. Cosmic Scientific Technologies. hs-CRP+CRP fast test kit (Immunofluorescence Assay). Available from: https://www.cosmic.net.in/product/hs-crpcrp/.
151. MyBioSource's Products. HS-CRP elisa kit :: Human high sensitivity C-reactive protein ELISA kit. Available from: https://www.mybiosource.com/hs-crp-human-elisa-kits/high-sensitivity-c-reactive-protein/40244?utm_source=TDSQR.
152. Bethune G, et al. Epidermal growth factor receptor (EGFR) in lung cancer: an overview and update. J Thorac Dis. 2010;2(1):48–51.
153. Roche Diagnostics. cobas® EGFR Mutation Test v2. Available from: https://diagnostics.roche.com/us/en/products/params/cobas-egfr-mutation-test-v2.html.
154. EntroGen, EntroGen's EGFR mutation analysis kit
155. Drain PK, Dorward J, Bender A, Lillis L, Marinucci F, Sacks J, Bershteyn A, Boyle DS, Posner JD, Garrett N. Point-of-care HIV viral load testing: an essential tool for a sustainable global HIV/AIDS response. Clin Microbiol Rev 2019;32:e00097–18. https://doi.org/10.1128/CMR.00097-18.3
156. Abbott. https://www.molecular.abbott/int/en/home. Available from: https://www.molecular.abbott/int/en/products/infectious-disease/realtime-hiv-1-viral-load.
157. altona Diagnostics GmbH, AltoStar® HIV RT-PCR Kit 1.5.
158. TRUPCR. TRUPCR ® HIV viral load kit. Available from: https://trupcr.com/products/infectious-disease/trupcr-hiv-viral-load-kit/.
159. Abdul Wahab A, et al. Anti-cyclic citrullinated peptide antibody is a good indicator for the diagnosis of rheumatoid arthritis. Pak J Med Sci. 2013;29(3):773–7.
160. Aiman AQ, et al. A new tool for early diagnosis of rheumatoid arthritis using combined biomarkers; synovial MAGE-1 mRNA and serum anti-CCP and RF. Pan Afr Med J. 2020;36:270.
161. Getein Biotech Inc. Anti-CCP fast test kit (Immunofluorescence Assay). Available from: https://www.getein.com/anti-ccp-fast-test-kit-immunofluorescence-assay_p106.html.
162. Diagnostic Automation/Cortez Diagnostics, I. Anti CCP ELISA kit. Available from: https://www.rapidtest.com/index.php?i=Anti-CCP-ELISA-kit&id=709&cat=8.
163. Cosmic Scientific Technologies. RF fast test kit (Immunofluorescence Assay). Available from: https://www.cosmic.net.in/product/rf/#:~:text=RF%20Fast%20Test%20Kit%20(Immunofluorescence,as%20rheumatoid%20arthritis%20(RA.
164. Charkhchi P, Cybulski C, Gronwald J, Wong FO, Narod SA, Akbari MR. CA125 and Ovarian cancer: a comprehensive review. Cancers (Basel). 2020;12(12):3730. https://doi.org/10.3390/cancers12123730. PMID: 33322519; PMCID: PMC7763876
165. Monobind Inc, CA-125 AccuBind ELISA kits.
166. Quadratech Diagnostics Ltd. Check-1 CA-125 quantitative rapid test for easy reader+® 15mins. Available from: https://www.quadratech.co.uk/product/check-1-ca-125-quantitative-rapid-test-for-easy-reader-15mins/.
167. Ltd, L.T. Ovarian cancer (CA125) test, CA125 rapid test cassette (whole blood). Available from: https://lifelabtesting.com/product/ovarian-cancer-test/.
168. Clinical Laboratory Improvement Amendments 2023 February 12, 2024 [cited 2024 29th May]. Available from: https://www.cdc.gov/clia/index.html.

Examining the Toxicological Landscape of New Molecular Entities and Biotherapeutics

Padmakumar Narayanan

Abstract Toxicity testing plays an essential role in drug development. The primary objective of toxicology studies at preclinical stages is to evaluate the safety of drug candidates before human trials. This chapter provides a comprehensive review of the principles, methodologies, and regulatory requirements that ensure the safety of new therapeutic agents. Beginning with basic concepts, the chapter reviews definitions of dose-response relationship, route of exposure, metabolism and bioactivation variability, and risk assessment. Most commonly conducted animal studies to support drug development, including acute, subacute, chronic, repeat dose toxicity studies, are also discussed. Key topics including basic principles of development and reproductive toxicity and carcinogenicity risk assessment will be discussed. In the second half of the chapter, in vitro technologies such as various liver, heart, skin, and kidney models that can potentially support early hazard identification and decrisking for drug candidates are discussed. Regulatory guidelines from major agencies like the FDA, EMA, and ICH are also discussed throughout the chapter, providing a framework for the design and interpretation of toxicology studies. Overall, the chapter illustrates strategies and comment challenges for mitigating potential toxicities, ultimately emphasizing the crucial contribution of toxicology to the safe and effective development of new drugs.

Keywords Toxicology studies · Preclinical safety assessment · Drug development · Dose-response relationship · Toxicity testing methodologies · In vitro toxicology models · Regulatory guidelines (FDA · EMA · ICH) · Reproductive toxicity · Carcinogenicity risk assessment · Hazard identification

P. Narayanan (✉)
Wave Life Sciences, Lexington, MA, USA
e-mail: pnarayanan@wavelifesciences.com

1   Introduction

1.1   Definition and Scope of Toxicology

Toxicology is the study of the adverse effects of chemical, physical, or biological agents on living organisms and the environment. It encompasses the evaluation of potential hazards posed by substances and the mechanisms through which these hazards manifest.

The scope of toxicology extends to various sectors, including pharmaceuticals, chemicals, environmental pollutants, occupational hazards, food safety, and consumer products. In drug development, toxicology plays a crucial role in assessing the safety profile of potential therapeutics [1].

1.2   Importance of Toxicology in Drug Development

Toxicology is an integral part of the drug development process, starting from early preclinical stages and continuing through clinical trials and post-market surveillance. Understanding the toxicological profile of a drug candidate helps in identifying potential risks and designing strategies to mitigate them, ensuring patient safety. Regulatory authorities such as the FDA (Food and Drug Administration) and EMA (European Medicines Agency) require comprehensive toxicology data to assess the safety and efficacy of new drugs before approval. Toxicology studies also contribute to the optimization of drug formulations, dosing regimens, and risk management plans, ultimately supporting successful drug commercialization.

1.3   Key Principles of Toxicology

1. *Dose-Response Relationship*: Toxicological effects often exhibit a dose-dependent relationship, where the severity of adverse outcomes correlates with the exposure level to a toxicant. Understanding dose-response curves helps in establishing safe exposure limits.
2. *Route of Exposure*: The route through which a substance enters the body influences its toxicological impact. Different routes, such as oral ingestion, inhalation, dermal contact, or injection, can lead to varied toxic effects.
3. *Metabolism and Bioactivation*: Metabolic processes can transform xenobiotics (foreign substances) into more toxic or less toxic metabolites. The balance between activation and detoxification pathways influences toxicity.
4. *Individual Variability*: Factors such as genetics, age, sex, health status, and concurrent medications contribute to individual variability in toxicological responses. Populations may exhibit different susceptibilities to toxicants.

5. *Cumulative Effects*: Some toxicants may cause cumulative damage over time, leading to chronic toxicity. Understanding cumulative effects is essential for long-term safety assessments.

6. *Risk Assessment*: Toxicology integrates risk assessment methodologies to estimate the likelihood of adverse effects in exposed populations. This involves hazard identification, dose-response assessment, exposure assessment, and risk characterization.

2   Preclinical Toxicology Studies

2.1   Overview of Preclinical Toxicology

Preclinical toxicology studies for small molecules, biopharmaceuticals, oligonucleotides (ODNs), cellular therapies are conducted during the early stages of drug development, prior to human clinical trials. These studies aim to assess the distribution and safety profile of a drug candidate and identify potential toxic effects. The data generated from preclinical toxicology studies are crucial for making informed decisions regarding the progression of a drug candidate to clinical trials and for designing appropriate safety monitoring protocols.

2.2   Species Selection

Selecting the appropriate animal species for toxicity testing is critical for assessing potential effects of drug candidates in humans. Although the objectives of the preclinical safety evaluation program are the same, rationale and experimental means by which relevant and appropriate species are selected for preclinical safety evaluations can be different among various drug modalities including new molecular entities (NMEs), biopharmaceuticals, ODNs, or cell and gene therapies [2]. Unlike protein therapeutics that remain in the extracellular space, small molecules are widely distributed and widely associated with off-target toxicities. Moreover, biodistribution and metabolic transformation on a given small molecule may alter the toxicity of the molecule. The question asked in this instance is which species most closely resembles humans in terms of metabolite profile (identity, number, quantity) in ex vivo or in vivo systems. On the contrary, protein-based therapeutics remain in the extracellular space and are not metabolized. They generally induce exaggerated pharmacologic/toxicological effects due to their on-target mechanism of action. Furthermore, the route and frequency of dosing in animal studies more closely parallel the proposed treatment regimen for humans. The default species for small molecules in toxicity testing are the rat and dog (as rodent and non-rodent species). The dog is by far the most common non-rodent species [3]. In cases when dog is not used (especially in the case of using primate instead of dog), the most common justification is the lack of efficacy (e.g., low pharmacological activity in dog tissues and/or similar activity/metabolic profiles between monkey and humans or the fact that the dog was known to be prone to emesis for the class of drug being investigated).

For biotechnology products such as monoclonal antibodies (mAbs) following the ICHS6 (R1) guideline [4–7], only studies in pharmacologically relevant species are expected, with studies in non-relevant species actively discouraged to prevent misleading results [8]. Pharmacological relevance is generally demonstrated by a species that expresses the target antigen and evokes a similar pharmacological response as that expected in humans. As many biotherapeutic products are highly selective, often there is only one relevant species (frequently the nonhuman primate (NHP), owing to higher genome sequence homology to humans and similarity in physiological systems, such as the immune system), and single species programs in NHP are common. However, if a rodent species is also pharmacologically relevant, toxicity studies in two species are recommended and this appears to be the case for 30–40% of mAbs [8–10]. In some instances, nonrelevant species may be used (e.g., when no relevant species is available or when one is assessing toxicity that results from contaminants). One example of this is keliximab, a human-cynomolgus monkey chimeric (primatized) monoclonal antibody with specificity for human and chimpanzee CD4 [11]. In order to conduct a comprehensive preclinical safety assessment of this antibody to support chronic treatment of rheumatoid arthritis in patients, a human CD4 transgenic mouse was used for chronic and reproductive toxicity studies and for genotoxic studies. In addition, immunotoxicity studies were conducted in these mice with *Candida albicans*, *Pneumocystis carinii*, and B16 melanoma cells to assess the effects of keliximab on host resistance to infection and immunosurveillance to neoplasia [11].

A collaboration between the NC3Rs and the ABPI aimed to review the species used for general toxicology studies of drug candidates within current pre-market portfolios [12]. The international working group reviewed anonymized data for 172 drug candidates received from 18 different organizations, consisting of 92 small molecules, 46 mAbs, 15 recombinant proteins, 13 synthetic peptides, and 6 antibody-drug conjugates (ADCs). Toxicology studies were conducted in both rodent and non-rodent species for the majority of small molecules, recombinant proteins, synthetic peptides, and ADCs (97%, 80%, 100%, and 83% of each drug modality, respectively), whereas a large number of mAbs (65%) were tested in single non-rodent species [13]. The species used for toxicity testing of the small molecules in the dataset were predominantly rat, dog, and NHP. The NHP was also used for all ADCs, for the majority of mAbs (96%) and recombinant proteins (87%), and half of the synthetic peptides (the dog being the non-rodent species used for these latter drug candidates). The rat was also used for testing 17% mAbs, 60% recombinant proteins, 92% synthetic peptides, and 66% ADCs. The mouse (both wild-type and transgenic models) or rabbits were also used for testing a small number of biologicals within the dataset (Fig. 1).

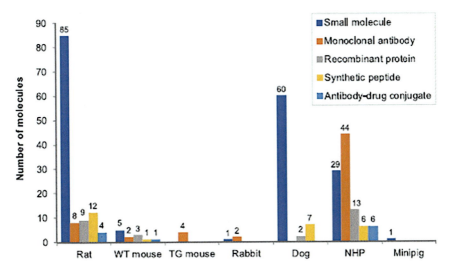

Fig. 1 Species used for different molecule types within the NC3Rs/ABPI "two species" working group dataset (Prior et al. 2020) [12]

Similar to small molecules, the general toxicity studies for ODNs, such as antisense oligonucleotides (ASOs) and small interfering RNA (siRNA), are performed in both a rodent and a non-rodent species [14]. For double-stranded siRNA and microRNA mimics, the rat is by far the most used rodent species, whereas the mouse is the rodent species of choice for most single-stranded PS backbone ASO candidates [15]. Although rat is the most common rodent for toxicity studies of small molecules, rat-specific lesions like chronic progressive nephropathy (CPN) [16] are aggravated by high kidney concentrations resulting from systemic delivery of PS backbone ASO. CPN is of no human relevance [17] but can become problematic in long-duration toxicity studies. Mice were generally used as the rodent species for early studies on the toxicity of ODNs because many of the pharmacology models are in mice, and therefore, surrogate molecules were readily available. Nonhuman primates (NHPs) are selected because the pharmacokinetics of ODN drugs appear to be very similar in monkeys and humans. At similar doses on a milligram per kilogram basis, plasma levels of a given sequence administered by the same route are nearly identical in monkeys and humans. Compared to rodents, the toxicity profiles in monkeys may be more predictive of the toxicity profiles in humans, especially for thrombocytopenia caused by systemically administered unconjugated ASOs [15].

Generally speaking, due to the highest likelihood of sequence-dependent crossover on activity and a robust historical background record, nonhuman primate (NHP) is by far the most common non-rodent species used for both ASO and siRNA candidates, but other non-rodents have been evaluated including the pig [18].

For cell therapy programs, generally more than one species may be necessary [19]. The species selected should be pharmacologically active, have relevant physiology and anatomy similar to humans, be immune tolerant to the cell therapy product, allow the ability to dose the appropriate anatomic site with a specified number of cells, and be permissive to viral infection (if a vector is needed). The age of the animals used should be representative of the target patient population. It is also important that the species used supports administration of the test article by the clinical route of administration. In some cases, a relevant species may not be available, so the use of a biologically active surrogate/analogous cell therapy product may be tested. However, in this case, analogy for the ability to harvest, identify, isolate, and expand cells; cell growth kinetics; and cell function and viability will likely need to be evaluated. Analogous cells should not be used to assess tumorgenicity.

In nonclinical studies, cross-species immunogenicity to a human cellular product may require the animal model to be modified so that it becomes immune tolerant for the administered human cells. This can be accomplished by administering immuno-suppressants to immune-competent animals, using genetically immunodeficient animals, using humanized animals, and/or administering the test article to an immune-privileged site. Genetically modified (i.e., transgenic or knockout/-in) animals are often used, when appropriate. For example, non-obese diabetic (NOD) severe combined immunodeficiency (scid) gamma (NSG) mice with interleukin 2 receptor gamma (IL2Rg) gene knockout and NOD-scid genetic background, lacking mature T, B, and NK cells, are recognized as a suitable immunodeficient model for cell xenografting. Similarly, NSG mice humanized with CD34+ cells or peripheral blood mononuclear cells (PBMC) can be used to assess human-relevant immune responses to human cell therapy administered to animals. The use of immunodeficient animals will often allow for an assessment of long-term safety.

The use of diseased/injured animals may also be acceptable, and in some cases preferred. For example, for potential treatments where there is a requirement for cell therapy to interact with a target that is only present in the disease state, then the use of a diseased animal models would provide more relevant information on the efficacy and safety of the test article. It should be acknowledged that the use of disease models also has some limitations that could interfere with the sponsor's ability to interpret data and/or achieve the objectives of the study. For example, there may be a lack of historical values for background findings, the untreated control animals may have a limited lifespan because of the disease, the characteristics of the disease in animals may not be fully reflective of what occurs in humans, or target organs for the disease model are similar to those observed with the test article.

Regardless of the animal species used, the nonclinical sections of regulatory submission documents (such as Investigational New Drug (IND)/Biologic License Application (BLA) or Clinical Trial Application (CTA)/Marketing Authorization Application (MAA)) should include a detailed justification for the species used to assess the pharmacology and toxicity of the test article. This justification should also include limitations of any non-standard animal models (when used).

2.3   Types of Preclinical Toxicology Studies

2.3.1   Acute Toxicity Studies

(a) *Acute (single dose) toxicity studies* evaluate the adverse effects of a single or short-term exposure to a drug candidate. They help determine the maximum tolerated dose (MTD) and identify acute toxic effects. The high dose level for single-dose toxicity studies is derived as a limit dose by notable clinical signs that may manifest rapidly or by deaths (as found in preliminary investigations with small numbers of animals) or by the limit of practicable dosing formulation [3, 20, 21]. In 1996, the Center for Drug Evaluation and Research (CDER) suggested a single-dose acute toxicity testing procedure for pharmaceutical substances that use a fixed safe dose that should not cause adverse events or threaten the life of an animal [20]. Assessment of acute toxicity can be performed in rodents as a stand-alone study, as part of acute toxicity screens, or as part of the dose-ranging process for the mouse in vivo micronucleus study.

Although acute toxicity studies have traditionally been used to assess tolerability for drug candidates, these studies may not be sensitive to some novel modalities. For instance, ODN therapeutics have a relatively low order of acute toxicity, with the exception of polyanionic phosphorothioate ODNs which have been known to activate complement through the alternative pathway in monkeys. Because ODNs do not cross the blood-brain barrier nor accumulate in cardiac muscle, and uptake of these compounds is dependent on saturable processes, rodents can tolerate very high doses of ODNs in acute toxicity studies [22]. Acute studies in non-rodent species are not performed for PS ODNNs, either, because of the complement cascade that can result in an anaphylactoid response.

(b) *Acute repeat dose* short-term exposure studies or range-finding work for conventional small molecules, to support the pivotal GLP study, can usually take the form of a 7 or 14-day investigation in both rodents and non-rodents, following a standard design. In the rodent, a 7-day range-finding study is the most performed study with clinical signs, bodyweight, food consumption, clinical pathology, organ weight, and histopathology. For small molecules in the non-rodent species, the most popular range-finding study involved dosing one or two pairs of male and female animals with ascending dose levels of the drug to establish an MTD and then a repeat fixed dose period ranging from 5 to 14 days in further one or two pairs of animals. As in the rat in-life studies, organ weight and macroscopic investigations were routinely performed. However, they were not usually performed as it can take many weeks for tissue processing.

Assessment in studies to support early clinical trials generally involves one vehicle-treated control group and three drug-treated (low, mid, and high) groups as shown in a recent evaluation [3]. Various regulatory guidance documents also highlight what parameters need to be measured, along with numbers of groups to be utilized [23–26]. Toxicology studies with biologicals also usually include anti-drug antibody (ADA) and sometimes immunophenotyping (flow cytometry) measurements. For oligonucleotides projects aiming to reach the clinic, it is useful to understand the expectations of health authorities approving clinical trials. In 2020, the Japanese Pharmaceuticals and Medical Devices Agency issued a guideline for preclinical safety assessment of oligonucleotide therapeutics [27]. In 2021 the US Food and Drug Administration (FDA) issued a draft guidance, "Nonclinical Testing of Individualized Antisense Oligonucleotide Drug Products for Severely Debilitating or Life-Threatening Diseases," providing a high-level overview of the key issues they will consider in the special situation of developing an OND to treat extremely rare mutations present on a single or very small number of patients (not for commercial development) [28]. No specific formal guidelines for preclinical testing of ONDs to treat more common conditions have been issued to date by the International Council for Harmonization of Technical Requirements for Pharmaceuticals for Human Use (ICH), FDA, or European Medical Agency (EMA), and so the testing principles normally follow the overall ICH M3(R2) guideline for small molecules. However, a series of consensus recommendations regarding regulatory considerations and expectations for OND have been generated by the Oligonucleotide Safety Working Group (OSWG) and other working groups in white papers [29].

2.3.2 Repeat-Dose Toxicity Studies

Repeat-dose toxicity study designs usually follow the principles followed in regulatory guidance [24–26]. These studies consist of subacute toxicity studies, subchronic toxicity studies, and chronic toxicity studies conducted in two animal species (one non-rodent). These studies are conducted to evaluate a drug's potential adverse effects over a longer period. The duration of the repeated-dose toxicity studies at any point in preclinical development should in principle be equal to or exceed the duration of the human clinical trials up to the maximum recommended duration of the repeated-dose toxicity studies [21]. The test compound, usually an NME, is administered daily for a specified duration. The target organs of toxicity should be clearly identified, and the reversibility of the toxic effects should be determined. The highest dose used in these studies is selected to deliberately induce toxicity. The lowest dose is selected to identify a NOAEL. When possible, the NOAEL dose should provide an adequate multiple of the projected maximum clinical dose. Subacute and subchronic toxicity studies are also used to determine dose levels for carcinogenicity studies, such as the MTD. The MTD, which is usually used as the highest dose for a carcinogenicity study, is a dose just high enough to elicit signs of minimal toxicity without significant alteration of the animal's normal lifespan due to effects other than carcinogenicity.

(a) *Subacute/Subchronic Toxicity Studies:* The duration of these good laboratory practice (GLP) studies range from 1 to 3 months in the beginning of a program for rodents and non-rodents, respectively. These studies involve repeated administration of a drug candidate to assess potential toxic effects that may occur with prolonged exposure. The aim is to identify adverse effects after repeated dosing, such as cumulative toxicity, organ-specific toxicity, and dose-dependent effects. These studies evaluate systemic toxicity, target organ toxicity, hematological parameters, biochemical markers, and histopathological changes. These studies usually follow acute toxicity studies and are conducted in at least two animal species—one rodent (rat or mouse) and one non-rodent (rabbit, dog, nonhuman primate, or other suitable species)—and are conducted as per regulatory requirements such as the FDA's good laboratory practices (GLP). For biologics, if the product is only pharmacologically active in one species, then studies only in that species are needed [4, 5]. If no pharmacologically responsive test species are available, then the FIH dose estimate may be based on pharmacology data alone (i.e., a minimal anticipated biological effect level (MABEL) approach can be used). Use of a MABEL is relatively uncommon and can necessitate starting with very low doses that may not provide clinical benefit and require protracted dose escalation [20]. It should be noted that test compounds, including oligonucleotides (ODNs), are not distributed equally throughout the body, even when systemically administered. The recently developed ODN products accumulate in specific tissues (e.g., accumulate in the kidney and liver) and induce injection site reactions [30, 31]. Therefore, for a locally administered ODN, it is necessary to evaluate toxicity, not only at the injection site but also in organs where the ODN is expected to accumulate through systemic circulation. Subchronic toxicity studies for ODNs have been performed primarily in (often mice) monkeys [22]. For ODNS, unlike small molecules, the catabolic pathways for degradation are very similar from species to species. Because metabolism is not an issue, the ability of the test species to mimic response in humans became the driving force for the selection of nonhuman primates.

Toxicity study parameters generally include clinical signs (including animal behavior), body weight and food consumption, clinical pathology, organ weights, gross pathology, histopathology, and toxicokinetics. Ideally, doses used in toxicity studies should identify a no observed adverse effect level (NOAEL) for toxicities and a maximum tolerated dose. Definitive animal toxicity studies establish the safety characteristics, including the no observed adverse effect level (NOAEL), of the NME. With very few exceptions, these studies are conducted under regulatory guidelines such as the FDA's good laboratory practices (GLP). The highest dose levels tested in the definitive animal toxicity studies are almost always based on the MTD determined from the range-finding studies. By the time an NME reaches definitive animal safety testing, a human trials' clinician is included in the drug development team to provide study details on the proposed first-in-human (FIH) clinical trial. In general, the route and frequency of drug administration to be used in initial human trials must be the same or similar to that used in the definitive animal toxicity studies.

2.3.3 Chronic Toxicity Studies

Chronic toxicity studies assess the effects of long-term exposure to a drug candidate, usually over a minimum of 6 months to 2 years in animal models [21]. They focus on detecting cumulative toxic effects, such as carcinogenicity, organ damage, reproductive toxicity, and immunotoxicity. These studies are designed to determine:

- The potential target organs of toxicity
- The reversibility of toxicities observed
- The NOAEL
- The potential clinical risk in relation to the anticipated clinical dose following long-term treatment

With regard to chronic toxicity testing, the ICH S4A guidance [32] recommends 6-month chronic toxicity studies in rodents and 9-month chronic toxicity studies in non-rodents. While the FDA considers 9-month studies in non-rodents acceptable for most drug development programs, shorter studies may be equally acceptable in some circumstances and longer studies may be more appropriate in others, as follows:

1. Six-month studies may be acceptable for indications of chronic conditions associated with short-term, intermittent drug exposure, such as bacterial infections, migraine, erectile dysfunction, and herpes.
2. Six-month studies may be acceptable for drugs intended for indications for life-threatening diseases for which substantial long-term human clinical data are available, such as cancer chemotherapy in advanced disease or in adjuvant use.
3. Twelve-month studies may be more appropriate for chronically used drugs to be approved based on short-term clinical trials employing efficacy surrogate markers where safety data from humans are limited to short-term exposure, such as some acquired immunodeficiency syndrome (AIDS) therapies.
4. Twelve-month studies may be more appropriate for NMEs acting at new molecular targets where post marketing experience is not available for the pharmacological class. Thus, therapeutic is the first in a pharmacological class for which there is limited human or animal experience on its long-term toxic potential.

Chronic toxicity studies should be initiated when Phase II clinical trials demonstrate the efficacy of an NME. These studies could be conducted concurrently with Phase III clinical trials and should be used to support the safety of long-term clinical trials and marketing approval. Different regulatory guidance for the design and duration of chronic toxicity studies are available, with flexibility in approaches often adopted for specific drug modalities. These guidance may provide opportunities to reduce time, cost, compound requirement, and animal use within drug development programs if applied more broadly and considered outside their current scopes of use. In an article summarizing presentations from a workshop at the 43rd Annual Meeting of the American College of Toxicology (ACT) in November 2022, different approaches for chronic toxicity studies were discussed. An industry collaboration between the Netherlands Medicines Evaluation Board (MEB) and the

UK National Centre for the Replacement, Refinement and Reduction of Animals in Research (NC3Rs) illustrated current practices and the value of chronic toxicity studies for monoclonal antibodies (mAbs). They evaluated a weight of evidence (WOE) model where a 3-month study might be adequate instead of a 6-month study. The data presented from the MEB/ Industry/NC3Rs consortia show that while new toxicities are identified with longer dosing durations than the FIH-enabling studies, these rarely translate to humans or modify clinical development. However, the provision of this data supports clinical development by confirming absence or presence of earlier findings and supports progression with appropriate monitoring, which could still be achieved using shorter studies. The data also provides useful information for dose setting for rat carcinogenicity studies (when required) or for WOE considerations (when appropriate) that mitigate the need for these studies [33]. Other topics included potential opportunities for single-species chronic toxicity studies for small molecules, peptides, and oligonucleotides and whether a 6-month duration non-rodent study can be used more routinely than a 9-month study (similar to ICH S6(R1) for biological products). Opportunities to reduce animal use for chronic toxicity studies were highlighted for mAbs; many companies perform 4-, 13-, and 26-week duration studies to support FIH and longer duration dosing trials, but for many mAbs, a 13-week (3-month) study may be sufficient for supporting long term. A unique problem encountered in a biopharmaceutical repeat-dose study is the development of antibodies to the drug, specifically neutralizing antibodies that diminish or eliminate its pharmacological activity, which may call into question the utility of conducting chronic toxicology studies.

As some oligonucleotides have successfully justified single-species programs (by retaining the capability to predict clinically translatable immune-mediated thrombocytopenia), more flexibility toward ICH S6(R1) [4] rather than ICH M3(R2) [34] approaches for these molecule types may be justified. Harmonizing non-rodent chronic toxicity studies to 6-month durations for small molecules and other drug modalities, following ICH regulatory guidelines (FDA, EMA, ICH) M3(R2) guidance, would be a refinement. This approach involves fewer procedures within a shorter dosing period, reducing test article requirements and shortening timelines. There could even be scope for some small molecules to use a single species for chronic toxicity studies when similar toxicities are identified in earlier FIH-enabling studies, and this will be evaluated further in a new phase of the NC3Rs project in 2024. Furthermore, for any modality, the inclusion of recovery animals in chronic toxicity studies for any drug modality only when necessary and scientifically justified, rather than in multiple studies/species across a package, is another means to potentially reduce animal use while also shortening timelines. Notably, the US FDA's pandemic-era guidance to minimize NHP use during COVID-19 provided some opportunities where scientifically justified. Encouragingly, the FDA appears to have continued interest in such approaches [35]. The door has been opened for regulatory agencies worldwide, and the industry as a whole, to consider broader flexibility and new approaches in the assessment of chronic toxicity potential/characterization for a variety of drug modalities.

2.3.4 Genotoxicity Studies

Genotoxicity studies evaluate the potential of a drug candidate to induce genetic damage, including mutations, chromosomal aberrations, and DNA damage. Standard genotoxicity assays include Ames test for mutagenicity, chromosomal aberration assay, micronucleus test, and in vivo comet assay. These tests enable hazard identification with respect to DNA damage and its fixation [21]. Fixation of damage to DNA in the form of gene mutations, larger-scale chromosomal damage, recombination, and numerical chromosome changes is generally essential for heritable effects and in the multistep process of malignancy, a complex process in which genetic changes may play only a part. NMEs that are positive in tests that detect such kinds of damage have the potential to be human carcinogens and/or mutagens. Preclinical drug development and registration of NMEs require a comprehensive assessment of their genotoxic potential. Since no single test can detect all relevant genotoxic agents, the usual approach is to carry out a battery of in vitro and in vivo tests for genotoxicity, which complement each other [36] and recommends the following standard test battery:

1. A test for gene mutation in bacteria (Ames test)
2. An in vitro test with cytogenetic evaluation of chromosomal damage with mammalian cells or an in vitro mouse lymphoma TK assay
3. An in vivo test for chromosomal damage using rodent hematopoietic cells

For NMEs that test negative, the completion of the standard test battery will usually provide a sufficient level of safety to demonstrate the absence of genotoxic potential. For NMEs that test positive in the standard test battery, depending on their therapeutic use, more extensive testing may be needed [37]. A gene mutation assay is generally considered sufficient to support all single-dose clinical development trials. To support repeated-dose clinical development trials, an additional assessment capable of detecting chromosomal damage in a mammalian system(s) should be completed [36]. The complete battery of genotoxicity tests should be completed before initiation of Phase II trials [37]. Companies perform this study early to avoid any delay in starting studies in patients, as required under current regulatory guidance (ICH, S2B). In cases in which a positive in vitro result was seen, a second in vivo study was performed (usually the rat liver micronucleus or unscheduled DNA synthesis test) in compliance with current guidance [37]. In recent years, the approach of backing up equivocal or real genotoxicity results with appropriate studies and a human risk assessment before commencing with clinical trial investigations has been further clarified [38]. Such an approach is important as positive genotoxicity results do occur but do not necessarily "kill" a drug. Thus, in the future, a different pattern of genotoxicity may occur compared to that found in the present evaluation (a new set of in vitro assays or even an Ames test only plus two in vivo tests).

2.3.5 Reproductive Toxicity Studies

Reproductive toxicity studies assess the effects of a drug candidate on fertility, pregnancy, embryonic development, and postnatal growth in animal models. These studies include assessments of mating behavior, fertility parameters, pregnancy outcomes, fetal development, and lactation. The FDA requires reproductive toxicity testing for any small molecule to be used in women of childbearing potential (WOCBP), regardless of whether the target population is pregnant women.

According to Colerangle 2017 [21], reproductive toxicity studies have been conducted in a three-segment testing protocol: testing follows guidelines established by agencies such as the FDA, EPA, and OECD. Key guidelines include OECD Test Guidelines 414 (Prenatal Development Toxicity Study), 421 (Reproduction/Developmental Toxicity Screening Test), and 422 (Combined Repeated Dose Toxicity Study with the Reproduction/Developmental Toxicity Screening Test).

Multi-Generational Studies: Tests often include multi-generational studies, such as the Two-Generation Reproductive Toxicity Study (OECD 416), assessing effects on parents and subsequent generations.

- Types of Studies:

    - Fertility and Early Embryonic Development Studies: Evaluate effects on mating behavior, conception, implantation, and early embryonic development.
    - Pre- and Postnatal Development Studies: Assess effects on prenatal development, including teratogenicity, and postnatal development through weaning.
    - Developmental Toxicity Studies: Focus on identifying potential teratogenic effects and developmental delays during gestation.

- Endpoints and Measurements

    - Fertility Parameters: Sperm count, motility, morphology, estrous cycle regularity, mating success, and conception rates.
    - Developmental Parameters: Embryo/fetal viability, growth, morphological abnormalities, and behavioral assessments in offspring.
    - Maternal Toxicity: Maternal body weight, food consumption, and clinical signs of toxicity.

- In Vivo and In Vitro Approaches

    - Animal Models: Commonly used species include rodents (rats and mice) and non-rodents (rabbits).
    - Alternative Methods: In vitro assays and computational models are increasingly used to complement in vivo studies, focusing on mechanisms such as endocrine disruption and genotoxicity.

As per ICH S5B [39], reproductive toxicity studies should be conducted as is appropriate for the population that is to be exposed to the NME. Men can be included in Phase I and II trials before the conduct of the male fertility study, since an evaluation of the male reproductive organs is performed in the repeated-dose toxicity studies. An assessment of male and female fertility by thorough standard histopathological examination of the testis and ovary in a repeated dose toxicity study (generally rodent) of at least 2-week duration is considered to be as sensitive as fertility studies for detecting toxic effects on male and female reproductive organs [39–41]. A male fertility study [39] should be completed before the initiation of large scale or long duration clinical trials (e.g., Phase III trials). According to ICH M3(R2) guidance [34] there is a high level of concern for the unintentional exposure of an embryo or fetus before information is available concerning the potential benefits versus potential risks. Therefore, for women of childbearing potential (WOCBP), it is important to characterize and minimize the risk of unintentional exposure of the embryo or fetus when including WOCBP in clinical trials.

The need for reproductive toxicology studies for biopharmaceuticals is dependent on the indication, the patient population for which it is intended, and the characteristics of the product. Because of the inherent nature of biopharmaceuticals with regard to issues related to species specificity, immunogenicity, biological activity, and/or long eliminating half-life, the study design and dosing schedule could be challenging. It is therefore recommended that consultation with the appropriate division of Center for Drug Evaluation and Research (CDER) is sought to discuss the need for and design of suitable reproductive toxicology studies.

Concerns regarding potential developmental immunotoxicity, which may apply particularly to certain biopharmaceuticals (e.g., monoclonal antibodies) with prolonged immunological effects, could be addressed in a study design modified to assess immune function of the neonate [4, 21]. For biopharmaceuticals, it is recommended that reproductive toxicology studies should be conducted in pharmacologically relevant animal species. If the biopharmaceutical is active in the commonly used species, then the studies should be conducted in compliance with guidance provided in ICH M3(R2) [34] and ICH S6 [4]. In many instances, the nonhuman primate is the only relevant species for conducting reproductive toxicology studies. The use of monkeys is not without challenges and limitations. These challenges and limitations include:

- The availability of sexually mature animals is limited.
- Nonhuman primates are markedly heterogeneous compared to in-bred laboratory animals.
- Nonhuman primates produce one offspring per pregnancy.
- Spontaneous abortions occur at high rates.
- Gestation periods are approximately 5–7 times longer relative to those of rats and rabbits.

It is recommended that "embryo-fetal development" study be conducted when necessary to assess risks to the conceptus with nonhuman primate [21, 42]. Effects on male and female fertility may be sufficiently addressed by incorporating appropriate endpoints in the nonhuman primate repeated-dose toxicology studies. However, if tissue cross-reactivity studies reveal binding of the biopharmaceutical to reproductive organs and/or repeated dose toxicology studies reveal concerns, "fertility to early embryonic development" studies should be considered in nonhuman primates. Due to the lengthy maturation and small number of offspring, complete "prenatal and postnatal development" studies cannot be expected in nonhuman primates. If the biopharmaceutical is an immune modulator and immune assessments of the offspring are desirable, relevant endpoints can be included in a modified "embryo-fetal development" study. When no relevant animal species is available, transgenic mice or analogous proteins should be considered.

ODNs, including ASOs, siRNA, and aptamers, represent a unique class of therapeutic agents that require specialized approaches for reproductive toxicology testing due to their distinct mechanisms and molecular properties. ODNs have attributes that are similar to small molecules and biopharmaceuticals. Both sets of guidelines can be useful when planning reproductive toxicity studies. The approach to DART testing should take into account both the potential effects of chemical structure and the intended pharmacology of the ODN. Standard reproductive toxicity species (rodent/rabbit) can often be used because most ODN toxicity is related to chemical structure. Animal-active surrogates can be useful to assess potential reproductive effects related to pharmacology in addition to effects related to the chemical backbone. The choice of the relevant animal model, the dosing regimen, and whether to use the clinical candidate or an animal surrogate will all need to be considered carefully based on the specific product attributes of the ODN. Information from general toxicity studies and previous experience with similar ONs can also help to inform these decisions. For various reasons outlined above, NHP studies should be used only as needed to answer specific questions or when surrogates are not practical. Dosing regimens should be tailored to ensure adequate exposure throughout the period of organogenesis without compromising the ability to assess PD- and PK-related effects [43].

Reproductive toxicology testing for biopharmaceuticals and ODNs requires a specialized approach that considers their unique molecular properties and mechanisms of action. Adapting standard reproductive toxicity study designs to address specific challenges associated with these platforms, such as off-target effects and immunogenicity, ensures a thorough assessment of their safety profile. Regulatory guidelines provide a framework for conducting these studies, emphasizing the importance of both in vivo and in vitro methods to comprehensively evaluate potential reproductive risks.

2.3.6 Carcinogenicity Studies

Carcinogenicity studies investigate the potential of a drug candidate to induce cancer development over prolonged exposure periods, typically 2 years in rodents. They involve monitoring for tumor incidence, tumor types, latency periods, metastasis, and histopathological characterization [21].

2.4   Study Design and Endpoints

- Preclinical toxicology studies are designed with specific protocols tailored to regulatory guidelines and study objectives.
- Key endpoints include clinical observations, body weight changes, hematology, clinical chemistry, organ weight ratios, histopathology, and toxicokinetic parameters.
- Study designs incorporate control groups (vehicle control and untreated control) to compare and interpret toxicological findings accurately.

The FDA follows the recommendations outlined in the ICH guidance for assessing genotoxic risk (ICH S2(R1), 2012) and overall cancer risk for small molecules (ICH S1A, 1996; ICH S1B, 1998; ICH S1C(R2), 2008) and biotechnology-derived products (ICH S6, 1997).

- Small molecules: A combination of genotoxicity tests and long-term studies in two rodent species, usually the rat and the mouse, are typically conducted for assessing carcinogenic risk of small molecules. One of the long-term bioassays may be substituted for by an alternative study, typically a transgenic mouse study, that involves less time and fewer animals than the standard mouse bioassay. Because carcinogenic risk may arise from chemical-specific attributes in addition to pharmacological properties, each small molecule has been assessed regardless of prior knowledge of the risk, or lack of risk, associated with the drug class.
- Biologics: A weight-of-evidence (WOE) approach is practiced for biotechnology-derived products when a carcinogenicity assessment is warranted. Assessment is tailored to the specific biologic and typically consists of an evaluation of known drug target pharmacology and compound-specific toxicology. Genotoxicity testing is generally inappropriate for biologic products and is not part of the assessment. When sufficient information is available, this type of WOE assessment can sometimes adequately address carcinogenic potential and preclude the need to conduct additional nonclinical studies, regardless of whether the biologic is active and testable in a rodent bioassay. Rodent bioassays are not warranted if the WOE clearly supports a concern regarding carcinogenic potential (e.g., immunosuppressives and growth factors). Similarly, if the WOE assessment does not suggest carcinogenic potential, a rodent bioassay is considered unlikely to add value, and no additional nonclinical testing of biologic products is recommended. Prior knowledge of target-based risks figures prominently in developing a strategy to assess carcinogenic risk of each new biologic because non-specific activity of such molecules is generally considered to be low.

Gap   Despite its central role in risk characterization, the predictivity of human cancer risk by the rodent bioassay has been questioned [44, 45]. Predictivity and human relevance are particularly problematic for non-genotoxic pharmaceuticals. Results of rodent carcinogenicity studies are frequently positive for tumor outcomes but are often concluded to be irrelevant to human risk for reasons of dose, mode of action, or species differences [46]. The value added by conducting rodent bioassays must be weighed against the extensive use of animals, resources, and time of conducting such assays. Given these considerations, the development of alternative and supplemental approaches that could improve assessment of human carcinogenic risk of new pharmaceuticals is of pressing interest. An ideal testing strategy would identify carcinogenic hazard by methods more applicable to human biology and establish dose response data which meaningfully informs risk assessment based on exposure and mechanism. A battery of approaches that extracts relevant information from the drug's intended target, drug-specific nonclinical data, and human carcinogenomics is more likely to meet this ideal than any single approach proposed to replace or supplement rodent bioassays.

The WOE strategy recommended for therapeutic proteins does not exclude the need for rodent bioassays, but in most cases, a rodent bioassay is not considered necessary to adequately assess carcinogenic risk, regardless of whether testing in rodents is feasible or not. As discussed below in FDA's current efforts, a similar strategy is being evaluated for assessing carcinogenic risk of small molecules as well. Experience with the WOE strategy with biologics has identified a need for additional testing methods that evaluate the hazard presented by immunomodulators.

2.4.1 Safety Pharmacology Studies

Safety pharmacology studies are conducted to determine whether a drug causes on- or off-target serious acute effects on critical organ systems (e.g., cardiovascular, respiratory, gastrointestinal, and central nervous system). Dose responses identified in these studies serve to establish a safety margin for the first-in-human (FIH) dose regimen. Follow-up experiments aimed at understanding the mechanism behind an observed specific toxicity can be undertaken to address the relevance of the finding to human risk. The risk to organs or tissues is demonstrated to be low. Safety pharmacology studies prior to the first administration in humans may not be needed for cytotoxic agents for treatment of end-stage cancer patients. However, for cytotoxic agents with novel mechanisms of action, there may be value in conducting safety pharmacology studies [21]. The effects of a drug candidate on the vital functions listed in the safety pharmacology core battery should be investigated prior to first in humans dosing. If there is a cause for concern not addressed by the safety pharmacology core battery, supplemental studies should be conducted (Table 1). Toxicology studies adequately designed and conducted to address safety pharmacology endpoints can reduce or eliminate the need for separate safety pharmacology studies.

Gaps According to Avila AM et al. 2020 [20], the current approach to assess the general toxicity of pharmaceuticals has been highly successful in allowing reasonably safe clinical trials to proceed [47], with generally good proficiency of animal toxicity data to predict potential adverse human effects [48, 49], and the absence of animal toxicity is particularly good at predicting an absence of toxicities in human phase 1 clinical trials [50]. However, there are areas in which traditional general

Table 1 An overview of current key challenges and needs in nonclinical safety assessments supporting development of human pharmaceuticals

Assessment	Key issues addressed	Key needs
Safety pharmacology	Identifies potential adverse pharmacological effects of the drug in vitro and/or in vivo Identifies specific parameters to monitor more closely in clinical trials	Refine approach to improve identification of proarrhythmic risk while reducing attrition of potentially useful drugs from clinical testing Refine Tier 1 neurological screens to achieve more quantitative and objective measures Standardize Tier 1 protocols and statistical approach to enable cross-laboratory comparisons Improve Tier 2 testing of higher cognitive functions and specificity of sensory tests
General toxicity	Estimates safe "first-in-human" (FIH) starting dose Estimates safe maximum doses in early clinical trials Identifies possible consequences of chronic exposure Identifies specific parameters to monitor more closely in clinical trials	Improve risk identification for rare and idiosyncratic toxicities Refine cardiovascular assessments to include functional endpoints Improve approach to evaluating human relevancy of toxicity findings in animals Identify best contexts-of-use for animal models of disease for toxicology testing Develop more human-relevant approaches to assessing risk of immune-directed therapies
Carcinogenicity	Predicts long-term risks that are difficult to assess or are unethical to assess in humans	Develop approaches more applicable to human biology and that provide dose response data based on anticipated human exposure and mechanism
Developmental and reproductive toxicity	Predicts risks that are difficult to assess or are unethical to assess in humans Identifies risks for special populations	Develop sensitive alternatives to detect potential teratogenicity earlier to facilitate clinical trials Develop sensitive alternatives for drug classes not tolerated by pregnant animals, such that the ability to conduct in vivo developmental and reproductive toxicity studies is limited
Special toxicity	Identifies specific parameters to monitor more closely in clinical trials Allows the mechanistic understanding of an adverse biological change observed in animals or humans	Develop validated non-animal methods for assessing human skin sensitization of mixtures

Avila AM, et al. *Regul Toxicol Pharmacol.* 2020;114:104662 [20]

toxicology studies have been less predictive of human toxicity, and improvements in these areas are desirable [48, 49]. The following describes these notable areas of need but is not meant to be inclusive of all attributes of general toxicology studies that could benefit from development of alternatives described in Table 2.

Table 2 Effects of the drug candidate onvarious organ systems and the parameters to be evaluated

Vital organs	Methods	Parameters to be evaluated
A.		
Cardiovascular system	In vivo, in vitro, and/ or ex vivo	Heart rate, blood pressure, electrocardiogram, repolarization, and conductance abnormalities
Central nervous system	In vivo	Behavioral changes, sensory/motor reflex responses, motor activity, coordination, and body temperature should be evaluated. The functional observational battery (FOB) [2], modified Irwin's test [3] and other tests [4] can be used to evaluate the effects of the drug
Respiratory system	In vivo	Respiratory rate as well as other measures of respiratory function (e.g., hemoglobin O_2 saturation, tidal volume) [5] should be evaluated
B.		
Cardiovascular system	In vivo, in vitro, and/ or ex vivo	Cardiac output, ventricular contractility, vascular resistance, the effects of endogenous and/or exogenous substances on the cardiovascular responses
Central nervous system	In vivo/in vitro	Behavioral pharmacology, learning and memory, ligand-specific binding, neurochemistry, visual, auditory, and/or electrophysiology examinations
Respiratory system	In vivo/in vitro/ex vivo	Airway resistance, compliance, pulmonary arterial pressure, blood gases, blood pH
Gastrointestinal system	In vivo/in vitro/ex vivo	Gastrointestinal injury potential, bile secretion, gastric secretion, transit time in vivo, ileal contraction in vitro, gastric pH measurement, and pooling may be considered
Renal/urinary system	In vivo/in vitro/ex vivo	Urinary volume, specific gravity, osmolality, pH, fluid/electrolyte balance, proteins, cytology, and blood chemistry determinations such as blood urea nitrogen, creatinine, and plasma proteins may be considered
Autonomic nervous system	In vivo/in vitro/ex vivo	Binding to receptors relevant for the autonomic nervous system, functional responses to agonists or antagonists in vivo or in vitro, direct stimulation of autonomic nerves and measurement of cardiovascular responses, baroreflex testing, and heart rate variability may be considered
Other organ systems	In vivo/in vitro/ex vivo	Effects of the drug candidate on organ systems not investigated elsewhere should be assessed when there is a concern for human safety. For example, dependency potential or skeletal muscle, immune, and endocrine functions can be investigated

Avila AM, et al. Supplemental safety pharmacology assays. *Regul Toxicol Pharmacol.* 2020;114:104662 [20]; Colerangle JB, Chapter 25: Preclinical development of nononcogenic drugs (small and large molecules). 2017;659-83 [21]

As a result of a better understanding of the underlying disease biology, it has become more apparent that many diseases are multi-factorial and treatment for these complex diseases remains limited. In addition, the development of resistance due to escape mechanisms, limited target expression, and sub-optimal effector mechanisms, among other reasons, continues to drive the need for more effective therapeutics with fewer dose-limiting toxicities. Therefore, the development of polyvalent biopharmaceuticals (PVBs) that target more than one epitope on the same or different target antigens has intensified over the past decade with the goal of overcoming some of the limitations of conventional mAbs. In general, similar challenges apply for the development of nonclinical safety strategies for monovalent mAbs and PVBs, except that engagement of two or more specific target pathways must be considered for PVBs. As one can envision, these challenges increase as the complexity of the desired attributes of PVBs increases. By design, PVBs target and modulate separate antigens/epitopes as part of their mechanism of action. Therefore, all of the binding regions of the molecule must be considered. Moreover, the ability to specifically introduce unique mechanisms of action (e.g., redirected immune effector cell lysis or enhanced transport to privileged sites) as part of the desired properties of multi-targeting therapeutics further increases the complexity for nonclinical safety evaluation [19].

As described in the previous sections, development of several novel multi-targeting formats and technologies has increased over the past decade as a result of advances in manufacturing and purification processes. Each technology or platform will create its own unique advantages and challenges that must be taken into consideration when developing the nonclinical safety assessment strategy to support various clinical indications. For example, certain platforms (e.g., antibody fragments) may allow for better tumor penetration yet may have rapid PK clearance properties that could limit exposure because of their small size. If prolonged exposure is required for a specific disease indication, one could potentially overcome this PK liability, through the addition of a "stabilizer" segment to the molecule, the encapsulation of the molecule in stabilizing material (i.e., liposomes), or continuous administration of the molecule through a portable infusion pump rather than a bolus IV infusion.

3   Principles for In Vitro Toxicity Assessment

In vitro assays are pivotal tools in toxicology assessment, utilizing biological models such as isolated organs, tissues, explants, cultured cells, organoids, and even subcellular fractions such as microsomes, mitochondria, to assess the toxic effects of various substances. In vitro toxicity assessment contributes to the 3Rs principle (reduction, refinement, and replacement) by reducing animal use [51]. Importantly, these assays can evaluate cellular responses, biochemical pathways, and molecular interactions upon the exposure of drug substances and, hence, offer the ability to evaluate events and endpoints directly associated with specific mechanisms of drug-induced toxicity [52]. Although limitations and challenges apply to many models, in vitro assays with human tissues mimicking human physiology can potentially improve the relevance and predictions of toxicity data for human health. Additionally, in vitro testing enables high-throughput screening of numerous compounds, facilitating the identification of drug liabilities early in the development process. This can significantly improve the timeline for drug discovery and lead optimization [27, 53].

Historically, a few in vitro assays designed for assessing adverse events that broadly applied to drug substances have been described in regulatory guidelines and should be considered to include in the preclinical safety assessment.

Genotoxicity assessment is probably one of the well-established and best known in vitro testing strategies. Several in vitro assays have been well recognized as indispensable components for determining potential hazard of a chemical structure for direct or indirect DNA interactions. These in vitro assessments provide tools for identifying drug candidates with the genotoxicity hazard early in development and contribute to regulatory decisions to define potential carcinogens. The standard in vitro test battery includes a bacterial reverse gene mutation test and several mammalian cell-based assays, which are summarized in this section.

The Ames test, developed by Dr. Bruce Ames, is a widely used in vitro assay for assessing the mutagenic potential of chemicals [54, 55]. This assay employs strains of the bacterium *Salmonella typhimurium* that carries mutations rendering them unable to synthesize the amino acid histidine. When exposed to a mutagenic substance, these bacteria can undergo reverse mutations that restore their ability to produce histidine, allowing them to grow on a histidine-free medium. The frequency of these reverted colonies provides a measure of the substance's mutagenicity. The Ames test is highly valued for its simplicity, cost-effectiveness, and rapid results, making it a standard initial screening tool in genetic toxicology.

Genotoxicity assessment in mammalian cells can include a metaphase chromosome aberration assay, or a micronucleus assay, or a thymidine kinase (TK) gene mutation assay [56–58]. The in vitro metaphase chromosome aberration assay is designed to identify the potential of a substance to cause chromosomal damage in cultured cells (typically human lymphocytes or Chinese hamster ovary [CHO] cells). Following the treatment of a drug candidate, cells were arrested in metaphase by mitotic inhibitors (e.g., colchicine). Chromosomes, which are most visible at metaphase, are examined microscopically for structural aberrations (such as breaks, fragments, and exchanges) [56]. The frequency and types of aberrations in cells were compared among treatment and the vehicle control group to evaluate the genotoxic potential of drug candidates. For in vitro micronucleus assays, cells were treated with test articles and then progress through mitosis. Micronuclei, which are small, extranuclear bodies containing chromosomal fragments or whole chromosomes that were not incorporated into the daughter nuclei during cell division, are then scored under a microscope [58]. The presence and frequency of micronuclei provide a measure of drug-induced chromosome breakage (clastogenicity) or whole chromosome loss (aneugenicity). For the TK gene mutation assay, L5178Y mouse lymphoma cells (carrying a functional TK gene) were treated with test articles. Chemicals that induce mutations that inactivate the TK gene make cells resistant to a toxic nucleoside analogue, trifluorothymidine (TFT). By measuring the frequency of resistant colonies, the mutagenic properties of the tested substance can be quantified [57].

The hERG (human Ether-à-go-go-Related Gene) encodes the pore-forming unit of the delayed rectifier potassium channel (IKr) crucial for maintaining normal electrical activity in the heart. A blockage of hERG channels by a diverse group of pharmaceuticals had been shown to cause QT interval prolongation which predisposes individuals to life-threating arrhythmias [59]. The in vitro hERG assays most commonly measure the channel's activity using electrophysiological techniques on isolated cardiomyocytes or cell lines expressing hERG protein in the presence of the test compounds. Other assay endpoints including drug-induced changes in ion flux across membranes, or plasma membrane potential, or radioligand binding assays have also been used [60]. The in vitro hERG assay is a cornerstone in cardiac safety pharmacology and widely used in the drug development process to screen out compounds with undesirable cardiac side effects early in development. Although delay of repolarization can occur through modulation of several types of ion channels, inhibition of IKr is the most common mechanism responsible for pharmaceutical-induced prolongation of QT interval in humans.

In vitro phototoxicity tests were also established for evaluating the potential of drug substances to cause skin irritation or damage when exposed to light, particularly ultraviolet (UV) radiation. The most widely used in vitro assay for phototoxicity is the 3T3 Neutral Red Uptake Phototoxicity Test (3T3 NRU-PT) which employs the 3T3 mouse fibroblast cells. Cells were treated with test articles and then exposed to UV light. The viability of cells is measured using the uptake of neutral red dye, which indicates plasma membrane integrity. A decrease in cell viability after light exposure, compared to controls, suggests phototoxic potential. The 3T3 NRU-PT are considered the most appropriate in vitro screen for soluble compounds [61].

Several in vitro skin models were established for assessing the potential dermatological effects of chemicals. Originally developed for cosmetic products, many of these assays had also been utilized to assess potential dermal effects of pharmaceutical gradients with topical formulations [62]. There are several commercially available versions of the in vitro human skin model including reconstructed human epidermis (RhE), EpiDerm, EST 1000, Labcyte, and Vitrolife skin models [63–65]. The general principles of all these models are based on the reconstruction of human epidermis or full-thickness skin models in vitro by keratinocytes or keratinocytes/fibroblasts co-culture. By mimicking the physiological and biochemical properties of human skin, these models can be used to evaluate skin corrosion, sensitization, permeability, and even metabolism under certain circumstances for topical drugs. Several of these models have been recommended by Organization for Economic Cooperation and Development (OECD) test guidelines and can be considered to replace animal irritations studies for supporting drug or formulation approval [66].

Besides those recommended by regulatory guidelines, many other in vitro methods are also widely used for understanding mechanisms of toxicity, ranking candidates, and, in some cases, contributing to risk assessment of drug substances. Most assays in this category are exploratory and typically developed in a fit for purpose mode. In the past decade, in vitro methods have evolved significantly in tandem with technological advancements. New technologies such as three-dimensional (3D) cell cultures, organ-on-a-chip, has enabled researchers to create more physiologically relevant models that closely mimic human tissues and organs [67, 68]. High-throughput screening (HTS), platforms, and automated systems have revolutionized the efficiency of in vitro testing, allowing for the rapid and simultaneous evaluation of thousands of compounds. Advancements in imaging technologies, such as high-content screening (HCS) and live-cell imaging, provide detailed, real-time insights into cellular responses and mechanisms of toxicity. Additionally, the integration of omics technologies, including genomics, proteomics, and metabolomics, allows for comprehensive analysis at the molecular level, uncovering biomarkers and pathways involved in toxic responses. Given the broad spectrum of in vitro toxicity assessment, it is difficult to thoroughly cover the types of assays nor the application scenarios in a book chapter. Therefore, in the next section we will focus on a few representative cellular assays that were developed to assess several drug toxicities.

3.1   Off-Target Profiling

Off-target effects are one of the main causes of drug-mediated adverse effects in humans and animals. For chemical drug entities, it is estimated that about 75% of all adverse drug reactions are dose-dependent and can be attributed to interactions with secondary targets (known as type A adverse drug reactions [ADRs]) [69]. Therefore, in vitro profiling of drug candidates against a broad range of known pharmacological targets (e.g., receptors, ion channels, or enzymes) distinct from the intended therapeutic target can identify candidates with potential unwanted effects. Although there was no regulatory guidance on targets that should be included in the screening (except hERG), several industrial-wise surveys have been published in the past which provided opportunities for companies to exchange experiences. Notably, a panel of 44 human targets proposed by Bowes et al. has been increasingly adopted in early pharmaceutical research [70]. The Bowes panel includes 24 G protein-coupled receptors (GPCRs), 8 ion channels, 6 enzymes, 3 transporters, 2 nuclear hormone receptors, and a protein kinase. Effects of drugs on these targets can be evaluated by either ligand bindings or functional assays. Similar to other in vitro assays, it is important to understand the concentration response relationship in the secondary pharmacology screening and generate quantitative data such as IC50 or EC50. Besides early hazard identification, these assays can also be used for understanding the molecular mechanisms of adverse effects in vivo.

Biologics are considered highly specific with a low risk of off-target effects. For monoclonal antibodies, a tissue-cross reactivity (TCR) assessment was recommended by ICHS6 for supporting first-in-human trial (FIH) if feasible. TCR employs immunohistochemical (IHC) techniques to characterize binding of monoclonal antibodies or antibody-like products in a panel of human tissues [71]. However, one of the most common challenges for this approach is that the biologic candidate is not a good IHC reagent and a TCR study is technically infeasible. Following an increasing case of off-target toxicity mediated by biologics, several new approaches, including in situ hybridization, protein chips, and cell-based off-target screening have been developed and be increasingly applied in drug development. For instance, retrogenix technology, which leverages a human cell microarray technology by individually expressing 6500 full-length human plasma membrane proteins in human cells, has been commonly used in evaluating off-target binding profile of chimeric antigen receptor (CAR)-based cell-therapy products [72].

3.2 In Vitro Models for Organ Toxicities

3.2.1 In Vitro Models for Hepatotoxicity

The liver is the primary site for drug metabolism in the body. Therefore it's not surprising that drug-induced livery injury (DILI) is one of the most commonly observed toxicities both in the clinic and in nonclinical studies for small molecule/chemical drug candidates. Numerous publications have demonstrated that in vitro hepatocyte models are crucial for derisking DILI by understanding the cellular, metabolic, and molecular mechanisms [73]. In vitro liver models typically employ human-derived cell systems, including primary human hepatocytes, hepatic cell lines like HepG2, or induced pluripotent stem cell (iPSC)-differentiated hepatocytes. In early days, hepatocytes were cultured in 2D alone or with stromal cells. More recently, several companies provided complex liver culture models including spheroids, organoids, bio-printed tissues, or liver-chips [74–76]. Given that DILI is a complex process that involves many biological and biochemical mechanism, the endpoints for most in vitro assays can be highly diverse. Common hepatotoxic endpoints evaluated in vitro can include cell viability, enzyme leakage (e.g., ALT, AST, LDH), reactive oxygen species (ROS) production, mitochondrial dysfunction, and bile acid accumulation [74, 75]. Several primary hepatocyte-based in vitro models have also shown the capability to detect metabolic activation of drugs by liver cytochrome P450 enzymes (CYP) and, hence, can inform potential toxicities mediated by drug metabolites [74–76].

3.2.2 In Vitro Models for Cardiotoxicity

Traditional approaches for in vitro cardiotoxicity assessment had been mainly focused on effects of drugs on ion channels (details described above). Obtaining and maintenance of high-quality primary cardiomyocytes remain challenging. In the past decade, one of the most important advancements in this area is the successful generation of functional cardiomyocytes from human stem cells or iPSCs. The iPSC-derived cardiomyocytes have been extensively characterized and various studies have demonstrated the value of the model to evaluate various types of drug-mediated cardiotoxicity [77]. A number of publications also reported that iPSC-derived cardiomyocytes retained the donor phenotype potentially allowing the evaluation of drug effects in specific patient populations [78]. Additionally, cultured cardiomyocytes expand the capability to detect drug-induced cardiotoxicity in vitro by enabling the measurement of cytotoxicity, contractility changes, mitochondrial toxicity, perturbed calcium regulation, and oxidative stress [77, 78].

3.2.3   In Vitro Models for Nephrotoxicity

As the primary site for drug or drug metabolite excretion, the kidney is another organ frequently involved in drug-induced adverse effects. Nephron, the functional unit of the kidney, is composed of different structural units including glomerulus, proximal tubule, loop of Henle, distal tubule, and the collecting ducts. The excretion of waste products is complex and typically involves glomerular filtration, tubular reabsorption, and tubular secretion. Most small molecule drug or chemical induces nephrotoxicity via proximal tubular injury as epithelial cells in this region are highly active in absorbing and excreting substances both in glomerular filtrate and in blood [72, 79]. It is not surprising that proximal tubular cells have been extensively used for nephrotoxicity screening. The primary human renal proximal tubule epithelial cell line RPTEC/hTERT is probably one of the most widely used in vitro models to assess drug-induced nephrotoxicity. Immortalized by human telomerase reverse transcriptase (hTERT), RPTEC/hTERT cells have been shown to express transporters including OATs, OCTs, OCTN2, MATE, and a few OATs, which are known to play important roles in drug uptake on both apical and on basolateral sides of tubular cells [80]. Another conditionally immortalized proximal tubule epithelial cell line (ciPTEC) has also been used [81]. One of the disadvantages of ciPTEC cells is the lack of OAT expression. However, a recent study showed that ciPTEC cells with OAT1 ectopically expressed can successfully predict nephrotoxicity by screening 62 benchmark drugs [80]. In recent years, efforts have been focused on the development of complex in vitro models for potentially improving the 2D cultures. These models include organoids and microphysiological systems that can create fluidic characteristics mimicking in vivo scenarios. Many of these models are in the early phases of development, and their value for predicting drug-induced nephrotoxicity is under investigation in most cases [81].

4   Conclusion

In summary, the field of toxicology in drug development is integral to ensuring the safety and efficacy of new pharmaceuticals. Preclinical toxicology studies provide essential data on the potential adverse effects of drug candidates, guiding safe dose selection and identifying risks before human trials. Compliance with regulatory guidelines and standards such as GLP, FDA, EMA, and ICH is crucial for the successful submission of IND, MAA, and NDA applications. The integration of in vitro and in vivo methods, along with advancements in high-throughput screening and omics technologies, enhances the predictive value of toxicology assessments and reduces animal usage.

Looking forward, predictive toxicology is poised to benefit significantly from emerging technologies like AI-driven modeling and multi-omics integration. However, challenges such as data integration and species differences remain. Regulatory harmonization and innovation, ethical considerations, and public trust are critical components that must be continually addressed. Ethical frameworks and transparency in data reporting are essential to maintain public trust and promote responsible conduct in toxicology research.

Continuous improvement in toxicology practices through adaptive strategies, risk management, and collaboration among stakeholders is vital. This ongoing process enhances patient safety and contributes to public health outcomes by ensuring the efficacy and safety of pharmaceutical products. The future of toxicology in drug development lies in leveraging technological advancements, embracing best practices, and addressing emerging challenges to deliver safe and effective therapies to patients worldwide.

By fostering a culture of innovation, regulatory compliance, and ethical consideration, the pharmaceutical industry can accelerate the development of new therapies while ensuring patient safety and maintaining public trust. The dynamic and evolving field of toxicology will continue to play a pivotal role in the advancement of medical science and the development of novel therapeutic agents.

Acknowledgments The author would like to thank Dr. Nianyu (Jason) Li, Merck Research Laboratories, Department of Safety Assessment and Laboratory Animal Resources, Merck & Co., Inc., 770 Sumneytown Pike, P.O. Box 4, West Point, Pennsylvania 19486, United States, for his review.

References

1. The Basic Science of Poison. Casarett and Doull's toxicology: the basic science of poisons. 8th ed. New York: McGraw-Hill Education; 2012.
2. Subramanyam M, Rinaldi N, Mertsching E, Hutto D. Selection of relevant species. In: Pharmaceutical sciences encyclopedia; 2010. p. 1–25. John Wiley and Sons, Inc.: Hoboken, NJ, USA.
3. Baldrick P. Safety evaluation to support First-In-Man investigations II: toxicology studies. Regul Toxicol Pharmacol. 2008;51(2):237–43.
4. S6(R1) FGD-I. Preclinical safety evaluation of biotechnology-derived pharmaceuticals. 2012. Available from: https://www.fda.gov/regulatory-information/search-fda-guidance-documents/s6r1-preclinical-safety-evaluation-biotechnology-derived-pharmaceuticals.
5. US-FDA. International conference on harmonisation; addendum to International conference on harmonisation guidance on S6 preclinical safety evaluation of biotechnology-derived pharmaceuticals; availability. Notice Fed Regist. 2012;77(97):29665–6.
6. Kendrick J, Stow R, Ibbotson N, Adjin-Tettey G, Murphy B, Bailey G, et al. A novel welfare and scientific approach to conducting dog metabolism studies allowing dogs to be pair housed. Lab Anim. 2020;54(6):588–98.
7. Prior H, Bottomley A, Champéroux P, Cordes J, Delpy E, Dybdal N, et al. Social housing of non-rodents during cardiovascular recordings in safety pharmacology and toxicology studies. J Pharmacol Toxicol Methods. 2016;81:75–87.
8. Prior H, Baldrick P, de Haan L, Downes N, Jones K, Mortimer-Cassen E, et al. Reviewing the utility of two species in general toxicology related to drug development. Int J Toxicol. 2018;37(2):121–4.
9. Brennan FR, Cauvin A, Tibbitts J, Wolfreys A. Optimized nonclinical safety assessment strategies supporting clinical development of therapeutic monoclonal antibodies targeting inflammatory diseases. Drug Dev Res. 2014;75(3):115–61.

10. Iwasaki K, Uno Y, Utoh M, Yamazaki H. Importance of cynomolgus monkeys in development of monoclonal antibody drugs. Drug Metab Pharmacokinet. 2019;34(1):55–63.
11. Bugelski PJ, Herzyk DJ, Rehm S, Harmsen AG, Gore EV, Williams DM, et al. Preclinical development of keliximab, a primatized anti-CD4 monoclonal antibody, in human CD4 transgenic mice: characterization of the model and safety studies. Hum Exp Toxicol. 2000;19(4):230–43.
12. Prior H, Haworth R, Labram B, Roberts R, Wolfreys A, Sewell F. Justification for species selection for pharmaceutical toxicity studies. Toxicol Res (Camb). 2020;9(6):758–70.
13. Prior H, Baldrick P, Beken S, Booler H, Bower N, Brooker P, et al. Opportunities for use of one species for longer-term toxicology testing during drug development: a cross-industry evaluation. Regul Toxicol Pharmacol. 2020;113:104624.
14. Andersson, P. (2022). Preclinical Safety Assessment of Therapeutic Oligonucleotides. In: Arechavala-Gomeza, V., Garanto, A. (eds) Antisense RNA Design, Delivery, and Analysis. Methods in Molecular Biology, vol 2434. Humana, New York, NY.
15. Tessier Y, Achanzar W, Mihalcik L, Amuzie C, Andersson P, Parry JD, et al. Outcomes of the European Federation of Pharmaceutical Industries and Associations Oligonucleotide Working Group Survey on nonclinical practices and regulatory expectations for therapeutic oligonucleotide safety assessment. 2020 (2159-3345 (Electronic)).
16. Henry SP, Kim T-W, Kramer-Stickland K, Zanardi TA, Fey RA, Levin AA, editors. Toxicologic properties of 2-O-methoxyethyl chimeric antisense inhibitors in animals and man. CRC Press; 2007.
17. Hard GC, Johnson KJ, Cohen SM. A comparison of rat chronic progressive nephropathy with human renal disease-implications for human risk assessment. Crit Rev Toxicol. 2009;39(4):332–46.
18. Braendli-Baiocco A, Festag M, Dumong Erichsen K, Persson R, Mihatsch MJ, Fisker N, et al. From the cover: the minipig is a suitable non-rodent model in the safety assessment of single stranded oligonucleotides. Toxicol Sci. 2017;157(1):112–28.
19. Narayanan P, Korgaonkar C, Staflin K. Nonclinical development of monovalent and polyvalent biopharmaceuticals. In: Plitnick LM, Fuller CL, editors. Nonclinical development of biologics, vaccines and specialty biologics. Elsevier; 2024.
20. Avila AM, Bebenek I, Bonzo JA, Bourcier T, Davis Bruno KL, Carlson DB, et al. An FDA/CDER perspective on nonclinical testing strategies: classical toxicology approaches and new approach methodologies (NAMs). Regul Toxicol Pharmacol. 2020;114:104662.
21. Colerangle JB. Chapter 25: Preclinical development of nononcogenic drugs (small and large molecules). In: Faqi AS, editor. A comprehensive guide to toxicology in nonclinical drug development. Second ed. Boston: Academic; 2017. p. 659–83. ISBN 9780128036204, https://doi.org/10.1016/B978-0-12-803620-4.00025-6
22. Levin AA, Henry SP. Toxicology of oligonucleotide therapeutics and understanding the relevance of the toxicities. In: Preclinical safety evaluation of biopharmaceuticals; 2008. p. 537–74.
23. Gennari A, Ban M, Braun A, Casati S, Corsini E, Dastych J, Descotes J, et al. The use of in vitro systems for evaluating immunotoxicity: the report and recommendations of an ECVAM workshop. (1547–6901 (Electronic)).
24. CHMP. Committee for Proprietary Medicinal Products (CPMP) Note for guidance on repeated dose toxicity. 2000. Available from: https://www.ema.europa.eu/en/documents/scientific-guideline/guideline-repeated-dose-toxicity-revision-1_en.pdf.
25. US-FDA RB. Short-term toxicity studies with rodents: IV.C.3.a. 2000.
26. US-FDA RB. Short-term toxicity studies with rodents: IV.C.3.b 2003.
27. Hirabayashi Y, Maki K, Kinoshita K, Nakazawa T, Obika S, Naota M, et al. Considerations of the Japanese research working group for the ICH S6 & related issues regarding nonclinical safety assessments of oligonucleotide therapeutics: comparison with those of biopharmaceuticals. Nucleic Acid Ther. 2021;31(2):114–25.
28. US-FDA. Nonclinical testing of individualized antisense oligonucleotide drug products for severely debilitating or life-threatening diseases guidance for sponsor-investigators. 2021.

Available from: https://www.fda.gov/regulatory-information/search-fda-guidance-documents/nonclinical-testing-individualized-antisense-oligonucleotide-drug-products-severely-debilitating-or.

29. Goyenvalle A, Jimenez-Mallebrera C, van Roon W, Sewing S, Krieg AM, Arechavala-Gomeza V, et al. Considerations in the preclinical assessment of the safety of antisense oligonucleotides. Nucleic Acid Ther. 2023;33(1):1–16.

30. Cavagnaro JA. Preclinical safety evaluation of biopharmaceuticals: a science-based approach to facilitating clinical trials. 2007.

31. Williams PD. Methods of production of biopharmaceutical products and assessment of environmental impact. In: Cavagnaro JA, editor. Preclinical safety evaluation of biopharmaceuticals: a science-based approach to facilitating clinical trials; 2007.

32. ICH-Q4A. Duration of chronic toxicity testing in animals (rodent and nonrodent toxicity testing). 1999. Available from: https://www.fda.gov/regulatory-information/search-fda-guidance-documents/s4a-duration-chronic-toxicity-testing-animals-rodent-and-nonrodent-toxicity-testing.

33. Baldrick P. Getting a molecule into the clinic: nonclinical testing and starting dose considerations. Regul Toxicol Pharmacol. 2017;89:95–100.

34. M3(R2) FGD-. M3(R2) nonclinical safety studies for the conduct of human clinical trials and marketing authorization for pharmaceuticals. 2010. Available from: https://www.fda.gov/regulatory-information/search-fda-guidance-documents/m3r2-nonclinical-safety-studies-conduct-human-clinical-trials-and-marketing-authorization.

35. Docket No. FDA-2021-D-1311-Nonclinical Considerations for Mitigating Nonhuman Primate Supply Constraints Arising From the COVID-19 Pandemic; Guidance for Industry; Availability. https://www.federalregister.gov/documents/2022/02/24/2022-03915/nonclinical-considerations-for-mitigating-nonhuman-primate-supply-constraints-arising-from-the

36. S2B FGD-. Genotoxicity: a standard battery for genotoxicity testing of pharmaceuticals. 1997. Available from: https://www.fda.gov/regulatory-information/search-fda-guidance-documents/s2b-genotoxicity-standard-battery-genotoxicity-testing-pharmaceuticals.

37. S2A FGD-. Specific aspects of regulatory genotoxicity tests for pharmaceuticals. 1997. Available from: https://www.fda.gov/regulatory-information/search-fda-guidance-documents/s2a-specific-aspects-regulatory-genotoxicity-tests-pharmaceuticals.

38. Q3B(R) FGD-I. Impurities in new drug products. 2006. Available from: https://www.fda.gov/regulatory-information/search-fda-guidance-documents/q3br-impurities-new-drug-products-revision-3.

39. S5(R3) FGD-. Detection of reproductive and developmental toxicity for human pharmaceuticals. 2010. Available from: https://www.fda.gov/regulatory-information/search-fda-guidance-documents/s5r3-detection-reproductive-and-developmental-toxicity-human-pharmaceuticals.

40. Sakai T, Takahashi M, Mitsumori K, Yasuhara K, Kawashima K, Mayahara H, et al. Collaborative work to evaluate toxicity on male reproductive organs by repeated dose studies in rats—overview of the studies. J Toxicol Sci. 2000;25 Spec No:1–21.

41. Sanbuissho A, Yoshida M, Hisada S, Sagami F, Kudo S, Kumazawa T, et al. Collaborative work on evaluation of ovarian toxicity by repeated-dose and fertility studies in female rats. J Toxicol Sci. 2009;34 Suppl 1:Sp1–22.

42. Johnson EM. Perspectives on reproductive and developmental toxicity. Toxicol Ind Health. 1986;2(4):453–82.

43. Cavagnaro J, Berman C, Kornbrust D, White T, Campion S, Henry S. Considerations for assessment of reproductive and developmental toxicity of oligonucleotide-based therapeutics. Nucleic Acid Ther. 2014;24(5):313–25.

44. Corvi R, Madia F, Guyton KZ, Kasper P, Rudel R, Colacci A, et al. Moving forward in carcinogenicity assessment: report of an EURL ECVAM/ESTIV workshop. Toxicol In Vitro. 2017;45:278–86.

45. Marone PA, Hall WC, Hayes AW. Reassessing the two-year rodent carcinogenicity bioassay: a review of the applicability to human risk and current perspectives. Regul Toxicol Pharmacol. 2014;68(1):108–18.

46. Bourcier T, Roy D. Addressing positive findings in carcinogenicity studies. In: Genotoxicity and carcinogenicity testing of pharmaceuticals; 2015. p. 159–82.
47. Butler LD, Guzzie-Peck P, Hartke J, Bogdanffy MS, Will Y, Diaz D, et al. Current nonclinical testing paradigms in support of safe clinical trials: an IQ Consortium DruSafe perspective. Regul Toxicol Pharmacol. 2017;87:S1–S15.
48. Clark M, Steger-Hartmann T. A big data approach to the concordance of the toxicity of pharmaceuticals in animals and humans. Regul Toxicol Pharmacol. 2018;96:94–105.
49. Olson H, Betton G, Robinson D, Thomas K, Monro A, Kolaja G, et al. Concordance of the toxicity of pharmaceuticals in humans and in animals. Regul Toxicol Pharmacol. 2000;32(1):56–67.
50. Monticello TM, Jones TW, Dambach DM, Potter DM, Bolt MW, Liu M, et al. Current nonclinical testing paradigm enables safe entry to First-In-Human clinical trials: the IQ consortium nonclinical to clinical translational database. Toxicol Appl Pharmacol. 2017;334:100–9.
51. Srivastava S, Mishra S, Dewangan J, Divakar A, Pandey PK, Rath SK. Chapter 2: Principles for in vitro toxicology. In: Dhawan A, Kwon S, editors. In Vitro toxicology. Academic; 2018. p. 21–43.
52. Eisenbrand G, Pool-Zobel B, Baker V, Balls M, Blaauboer BJ, Boobis A, et al. Methods of in vitro toxicology. Food Chem Toxicol. 2002;40(2–3):193–236.
53. Roggen EL. In vitro toxicity testing in the twenty-first century. Front Pharmacol. 2011;2:3.
54. Ames BN, Durston WE, Yamasaki E, Lee FD. Carcinogens are mutagens: a simple test system combining liver homogenates for activation and bacteria for detection. Proc Natl Acad Sci USA. 1973;70(8):2281–5.
55. Ames BN, Lee FD, Durston WE. An improved bacterial test system for the detection and classification of mutagens and carcinogens. Proc Natl Acad Sci USA. 1973;70(3):782–6.
56. Clare G. The in vitro mammalian chromosome aberration test. Methods Mol Biol. 2012;817:69–91.
57. Clements J. The mouse lymphoma assay. Mutat Res. 2000;455(1–2):97–110.
58. Kuo B, Beal MA, Wills JW, White PA, Marchetti F, Nong A, et al. Comprehensive interpretation of in vitro micronucleus test results for 292 chemicals: from hazard identification to risk assessment application. Arch Toxicol. 2022;96(7):2067–85.
59. Garrido A, Lepailleur A, Mignani SM, Dallemagne P, Rochais C. hERG toxicity assessment: useful guidelines for drug design. Eur J Med Chem. 2020;195:112290.
60. Priest BT, Bell IM, Garcia ML. Role of hERG potassium channel assays in drug development. Channels (Austin). 2008;2(2):87–93.
61. Kim K, Park H, Lim KM. Phototoxicity: its mechanism and animal alternative test methods. Toxicol Res. 2015;31(2):97–104.
62. Danilenko DM, Phillips GD, Diaz D. In vitro skin models and their predictability in defining normal and disease biology, pharmacology, and toxicity. Toxicol Pathol. 2016;44(4):555–63.
63. Agonia AS, Palmeira-de-Oliveira A, Cardoso C, Augusto C, Pellevoisin C, Videau C, et al. Reconstructed human epidermis: an alternative approach for in vitro bioequivalence testing of topical products. Pharmaceutics. 2022;14(8):1554.
64. Hoffmann J, Heisler E, Karpinski S, Losse J, Thomas D, Siefken W, et al. Epidermal-skin-test 1,000 (EST-1,000)—a new reconstructed epidermis for in vitro skin corrosivity testing. Toxicol In Vitro. 2005;19(7):925–9.
65. Kojima H, Ando Y, Idehara K, Katoh M, Kosaka T, Miyaoka E, et al. Validation study of the in vitro skin irritation test with the LabCyte EPI-MODEL24. Altern Lab Anim. 2012;40(1):33–50.
66. OECD. Guidelines for the testing of chemicals, section 4, OECD series on testing and assessment; 2021. p. 331.
67. Cacciamali A, Villa R, Dotti S. 3D cell cultures: evolution of an ancient tool for new applications. Front Physiol. 2022;13:836480.
68. Ingber DE. Human organs-on-chips for disease modelling, drug development and personalized medicine. Nat Rev Genet. 2022;23(8):467–91.
69. Iasella CJ, Johnson HJ, Dunn MA. Adverse drug reactions: type A (intrinsic) or type B (idiosyncratic). Clin Liver Dis. 2017;21(1):73–87.

70. Bowes J, Brown AJ, Hamon J, Jarolimek W, Sridhar A, Waldron G, et al. Reducing safety-related drug attrition: the use of in vitro pharmacological profiling. Nat Rev Drug Discov. 2012;11(12):909–22.

71. Leach MW, Halpern WG, Johnson CW, Rojko JL, MacLachlan TK, Chan CM, et al. Use of tissue cross-reactivity studies in the development of antibody-based biopharmaceuticals: history, experience, methodology, and future directions. Toxicol Pathol. 2010;38(7):1138–66.

72. Freeth J, Soden J. New advances in cell microarray technology to expand applications in target deconvolution and off-target screening. SLAS Discov. 2020;25(2):223–30.

73. Soldatow VY, Lecluyse EL, Griffith LG, Rusyn I. In vitro models for liver toxicity testing. Toxicol Res (Camb). 2013;2(1):23–39.

74. Ali ASM, Berg J, Roehrs V, Wu D, Hackethal J, Braeuning A, et al. Xeno-free 3D bioprinted liver model for hepatotoxicity assessment. Int J Mol Sci. 2024;25(3), 1811; https://doi.org/10.3390/ijms25031811

75. Jang KJ, Otieno MA, Ronxhi J, Lim HK, Ewart L, Kodella KR, et al. Reproducing human and cross-species drug toxicities using a Liver-Chip. Sci Transl Med. 2019;11(517), eaax5516. https://doi.org/10.1126/scitranslmed.aax5516

76. Kammerer S. Three-dimensional liver culture systems to maintain primary hepatic properties for toxicological analysis in vitro. Int J Mol Sci. 2021;22(19), 10214. https://doi.org/10.3390/ijms221910214

77. Prondzynski M, Berkson P, Trembley MA, Tharani Y, Shani K, Bortolin RH, et al. Efficient and reproducible generation of human iPSC-derived cardiomyocytes and cardiac organoids in stirred suspension systems. Nat Commun. 2024;15(1):5929.

78. Paik DT, Chandy M, Wu JC. Patient and disease-specific induced pluripotent stem cells for discovery of personalized cardiovascular drugs and therapeutics. Pharmacol Rev. 2020;72(1):320–42.

79. Soo JY, Jansen J, Masereeuw R, Little MH. Advances in predictive in vitro models of drug-induced nephrotoxicity. Nat Rev Nephrol. 2018;14(6):378–93.

80. Faria J, Ahmed S, Gerritsen KGF, Mihaila SM, Masereeuw R. Kidney-based in vitro models for drug-induced toxicity testing. Arch Toxicol. 2019;93(12):3397–418.

81. Nieskens TT, Peters JG, Schreurs MJ, Smits N, Woestenenk R, Jansen K, et al. A human renal proximal tubule cell line with stable organic anion transporter 1 and 3 expression predictive for antiviral-induced toxicity. AAPS J. 2016;18(2):465–75.

PadmaKumar "Padma" Narayanan BVSc (DVM equivalent), MS, PhD is currently Vice President and Head of Preclinical and Clinical Development Sciences at Wave Life Sciences, Boston, MA. Padma has spent 27 years in the Pharmaceutical Industry, with previous stints as Senior Director and Global Head of Immunology and Translational Toxicology in the Nonclinical Safety Department at Janssen Pharmaceuticals R&D, La Jolla, CA, Executive Director of Pathology and Toxicology at Ionis Pharmaceuticals, Carlsbad, CA, Director of Cell Signaling and Immunotoxicology Group of Discovery Toxicology, Amgen, Inc, Seattle, WA, and Senior Pathologist, SmithKline Beecham/GlaxoSmithKline, Philadelphia, PA. He obtained his BVSc degree from College of Veterinary & Animal Sciences, Kerala, India; a Masters' in Veterinary Surgery from Madras Veterinary College, Tamil Nadu, India, a Ph.D in Immunopharmacology from School of Veterinary Medicine, Purdue University, Indiana, USA and completed his post-doctoral training at the Life Sciences Division, Los Alamos National Laboratories, Los Alamos, NM. Padma not only leads and contributes to the overall strategic direction of preclinical development programs supporting development of novel RNA therapeutics to combat neurodegenerative diseases but is also deeply interested in understanding immune-mediated adverse events to drugs in nonhuman primates and its relevance in a clinical setting.

Perspective on Clinical Trials: What Researchers Need to Know

Prajak Barde and Mohini Barde

Abstract This chapter provides a comprehensive overview of clinical research, emphasizing its pivotal role in modern medical practice, including treatment, management, diagnosis, and prevention strategies. The evolution of clinical trials has been marked by significant advancements in design and methodology, greatly enhancing the efficiency and reliability of research outcomes. Initially, clinical trials were naive and lacked controls, but the introduction of randomized controlled trials (RCTs) marked a crucial turning point, establishing a robust framework to minimize bias and ensure more reliable results. Recent innovations, such as adaptive trial designs, allow modifications based on interim results, improving both efficiency and patient safety. This chapter discusses various types of clinical trials, different study designs, and the phases of trials, providing examples to illustrate when their implementation is appropriate and feasible. The implementation of information obtained from well-conducted clinical research is fundamental to evidence-based medicine, making it essential to fully understand the scope and importance of clinical research in view of increasingly sophisticated study designs, rigorous ethical standards, and technological innovations.

Keywords Clinical Trials · Historical Perspective · Ethical Perspective · Regulatory Framework · Treatment Trials · Prevention Trials · Diagnostic Trials · Screening Trials · Device Trials · Phases of Clinical Trials · Study Designs · Blinding · Randomization · Adaptive Study Designs · Precision Medicine · Basket Trials · Umbrella Trials · Platform Trials · Biomarkers · Artificial Intelligence

P. Barde (✉) · M. Barde
Med Indite Communication (MIC) Private Limited, Pune, Maharashtra, India
e-mail: drprajaktb@medindite.com

1 Introduction

The evolution of clinical trials has been instrumental in shaping the landscape of modern medicine, offering insights into treatment efficacy, safety, and patient outcomes. This chapter explores the intricate history and evolving landscape of clinical trials. Starting with a historical and ethical perspective, it explores the transformation of clinical trials over the years, guided by stringent regulatory frameworks. It elucidates the various types of clinical trials with an emphasis on treatment, preventive, diagnostic, screening, and device trials, providing comprehensive insights into each category.

The chapter highlights the different phases of clinical trials, from early exploratory phase 0 to post-marketing surveillance phase IV studies, emphasizing the objectives and methodologies distinctive to each phase. Phase I typically involves dose-escalation studies (e.g., 3 + 3 design, Single Ascending Dose (SAD), Multiple Ascending Dose (MAD)); phase IIa focuses on efficacy, patient population selection, and use of biomarkers; and phase IIb determines optimal dosing using traditional or modeling approaches. Phase III is critical for confirming efficacy and safety. Phase IV is done to monitor long-term effects and efficacy.

Study designs are critically analyzed, encompassing blinding techniques, randomization methods, and advanced study designs like adaptive study designs, Bayesian designs, model-based, and model-assisted designs. The discussion extends to precision medicine, explaining master protocols and specific trial types such as basket, umbrella, and platform trials.

Special considerations in clinical trials are thoroughly examined, including patient enrichment, sample size determination, the role of biomarkers, and the significance of N-of-1 trials, as these trials focus on individual patients, tailoring treatment to specific needs and evaluating efficacy through repeated crossover comparisons. The chapter also addresses drug designations, approvals, and real-world evidence (RWE) studies, alongside an analysis of clinical trial failures during development. RWE trials utilize data from real-world settings to assess treatment outcomes. Although they provide valuable insights into everyday clinical practice, challenges include variability in data quality and potential confounding factors. The evolving paradigm section discusses the integration of artificial intelligence (AI) and machine learning (ML) in clinical trials as they are increasingly integrated into drug development, enhancing data analysis, patient recruitment, and trial design. AI/ML can predict trial outcomes, optimize dosing, and also can personalize treatments.

In summary, the evolution of clinical trials from ethical and regulatory perspectives, coupled with advancements in trial types, phases, and designs, underscores the complexity and importance of clinical research. Addressing the challenges and limitations inherent in each aspect is crucial for advancing medical science and improving patient outcomes.

2   History and Background of Clinical Trials

2.1   *Historical Perspective*

Clinical trials and their advancement form a very crucial part of the medical land-scape. It has covered a very exciting path starting from 500 BC when the trial of legumes was first recorded in biblical times till the era of randomized controlled trials (RCTs). The famous surgeon Ambroise Parè accidentally conducted the first clinical trial of a novel therapy in 1537 when he applied digestives made of egg yolks, oil of roses, and turpentine to the wounded soldiers instead of conventional oils.

Of the current time, James Lind (1716–1794), a Scottish physician, is considered as the first physician who had conducted a controlled clinical trial. While working as a surgeon on the ship, he faces a situation of high mortality among the sailors. On May 20, 1747, he begins his scurvy trial. He divided the sick sailors into groups and had each of them follow a particular regimen. He concluded in this trial that oranges and lemons (sources of Vitamin C) had visibly good effects in just 6 days, paving the way to keep the sailors free of scurvy in the following years. This achievement leads to the celebration of "International Clinical Trials Day" on May 20 every year to mark the day he began his trial.

The word placebo first entered the medical literature in the early 1800s, including the 1811 edition of Hooper's Medical Dictionary. In 1863, American physician Austin Flint (1812–1886) planned the first clinical study where a dummy remedy was compared to an active treatment. This was the first known use of placebo in clinical trials. He treated 13 patients suffering from rheumatism with an herbal extract instead of an established remedy. In his 1886 book, *A Treatise on the Principles and Practice of Medicine,* Flint wrote that "this was given regularly and became well known in my wards as the 'placeboic remedy' for rheumatism" [1].

In 1943–44, the Medical Research Council (MRC) in the UK conducted the first double-blind comparative clinical trial to investigate patulin treatment for the common cold. The study was strictly controlled by ensuring that the treating physician and patient were blinded to the treatment [2]. Though the results of the trial did not show the effectiveness of Patulin, this trial serves as a base for future double-blind clinical trials.

The concept of randomization was initially introduced in the year 1923; however, the first randomized controlled trial was carried out in 1946 by MRC of the UK for testing streptomycin in pulmonary tuberculosis. This trial was planned very meticulously by using proper enrollment criteria, allocation concealment, objective measures, and data collection methods to improve the quality of data collected. This trial of Sir Bradford Hill was a model of design and implementation, and in years to come, random allocation in clinical trials becomes the universal norm [1].

The streptomycin trial certainly left a legacy and provided much of the evidence upon which rational treatment policies came to be based. In addition to testing new drugs, new combinations of drugs, and new drug regimens, trials were also designed to assess the value of bed rest and treatment in hospitals and the preventive potential of vaccines [3].

2.2 Ethical and Regulatory Perspective

The evolution of clinical trials from an ethical and regulatory perspective has been a dynamic process shaped by historical events, scientific advancements, and growing awareness of the need to protect human subjects participating in research. It's a concerted effort to prioritize the rights, safety, and well-being of participants.

Emerging from the atrocities of World War II, the Nuremberg Code, in 1947, established fundamental principles of informed consent, voluntary participation, and avoidance of harm to human subjects. This field has progressed through tragedies like the thalidomide disaster in the 1960s, highlighting the critical need for robust regulatory oversight and stringent safety measures. In response, regulatory reforms such as the FDA Amendments Act of 1962 empowered regulatory agencies to enforce stricter requirements for drug approval, including the establishment of institutional review boards (IRBs). Subsequently, the Declaration of Helsinki in 1964 provided a global ethical framework, which reinforced the importance of protecting participants' rights and ensuring scientific integrity in medical research. The latter part of the twentieth century and the early twenty-first century witnessed an increasing emphasis on international collaboration and regulatory harmonization. The International Conference for Harmonization (ICH), formed in the early 1990s, aimed to standardize regulatory requirements globally, ensuring consistency in the design, conduct, and reporting of clinical trials.

The advent of clinical trial registries and increased transparency initiatives has further enhanced accountability and public trust. Recent revisions to regulations, like the ICH E6 (R2) and the EU Clinical Trials Regulation, reflect ongoing efforts to adapt to technological advancements, streamline processes, and harmonize standards globally. Throughout this evolution, the overarching goal remains the same: to conduct ethical and rigorous research that promotes patient welfare and contributes to medical progress.

3 Types of Clinical Trials

As per ICH guidelines, a clinical trial is defined as any investigation in human subjects intended to discover or verify the clinical, pharmacological, and/or other pharmacodynamic effects of a drug and/or to identify any adverse reactions to a drug and/or to study absorption, distribution, metabolism, and excretion of a drug with the object of ascertaining its safety and/or efficacy. The terms clinical trial and clinical study are synonymous [4].

Clinical trials are categorized into different types, such as the trials for treatment, prevention, early detection/screening, and diagnosis, depending on whether the investigational drug can treat the disease, or can prevent the disease/condition, or test the ability of a device to detect/screen the disease, or test the efficacy of a medical test to diagnose the disease/condition [5].

3.1 Treatment Trials

Treatment trials, also known as therapeutic trials, are designed to evaluate the safety, efficacy, and effectiveness of new/improved drugs, therapies, or treatment approaches for specific diseases or conditions in human subjects. These trials aim to assess whether the investigational treatment is effective in improving patient outcomes, such as treating or managing the disease, reducing the symptoms, slowing disease progression, or achieving remission.

These trials are pivotal in expanding treatment options, especially for conditions where existing therapies are limited or ineffective. These trials can evaluate the impact of experimental treatments by determining their overall benefit-risk profile. By comparing the experimental intervention with standard treatments or placebo, treatment trials provide valuable insights into the relative effectiveness and safety of novel therapeutic approaches.

Examples: A clinical study of Trastuzumab, a recombinant monoclonal antibody against HER2, in patients undergoing surgery and chemotherapy for early-stage breast cancer, showed that 1 year of treatment with trastuzumab significantly improves disease-free survival among women with HER2-positive breast cancer [6]. The evaluation of double-dose oseltamivir on clinical and virological outcomes in children and adults admitted to the hospital with severe influenza showed that there were advantages with double dose of oseltamivir compared with the standard dose [7].

3.2 Prevention Trials

Prevention trials, also known as prophylactic trials, investigate interventions aimed at preventing the development of a disease or condition in individuals who are at risk. These trials evaluate the effectiveness of vaccines, medications, dietary supplements, lifestyle modifications, behavioral interventions, or public health interventions in reducing the risk of disease occurrence. Prevention trials prove to be helpful in developing strategies to combat diseases before they manifest. These strategies are based on the underlying disease process and various elements of the intervention, like sustainability of behavior change, the timing of the intervention relative to the disease progression, the necessary dose and duration of exposure to achieve risk reduction, and the durability of the impact of an intervention [8].

Examples: The findings from the P-1 study, which tested the hypothesis that "tamoxifen-treated women had a lower incidence of contralateral breast cancer than did women who received a placebo" demonstrated that tamoxifen administration resulted in a significant reduction in the risk of invasive breast cancer [9].

Other examples included the development of vaccines for new and emerging infectious diseases, such as Ebola and Zika virus, as well as the development of a universal influenza vaccine and COVID-19 vaccine. The PREDIMED trial assessing the long-term effects of the Mediterranean diet (MeDiet) on clinical events of cardiovascular disease (CVD) showed significant improvements in CVD risk factors with strong evidence that a MeDiet can prevent CVD [10].

3.3   Diagnostic Trials

Diagnostic trials, also known as diagnostic test evaluation studies, are designed to assess the performance and accuracy of new diagnostic tests, biomarkers, imaging techniques, or screening protocols for detecting specific diseases or conditions. Diagnostic trials are of two types, Type I trials aim to evaluate the accuracy of diagnostic tests or tools by measuring the sensitivity, specificity, and predictive value of a diagnostic tool in detecting a particular disease or health condition and its severity. Type II trials evaluate the value of test or tool results to guide or determine treatment decisions in the management of patients. These trials assess the sensitivity, specificity, predictive value, and reliability of diagnostic tools compared to standard diagnostic methods to aid in timely and accurate disease diagnosis. Example: "Spirokit," an inexpensive, smart device-based handheld spirometer, when compared with conventional spirometry in patients with chronic obstructive pulmonary disease, showed high validity and portability, with a high potential for replacing PC-based pulmonary function test equipment [11]. The MINDACT trial validated the performance of a 70-gene gene expression profile (MammaPrint™), based on DNA microarray technology that accurately determines the risk of relapse for breast cancer patients [12].

3.4   Screening Trials

Screening trials, also known as early detection trials, assess the ability of a diagnostic test, device, or screening protocols or strategies to identify a disease or condition in asymptomatic individuals at an early stage. These trials aim to validate the sensitivity, specificity, and accuracy of screening methods and assess whether early detection through screening leads to improved health outcomes, such as reduced disease-specific mortality, improved survival rates, or better treatment outcomes. Screening trials are conducted in the general population or in individuals with a higher-than-normal risk of developing a certain disease [13].

Example: The European randomized study of screening for prostate cancer assesses the usefulness of prostate-specific antigen (PSA)-based screening on prostate cancer mortality [14]. Results showed that this screening reduced the rate of death from prostate cancer by 20% but was associated with a high risk of overdiagnosis [15]. Other examples included a study assessing routine testing at birth for glucose-6-phosphate dehydrogenase (G6PD) deficiency to avoid dose-dependent acute hemolytic anemia caused by primaquine in patients having G6PD deficiency [16].

3.5 Device Trials

Device trials, also known as medical device clinical trials, are designed to evaluate the safety, efficacy, and performance of medical devices intended for use in the diagnosis, treatment, monitoring, or prevention of diseases or health conditions. Medical devices encompass a broad range of products, including diagnostic tools, therapeutic devices, surgical instruments, implants, wearable technologies, and medical software. Medical device trials are often conducted in controlled clinical settings and are compared to standard or existing devices or procedures. These trials generate clinical evidence supporting the device's performance and its impact on patient outcomes.

Example: Spectrophotometric tests for G6PD screening are complex and error-prone. Two newly developed kits, the CareStart™ Biosensor kit and the STANDARD G6PD, had promising performance with high sensitivity and specificity. Both kits showed high reliability and performed well in comparison to the spectrophotometric reference standard [17].

4 Phases of Clinical Trials

During clinical development, the drug needs to be tested pre-clinically and then in a series of clinical study trials. This testing is done in phases (Table 1). The phases of clinical trials offer a structured approach to evaluate the safety, efficacy, and feasibility of new medical interventions, ultimately advancing medical knowledge and improving patient care (Fig. 1).

4.1 Phase 0: Navigating the Frontier of Early Exploration

Phase 0 trials, also called ExpIND studies, represent an innovative approach in the early stages of drug development. These studies are conducted before phase I studies with very few subjects at the micro-dose level to assess the pharmacokinetics

Table 1 Phases of clinical trial

	Phase 0 "Exploratory"	Phase I	Phase II	Phase III	Phase IV
Description	First-in-human early trial to assess the feasibility of further clinical development of a drug	Initial safety evaluations, determine safe dosage range, identify common side effects, study toxicity profile of the drug	Begin to explore efficacy while maintaining safety	Final confirmation of safety and efficacy	Any trials conducted after FDA approval of the drug
Objectives	Pk, PD, MOA and if drug engages its expected target	Safety and PK	Clinical activity and safety, exploratory biomarkers	Efficacy and safety	Efficacy and Safety
Number of subjects	10–15 healthy volunteers	20–80 healthy volunteers and patients in case of anti-cancer drugs	100–300 volunteers with the targeted medical condition	1000–3000 subjects with the targeted medical condition	Number of subjects depends on trial endpoints
Study design	Open-label, single, low dose (<1% of the dose calculated to produce a clinical effect)	Open-label, randomized/non-randomized study. [Single ascending dose (SAD)/ Multiple ascending dose (MAD)] 3 + 3 dose escalation followed by dose expansion (Oncology).	Proof of concept study: (Exploratory studies) Dose-finding study: Open-label, placebo-controlled, randomized study.	Open-label, placebo-controlled, randomized study.	Variable
Endpoints	Not expected to show clinical effects or significant adverse effects. Helps to choose between competing chemical analogs for further study	Incidence and severity of adverse events, changes in vital signs, laboratory abnormalities, any dose-limiting toxicities (DLTs), and pharmacokinetic parameters.	Measures of treatment efficacy, such as disease response rates, progression-free survival, overall survival, or biomarker changes. Also include safety assessments, patient-reported outcomes, quality-of-life measures, and PK parameters.	Confirms clinical efficacy of the drug. Efficacy measurements such as Disease-free survival, overall survival, EFS, PFS, Quality of life (QoL), and adverse events.	Safety outcomes such as adverse events, drug interactions, and long-term side effects, as well as effectiveness outcomes such as disease recurrence, treatment response rates, and patient-reported outcomes.
Timing	Can be conducted with prior approval while the final IND review is pending	Together with phase 0 trials, the first clinical trials conducted in an IND process	Conducted after results of phase I trials are reported to regulators	Conducted after results of phase II trials are reported to regulators	Conducted after the approval of the drug by regulators for marketing

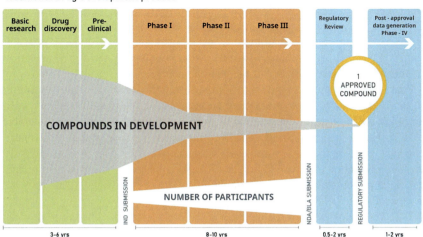

Fig. 1 Traditional Drug Development Process

(PK), target engagement, MOA, and pharmacodynamics of new molecules. These studies have no therapeutic intent and are not intended to study drug safety or tolerance. On the contrary, these studies help in selecting one of the preclinical candidates that can be considered for further clinical development, resolving conflict or inadequacy of the preclinical data, and providing a valuable bridge between laboratory research and traditional clinical trial phases [18].

Phase 0 trials may help to address some of the most challenging issues for new drug development by helping to prioritize potential agents for future study, reducing development timelines, and demonstrating proof-of-concept target inhibition.

Examples: The development of an anti-malarial drug (GSK3191607) was terminated using a microdosing study approach where it was determined that the elimination half-life (17 h) of the drug was too short to be consistent with developmental objectives [19]. Microdose studies were conducted by Pfizer to explore the human PK of four selective inhibitors of Nav1.7 channel blockers. It was seen that differences were related to the organic anion-transporting polypeptide (OATP)-mediated hepatic uptake, a mechanism for which few validated human modeling approaches were available [20].

Thus, phase 0 trials represent an innovative approach in the early stages of drug development; however, these trials are not commonly performed.

4.2   Phase I: Safety and Dose Selection

Phase I clinical trials represent the first stage of clinical testing of an investigational product in humans (also called first-in-human studies), typically a drug or therapeutic agent. The primary objective of phase I trials is to assess the safety profile of the investigational treatment and determine the maximum tolerated dose (MTD) or recommended phase II dose (RP2D) for further investigation in subsequent phases [21, 22]. These trials also aim to characterize the drug's pharmacokinetic properties (including its absorption, distribution, metabolism, and excretion (ADME)) in humans. Phase III trials typically involve a small number of healthy volunteers or patients with the target disease for the oncology drugs.

Phase I trials commonly employ dose-escalation designs (Single Ascending Doses (SAD) and Multiple Ascending Doses (MAD), where participants receive escalating doses of the investigational treatment to evaluate its safety and tolerability [23]. These designs typically follow a traditional 3 + 3 or modified Fibonacci sequence, with predefined dose levels and criteria for dose escalation or de-escalation based on observed toxicities [22]. Some phase I trials may also include cohort expansion or dose expansion cohorts to further characterize the safety and pharmacokinetics of the treatment.

The dose range needs to be determined before starting the phase I trial of any new drug candidate. Preclinical data generated in safety pharmacology studies, pharmacodynamics studies, and acute and repeated dose toxicity studies serve as a guide for dose selection and safety monitoring in phase I studies. The no-observed adverse effect level (NOAEL) observed in the most sensitive animal species is used to calculate the starting dose in the study after applying the human equivalent dose (HED) and safety factor (a default safety factor of 10). Minimally anticipated biological effect level (MABEL) is used to calculate the starting dose for biologics with agonistic properties.

There is no single approach for selecting a starting dose for FIH studies due to the uniqueness of each new drug candidate and the many different assumptions that have to be made for translation across species. For these reasons, it is difficult to establish a standard uniform approach. The EMA and the FDA have published guidance documents for investigators to follow when determining an appropriate starting dose [24].

An appropriate starting dose is also important to ensure patient safety. About 20–40% of phase I trials fail in further development because of safety issues. Noteworthy examples here are a study of TGN1412 (CD28 superagonist antibody), where six human volunteers faced life-threatening conditions involving multiorgan failure after the very first infusion of a dose 500 times smaller than that found safe in animal studies [25]. Another example includes a study in France of BIA10-2474, a fatty acid amide hydrolase (FAAH) inhibitor, leaving one healthy volunteer dead and four others seriously injured [26].

Food effects BA studies are usually conducted for new drugs and drug products during the IND period to assess the effects of food on the rate and extent of absorption of a drug when the drug product is administered shortly after a meal (fed conditions), as compared to administration under fasting conditions [27]. These assessments are incorporated into the SAD arm of the study when dose escalation has reached an anticipated therapeutically relevant level. This is particularly important for drugs that exhibit nonlinear PK. These studies should be conducted with a crossover design to draw meaningful conclusions [24].

Overall, phase 1 trials play a critical role in the early evaluation of new medical interventions, providing essential data to guide dose selection, inform subsequent clinical development plans, and ensure the safety of participants in later-phase trials.

4.3   Phase II: Efficacy and Expanded Patient Population

Phase II trials, also referred to as "therapeutic exploratory" trials, are usually larger than phase I studies and are conducted in a large number of participants (20–300) who have the disease of interest [23]. These trials lie at the critical juncture between the safety and established efficacy of a drug in development.

The phase II design depends on the quality and adequacy of phase I studies. The objective of phase II trials is to assess the preliminary efficacy of the investigational drug and further characterize its safety profile. These trials also aim to provide informed dose selection and regimen for further investigation in phase III trials [28].

Phase II trials may employ various study designs, including randomized controlled trials (RCTs), non-randomized trials, single-arm trials, or adaptive designs, depending on the research objectives and practical considerations. These trials may also include multiple treatment arms to evaluate different doses, dosing regimens, or treatment combinations to identify the most promising approach for further investigation.

Phase II trials are subdivided into phase IIa and phase IIb trials. Phase IIa is a proof of concept (PoC) or pilot or exploratory study involving around 50 participants. These trials are usually conducted to explore if the drug has some initial evidence of clinical efficacy or not. PoC studies usually use surrogate markers as endpoints. These trials facilitate the successful conduct of large clinical trials by informing study design and streamlining protocol implementation. These studies often link phase I (first in human) and dose-ranging phase II studies and allow drug developers to make "Go/No Go" decisions about proceeding with larger, more expensive studies [29].

Phase IIb trials are dose-finding studies that are conducted to identify the most promising dose or doses of the drug to be used in later studies. These studies are conducted in several hundred patients, and treatment time is commonly longer than in a phase IIa trial. The objective of these studies is to find the minimum effective dose or a dose with the best adverse event–efficacy trade-off. If a low dose gets selected, then the treatment effect will be insignificant to yield a positive result, and if a high dose gets selected, adverse effects may threaten the further development program in phase III trials. A well-designed dose-finding trial can establish the optimal dose of medication and facilitate the decision to proceed with a phase III trial. Selection of a dose for further testing requires an understanding of the relationships between dose and both efficacy and safety. These relationships can be assessed by comparing the data from each dose group with placebo, or with the other doses, in a series of pairwise comparisons [30]. As the drug can also be given together with other medications, interactions between them may be explored in these trials.

Phase II trials are considered to be successful if a statistically significant difference in the primary outcome favoring the experimental treatment is identified. These results are used to justify moving on to phase III trials and to help calculate their sample size.

Some phase II trials may include interim analyses to assess early treatment efficacy or safety signals and make informed decisions regarding dose selection, treatment continuation, or modification of study protocols. These analyses may involve data monitoring committees (DMCs) or independent review boards to ensure the integrity and validity of the trial results. Many times, decisions taken based on the phase II studies are not perfect because of the limited size and assumptions used for statistical calculation. However, these decisions can be strengthened by (i) the observation of a predefined threshold of activity, and (ii) pharmacodynamic (PD) studies to support the target engagement [31]. As the drug can also be given together with other medications, interactions between them may be explored in these trials.

In summary, phase II trials play a critical role in providing preliminary evidence of a new medical intervention's efficacy and safety in patients with the target disease. These trials help optimize treatment strategies, inform subsequent clinical development plans, and guide decision-making regarding the advancement of the intervention to phase III trials.

Examples:

In a randomized placebo-controlled dose-finding SOCRATES-REDUCED trial by Gheorghiade et al., four different target doses of vericiguat were tested in patients with worsening chronic heart failure and reduced LVEF, compared with placebo. In this trial, vericiguat did not have a statistically significant effect on the change in NT-proBNP level at 12 weeks but was well tolerated. Further studies were required to determine the potential role of this drug for patients with worsening chronic HF [32].

In a first phase IIa multicenter, double-blind, placebo-controlled study of dupilumab in 109 adult patients with moderate-to-severe atopic dermatitis (AD), it was observed that, per PK parameters, dupilumab results in a rapid and dose-dependent improvement in all measured clinical parameters. In another phase IIa study, the efficacy and safety of dupilumab were evaluated in combination with topical glucocorticoids. In this study, all patients treated with combination therapy achieved desirable clinical index values, indicating efficiency in treating AD [33].

A phase IIb multicenter, double-blind, placebo-controlled, randomized, parallel study of dupilumab was conducted in a significantly higher number of adult AD patients [34]. The main objective of this phase was to assess the efficacy and safety of several dose regimens of dupilumab. Results showed Eczema Area and Severity Index (EASI) score improvements with all dupilumab regimens versus placebo. A favorable benefit–risk ratio was derived with a recommended dosing of 300 mg QW or Q2W. Thus, dupilumab improved clinical responses in adults with moderate-to-severe AD in a dose-dependent manner, without significant safety concerns. These benefits were to be further confirmed in phase III studies.

4.4   Phase III: Confirmatory Trials and Regulatory Approval

Phase III trials, also known as therapeutic confirmatory trials, are a crucial stage in clinical development and are designed to provide definitive evidence of the intervention's efficacy, safety, and overall benefit-risk profile. Phase III trials aim to confirm the preliminary efficacy observed in phase II trials and provide robust evidence of the treatment's effectiveness. These trials typically employ randomized, controlled designs to compare the investigational treatment with standard treatments, placebo, or alternative interventions, allowing researchers to assess its superiority, non-inferiority, or equivalence. The most common type of phase III trials are comparative efficacy trials (often referred to as "superiority" or "placebo-controlled trials"), and another type of phase III trial, the equivalency trial (or "positive-control study"), is designed to ascertain whether the experimental treatment is like the chosen comparator within some margin prespecified by the investigator [21, 22, 28, 35].

These trials are powered to detect clinically meaningful treatment effects with a high degree of statistical confidence. These trials typically require a sufficient sample size to achieve adequate statistical power, ensuring the reliability and validity of the study results. Positive results from phase III trials are often a prerequisite for regulatory approval or marketing authorization of the investigational treatment.

More than 50% of phase III studies failed to demonstrate the therapeutic efficacy. The reason for this failure could be the overestimated results in phase II studies, which do not get replicated in a large-scale phase III trials. This pattern of overestimation has been consistently observed in cancer and other indications and is related to factors such as study design, sample size estimation, and the eligibility criteria of the study population [36].

Estimating sample size requirements based on phase II trial results poses several challenges. First, the phase II primary outcome may not be the clinically relevant primary outcome used for the phase III trial. Secondly, the effect size estimated from a small population is imprecise and may lead to an overestimation of the treatment effect. Third, phase III clinical trials include more diverse populations and treatment settings than phase II trials. All these factors lead to potentially smaller treatment effects in phase III trials than in phase II. Furthermore, phase III trials must be adequately powered to detect minimally important differences in clinically relevant outcomes, as these outcomes have widespread implications on physician practices.

For instance, in the Swedish Post-term Induction Study (SWEPIS) randomized control trial, which compared induction of labor at 41 weeks with expectant management and induction of labor at 42 weeks, a 0.4% increase in perinatal mortality in the expectant management group led to early trial termination and a recommendation to induce labor at 41 weeks [37].

Four major phase III clinical trials (SOLO 1, SOLO 2, SOLO-CONTINUE, and CHRONOS) were conducted worldwide in adults with moderate-to-severe AD. SOLO 1 and SOLO 2 had an identical study design. Patients who received dupilumab treatment and achieved a response were rerandomized in

SOLO-CONTINUE to continue their original regimen of dupilumab (300 mg dupilumab qw or q2w) or to receive dupilumab 300 mg every 4 weeks (q4w) or 8 weeks (q8w) or placebo for 36 weeks. Based on these study results, Dupilumab is approved in the United States for the treatment of moderate-to-severe AD in adults at the dose of 300 mg q2w and, adolescents, and children at the doses 300 mg q4w (weight ≥ 15 to < 30 kg), 200 mg q2w (weight ≥ 30 to < 60 kg), and 300 mg q2w (weight ≥ 60 kg) [38].

Overall, phase III trials play a pivotal role in advancing the clinical development of new medical interventions, providing definitive evidence of efficacy, safety, and clinical utility to support regulatory approval and clinical adoption.

4.5 Phase IV: Post-marketing Surveillance

Phase IV (post-marketing trial) is performed after the approval of the drug and is related to the approved indication. The primary objective of phase IV trials is to evaluate the safety and effectiveness of the intervention in real-world clinical settings, beyond the controlled conditions of clinical trials conducted during earlier phases. Phase IV trials help generate additional safety and efficacy data on drug-drug interaction and dose response and support use under the approved indication, for example, mortality or morbidity studies, epidemiological studies, etc. [39].

Phase IV trials evaluate a wide range of endpoints, including safety outcomes such as adverse events, drug interactions, and long-term side effects, as well as effectiveness outcomes such as disease recurrence, treatment response rates, and patient-reported outcomes. These trials may also assess healthcare utilization, cost-effectiveness, and quality-of-life measures to provide a comprehensive assessment of the intervention's real-world impact. These trials often involve large patient populations, extended follow-up periods, and data collection from multiple healthcare settings to capture a broad range of treatment outcomes and patient experiences. While phase IV trials are not always required for regulatory approval, regulatory agencies may request or require post-marketing surveillance studies as a condition of approval, particularly for interventions with limited pre-approval safety data or if there are ongoing safety concerns. These trials may also be mandated to fulfil post-marketing commitments or requirements specified in regulatory approvals.

Overall, phase IV trials play a critical role in ensuring the continued safety and effectiveness of medical interventions post-approval, guiding evidence-based decision-making, and enhancing patient care and public health outcomes. For example, adverse reactions revealed during phase IV studies may result in the drug being discontinued or restricted to specific uses [39, 40].

Examples: (1) Rofecoxib, an NSAID, was approved for pain relief but was voluntarily withdrawn from the market due to increased cardiovascular risks, including heart attacks and strokes, observed in a phase IV trial APPOVe (Adenomatous Polyp Prevention on Vioxx) [41]. (2) Troglitazone, an oral anti-diabetic medication, was withdrawn from the market due to reports of severe liver toxicity, including cases of liver failure, which were identified during phase IV surveillance and post-marketing reports [42]. (3) Rosiglitazone, another oral anti-diabetic medication, came under scrutiny due to safety concerns related to an increased risk of cardiovascular events. The drug's use was restricted in some countries after a meta-analysis of phase IV studies raised concerns, and the US FDA imposed restrictions on its use [43]. (4) Cisapride was a medication used to treat gastrointestinal reflux disease. It was withdrawn from the market in 2000 due to safety concerns regarding serious cardiac arrhythmias due to potential drug-drug interactions [44].

5 Study Designs

Study design is a framework, or the set of methods and procedures used to collect and analyze data on variables specified in a particular research problem [45]. The study designs broadly include two major types: non-interventional/observational and interventional/experimental studies. Non-interventional studies are hypothesis-generating studies wherein a naturally occurring relationship between the exposure and the outcome is documented without any active intervention. These studies may have a comparator group that provides a measurement of the association between the exposure and the outcome (e.g., case-control and cohort studies) or are without it (descriptive studies) where a description of the exposure and/or the outcome is provided.

Interventional studies are hypothesis-testing studies and are divided broadly into two main types: (i) "clinical trials" (or "controlled clinical trials"), in which individuals are assigned to one of two or more competing interventions such as a drug, medical device, vaccine, or procedure, and (ii) "community trials" (or field trials), in which entire groups, for example, villages, neighborhoods, schools, or districts, are assigned to different interventions.

Interventional studies can either be randomized or non-randomized (based on the process of assigning treatment), blinded or open-label (based on the concealment of treatment), and parallel, crossover, or factorial (depending on the sequence of the treatment), with placebo or active comparator (based on the type of comparator) and superiority or non-inferiority or equivalent (based on the assumptions for comparison) (Table 2).

5.1 Blinding

Blinding refers to the concealment of group allocation from one or more individuals involved in a clinical research study, most commonly done in a randomized controlled trial (RCT). If patients, clinicians, or assessors are aware of treatment assignments, this may influence the reporting or measurement of the outcome and introduce bias; for example, if participants are not blinded to the treatment group,

Table 2 Different types of study designs with or without randomization

Study design	Randomized or non-randomized	Description	Comment
Parallel group	Randomized/non-randomized	Enroll subjects with similar characteristics to two (or multiple groups): Group N1: Receives the intervention/experimental therapy Group N2: Receive the placebo (or standard of care)	The placebo arm does not receive the trial drug, so may not get the benefit of it
Crossover	Randomized	Each group serves as a control while the other group is undergoing the intervention/experiment Depending on the intervention/experiment, a "washout" period is recommended Most appropriate for studies where the effects of the treatment(s) are short-lived and reversible Best suited for symptomatic but chronic conditions or diseases	Avoids participant bias in treatment and requires a small sample size
Factorial	Randomized	Two or more interventions on the participants and the study can provide information on the interactions between the drugs Address two (or more) intervention comparisons carried out simultaneously In 2 × 2 factorial design, participants are randomized to one of four groups: Group A: Receives both treatments A and B (AB) Group B: Receives only treatment A (A0) Group C: Receive only treatment B (B0) Group D: Receives neither treatment A nor B (00)	Not suited for drugs or interventions, overlaps with each other on modes of action or effects, as the results obtained would not attribute to a particular drug or intervention
Cross-sectional	Non-randomized	These studies are carried out at one point in time, or over a short period. They find out who has been exposed to a risk factor or who has the disease	Quicker and cheaper but the results can be less useful

knowledge of group assignment may affect their behavior in the trial (less likely to comply with the trial protocol and can seek additional treatment) and their responses to subjective outcome measures. Similarly, blinded clinicians are less likely to transfer their attitudes to participants or to provide differential treatment to the active and placebo groups than unblinded clinicians [46].

Hence, a methodology of blinding is adopted in a study design to intentionally not provide information related to the allocation of the groups to the subject participants, investigators, and/or data analysts.

There are three forms of blinding: single-blind, double-blind, and triple-blind. In single-blind studies, the subjects or participants are not revealed which group they have been allocated to. In double-blind studies, both the study participants and the investigator are unaware of the group allocation. In triple-blind studies, the subject participants, investigators, and data analysts (data collectors, outcome adjudicators, and data analysts) are not aware of the group allocation. Triple-blinded studies are more difficult and expensive to design, but the results obtained will exclude confounding effects from knowledge of group allocation.

Blinding is especially important in studies where subjective responses are considered as outcomes, as the treatment effect may be exaggerated for such outcomes when outcome assessors are not blinded. This bias can be prevented by using independent clinicians who are not otherwise involved in the trial to assess patients or using a blinded adjudication committee to determine the outcome.

However, in some cases, the nature of the treatments under investigation may not permit blinding. This can be an issue in trials assessing surgical interventions, device trials, or other non-pharmacologic interventions, which are more difficult to blind. Many such trials are therefore open-label, where patients, clinicians, and care providers are aware of treatment allocations [47].

5.2   Randomization

Randomization is the process of assigning participants to treatment and control groups, assuming that each participant has an equal chance of being assigned to any group. Randomization in trials is needed for several reasons. First, it is needed so that participants in various groups do not differ in any systematic way. If treatment groups differ systematically, trial results will be biased. For example, if a greater proportion of older adults are assigned to the treatment group than the younger population in a walking intervention study, then the outcome will be influenced by this imbalance. Second, proper randomization helps in allocation concealment (group assignment). Knowledge of group assignments creates a potential selection bias that may taint the data. It is stated that trials with inadequate or unclear randomization can overestimate treatment effects up to 40% compared with those that used proper randomization, impacting the trial outcome negatively [48].

Randomized trials are prospective trials that measure the effectiveness of a new intervention or treatment. Randomization reduces bias and provides a rigorous tool to examine cause-effect relationships between an intervention and outcome [49]. This is typically performed by using computer software or can be done manually. It helps to measure the outcomes without bias as subjects are randomized to their respective groups with similar baseline characteristics.

A non-randomized clinical trial involves an approach to selecting controls without randomization. Participants are not assigned by chance to different treatment groups. They may be assigned to the groups by the researchers/clinicians conducting the study. By this method, the selection of the subjects becomes predictable, and therefore, the validity of the results obtained may be questioned.

Randomization is one of the fundamental principles of an experimental study design and ensures scientific validity.

There are four types of randomizations: simple, block, stratified, and cluster randomization. Randomization based on a single sequence of random assignments is known as simple randomization. The block randomization method is designed to randomize participants within blocks such that an equal number are assigned to each treatment. This method is used to ensure a balance in sample size across groups over time. The stratified randomization method addresses the situation where the strata are based on the level of prognostic factors or covariates. This method can be used to achieve balance among groups in terms of participants' baseline characteristics (covariates). In cluster randomization, pre-existing groups or clusters of individuals are randomly allocated to treatment arms. Cluster randomization is carried out to evaluate the complex interventions that are increasingly being adopted in health research. Simple randomization works well for a large trial (e.g., $n > 200$) but not for a small trial ($n < 100$). To achieve balance in sample size, block randomization is desirable. To achieve balance in baseline characteristics, stratified randomization is widely used [49] (Table 3).

5.3   3 + 3 Dose Escalation Study Designs

The traditional "3 + 3 design" is being used for evaluating oncology molecules in first-in-human trials as well as for first-time use of experimental or approved drugs in novel combinations. The traditional 3 + 3 design is the most widely adapted design for phase I development.

In this design, three patients are initially enrolled in a given dose cohort. If there is no DLT seen in any of these participants, the trial proceeds to enroll an additional three patients into the next higher-dose cohort. Dose-limiting toxicities are generally defined as clinically relevant toxicities or grade 3 or higher toxicities that occur in the first cycle of the therapy. If one patient manifests a DLT in a given dose cohort, an additional three patients are enrolled into that same dose cohort. Development of DLTs in two or more out of six patients at a specific dose level indicates that the MTD has been exceeded; further dose escalation is not pursued, and the prior dose level is considered as the MTD.

Table 3 Types of randomizations

Randomization type	Description	Comments
Simple randomization	Random allocation to experiment/intervention groups based on a constant probability	Advantage: It eliminates selection bias. Limitation: May introduce an imbalance in the number allocated to each group as well as the prognostic factors between groups
Block randomization	Subjects of similar characteristics are classified into blocks. Balance the number of subjects allocated to each experiment/intervention group	Advantage: Control the balance between the experiment/intervention groups. Limitation: There is still a component of predictability in the selection of subjects
Stratified randomization	Subjects are defined based on certain strata, which are covariates. For example, prognostic factors like age can be considered as a covariate	Advantage: enables comparability between experiment/intervention groups Limitation: The covariates need to be measured and determined before the randomization process
Clustered randomization	Intervention is administered to clusters/groups by randomization	Evaluate the complex interventions

In dose escalation, escalating doses are predetermined or preferably adjusted to toxicity. The most frequently used predetermined escalation rules use a modified Fibonacci method for deciding the amount of dose increase for cohorts of sequentially enrolled patients. This is the sequence in which each number is the sum of the two numbers that precede it; for example, a phase-I dose-escalation trial of ABT-199, a BCL-2 inhibitor, is planned using a modified Fibonacci design in patients with relapsed/refractory NHL. A strategy of starting with a 2- to 3-week lead-in period with stepwise increases to the target cohort dose was used here. In the first four cohorts, the starting dose increased from 50 to 200 mg (50, 100, 200, and 200 mg, respectively), with target cohort doses of 200 mg [$n = 3$], 300 mg [$n = 3$], 400 mg [$n = 4$], and 600 mg [$n = 7$] [50].

Another dose escalation way is based on adjusted toxicity. If there are no toxicities within a 3 + 3 design in the prior dose level, then the next escalation is by 100%; if there is grade 1 toxicity in the prior dose level, then the next escalation is by 50%; if the grade of toxicity was grade 2, then the next dose escalation is by 25%; if the highest grade of toxicity was grade 3 (or higher), dose escalation is held, and the dose level is expanded; and if less than one-third of the patients have a toxicity of grade 3 or higher at the expanded dose level, a further dose escalation of 25% occurs [51].

For improving phase I 3 + 3 design, toxicity-adjusted dose escalation should be used in place of a predetermined Fibonacci-related schema, thus shifting dose escalation based on the toxicities rather than on a predetermined formula.

5.4   Adaptive Study Design

In an adaptive design, trials and/or statistical procedures can be modified or changed (adaptations are allowed) during the clinical trials based on the review of the interim data. This is done without undermining the validity and integrity of the trial.

The goal of the modifications is to improve upon the probability of success of the trial as well as to correctly identify the clinical benefits of the investigational agent. Different types of modifications/changes can be made during an adaptive design. Of these, prospective adaptations are changes such as stopping a trial early for safety or lack of efficacy reasons, dropping the loser, or sample size re-estimation. Modifications in hypotheses, inclusion/exclusion criteria, dose/ regimen, treatment duration, and endpoints are examples of ad-hoc modifications that are usually not initially recognized or anticipated for modification but become necessary as the trial progresses. Changes that are made to the statistical analysis plans before database lock or un-blinding of treatment codes are known as retrospective adaptations.

Adaptive dose-finding design is used during phase I studies to determine the minimum effective dose (MED) and/or the MTD. In this design, the continual re-assessment method with the Bayesian approach is usually used to estimate the dose-response curve.

5.5   Bayesian Study Design

Bayesian designs are based on the principle of Bayes theorem, proposed by Reverend Thomas Bayes in 1763. The theorem proposes that the probability of an event is based upon prior knowledge of conditions related to the event [51]. For example, if the prevalence of the disease in the population is known, the occurrence rate of a symptom or of a laboratory finding among the diseased, and the sensitivity of a diagnostic test, one can estimate the occurrence of the disease (prevalence) in that population. According to the ways of incorporating statistical models in clinical trials, there are two categories of phase I Bayesian design: model-based and model-assisted designs.

According to previous knowledge or experience, like the occurrence rate of a symptom or a laboratory finding among the diseased, or the sensitivity of a diagnostic test, one can estimate the occurrence of the disease (prevalence) in that population. According to the ways of incorporating statistical models in clinical trials, there are two categories of phase I Bayesian design: model-based and model-assisted designs.

5.5.1 Model-Based Designs

Model-based designs are earlier Bayesian designs. These are a class of novel adaptive designs that use a prespecified statistical model (e.g., a logistic model) and do not follow a predetermined algorithm to describe the dose-toxicity curve and guide dose transition. Examples include the continuous reassessment method (CRM) and its various extensions and the Bayesian logistic regression model. As the data gets collected during the trial, the CRM updates the estimate of the model after each cohort and then uses these estimates to determine the dose for the next cohort. Thus, the dose level of the next to-be-enrolled patient remains unknown unless the information about previous patients can be integrated into the model and the designs and parameters are updated. Studies have shown that the CRM can outperform the 3 + 3 design, with higher accuracy to identify and allocate more patients to the MTD as well as the ability to target any prespecified DLT rates [52]. Other key benefits include relatively more patients being treated at the optimal dose and fewer patients being treated at sub-therapeutic doses and efficient utilization of available data.

Despite several advantages of model-based designs, the high requirement of expertise and infrastructure, as this design requires intensive help from statisticians, and the lack of predetermined algorithms to be followed hindered the wide adoption of model-based designs.

5.5.2 Model-Assisted Design

In model-assisted designs, the advantages of algorithm-based designs and model-based designs are combined together. Model-assisted design also uses a statistical model (e.g., the binomial model) to derive the design for efficient decision-making. Its dose escalation and de-escalation rule can be predetermined before the onset of the trial, and, thus, it can be implemented in a simple way like the algorithm-based designs.

Examples of model-assisted designs include the modified toxicity probability interval (mTPI) design and its variation, mTPI-2; Bayesian optimal interval (BOIN) design; and keyboard design. Among the model-assisted designs, BOIN outperforms the mTPI with higher accuracy identifying the MTD and a lower risk of overdosing patients.

Despite several publications demonstrating the superiority of model-based or model-assisted designs compared with traditional 3 + 3 designs, the application of novel Bayesian designs in drug development is still limited. Although many pharmaceutical companies have overcome practical barriers to implementing model-based designs, the effect of these models on the drug development has not yet fully expanded [51].

5.6 *Master Protocol*

Master protocols refer to a single overarching protocol developed to evaluate multiple questions (Fig. 2). The general goals of having a master protocol are to improve efficiency and to establish uniformity through standardization of procedures in the development and evaluation of different interventions. In a way, the master protocol allows the testing of multiple drugs and/or multiple subpopulations in parallel under a single protocol, without the need to develop new protocols for every trial. This is particularly applicable to the field of "precision oncology," in which therapies are selected to specifically target cancers based on their genetic mutations. They can also be used to incorporate biomarker development, genetic subtyping, and to test therapies with different mechanisms of action.

The imatinib story is a good example of a precision medicine trial designed using the master protocol. In 2006, imatinib was approved by the US FDA in 2006 for five indications based on a single-phase 2 clinical trial [53]. The approval was given without the customary two pivotal phase 3 trials, and it was also cleared to treat a number of cancers based on an open-label, single-arm prospective study that enrolled just 186 patients with 40 malignancies.

Master protocols offer a powerful new approach to drug development, allowing for flexibility and creativity in the highly regulated clinical trial sector. Under the term "master protocol" are included three distinct entities called "umbrella," "basket," and "platform" trials. They constitute a collection of trials or substudies that share key design components and operational aspects to achieve better coordination than can be achieved in single trials designed to conduct independently.

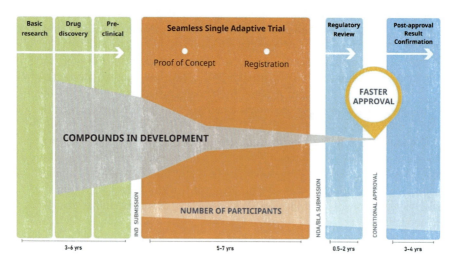

Fig. 2 Emerging drug development process

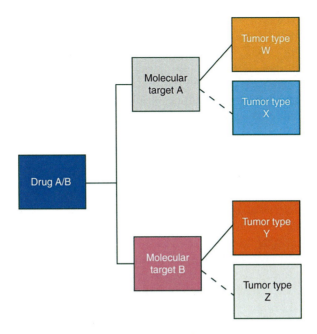

Fig. 3 Basket trial

5.6.1 Basket Trials

A basket trial refers to designs in which the single investigational drug or single drug combination is used to treat patients with multiple diseases who share a common characteristic, such as a genetic or molecular alteration or a specific biomarker. The basket trial design embodies the core theme "treating different diseases harboring similar genetic/molecular alterations with the same treatment" (Fig. 3). A basket trial is a master protocol for which patient eligibility is defined by the presence of a particular biomarker or molecular alteration rather than a particular disease type.

The aforementioned imatinib trial was a basket design trial where multiple cancers, both hematologic and solid, were treated with one drug.

Basket trials have several advantages. First, basket trials help in accessing molecularly targeted agents for patients across a broad range of disease or tumor types, potentially including those not otherwise studied in clinical trials of targeted therapies. Second, a basket trial targets specific oncogenic pathways responsible for the growth and metastasis of tumors, resulting in a higher response rate in patients with mutation as compared to patients without mutation. Third, cohorts within basket trials are often small and utilize single-stage or two-stage designs, which help in getting quick results [54].

Examples: A trial of trastuzumab emtansine to evaluate an antitumor response in HER2 -amplified/-mutant cancers of multiple histologies (e.g., lung, endometrial, and salivary gland). In this case, HER2 amplification was the common predictive biomarker risk factor that can predict whether different cancers would respond to this targeted therapy [55]. A study of Vemurafenib in participants with BRAF V600 mutation-positive cancers (solid tumors and multiple myeloma) [56], a KEYNOTE-158 trial evaluating Pembrolizumab in patients with various types of advanced solid tumors having PD1/PD-L1 expression [57], and NAVIGATE trial evaluating Larotrectinib for the treatment of NTRK-fusion-positive advanced solid tumors [58] are other examples.

5.6.2 Umbrella Trials

An umbrella trial refers to designs where multiple therapies or interventions are evaluated for patients with a single disease that are stratified into subgroups according to clinical features and molecular alterations. The umbrella trial design embodies the core theme of "treating the same disease with different treatments" due to the different molecular phenotypes of a certain disease (Fig. 4). In an umbrella trial, patients are screened centrally for tumors with specified cancer types and are assigned to one of several molecularly defined sub-trials where they receive a matched targeted treatment [54]. The principle of the umbrella trial design stems from a deep understanding of disease heterogeneity, including genomic heterogeneity and clinical phenotypic diversity [59].

The advantage of umbrella trials is the ability to draw meaningful conclusions that are specific to a tumor type and therefore less prone to chance tumor heterogeneity present within a given trial cohort [54]. In a UK plasma MATCH trial, 5 different targeted therapies were evaluated for advanced breast cancer. Patients were assigned to 1 of the 5 subgroups based on their biomarker status/ molecular signatures. Subgroup A patients had an ESR1 (estrogen receptor gene 1) mutation and received an extended dose of *fulvestrant*. Subgroup B included patients with an HER2 mutation who received an HER tyrosine kinase inhibitor (neratinib) and fulvestrant if they had an estrogen receptor co-mutation. Subgroup C had an AKT (serine/threonine-specific protein kinase B) mutation and received the AKT inhibitor AZD5364 plus fulvestrant. Subgroup D patient had AKT activation and received AZD5364 only. Subgroup E had patients with triple-negative status and received the PARP inhibitor olaparib plus AZD5364 [60].

5.6.3 Platform Trials

Platform trials, also referred to as multi-arm, multi-stage (MAMS) design trials, evaluate several interventions against a common control group or several diseases and can be perpetual. This design has pre-specified adaptation rules to allow the dropping of ineffective intervention(s) and flexibility of adding new intervention(s) during the trial depending on the futility/efficacy assessment, often according to Bayesian decision rules. The key features of platform trials are their scope and adaptability. They are designed to evaluate multiple therapies. In these trials, rather

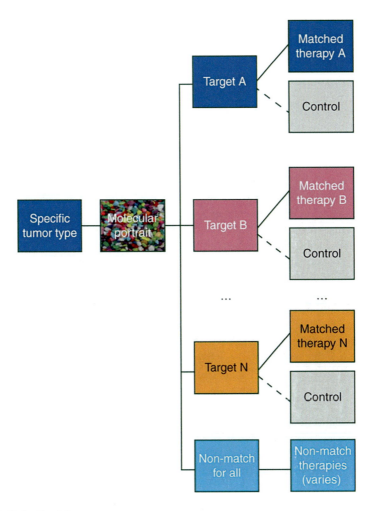

Fig. 4 Umbrella trial

than testing drugs one at a time, multiple drugs, including the combination, can be tested all at the same time, giving a quicker result [54].

Examples: A study of multiple immunotherapy-based treatment combinations in participants with metastatic pancreatic ductal adenocarcinoma, which included a few phase Ib/II trials in different cancers [61]. I-SPY 2 was another exploratory phase platform trial where neoadjuvant and novel agents were evaluated for early breast cancers. The goal was to identify treatment strategies for subsets based on molecular characteristics (biomarker signatures) of their disease with a high estimated pCR rate [62].

LUNG-MAP was a phase II/III study of targeted therapies in patients with previously treated advanced squamous NSCLC. In this trial, four randomized phase II trials of targeted therapy versus standard of care were conducted in parallel within mutation-enriched cohorts, with the graduation of a cohort to a phase III registration study if progression-free survival (PFS) crosses an efficacy boundary during phase II [54].

5.7 Study Endpoints

Endpoints are measures designed to reflect the intended effects of a drug and include clinical events (e.g., mortality, stroke, disease progression, venous thromboembolism), symptoms (e.g., pain, dyspnea, symptoms of depression), measures of function (e.g., ability to walk or exercise), or surrogate endpoints that are reasonably likely or expected to predict a clinical benefit. The endpoints determine whether the clinical trial has been successful in finding out if an intervention/drug has proven beneficial to patients. It determines the validity of the clinical trial results.

Most diseases can potentially cause more than one clinical event, symptom, and/ or altered function; hence, trials are designed to examine the effect of a drug on more than one aspect of the disease.

As single endpoint may not capture the important effects of an intervention to the satisfaction of all end-user groups, multiple endpoints are usually selected, which are categorized as *primary, secondary*, or *tertiary*. The primary endpoint(s) are typically measures that address the main study question. These endpoints are the most important outcomes of the trial that directly correspond to the trial's scientific objective and provide the direct evidence of a treatment's effectiveness. Secondary endpoints are generally not sufficient to influence decision-making alone but may support the claim of clinical activity by demonstrating additional effects or by supporting a causal mechanism. For example, secondary outcomes of the trials include measurements of safety and tolerability of the intervention, effects on quality of life, or subgroup analyses to identify which patient populations benefit most from the intervention. Secondary outcomes directly align with secondary objectives [63]. Tertiary endpoints are likely to be more exploratory and typically capture outcomes that occur less frequently or that may be useful for exploring novel hypotheses [64]. Tertiary endpoints could be used to explore disease processes or the mechanisms by which an intervention is effective. Examples of these endpoints include laboratory measurements, biomarkers, or physical signs. By themselves, tertiary endpoints are unlikely to modify the key conclusions in a clinical trial, but they may still be of interest for guiding future research [65].

The selection of proper endpoints for clinical trials is imperative to the accurate interpretation of trial results and to achieving the goal of novel therapies to prolong and/or improve the quality of patients' lives [66].

There are two types of endpoints: true endpoints and surrogate endpoints. A true endpoint is a clinically meaningful endpoint that usually reflects "how a patient feels, functions, or survives" (e.g., endpoints such as mortality and strokes). A surrogate endpoint is a biomarker intended for substituting a clinical endpoint and expected to predict clinical benefit, harm, or lack of these (e.g., bone mineral density for bone fracture, fetal heart rate for fetal brain oxygenation, and PFS for OS).

In clinical studies, biomarkers are used as surrogate endpoints with the assumption that there exists a correlation between the measured level of a biomarker and a clinical outcome. For a biomarker to qualify as a surrogate, the biomarker must not only be correlated with the outcome, but the change in the biomarker must "explain" the change in the clinical outcome.

There is a common misconception that if an outcome is correlated (that is, correlated with a true clinical outcome), it can be used as a valid surrogate endpoint. For the valid surrogate, the effect of the intervention on the potential surrogate endpoint should predict the effect on clinical outcome—a much stronger condition than correlation." The figure below shows why a potential surrogate endpoint can fail to provide correct inference about the true endpoint.

Figure 5 delineates the reasons for the failure of surrogate endpoints. In Fig. 5a, the disease affects the surrogate endpoint and the true clinical outcome via different mechanisms so that any correlation between the two is not causal. In Fig. 5b, the intervention affects the surrogate endpoint, which has some impact on the true clinical outcome. The disease also affects the true clinical outcome by other mechanisms, which makes the change in the putative surrogate an unreliable measure of change in the true clinical outcome. In Fig. 5c, the intervention affects the surrogate endpoint through mechanisms independent of its effect on the true clinical outcome. Thus, the change in the surrogate endpoint is not a reliable measure of the change in the true clinical outcome. In Fig. 5d, all of the above issues are in play.

Examples: High-density lipoprotein (HDL) cholesterol, when used as a surrogate, has failed multiple times across many classes of drugs. People with low levels of HDL are susceptible to developing atherosclerosis and thus are more likely to have poor outcomes, but drugs that raise levels of HDL cholesterol have had either no effect or detrimental effects on clinical outcomes [67].

In oncology, PFS is frequently used as a primary endpoint in place of overall survival (OS). More than one-third of trials received marketing approval between 2009 and 2013 using PFS as a primary endpoint. However, the approval (FDA-accelerated approval) of bevacizumab in 2008 for metastatic breast cancer was based on PFS improvement and was withdrawn in 2011 after several trials showed that there was no OS improvement. However, it is argued that the reason for the failure to see a survival gain after an improvement in PFS is the influence of post-progression therapy [68, 69].

Accelerated approval was granted to atezolizumab in March 2019 based on data from the phase III IMpassion130 trial, which demonstrated a statistically significant benefit to progression-free survival [70]. However, the results of the trial conducted for continued approval indicated that the trial failed to meet the primary endpoint of PFS superiority in the frontline treatment of patients with PD-L1 positivity. Also, there was no difference in survival advantage in the PD-L1–positive or the intention to treat population [71].

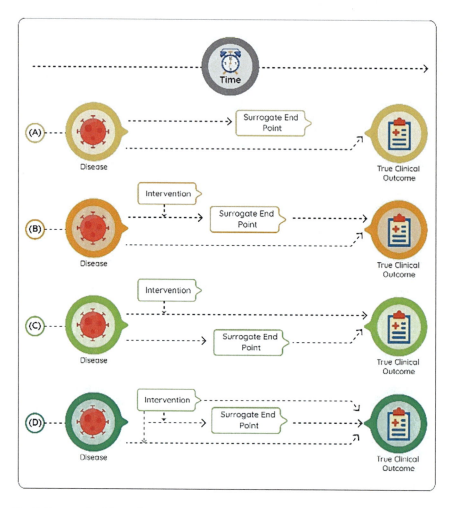

Fig. 5 Reasons for failure of surrogate endpoints

6 Special Considerations

6.1 Patient Enrichment

In clinical trials, numerous examples showed a benefit in the subgroup of the patient population who have a certain marker characteristic or profile. For instance, gefitinib showed significant improvement in patients with advanced pulmonary adenocarcinoma, who tested positive for epidermal growth factor receptor (EGFR). Whereas no improvement was observed in patients who tested negative. This led to the concept of patient enrichment, which refers to the inclusion of patients with the presence or absence of a certain marker characteristic or profile. Patient enrichment is based on the paradigm that not all patients will benefit from the study treatment under consideration but rather that the benefit is restricted to a subgroup of patients who either express or do not express a specific molecular feature [72].

"Enrichment" can be (i) practical enrichment to reduce noise (variability of measurement) and heterogeneity; (ii) prognostic enrichment to find patients who are likely to have the event of interest when enrolling for risk-reduction studies; and (iii) predictive enrichment to find the individuals who are more likely to respond [73].

Examples of enrichment include the inclusion of patients with advanced anaplastic lymphoma kinase (ALK)-positive non-squamous non-small cell lung cancer (NSCLC) in a clinical study of crizotinib [74].

Similarly, clinical trials have demonstrated that PARP inhibitors (PARPi) significantly prolong the time to recurrence and progression in all women with ovarian cancer, with the most robust antitumor activity observed in those women harboring a somatic or germline *BRCA* mutation. These clinical data have led to FDA approval of multiple PARPi for ovarian cancer [75].

6.2   Sample Size

The sample size is important for a clinical trial for two reasons: (a). it should be large enough to provide a reliable answer to the questions addressed and (b). it should be small enough to be financially feasible. An appropriate sample size relies on the choice of certain factors and estimates (Table 4), which influences the sample size independently.

Study Design, Outcome Variable, and Sample Size   The design of the study has a significant effect on the size of the sample. Descriptive study design requires hundreds of subjects to provide confidence limits for small effects. On the contrary, experimental studies require smaller samples. Cross-over designs require a quarter of the number required for a controlled study because every subject in the cross-over study receives the experimental treatment. An evaluation study conducted in a single group using a pre-post design requires half the number of subjects for a similar study conducted with a controlled study. A one-tailed hypothesis design requires 20% fewer subjects than a two-tailed study. A non-randomized study requires 20% more subjects than a randomized study to account for other confounders. An additional 10–20% of subjects are needed to allow for adjustments for withdrawals, missing data, and lost to follow-up.

Alpha (α) Level   In most of the studies, the alpha level used to determine the sample size is 0.05. The lower the alpha level, the larger the sample size is. For instance, a study with a 0.01 alpha level requires more subjects than a study with a 0.05 alpha level for a similar outcome variable. A lower alpha viz. of 0.01 or lower is used when the decision-making based on the research is critical and the errors are likely to cause significant economic damage.

Table 4 Key parameters for sample size calculations

Component	Definition
α (type I error)	The probability of falsely rejecting H_0 and detecting a statistically significant difference when the groups in reality are not different, i.e. the chance of a false-positive result.
β (type II error)	The probability of falsely accepting H_0 and not detecting a statistically significant difference when a specified difference between the groups in reality exists, i.e. the chance of a false-negative result.
Power $(1-\beta)$	The probability of correctly rejecting H_0 and detecting a statistically significant difference when a specified difference between the groups in reality exists.
Minimal detectable difference (effect size)	It's the magnitude of the difference between groups.
Variance	It's a statistical measure to measure the dispersion of data points from the center or mean.

Power of the Study The ideal study is one that has high power, which means that the study has a high chance of detecting a difference between groups if it exists; consequently, if the study demonstrates no difference between the groups, the researcher can reasonably be confident in concluding that none exists. The ideal power for any study is considered to be 80%.

Statistical power is positively correlated with the sample size; a larger sample size gives greater power. However, the "statistical difference" is different from the "scientific difference." Although a larger sample size enables the study to find smaller differences statistically significant, the difference found may not be scientifically meaningful. Therefore, a prior idea of the expected scientifically meaningful difference is important before performing a power analysis.

Variance or Standard Deviation The variance is the variability of the outcome measure and is generally obtained either from previous studies or from pilot studies to understand the preliminary effect of the drug. The larger the standard deviation, the larger the sample size required in a study. For example, for the sample size calculation, in the Australian Initiating Dialysis Early and Late (IDEAL) Study, which aimed to determine the effect of the timing of dialysis initiation on left ventricular mass, the author used data from another laboratory indicating that the mean left ventricular mass in renal failure patients in Australia was 140 g/m² with an SD of 60 g/m² [76].

Minimum Detectable Difference This is the smallest effect of interest or minimal difference (also known as the effect size) between the intervention and control groups that are being detected by the study. This should be a difference that the investigator believes to be clinically relevant and biologically plausible. For example, in a study to evaluate the effect of the intervention on body weight, a study should be powered to detect the minimal difference of 5 kg, which should be the minimally clinically relevant difference for change in body weight [77]. In a myocardial infarction study, an estimated minimally relevant difference between the event rates in both treatment groups could be 10%.

The effect size must be clinically relevant, and a mere significant difference should not be considered for the successful outcome of the study. For example, in a Physicians Health Study of aspirin to prevent myocardial infarction (MI), aspirin was found to be associated with a reduction in MI ($P < 0.00001$). From the result of this study, aspirin was recommended for general prevention of MI. However, the effect size of this study was very small (a risk difference of 0.77% with $r^2 = 0.001$— an extremely small effect size). As a result, many people were advised to take aspirin who would not experience benefits but were at risk for adverse effects. Further studies found even smaller effects, and the recommendation to use aspirin has since been modified [78].

A prior idea of the expected scientifically meaningful difference is important before performing a power analysis. Effect size is generally estimated from prior studies. Where no previous study exists, the effect size is determined from a literature review, logical assertion, and conjecture [79].

6.3   Role of Biomarker

The Biomarkers, EndpointS and Other Tools (BEST) glossary by the FDA-NIH Joint Leadership Council defines a biomarker as a characteristic that is objectively measured and evaluated as an indicator of normal or pathological biological processes, or responses to an exposure or intervention. Biomarkers are distinct from a clinical outcome assessment (COA), which are direct measures of how a person feels, functions, or survives, and from the endpoints, which are precisely defined variables designed to indicate an outcome of interest [67]. In comparison to current measurements, biomarkers are expected to provide tests with greater sensitivity and specificity, improve the decision-making process, and facilitate the development of therapies [80]. With the advancement in genomics and molecular biology, the use of biomarkers in clinical trials has tremendously increased.

Based on the application, seven biomarker categories are defined: diagnostic, monitoring, susceptibility/risk, pharmacodynamic/response, predictive, prognostic, and safety. A diagnostic biomarker detects or confirms the presence of a disease or condition of interest or identifies an individual with a subtype of the disease [67]. These biomarkers are used to identify people with diseases; for example, defect in the "cystic fibrosis transmembrane conductance regulator (CFTR) gene" confirms the diagnosis of cystic fibrosis. Similarly, carcinoembryonic antigen (CEA) helps in the diagnosis of colon cancer and cancer antigen 125 (CA-125) in the ovarian cancer. These biomarkers may also be used to redefine the classification of the diseases.

Markers that are measured serially to monitor disease progression, therapeutic response, and safety are called monitoring biomarkers. Monitoring biomarkers has important applications in clinical care. For example, in the case of blood pressure-lowering agents and cholesterol-lowering drugs, systolic/diastolic blood pressure and low-density lipoprotein (LDL) cholesterol levels are monitored routinely to understand the effect of drugs. When the level of a marker changes in response to exposure to a medical product or an environmental agent, it can be called a pharmacodynamic/response biomarker. These biomarkers are useful both in clinical practice and early therapeutic development. For example, in the case of the clinical study of antihypertensive or diabetic drugs, the reduction in blood pressure or glucose levels is a good reason to continue the intervention and plan for confirmatory trials.

A predictive biomarker can predict which individual or group of individuals is more likely to experience a favorable or unfavorable effect from exposure to an investigational or environmental agent. The development of genetic and genomic markers for precision medicine is a good example of predictive biomarkers [67]. Human epidermal growth factor receptor −2 (HER-2) protein overexpression in patients with breast cancer correlated with decreased relapse-free and overall survival. Subsequent research showed that HER2 amplification/overexpression can predict a greater or lesser response to available chemotherapy and hormonal therapy and a promising response to trastuzumab (Herceptin) [81].

Prognostic biomarkers indicate an increased (or decreased) likelihood of a future clinical event, disease recurrence, or progression in an identified population. These biomarkers can predict disease trajectory, which can guide prevention and treatment efforts and can stratify patients, thereby redefining disease categories to align with pathophysiology [67]. Examples of prognostic biomarkers are PSA level at the time of a prostate cancer diagnosis or the PIK3CA mutation status of tumors in women with human epidermal growth factor receptor 2 (HER2)–positive metastatic breast cancer [82]. In clinical trials, prognostic biomarkers are routinely used to set trial entry and exclusion criteria to identify higher-risk populations. Susceptibility/risk biomarkers identify risk factors and individuals at risk. These biomarkers indicate the potential for developing a disease or medical condition in an individual who does not currently have the clinically apparent disease or the medical condition [67]. For example, BReast CAncer genes 1 and 2 (BRCA1/2) mutations may be used as a susceptibility/risk biomarker to identify individuals with a predisposition to develop breast cancer. Factor V Leiden may be used as a susceptibility/risk biomarker to identify individuals with a predisposition to develop deep vein thrombosis (DVT) [83].

Safety biomarkers reflect the potential or presence of toxicity related to a therapeutic agent. Safety biomarkers are useful for identifying patients who are experiencing adverse effects from a treatment. When antiarrhythmic drugs are prescribed, prolongation of the QT interval on the electrocardiogram is used as a safety biomarker because it predicts the risk of developing the lethal arrhythmia torsades de pointes and can be used to identify patients in need of countermeasures for effective therapy [67]. Patients with UGT1A1 gene mutation have reduced ability to clear drugs or endogenous substances (e.g., irinotecan and bilirubin) and thus are prone to irinotecan-related toxicities such as severe diarrhea. Thus, dose reductions are recommended in these patients to minimize the toxicity of irinotecan. Hepatic

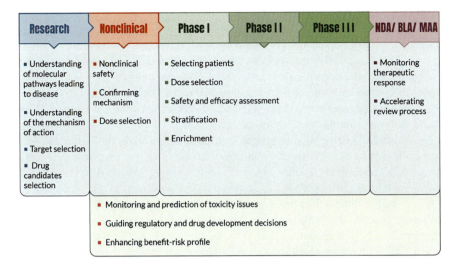

Fig. 6 Role of biomarkers in drug discovery and development processes

aminotransferases (AST/ALT) and bilirubin are used as safety biomarkers when evaluating potential hepatotoxicity.

Biomarkers have gained prominence in drug discovery, development, and approval processes. The development of suitable biomarkers can contribute to understanding the mechanism of action of a drug, selecting the right patients for a clinical trial, monitoring and predicting toxicity issues, and guiding regulatory as well as drug development decisions (Fig. 6). Furthermore, biomarkers facilitate more adaptive clinical studies.

Overall, biomarkers have the potential to make development more sustainable, improve the quality and safety of a drug, reduce development costs, and accelerate the approval process significantly [67].

6.4 N-of-1 Trial

An N-of-1 clinical trial is also known as a single-subject trial. These trials consider an individual patient as the sole unit of observation in a study and act as his or her control. The goal of an n-of-1 trial is to determine the optimal or best treatment for an individual patient using objective criteria [84]. N-of-1 trial designs are primarily used in chronic and stable diseases with different treatments that have comparable efficacy.

In N-of-1 designs, patients usually receive different treatment options or a placebo in sequential order with a possible washout period in between. This allows for intraperson comparison and overcomes the possibility of interpatient heterogeneities observed in classic clinical trial designs by eliminating treatment-unrelated differences between study participants.

A classic example of an N-of-1 trial is the study by Molloy and colleagues wherein patients with Alzheimer's disease received 3 weeks of tetrahydroaminoacridine and 3 weeks of placebo for three treatment periods in a randomized, double-blind fashion after an adequate washout period [85]. Similarly, the assessment of the effectiveness of traditional Chinese medicine (TCM) on chemotherapy-induced leukopenia is another example of N-of-1 trial design where patients with gastric cancer after gastrectomy received 3 weeks of standard chemotherapy and 3 days of TCM in a randomized, double-blinded fashion [86].

N-of-1 studies can be used in precision oncology. In a study by Drilon and colleagues, two patients who received larotrectinib for *TRK* fusion-positive cancers developed an acquired resistance to larotrectinib that was attributed to G595R and G623R mutations. Subsequently, these two patients were treated with selitrectinib based on the preclinical data that demonstrated its activity against these mutations. Both these patients showed a response to the treatment, indicating the usefulness of selitrectinib in NTRK mutations [87].

In a similar study, two patients with *RET* alterations (one with a RET^{M918T} mutation and the other with a *KIF5B–RET* fusion) were treated with selpercatinib. In this study, responses to selpercatinib were very encouraging, giving a quick understanding of the role of selpercatinib in these mutations [88].

However, the implementation of N of 1 study design in oncology studies poses multiple challenges, such as 1. Patients with metastatic cancer are difficult to keep on washout period for the sake of study design. as these patients have a high disease burden and dynamic disease progression. 2. Time to response may vary between different treatment options with variable mechanisms of action, making the response assessment difficult [88]. These challenges hamper N of 1 trial design adoption during drug development.

6.5 Drug Designations/Approvals

The US FDA grants special designations to expedite drug development and regulatory review for promising drugs treating diseases with unmet medical needs. To date, five special FDA review programs exist mostly for cancer drugs: orphan drug designations, fast-track approvals, accelerated approval, priority review, and breakthrough therapy designation (Fig. 7) [89, 90].

6.5.1 Orphan Designation

Orphan drug designation was introduced through the Orphan Drug Act (ODA) in 1983 to motivate the pharma companies who are addressing the unmet need for rare diseases through their R&D efforts (conditions that affect fewer than 200,000 people in the United States). This is mainly to offset the high R&D costs associated with bringing a new rare disease drug to market.

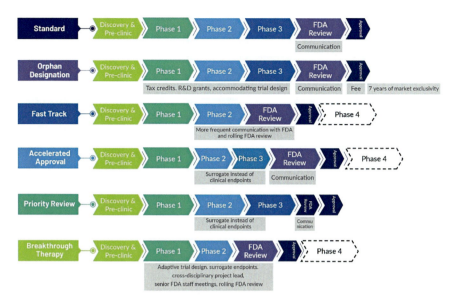

Fig. 7 An overview of the FDA's special review pathways and designations

Orphan Drug Designation provides certain benefits, including financial incentives, to support clinical development and the potential for up to 7 years of market exclusivity for the designated orphan indication in the United States. The FDA even encourages the use of "innovative clinical trial methods such as adaptive and seamless trial designs, modeling and simulations, and basket and umbrella trials" to conduct this research [90]. As of 31st December 2022, 6340 orphan drug designations are granted, representing 1079 diseases [91]. Some of the examples included AUM302, a PI3K, PIM, and mTOR inhibitor for neuroblastoma, BVX001, CD33/CD7-directed antibody-drug conjugate, for acute myeloid leukamia, Tenalisib, a PI3k δ/γ and SIK-3 inhibitor for PTCL and CTCL, Tirabrutinib, a Bruton's tyrosine kinase inhibitor for primary central nervous system lymphoma (PCNSL).

6.5.2 Fast Track Designation

Fast track is a process that was introduced in 1988 to "facilitate the development, and expedite the review of drugs to treat serious conditions (e.g. AIDS, dementia, cancer, heart failure, and even epilepsy, depression, and diabetes) and fill an unmet medical need." The purpose is to get important new drugs to the patient earlier. This program was conceptualized based on the emerging threat of acquired immunodeficiency syndrome (AIDS) in the 1980s, which needed new antiretroviral treatments to quickly combat the emerging virus.

Under this process, a drug showing some advantage over available therapy, such as showing superior effectiveness, effect on serious outcomes, or improved effect on serious outcomes, avoiding serious side effects of available therapy, improving the diagnosis of a serious condition where early diagnosis results in an improved outcome, decreasing clinically significant toxicity of an available therapy that is common and causes discontinuation of treatment, and the ability to address emerging or anticipated public health needs are considered.

A drug that receives Fast Track designation is eligible for more frequent meetings with the FDA to discuss the drug's development plan, the design of the proposed clinical trials, and the use of biomarkers. For example, after a successful phase I trial, a meeting with FDA can help to design a phase II trial that can be used for drug approval. If successful, this phase III, instead of phase III, trial would, therefore, be sufficient to prove a drug's safety and efficacy [90].

The FDA granted Fast Track status to 12 of the 37 novel drugs (32%) approved in 2022. Some of the examples included "Amvuttra" to treat polyneuropathy in adults with hereditary transthyretin-mediated amyloidosis, "Elahere" to treat patients with recurrent ovarian cancer that is resistant to platinum therapy, and "Adagrasib" (Krazati) to treat KRAS G12C-mutated locally advanced or metastatic non-small cell lung cancer in adults who have received at least one prior systemic therapy.

6.5.3 Accelerated Approval

In 1992, the US further expedited drug development by introducing the accelerated approval program. Similar to the fast-track program, accelerated approval can be obtained for drugs treating a serious condition and hence fill an unmet clinical need [92]. Accelerated approval enables the FDA to evaluate a drug's efficacy based on surrogate or intermediate endpoints rather than clinical endpoints. This can save valuable time in the drug approval process. For example, overall response rate (ORR) or progression free survival can be used for efficacy assessment rather than patient survival which may reduce clinical trial duration by 11 and 19 months, respectively [93].

For example, Adagrasib (Krazati) received accelerated approval for adult patients with KRAS G12C—mutated metastatic non-small cell lung cancer (NSCLC) based on a single-arm, open-label, Phase 1/2 clinical trial in 112 patients. ORR was the primary endpoint of the study. Vonjo (pacritinib) received accelerated approval for intermediate or high-risk myelofibrosis based upon the proportion of patients who had a 35% or greater spleen volume reduction from baseline to week 24 rather than clinical benefit and overall survival. Thus, the accelerated approval program expedited drug development by shortening clinical trial durations and enabling approvals based on trial designs with fewer enrolled patients. However, the companies are still needed to conduct confirmatory studies to verify clinical benefits [93].

6.5.4 Priority Review

For the drugs marketed in the United States, a detailed FDA review process is done prior to the approval. In 1992, under the Prescription Drug User Fee Act (PDUFA), the US introduced a priority review under which the FDA agreed to specific goals for improving the drug review time and prioritized and allocated more resources to new drug approvals based on their therapeutic gain. A drug receives a priority review designation if it is determined that the drug treats a serious condition and, if approved, would provide a significant improvement in the safety or effectiveness of the treatment, diagnosis, or prevention of serious conditions [89]. The FDA states that a significant therapeutic improvement constitutes 'evidence of increased effectiveness,' 'elimination or substantial reduction of a treatment-limiting drug reaction,' improved patient compliance leading to better outcomes, or 'evidence of safety and effectiveness in a new subpopulation' [89]. A priority review ought to be completed within 6 months of filing (compared to a target of 10 months under standard review).

Some of the examples of priority review included "Umbralisib" (Ukoniq) for treatment of marginal zone lymphoma and follicular lymphoma, "Sotarasib" (Lumakras) for non-small cell lung cancer, "Melphalan flufenamide" (Pepaxto) for relapsed or refractory multiple myeloma, and "Loncastuximab" (Zynlonta) for relapsed or refractory large B-cell lymphoma.

6.5.5 Breakthrough Therapy Designation

In 2012, USFDA introduced the Breakthrough Therapy Designation process under which drugs that treat a serious or life-threatening condition and preliminary clinical evidence indicates that the drug may demonstrate substantial improvement on a clinically significant endpoint(s) over available therapies receive breakthrough designation [89]. A Breakthrough Therapy designation includes all the Fast Track program features and also offers intensive FDA guidance during drug development and review, including involvement from senior managers.

Breakthrough Therapy designation is usually requested by the sponsors. In some cases, after reviewing the submitted data and information, if USFDA thinks that the drug development program meets the criteria for Breakthrough Therapy designation and if the drug development program can benefit from this designation, then the FDA suggests the sponsor consider submitting a breakthrough designation request.

By 2023, this designation has helped sponsors and the FDA streamline the development and approval of 225 drugs, over 56% of which were oncology indications [94]. The USFDA has granted Breakthrough Therapy Designation to "Pracinostat" (HDACs inhibitor) in combination with azacitidine for the treatment of elderly patients with newly diagnosed acute myeloid leukemia (AML), "Rezurock" (belumosudil) for chronic graft-versus-host disease after failure of at least two prior lines of systemic therapy, and "Rybrevant" (amivantamab) for non-small cell lung cancer.

Overall, these special designations incentivize and facilitate the expedited development of drugs for rare and severe medical conditions. Over the past decades, these designations provided pharmaceutical companies with more flexibility in conducting clinical trials to provide patients with timely access to promising new drugs.

6.6 Real-World Evidence (RWE) Studies

Real-world evidence (RWE) is the clinical evidence about the usage and potential benefits or risks of a medical product/drug derived from the analysis of real-world data (RWD). RWD are data relating to patient health status and/or the delivery of health care, routinely collected from a variety of sources like electronic health records, insurance data, product and disease registries, and billing data. The analysis of RWD generates real-world evidence (RWE) [95]. RWE studies are used to explore different aspects of health and disease, such as epidemiology, disease burden, safety, treatment patterns, treatment outcomes, long-term outcomes, and patient-reported outcomes such as satisfaction, quality of life, medication adherence, and patient experience. RWE studies can be conducted as prospective and retrospective cohort studies, case-control studies, and pragmatic clinical trials where the data can be collected and analyzed. Implementing RWE in the early stages of drug development can result in a shorter duration of trials and cost savings. It can also strongly complement the evidence gathered from RCTs, thus filling gaps in existing clinical knowledge. Of late, RWE is increasingly being utilized to support clinical and regulatory decisions, including approval of medical products. For example, Palbociclib, a CDK4/6 inhibitor, which was approved for women with ER+/HER2− breast cancer, was approved in men based on EHR outcomes data related to its off-label use among men. In 2017, RWE data was used in the form of a historical control arm to grant marketing approval of avelumab for treating Merkel cell carcinoma, marking the first instance of the use of RWE for original drug approvals [96].

In addition, historical information or data from EHRs and other RWD sources can be used as a historical control, especially the RCTs conducted for rare diseases. The most commonly reported challenge in using RWE is missing data and lack of randomization, and the resulting bias in patient selection. The inherent principle of using RWE is based on the analysis of routinely collected patient data, which, if not managed appropriately, can lead to bias. Real-world evidence studies have limitations such as information bias (e.g., ad hoc collection of unstandardized data), recall bias (i.e., data resulting from a selective recollection of events by patients/caregivers), and detection bias (i.e., the likelihood of an event being captured in one treatment group rather than in another). These challenges continue to limit the applicability of RWE despite its advantages and noteworthy data capabilities.

6.7 Clinical Trial Failures During Clinical Development

Drug discovery and development process takes over 10–15 years for each new drug to be approved for clinical use. Nine out of ten drug candidates that enter clinical development would fail during phases I, II, and III clinical trials and drug approval. If drug candidates in the preclinical stage are also counted, the failure rate of drug discovery/development is even higher than 90%. Data also showed that the probability of success is 63% in phase I trials, 31% in phase II trials, 58% in phase III trials, and 85% during the regulatory review process, for an overall success rate of 9.6% [97, 98].

There are four possible reasons attributed to the 90% clinical failures of drug development: lack of clinical efficacy (40–50%), unmanageable toxicity (30%), poor drug-like properties (10–15%), and lack of commercial needs and poor strategic planning (10%) [97].

Lack of clinical efficacy: Many times, cell lines and tissues fail to accurately represent the complexity of the human body. Likewise, animal models are poor predictors of human efficacy as animals fail to replicate the disease accurately, particularly when the pathogenesis of the disease is unknown or in the case of the new mechanism of the drug. This is reflected in failure rates, which are highest for drugs with a new mechanism of action and for diseases (e.g., Alzheimer's disease) where the pathogenesis is poorly understood [99].

Lack of clinical efficacy can be addressed by improving drug efficacy in preclinical and clinical studies by rigorously validating targets using genetic, genomic, and proteomic studies in cell lines, tissues, preclinical models, and human disease models. However, true validation of any new molecular target in human disease will always be challenging before a drug can be successfully developed since biological discrepancies among in vitro, in-disease animal models, and human disease may hinder the true validation of the molecular target's function [97].

It is seen that 55% of the total failures due to therapeutic efficacy occur in phase III trials. This implies that the results were overestimated in phase II studies but could not be reproduced on a larger scale in phase III trials. This pattern of overestimation has been consistently observed in cancer and other indications. In the case of cancer chemotherapy and targeted therapy, factors related to study design and sample size are implicated in failure [36].

It is also seen that many phase III trials proceed despite a threshold of activity not being reached in phase II evaluation. For example, nine phase III studies were initiated for targeted agents (e.g., bevacizumab and aflibercept) in combination with chemotherapy docetaxel without having strong evidence of activity from early-phase clinical trials for metastatic castrate-resistant prostate cancer [100].

In some cases, the potentially efficacious drugs fail to demonstrate efficacy due to a flawed study design, an inappropriate statistical endpoint, an inadequate sample size, or a reduced sample size that may result from patient dropouts and insufficient enrollment [101].

Drug toxicity: Drug safety is an utmost important consideration during the process of drug discovery and development. A large proportion of failures in drug discovery and development are due to toxicity. Drug toxicity is related to multiple factors, including pharmacokinetic factors (such as enhanced absorption, varied distribution into tissues, altered metabolism, and decreased excretion), drug overdose, genetics, comorbidities, drug interactions, and additive adverse effects from concomitant therapies. The direct pharmacodynamic effects of drugs at the receptor site also can lead to increased toxicity.

Failure due to unacceptable toxicity can be minimized by better understanding off-target or on-target inhibition of the molecular targets. For instance, the kinase inhibitor needs to be screened against other several hundred kinases for selectivity of disease. Additionally, several known toxicity targets in the major vital organs (e.g., in vitro and in vivo hERG assays) are usually performed as a predictive marker of cardiotoxicity. The structure of drug candidates may need to be further modified to minimize drug-protein adducts or drug–DNA adducts for reducing potential organ toxicity. In vitro/in vivo animal studies are always conducted for potential genotoxicity and carcinogenicity. Unfortunately, there is no well-developed strategy to optimize drug candidates to reduce tissue accumulation in the major vital organs to minimize the potential toxicity.

Drug-like properties: Drug-like properties, such as solubility, permeability, metabolic stability, and transporter effects, are of critical importance for the success of drug candidates. They affect oral bioavailability, metabolism, clearance, and toxicity, as well as in vitro pharmacology. Improving the drug optimization and selection process may significantly improve success of a given candidate. Poor drug-like properties contributed to the 30–40% drug development failures in the 1990s, but they only account for 10–15% of drug development failures today. This improvement benefits from rigorous selection criteria for drug-like properties during drug optimization, including solubility, permeability, protein binding, metabolic stability, and in vivo pharmacokinetics.

Lastly, poor strategic planning, which may include a change in therapeutic focus, company mergers, or poor clinical study conduct, accounts for 10% of drug development failures. To overcome this, the companies should develop a meticulous development plan with a detailed roadmap and milestones to advance new compounds from the lab through each stage of development [97].

7 Evolving Paradigm

7.1 Role of AI/ML

In the rapidly evolving field of technology, artificial intelligence (AI) is transforming various sectors and is also making significant strides in healthcare, enhancing capabilities in patient care and medical diagnostics. This technological revolution has deep historical roots, dating back to 1943 when the model of artificial neurons was first proposed by Warren McCulloch and Walter Pits.

The use of AI in healthcare began in the early 1970s with systems like MYCIN, which provided diagnostic support by using AI to identify pathogens and recommend antibiotics. These early systems depended on extensive medical knowledge encoded as logic rules [102].

In clinical research, AI has emerged as a transformative force, revolutionizing how we approach drug discovery, patient care, and the conduct of clinical trials. One of the most significant contributions of AI to clinical research is its ability to streamline and accelerate drug discovery processes. Traditional drug development is a lengthy and expensive journey, often taking years to bring a new drug to the market. AI-powered algorithms are changing this landscape by rapidly analyzing vast amounts of biomedical data to identify promising drug candidates and by predicting their efficacy and safety profiles. Machine learning models can sift through complex genetic, molecular, and patient data to pinpoint potential drug targets, significantly shortening the time it takes to identify lead compounds and move them into preclinical and clinical testing phases.

AI is also revolutionizing clinical trials by enhancing their design and execution, making them more efficient and cost-effective. AI algorithms use historical data to better identify and recruit patient groups most likely to benefit from specific treatments, boosting recruitment and patient retention. AI also optimizes trial protocols by recommending appropriate sample sizes, treatment doses, and relevant endpoints, which ultimately help in reducing duration and costs of the trial. Real-time data monitoring via AI safeguards data quality, swiftly identifies safety issues, and supports adaptive trials that evolve based on emerging new data. Furthermore, AI facilitates personalized medicine and patient-centered care by analyzing existing patient data to customize treatments, thereby improving outcomes and reducing side effects. The use of AI ensures treatments are precisely tailored to individual genetic and clinical profiles, leading to more targeted and effective healthcare interventions.

AI is significantly transforming clinical research and healthcare operations through the application of natural language processing (NLP) and predictive analytics. NLP is utilized to parse unstructured clinical data from electronic health records, physician notes, and medical literature, identifying trends and patterns that aid in clinical decision-making and risk prediction. This capability enhances the efficiency of literature reviews and supports systematic reviews and meta-analyses by spotting knowledge gaps and emerging research areas and informing potential research design and cohort identification for clinical trials.

Despite these remarkable advancements, integrating AI into clinical research presents challenges related to data privacy, regulatory compliance, and algorithm transparency. Addressing these challenges requires collaboration between researchers, clinicians, data scientists, and regulatory agencies to establish standards for data governance, algorithm validation, and ethical use of AI technologies in healthcare.

These issues must be addressed to leverage its maximum potential and successful implementation of artificial intelligence in clinical research.

AI is reshaping the landscape of clinical research by offering unprecedented opportunities to accelerate drug discovery, optimize clinical trials, and deliver personalized healthcare solutions. By harnessing the power of AI, we can unlock new insights into disease mechanisms, improve patient outcomes, develop personalized treatment, detect the diseases earlier, and pave the way for a future where precision medicine is the standard of care. As AI continues to evolve, its impact on clinical research will undoubtedly continue to grow.

Acknowledgments We wish to extend our heartfelt thanks to Mrs. Varsha Shitole and Dr. Jay Verma for their valuable support and contributions during the writing of this chapter. Their assistance was instrumental in shaping the final content.
Thank you all for your unwavering support and belief in our work.

References

1. Bhatt A. Evolution of clinical research: a history before and beyond james lind. Perspect Clin Res. 2010;1(1):6–10.
2. Hart PD. A change in scientific approach: from alternation to randomised allocation in clinical trials in the 1940s. BMJ. 1999;319(7209):572–3.
3. Crofton J. The MRC randomized trial of streptomycin and its legacy: a view from the clinical front line. J R Soc Med. 2006;99(10):531–4.
4. ICH: International Council for Harmonisation of technical requirements for pharmaceuticals for human use. 2016. https://database.ich.org/sites/default/files/ICH_E6%28R3%29_DraftGuideline_2023_0519.pdf [Apr; 2024].
5. Kandi V, Vadakedath S. Clinical trials and clinical research: a comprehensive review. Cureus. 2023;15(2):e35077.
6. Piccart-Gebhart MJ, et al. Herceptin Adjuvant (HERA) Trial Study Team. Trastuzumab after adjuvant chemotherapy in HER2-positive breast cancer. N Engl J Med. 2005; 353(16):1659–72.
7. South East Asia Infectious Disease Clinical Research Network. Effect of double dose oseltamivir on clinical and virological outcomes in children and adults admitted to hospital with severe influenza: double blind randomised controlled trial. BMJ. 2013;346:f3039.
8. Colditz GA, Taylor PR. Prevention trials: their place in how we understand the value of prevention strategies. Annu Rev Public Health. 2010;31:105–20.
9. Fisher B, Costantino JP, Wickerham DL, Cecchini RS, Cronin WM, Robidoux A, et al. Tamoxifen for the prevention of breast cancer: current status of the National Surgical Adjuvant Breast and Bowel Project P-1 study. J Natl Cancer Inst. 2005;97(22):1652–62.
10. Martínez-González MA, Salas-Salvadó J, Estruch R, Corella D, Fitó M, Ros E, PREDIMED INVESTIGATORS. Benefits of the Mediterranean diet: insights from the PREDIMED study. Prog Cardiovasc Dis. 2015;58(1):50–60.
11. Kim BS, Park SH, Jung SS, Kim HJ, Woo SD, Lee MM. Validity study for clinical use of hand-held spirometer in patients with chronic obstructive pulmonary disease. Healthcare (Basel). 2024;12(5):507.

12. Cardoso F, Piccart-Gebhart M, Van't Veer L, Rutgers E, TRANSBIG Consortium. The MINDACT trial: the first prospective clinical validation of a genomic tool. Mol Oncol. 2007;1(3):246–51.

13. Lee JM, Lehman CD, Miglioretti DL. Screening trials and design. In: Handbook for clinical trials of imaging and image-guided interventions; 2016. p. 76–90.

14. Eckersberger E, Finkelstein J, Sadri H, Margreiter M, Taneja SS, Lepor H, Djavan B. Screening for prostate cancer: a review of the ERSPC and PLCO trials. Rev Urol. 2009;11(3):127–33.

15. Schröder FH, et al. ERSPC Investigators. Screening and prostate-cancer mortality in a randomized European study. N Engl J Med. 2009;360(13):1320–8.

16. Anderle A, Bancone G, Domingo GJ, Gerth-Guyette E, Pal S, Satyagraha AW. Point-of-care testing for G6PD deficiency: opportunities for screening. Int J Neonatal Screen. 2018;4(4):34.

17. Zailani MAH, Raja Sabudin RZA, Ithnin A, Alauddin H, Sulaiman SA, Ismail E, Othman A. Population screening for glucose-6-phosphate dehydrogenase deficiency using quantitative point-of-care tests: a systematic review. Front Genet. 2023;14:1098828.

18. Burt T, Roffel AF, Langer O, Anderson K, DiMasi J. Strategic, feasibility, economic, and cultural aspects of phase 0 approaches: is it time to change the drug development process in order to increase productivity? Clin Transl Sci. 2022;15(6):1355–79.

19. Okour M, Derimanov G, Barnett R, et al. A human microdose study of the anti-malarial GSK3191607 in healthy volunteers. Br J Clin Pharmacol. 2017;84:482–9.

20. Jones HM, Butt RP, Webster RW, et al. Clinical micro-dose studies to explore the human pharmacokinetics of four selective inhibitors of human Nav1.7 voltage-dependent sodium channels. Clin Pharmacokinet. 2016;55:875–87.

21. Wright B. Clinical trial phases. In: Shamley D, Wright B, editors. A comprehensive and practical guide to clinical trials. Academic Press; 2017. p. 11–5.

22. Mahan VL. Clinical trial phases. Int J Clin Med. 2014;05(21):1374–83.

23. Thorat SB, Banarjee SK, Gaikwad DD, Jadhav SL, Thorat RM. Clinical trial: a review. Int J Pharm Sci Rev Res. 2010;1(2):Article 019.

24. Shen J, Swift B, Mamelok R, Pine S, Sinclair J, Attar M. Design and conduct considerations for first-in-human trials. Clin Transl Sci. 2019;12(1):6–19.

25. Attarwala H. TGN1412: from discovery to disaster. J Young Pharm. 2010;2(3):332–6.

26. Singh H, Sarangi SC, Gupta YK. French Phase I clinical trial disaster: issues, learning points, and potentialsafety measures. J Nat Sc Biol Med. 2018;9:106–10.

27. Small DS, Zhang W, Royalty J, Cannady EA, Downs D, Friedrich S, Suico JG. A multidose study to examine the effect of food on Evacetrapib exposure at steady state. J Cardiovasc Pharmacol Ther. 2015;20(5):483–9.

28. Umscheid CA, Margolis DJ, Grossman CE. Key concepts of clinical trials: a narrative review. Postgrad Med. 2011;123:194–204.

29. Friedman LG, McKeehan N, Hara Y, Cummings JL, Matthews DC, Zhu J, et al. Value-generating exploratory trials in neurodegenerative dementias. Neurology. 2021;96(20):944–54.

30. Ivanova A, Xiao C, Tymofyeyev Y. Two-stage designs for Phase 2 dose-finding trials. Stat Med. 2012;31(24):2872–81.

31. Goodwin R, Giaccone G, Calvert H, Lobbezoo M, Eisenhauer EA. Targeted agents: how to select the winners in preclinical and early clinical studies? Eur J Cancer. 2012;48:170–8.

32. Gheorghiade M, Greene SJ, Butler J, Filippatos G, Lam CS, Maggioni AP, et al. Effect of vericiguat, a soluble guanylate cyclase stimulator, on natriuretic peptide levels in patients with worsening chronic heart failure and reduced ejection fraction: the SOCRATES-REDUCED randomized trial. JAMA. 2015;314(21):2251–62.

33. Beck LA, Thaçi D, Hamilton JD, Graham NM, Bieber T, Rocklin R, et al. Dupilumab treatment in adults with moderate-to-severe atopic dermatitis. N Engl J Med. 2014;371(2):130–9.

34. Thaçi D, Simpson EL, Beck LA, Bieber T, Blauvelt A, Papp K, et al. Efficacy and safety of dupilumab in adults with moderate-to-severe atopic dermatitis inadequately controlled by topical treatments: a randomised, placebo-controlled, dose-ranging phase 2b trial. Lancet. 2016;387(10013):40–52.

35. Rohilla A, Singh RK, Sharma D, Keshari R, Kushnoor A. Phases of clinical trials: a review. IJPCBS. 2013;3(3):700–3.

36. Li X, Zhou Y, Xu B, et al. Comparison of efficacy discrepancy between early-phase clinical trials and phase III trials of PD-1/PD-L1 inhibitorsJournal for ImmunoTherapy of. Cancer. 2024;12:e007959.

37. Wennerholm UB, Saltvedt S, Wessberg A, Alkmark M, Bergh C, Wendel SB, et al. Induction of labour at 41 weeks versus expectant management and induction of labour at 42 weeks (SWEdish Post-term Induction Study, SWEPIS): multicentre, open label, randomised, superiority trial. BMJ. 2019;367:16131. Erratum in: BMJ. 2021 Dec 15;375:n3072.

38. Cather J, Young M, DiRuggiero DC, Tofte S, Williams L, Gonzalez T. A review of phase 3 trials of dupilumab for the treatment of atopic dermatitis in adults, adolescents, and children aged 6 and up. Dermatol Ther (Heidelb). 2022;12(9):2013–38.

39. Suvarna V. Phase IV of drug development. Perspect Clin Res. 2010;1(2):57–60. PMID: 21829783; PMCID: PMC3148611.

40. Trailokya A, Srivastava A, Mukaddam Q, Chaudhry S, Patel K, Suryawanshi S, Naik M. Phase IV studies: what clinicians need to know! J Assoc Physicians India. 2024;72(1):85–7.

41. Baron JA, et al. A randomized trial of aspirin to prevent colorectal adenomas. N Engl J Med. 2003;348(10):891–9.

42. Graham DJ, Green L, Senior JR, Nourjah P. Troglitazone-induced liver failure: a case study. Am J Med. 2003;114(4):299–306.

43. Xu B, Xing A, Li S. The forgotten type 2 diabetes mellitus medicine: rosiglitazone. Diabetol Int. 2021;13(1):49–65.

44. Jones JK, Fife D, Curkendall S, Goehring E Jr, Guo JJ, Shannon M. Coprescribing and codispensing of cisapride and contraindicated drugs. JAMA. 2001;286(13):1607–9.

45. Ranganathan P, Aggarwal R. Study designs: part 1 – an overview and classification. Perspect Clin Res. 2018;9(4):184–6.

46. Karanicolas PJ, Farrokhyar F, Bhandari M. Practical tips for surgical research: blinding: who, what, when, why, how? Can J Surg. 2010;53(5):345–8.

47. Sil A, Kumar P, Kumar R, Das NK. Selection of control, randomization, blinding, and allocation concealment. Indian Dermatol Online J. 2019;10(5):601–5.

48. Kang M, Ragan BG, Park JH. Issues in outcomes research: an overview of randomization techniques for clinical trials. J Athl Train. 2008;43(2):215–21.

49. Hariton E, Locascio JJ. Randomised controlled trials – the gold standard for effectiveness research: study design: randomised controlled trials. BJOG. 2018;125(13):1716.

50. Seymour JF, Davids MS, Anderson MA, Kipps TJ, Wierda WG, Pagel JM, et al. The BCL-2-specific BH3-mimetic ABT-199 (GDC-0199) is active and well-tolerated in patients with relapsed/refractory chronic lymphocytic leukemia: interim results of a phase I first-in-human study. Blood. 2012;120(21):3923.

51. Kurzrock R, et al. Moving beyond 3+3: the future of clinical trial design. Am Soc Clin Oncol Educ Book. 2021;41:e133–44.

52. Yuan Y, Lee JJ, Hilsenbeck SG. Model-assisted designs for early-Phase clinical trials: simplicity meets superiority. JCO Precis Oncologia. 2019;3:PO.19.00032.

53. Bogin V. Master protocols: new directions in drug discovery. Contemp Clin Trials Commun. 2020;18:100568.

54. Renfro LA, Sargent DJ. Statistical controversies in clinical research: basket trials, umbrella trials, and other master protocols: a review and examples. Ann Oncol. 2017;28(1):34–43.

55. Li BT, Shen R, Buonocore D, Olah ZT, et al. Ado-Trastuzumab Emtansine for patients with HER2-mutant lung cancers: results from a phase II basket trial. J Clin Oncol. 2018;36(24):2532–7.

56. A study of Vemurafenib in participants with BRAF V600 mutation-positive cancers. ClinicalTrials.gov ID NCT01524978. Accessed on 13 Apr 2024. https://clinicaltrials.gov/study/NCT01524978.

57. Study of Pembrolizumab (MK-3475) in participants with advanced solid tumors (MK-3475-158/KEYNOTE-158). ClinicalTrials.gov ID NCT02628067. Accessed on 13 Apr 2024. https://clinicaltrials.gov/study/NCT02628067.

58. A study to test the effect of the drug Larotrectinib in adults and children with NTRK-fusion positive solid tumors (NAVIGATE). ClinicalTrials.gov ID NCT02576431. Accessed on 13 Apr 2024. https://clinicaltrials.gov/study/NCT02576431.

59. Duan XP, Qin BD, Jiao XD, et al. New clinical trial design in precision medicine: discovery, development and direction. Sig Transduct Target Ther. 2024;9:57.

60. The UK plasma based molecular profiling of advanced breast cancer to inform Therapeutic CHoices (plasmaMATCH) Trial (plasmaMATCH). ClinicalTrials.gov ID: NCT03182634. Accessed on 13 Apr 2024. https://classic.clinicaltrials.gov/ct2/show/NCT03182634.

61. A study of multiple immunotherapy-based treatment combinations in participants with metastatic pancreatic ductal adenocarcinoma (Morpheus-Pancreatic Cancer). ClinicalTrials.gov ID NCT03193190. https://clinicaltrials.gov/study/NCT03193190.

62. I-SPY TRIAL: neoadjuvant and personalized adaptive novel agents to treat breast cancer (I-SPY). ClinicalTrials.gov Identifier: NCT01042379. Accessed on 13 Apr 2024. https://classic.clinicaltrials.gov/ct2/show/NCT01042379.

63. Kapoor MC, Goyal R. Objectives and outcomes of a clinical trial. Indian J Anaesth. 2023;67(4):328–30.

64. McLeod C, Norman R, Litton E, Saville BR, Webb S, Snelling TL. Choosing primary endpoints for clinical trials of health care interventions. Contemp Clin Trials Commun. 2019;16:100486.

65. Parker RA, Weir CJ. Multiple secondary outcome analyses: precise interpretation is important. Trials. 2022;23(1):27.

66. Ghali F, Zhao Y, Patel D, Jewell T, Yu EY, Grivas P, et al. Surrogate endpoints as predictors of overall survival in metastatic urothelial cancer: a trial-level analysis. Eur Urol Open Sci. 2022;47:58–64.

67. Califf RM. Biomarker definitions and their applications. Exp Biol Med (Maywood). 2018;243(3):213–21.

68. Belin L, Tan A, De Rycke Y, et al. Progression-free survival as a surrogate for overall survival in oncology trials: a methodological systematic review. Br J Cancer. 2020;122:1707–14.

69. Booth CM, et al. Progression-free survival: meaningful or simply mseasurable? JCO. 2012;30:1030–3.

70. FDA approves atezolizumab for PD-L1 positive unresectable locally advanced or metastatic triple-negative breast cancer. FDA. March 8, 2019. Accessed 13 Apr 2024. https://bit.ly/3Bn1vP1.

71. Miles D, Gligorov J, André F, et al. Primary results from IMpassion131, a double-blind, placebo-controlled, randomised phase III trial of first-line paclitaxel with or without atezolizumab for unresectable locally advanced/metastatic triple-negative breast cancer. Ann Oncol. 2021;32(8):994–1004.

72. Mandrekar SJ, Sargent DJ. All-comers versus enrichment design strategy in phase II trials. J Thorac Oncol. 2011;6(4):658–60.

73. Temple R. Enrichment of clinical study populations. Clin Pharmacol Ther. 2010;88(6):774–8.

74. Uozumi R, Yada S, Kawaguchi A. Patient recruitment strategies for adaptive enrichment designs with time-to-event endpoints. BMC Med Res Methodol. 2019;19:159.

75. Bellio C, DiGloria C, Foster R, James K, Konstantinopoulos PA, Growdon WB, Rueda BR. PARP inhibition induces enrichment of DNA repair-proficient CD133 and CD117 positive ovarian cancer stem cells. Mol Cancer Res. 2019;17(2):431–45.

76. Noordzij M, Tripepi G, Dekker FW, Zoccali C, Tanck MW, Jager KJ. Sample size calculations: basic principles and common pitfalls. Nephrol Dial Transplant. 2010;25(5):1388–93.
77. Koeder C, Kranz RM, Anand C, Husain S, Alzughayyar D, Schoch N, Hahn A, Englert H. Effect of a 1-year controlled lifestyle intervention on body weight and other risk markers (the healthy lifestyle community programme, cohort 2). Obes Facts. 2022;15(2):228–39.
78. Sullivan GM, Feinn R. Using effect size-or why the P value is not enough. J Grad Med Educ. 2012;4(3):279–82.
79. Zodpey SP. Sample size and power analysis in medical research. Indian J Dermatol Venereol Leprol. 2004;70(2):123–8.
80. Bodaghi A, Fattahi N, Ramazani A. Biomarkers: promising and valuable tools towards diagnosis, prognosis and treatment of Covid-19 and other diseases. Heliyon. 2023;9(2):e13323.
81. Swain SM, Shastry M, Hamilton E. Targeting HER2-positive breast cancer: advances and future directions. Nat Rev Drug Discov. 2023;22(2):101–26.
82. Zhou Y, Tao L, Qiu J, Xu J, Yang X, Zhang Y, et al. Tumor biomarkers for diagnosis, prognosis and targeted therapy. Signal Transduct Target Ther. 2024;9(1):132.
83. FDA-NIH Biomarker Working Group. BEST (Biomarkers, EndpointS, and other Tools) Resource [Internet]. Silver Spring (MD): Food and Drug Administration (US); 2016-. Susceptibility/Risk Biomarker. 2016 Dec 22 [Updated 2020 Aug 27].
84. Lillie EO, Patay B, Diamant J, Issell B, Topol EJ, Schork NJ. The n-of-1 clinical trial: the ultimate strategy for individualizing medicine? Per Med. 2011;8(2):161–73.
85. Molloy DW, Guyatt GH, Wilson DB, Duke R, Rees L, Singer J. Effect of tetrahydroaminoacridine on cognition, function and behaviour in Alzheimer's disease. CMAJ. 1991;144(1):29–34.
86. Li J, Niu J, Yang M, Ye P, Zhai J, Yuan W, et al. Using single-patient (n-of-1) trials to determine effectiveness of traditional Chinese medicine on chemotherapy-induced leukopenia in gastric cancer: a feasibility study. Ann Transl Med. 2019;7(6):124.
87. Harada G, Drilon A. TRK inhibitor activity and resistance in TRK fusion-positive cancers in adults. Cancer Genet. 2022;264-265:33–9.
88. Gouda MA, Buschhorn L, Schneeweiss A, Wahida A, Subbiah V. N-of-1 trials in cancer drug development. Cancer Discov. 2023;13(6):1301–9.
89. FDA. Fast track, breakthrough therapy, accelerated approval, priority review. In: US Food Drug Adm. https://www.fda.gov/patients/learn-about-drug-and-device-approvals/fast-track-breakthrough-therapy-accelerated-approval-priority-review (2018). Accessed 13 Apr 2024.
90. Michaeli DT, Michaeli T, Albers S, et al. Special FDA designations for drug development: orphan, fast track, accelerated approval, priority review, and breakthrough therapy. Eur J Health Econ. 2024;25(6):979–97.
91. Fermaglich LJ, Miller KL. A comprehensive study of the rare diseases and conditions targeted by orphan drug designations and approvals over the forty years of the Orphan Drug Act. Orphanet J Rare Dis. 2023;18(1):163.
92. FDA. Accelerated approval program. In: US Food Drug Adm. https://www.fda.gov/drugs/information-health-care-professionals-drugs-accelerated-approval-program. 2020. Accessed 13 Apr 2024.
93. Chen EY, Joshi SK, Tran A, Prasad V. Estimation of study time reduction using surrogate end points rather than overall survival in oncology clinical trials. JAMA Intern Med. 2019;179:642–7.
94. Collins G, Stewart M, McKelvey B, Stires H, Allen J. Breakthrough therapy designation criteria identify drugs that improve clinical outcomes for patients: a case for more streamlined coverage of promising therapies. Clin Cancer Res. 2023;29(13):2371–4.
95. Bhatt A. Conducting real-world evidence studies in India. Perspect Clin Res. 2019; 10(2):51–6.
96. Feinberg BA, Gajra A, Zettler ME, Phillips TD, Phillips EG Jr, Kish JK. Use of real-world evidence to support FDA approval of oncology drugs. Value Health. 2020; 23(10):1358–65.

97. Sun D, Gao W, Hu H, Zhou S. Why 90% of clinical drug development fails and how to improve it? Acta Pharm Sin B. 2022;12(7):3049–62.

98. Mullard A. Parsing clinical success rates. Nat Rev Drug Discov. 2016;15(7):447.

99. Hingorani AD, Kuan V, Finan C, et al. Improving the odds of drug development success through human genomics: modelling study. Sci Rep. 2019;9:18911.

100. Seruga B, Ocana A, Amir E, Tannock IF. Failures in Phase III: causes and consequences. Clin Cancer Res. 2015;21(20):4552–60.

101. Fogel DB. Factors associated with clinical trials that fail and opportunities for improving the likelihood of success: a review. Contemp Clin Trials Commun. 2018;11:156–64.

102. Stafie CS, Sufaru IG, Ghiciuc CM, Stafie II, Sufaru EC, Solomon SM, Hancianu M. Exploring the intersection of artificial intelligence and clinical healthcare: a multidisciplinary review. Diagnostics (Basel). 2023;13(12):1995.

From Orphan Drugs to Inclusive Research: Bridging Global Gaps in Rare Disease Treatment

Harsha K. Rajasimha, Mohua Chakraborty Choudhury, and Partha Sarathi Mukherjee

Abstract This chapter explores the global disparities in rare disease research and drug development, highlighting that over 90% of efforts are concentrated in less than 10% of the world's population, primarily in the US and EU. Despite significant progress with over 1100 FDA-approved orphan drugs, 95% of rare diseases still lack approved treatments. The unique challenges in rare disease research are discussed, including limited scientific literature, small patient populations, and difficulties in patient identification. The critical role of patient registries and natural history studies in understanding disease progression and designing clinical trials is emphasized. The chapter examines the impact of the Human Genome Project on rare disease diagnosis and treatment, alongside emerging treatment modalities like gene therapy and innovative approaches using CRISPR and mRNA technologies. It addresses the potential and pitfalls of AI and machine learning in rare disease research, highlighting the risk of bias due to limited global data representation. The stark inequality in clinical trial distribution is noted, with over 80% of active rare disease trials concentrated in the US and EU. The urgent need for unified, patient-centric research platforms and strategies to bridge the global divide in rare disease research and drug development is underscored, calling for more inclusive approaches to ensure equitable progress in this critical field. This chapter also focuses on an overview of Chemistry, Manufacturing, and Controls (CMC) and its critical role in drug development for all modalities (small molecules, biologics, oligonucleotides,

H. K. Rajasimha (✉)
Jeeva Clinical Trials Inc, Manassas, VA, USA

Indo US Organization for Rare Diseases (IndoUSrare), Herndon, VA, USA

School of Systems Biology, George Mason University, Fairfax, VA, USA
e-mail: harsha@jeevatrials.com

M. C. Choudhury
DST Center for Policy Research, Indian Institute of Science, Bengaluru, Karnataka, India

P. S. Mukherjee
Indo US Organization for Rare Diseases (IndoUSrare), Herndon, VA, USA

© The Author(s), under exclusive license to Springer Nature Switzerland AG 2025
N. Chirmule, V. V. Ghalsasi (eds.), *Approved: The Life Cycle of Drug Development*,
https://doi.org/10.1007/978-3-031-81787-8_12

and gene therapy) with a focus on rare disease. The key elements of CMC requirements in regulatory dossier submission and approval are highlighted along with the latest trends of using AI in CMC.

Keywords Rare diseases · Orphan diseases · Drug development · Global disparities · Orphan Drug Act · Patient registries · Natural history studies · Genetic mutations · Newborn screening · Genome sequencing · Clinical trials · Patient identification · Human Genome Project · Gene therapy · CRISPR · Artificial intelligence (AI) · Machine learning (ML) · Low- and middle-income countries (LMICs) · Diversity, equity, inclusion, and accessibility (DEIA) · Personalized medicine · Chemistry, Manufacturing, and Controls (CMC) · Drug Substance (DS) · Drug Product (DP)

1   Introduction

Rare diseases, also known as orphan diseases, represent a unique and challenging frontier in medical research and drug development. These conditions, while individually uncommon, collectively affect a significant portion of the global population. In the United States, a rare disease is defined as one that affects fewer than 200,000 individuals, while the European Union considers a disease rare when it affects fewer than 1 in 2000 people. With over 10,000 known rare diseases impacting an estimated 350 million people worldwide, the collective burden of these conditions is substantial [1–3].

The field of rare disease research and drug development is characterized by significant challenges and inequities. Despite the progress made since the implementation of the Orphan Drug Act in 1983, which has led to over 1100 FDA-approved orphan drugs, almost 95% of rare diseases still lack a single approved treatment option [4]. This stark statistic underscores the critical need for continued efforts and innovation in this field.

One of the most pressing issues in rare disease research is the massive global disparity in research efforts and clinical trials. Over 90% of rare disease research and drug development is concentrated in less than 10% of the world's population, primarily in the United States and European Union. This inequity leaves the majority of the world's population, particularly in low- and middle-income countries (LMICs), disconnected from crucial advancements and potential treatments. As the orphan drug industry grows in Western countries, this divide is further widening, necessitating urgent attention and strategies to bridge these gaps.

The unique aspects of rare disease research present several challenges. Limited scientific literature, small patient populations that are often geographically dispersed, and difficulties in patient identification all contribute to the complexity of studying and treating these conditions. The average time to diagnosis for a rare disease patient is 5–7 years, highlighting the critical need for improved screening and diagnostic methods [5]. While newborn screening programs in countries like the United States can identify about 80 diseases at birth, technologies capable of screening or diagnosing thousands of genetic diseases, such as low-cost genome and exome sequencing, remain underutilized globally.

Patient registries and natural history studies play a vital role in rare disease research. These resources are crucial for understanding disease progression, identifying potential therapeutic targets, and designing effective clinical trials. However, the lack of standardized, digitized patient data that is properly consented for research use presents a significant hindrance to progress. Moreover, most patient registries and natural history studies are initiated and funded in Western countries, leaving patients in other parts of the world underrepresented and often unaware of the importance of their participation in these research databases.

The Human Genome Project has revolutionized our understanding of rare diseases, many of which have a genetic basis. Approximately 80% of rare diseases have a causal genetic mutation, making the era of low-cost whole genome and exome sequencing particularly promising for high-throughput screening, diagnosis, and personalized medicine approaches. This genetic understanding has paved the way for innovative treatment modalities, including gene and cell therapies, enzyme replacement therapies, and repurposed drugs. Cutting-edge approaches such as CRISPR gene editing and mRNA technologies are also being explored to expand treatment possibilities for rare diseases.

The application of artificial intelligence (AI) and machine learning (ML) in rare disease research holds great potential but also presents challenges [6]. The primary concern is the risk of bias in AI/ML algorithms due to the predominance of data from Western populations. To address this issue, there is an urgent need to include global patients in research databases and support LMICs in digitizing their rare disease populations through high-quality patient registries and natural history studies.

Clinical trials for rare diseases face significant geographical disparities. As of June 2024, over 7600 active clinical trials for rare diseases are ongoing, with more than 80% concentrated in the United States and European Union. This imbalance is particularly stark when considering that countries with large populations, such as India, host less than 1% of these trials. Initiatives like the FDA's Project Asha, which facilitates the inclusion of Indian sites in US-based cancer clinical trials, represent steps towards addressing this disparity.

This chapter explores these multifaceted challenges in rare disease research and drug development, emphasizing the need for global collaboration, innovative approaches, and inclusive strategies. It delves into the role of patient advocacy, the impact of technological advancements, and the importance of diversity, equity, inclusion, and accessibility (DEIA) in clinical trials. By examining these aspects, we aim to provide a comprehensive overview of the current landscape of rare disease research and highlight potential pathways to democratize and advance this critical field of medicine globally.

This chapter also includes a high level of the critical role played by CMC in drug development, with specific emphasis on the rare disease arena. The various key elements of CMC in terms of drug substance and drug product identity, potency, quality, safety, and efficacy are highlighted, along with the required information from a drug or biologics marketing authorization holder (MAH) to file regulatory dossier for approval of medicines. The key attributes of AI in CMC and the critical roles played by contract manufacturers for rare disease drug development are also mentioned.

2 Rare Diseases: Definition

Rare diseases, also known as orphan diseases, are a diverse group of conditions that affect a small percentage of the population. While there is no universally accepted definition, the United States defines a rare disease as one that affects fewer than 200,000 individuals, while the European Union considers a disease rare when it affects fewer than 1 in 2000 people. Despite their individual rarity, rare diseases collectively impact a significant portion of the global population. There are over 10,000 known rare diseases, affecting an estimated 25–30 million people in the United States alone. Worldwide, it is estimated that 350 million people suffer from rare diseases. The majority of these conditions are genetic in origin and often chronic, progressive, and life-threatening. Unfortunately, only a small fraction of rare diseases has approved treatments, with fewer than 5% of the known rare diseases having an FDA-approved drug. This highlights the significant unmet medical need and the importance of ongoing research and drug development efforts in the rare disease space.

3 Patient Advocacy

The role of patient advocacy has been pivotal in advancing orphan drug development and improving the lives of individuals affected by rare diseases. In the United States, the Orphan Drug Act (ODA) of 1983 was a landmark legislation that provided incentives for pharmaceutical companies to develop treatments for rare diseases. The ODA was the result of tireless efforts by patient advocates, who lobbied Congress to address the unmet medical needs of the rare disease community. Since then, patient advocacy organizations have evolved and grown in number and influence, both in the United States and globally.

Notable patient advocacy organizations in the rare disease space include the National Organization for Rare Disorders (NORD) in the United States, the European Organization for Rare Diseases (EURORDIS), Global Genes, Every Life Foundations for Rare Diseases, and many others. These organizations provide support, education, and advocacy for patients and families affected by rare diseases. They also collaborate with industry partners, researchers, and policymakers to advance rare disease research and drug development.

In recent years, patient advocates have become increasingly involved in the orphan drug industry, serving as key partners in the drug development process. They provide valuable insights into the patient experience, help to identify and prioritize research objectives, and assist in the design and implementation of clinical trials. Patient advocates also play a critical role in raising awareness about rare diseases and the importance of developing treatments. At the global level, driven by global patent advocacy movements, the United Nations has recognized the importance of addressing rare diseases as part of its Universal Health Coverage (UHC) agenda.

Some of the key achievements of patient advocacy groups have been:

- Engaging in research and development of drugs and prioritizing it as a key goal
- Rare disease patients and their advocates have often had to take on the herculean task of driving drug development forward themselves, in the face of limited interest from industry and government.

In one such early instance, in the 1950s, when very little was known about cystic fibrosis and there were no available treatments, a group of concerned parents came together to form what would become the Cystic Fibrosis Foundation (CFF). Over the following decades, the CFF would go on to invest hundreds of millions of dollars in research and drug development, ultimately leading to the approval of several groundbreaking therapies that have dramatically improved the lives of cystic fibrosis patients.

The CFF's success was built on an innovative "venture philanthropy" model, in which the foundation provided early-stage funding and scientific expertise to biotech companies working on cystic fibrosis treatments. In exchange, the CFF received a share of royalties from any approved drugs, which it could then reinvest into further research. This approach helped de-risk the development process and incentivize industry partners to take on the challenge of developing drugs for a relatively small patient population.

The CFF's story is just one example of how rare disease patients and their advocates have had to become drug developers out of necessity. Many other PAGs over the years have followed a similar model, establishing research funding programs, scientific advisory boards, and industry partnerships to advance treatments for their disease. Many PAGs are directly engaging in research activities to help drive progress for their rare diseases. The survey by Patterson et al. [7] found that 79% of responding PAGs engage in research in some capacity, such as creating patient registries (64%), conducting natural history studies (43%), and initiating translational research (40%). "Ultra-rare" disease PAGs were even more likely than "rare" disease PAGs to cite research as their top priority.

NORD is a key organization that collaborates with industry players to support the discovery and financing of rare disease therapies (https://rarediseases.org/). They also foster patient-led activism and advocacy for important legislation like the Orphan Drug Act, which provides incentives for developing treatments for rare conditions. The Chan Zuckerberg Initiative's "Rare As One" project (https://chanzuckerberg.com/science/programs-resources/rare-as-one/) provides research grants and a collaborative network to help existing nonprofit PAGs drive progress for their disease areas.

Global Genes offers resources like the "Data DIY" guidebook to empower PAG leaders to collect and manage patient data, which can be valuable for advancing research.

Many PAGs focus on educating patients about their condition using vetted, credible information across various direct-to-consumer channels. This allows patients to become well-informed about their disease and potential treatments before engaging with healthcare providers. PAGs often serve as a liaison between patients and trusted clinical researchers, helping to overcome barriers like lack of awareness or mistrust that can hinder participation in clinical trials, especially among underserved populations. The Rare Diseases Clinical Research Network's "Coalition of Patient Advocacy Groups" brings together mature PAGs interested in research to share education and expertise.

PAGs often contribute to clinical trials run by external researchers/companies. 81% of research-engaged PAGs in the Patterson survey have contributed to a clinical trial (Emerging roles and opportunities for rare disease patient advocacy groups). They support trials by providing data/biospecimens, encouraging patient participation, assisting in study design, and supporting enrolled patients. Sixteen percent have even initiated their own clinical trial. NORD has collaborated with industry partners to support the development and execution of clinical trials. NORD leverages its extensive patient network to help with trial recruitment and retention, ensuring that studies have adequate representation from the patient community.

It is important to understand the patient journey with a rare disease across the continuum from early onset of symptoms to end of life with the limitations of the healthcare ecosystems across the world in screening, diagnosing, treating, and developing new products to help patients with rare diseases.

3.1   Building Patient Registries and Biorepositories

Patient registries and natural history studies are foundational resources for rare disease research and orphan drug development to advance. These are almost prerequisite for rare and genetic diseases with limited funding and scientific literature published about them compared to common indications. The full gamut of the disease symptoms and how it manifests in a diverse pool of patients is critical to gain a comprehensive scientific and clinical understanding of the natural history and progression of the disease in the affected population. Historically, most of these patient registries have been created and maintained in the USA and EU (Fig. 1).

Many PAGs have created patient registries (64%) and biobanks (31%) to provide valuable data and biospecimen resources to researchers. This helps attract research interest and enables studies in rare disease patient populations. The systematic review cited by Patterson et al. found that having a registry/biobank was associated with increased research activity, publications, and clinical trial participation for a

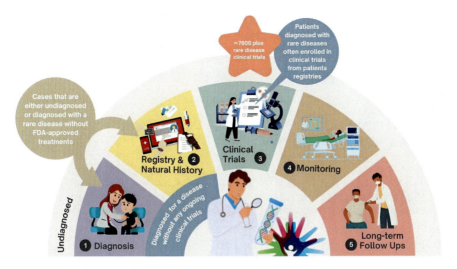

Fig. 1 The patient journey

PAG. One PAG that has been a leader in this area is the Chordoma Foundation, which supports research on chordoma, a rare type of bone cancer. In 2007, the Foundation launched the Chordoma Biobank (https://www.chordomafoundation.org/researchers/biospecimens/), a centralized repository of chordoma tumor tissue and associated clinical data from patients around the world. The Biobank operates through a network of partnering institutions, including hospitals, clinics, and research centers, that collect and process samples according to standardized protocols. Patients can consent to donate their tissue samples and clinical information through their treating physician or by contacting the Foundation directly. The samples and data in the biobank are then made available to qualified researchers who are working to develop new treatments and diagnostic tests for chordoma. Researchers can apply to access the samples through a formal request process, which involves a scientific review to ensure that the proposed research is of high quality and aligns with the Foundation's mission. To date, the Chordoma Biobank has collected samples from over 1000 patients, making it the largest repository of chordoma tissue in the world. The Biobank has supported numerous research projects, including studies to identify new drug targets, develop more accurate diagnostic tests, and understand the genetic factors that contribute to chordoma risk and progression. One of the key advantages of the Biobank is that it allows researchers to access a large, diverse collection of samples that would be difficult or impossible to obtain through individual institutions. This is particularly important for rare diseases like chordoma, where any given hospital or clinic may only see a handful of cases per year.

The Chordoma Foundation's Biobank is just one example of how PAGs are taking a proactive role in driving research progress through biospecimen collection and sharing. By creating these vital research resources, PAGs are helping to accelerate the pace of discovery and bring new treatments to patients faster. However, establishing and maintaining a biobank is a complex and resource-intensive undertaking, requiring robust infrastructure, expertise, and ongoing funding. As such, many PAGs are also partnering with academic institutions, industry partners, and other stakeholders to build collaborative biobanking initiatives that can maximize impact and sustainability over the long term.

3.2 Collaborating with Industry on Patient-Centric Drug Development

PAGs provide the patient perspective to help biopharmaceutical companies develop meaningful treatments. Over half the surveyed PAGs receive some industry funding. The GENE TARGET framework proposed by Chopra et al. [8] emphasizes the need to engage patient populations and PAGs to enable successful gene therapy development. Guidelines like the Stein et al. [9] recommendations aim to facilitate productive, ethical PAG-industry collaborations.

3.3 Assisting Patients in Accessing Approved Therapies

30% of PAGs with an approved therapy, over 60% educate healthcare providers and patients about access [10]. They teach patients self-advocacy skills, write letters of medical necessity, and advocate to insurers to help enable access to expensive rare disease therapies.

The PAGs play vital roles in enabling access to approved rare disease therapies. Of PAGs associated with a disease that has an approved treatment, over 60% actively educate healthcare providers and patients about access (*Patterson Survey* [7]). This involves teaching patient's self-advocacy skills to navigate the complex reimbursement landscape and empowering them to communicate effectively with insurers and healthcare providers. PAGs also directly advocate on behalf of patients by writing letters of medical necessity to payers and engaging with insurance companies to push for coverage of costly rare disease treatments. Securing reimbursement often requires "unwavering tenacity" from PAGs due to the fragmented payer landscape and potential coverage barriers, even for life-saving therapies. By educating regulators and payers about the disease burden and real-world benefits of treatments, PAGs can influence coverage decisions and ensure fair valuation of innovative therapies. For example, the National Organization for Rare Disorders (NORD) leverages its extensive patient network to gather insights on the patient experience and socialize these perspectives with key access stakeholders. Ultimately, proactive and persistent PAG advocacy with payers, in tandem with their educational efforts targeting the healthcare and patient communities, helps surmount reimbursement hurdles to make expensive rare disease drugs accessible to the patients who need them.

3.4   Advocating for Research and Policy

Beyond direct research, PAGs often advocate for government and other stakeholders on key issues. About 30% of surveyed PAGs advocate for research funding, resources, and policies. Most PAGs actively advocate for research funding, resources, and policies. The National Organization for Rare Disorders (NORD) collaborates with industry partners to support pivotal legislation like the Orphan Drug Act. This act provides key incentives for developing treatments for rare diseases, such as tax credits for clinical trial costs, waived FDA fees, and 7 years of market exclusivity. By working together to champion this legislation, NORD and its industry allies helped create a more favorable environment for rare disease drug development. Similarly, the EveryLife Foundation's "Guide to Patient Involvement in Rare Disease Therapy Development" (https://everylifefoundation.org/) serves as a roadmap for PAGs looking to engage in policy advocacy. The Foundation also offers programs like the "YARR Leadership Academy" to empower young adult rare disease advocates to become effective leaders and change agents. These examples underscore the multifaceted ways in which PAGs work beyond the research arena to shape policies that benefit the rare disease community. By harnessing their patient networks, deep disease expertise, and passion, PAGs can be formidable advocates on Capitol Hill and beyond.

3.5   Leveraging New Technologies and Collaborating to Enable Research

PAGs are leveraging emerging technologies and collaborative platforms to accelerate research and care for rare diseases. One such example is the RARE-X platform (https://rare-x.org), which provides PAGs with secure data storage, sharing, and analysis tools to facilitate patient data collection and research collaborations. Similarly, AllStripes (https://www.linkedin.com/company/allstripes/) offers a platform for patients to directly contribute their health records to a centralized database that researchers can access. These innovative data infrastructure solutions enable PAGs to more efficiently gather and manage patient data, which is critical for understanding disease natural history and developing targeted therapies.

PAGs are also actively participating in collaborative research networks to advance progress for their specific disease areas. The NIH's Rare Diseases Clinical Research Network (RDCRN) (https://ncats.nih.gov/research/research-activities/rdcrn) is one such initiative that brings together patient advocates, researchers, and clinicians to facilitate studies across over 200 rare diseases. The Chan Zuckerberg Initiative's "Rare as One" project is another example, providing selected PAGs with research grants and access to a collaborative network of experts and resources. These partnerships allow PAGs to tap into a broader ecosystem of expertise, infrastructure, and funding to drive research forward.

Moreover, PAGs are harnessing telemedicine and remote data collection technologies to break down geographic barriers and improve access to care and research participation. The COVID-19 pandemic has accelerated the adoption of these tools, enabling patients to connect with specialists, participate in clinical trials, and share health data from the comfort of their homes. By embracing these digital innovations, PAGs are not only enhancing the patient experience but also expanding the reach and efficiency of rare disease research efforts.

In conclusion, patient advocacy has been a driving force behind the progress made in rare disease research and orphan drug development. As the rare disease community continues to grow and evolve, patient advocates will remain essential partners in the ongoing effort to improve the lives of those affected by rare diseases worldwide.

Patient advocacy efforts have also been growing in low- and middle-income countries (LMICs), where the challenges of rare diseases are compounded by limited resources and access to healthcare. In India [4], the National Policy for Rare Diseases (NPRD) was introduced in 2017 to provide a framework for the management and treatment of rare diseases. However, the policy faced challenges in implementation and was later withdrawn. In 2021, the Indian government announced the formation of a National Rare Disease Policy to provide financial assistance for the treatment of rare diseases. Organizations such as the Organizations for Rare Disease India [11], the Indo-US Rare Diseases Initiative (IndoUSrare, https://www.indousrare.org/) [12] and Indian Organization for Rare Diseases (IORD, https://www.rarediseases.in/) [13] are umbrella organizations covering all rare diseases, working to raise awareness, provide support to patients and families, and advocate for policies to address the needs of the rare disease community in India and other LMICs.

However, it's important to note that not all rare diseases have the same level of resources or organizational infrastructure as the CFF. Many smaller, less well-known rare diseases may struggle to attract industry interest or secure adequate funding for research. This is where collaborative initiatives like the NIH RDCRN and the "Rare as One" project can play a crucial role in providing support and resources to help level the playing field.

Ultimately, while rare disease drug development often starts with the tireless efforts of patients and advocates, it requires the engagement and collaboration of all stakeholders—including industry, government, and academia—to truly thrive. By working together and leveraging the unique strengths and perspectives of each group, we can accelerate progress and bring new hope to the millions of people living with rare diseases worldwide.

4 Unique Aspects of Rare Diseases Research and Drug Development

Rare disease research and drug development face significant global disparities, with over 90% of efforts concentrated in less than 10% of the world's population, primarily in the US and EU. Despite 40 years of the Orphan Drug Act and over 1100

FDA-approved orphan drugs, 95% of rare diseases lack approved treatments. This chapter explores the unique challenges in rare disease research, including limited scientific literature, small patient populations, and difficulties in patient identification. It emphasizes the critical role of patient registries and natural history studies in understanding disease progression and designing clinical trials. The impact of the Human Genome Project on rare disease diagnosis and treatment is discussed, alongside emerging treatment modalities like gene therapy and innovative approaches using CRISPR and mRNA technologies. The chapter also addresses the potential and pitfalls of AI and machine learning in rare disease research, highlighting the risk of bias due to limited global data representation. It calls for more inclusive approaches to clinical trials, noting that over 80% of active rare disease trials are concentrated in the US and EU. The urgent need for unified, patient-centric research platforms and strategies to bridge the global divide in rare disease research and drug development is underscored.

There are over 1100 FDA-approved orphan drugs over the 40 years of the Orphan Drug Act of 1983. Yes, almost 95% of rare diseases still remain without a single approved treatment option. A majority of the approved orphan drugs are targeting rare forms of cancers. The hallmark of rare disease research is limited scientific literature on most rare diseases and very small patient populations that are sparsely distributed geographically.

The massive inequities in rare disease research are that over 90% of the research and drug development is restricted to less than 10% of the global population residing in the US and EU. The rest of the world, with most of the world's population, remains disconnected. The growing orphan drugs industry in the Western world is increasing this divide, which needs urgent attention in bridging the gaps.

Identification of patients with rare diseases remains a challenge everywhere. Newborn screening programs are helping identify about 80 diseases included in the universal screening panel (USP) at birth in the US. However, technologies capable of screening or diagnosing thousands of genetic diseases such as low-cost genome and exome sequencing, remain underutilized everywhere. As a result, millions of people suffering from rare diseases remain undiagnosed for 5–7 years on an average [5]. It is critical to screen, identify, diagnose, and engage patients with rare diseases early to ensure better patient outcomes.

4.1 Role of Patient Registries and Natural History Studies

Patient registries and natural history studies are vital for understanding the progression of rare diseases and identifying potential therapeutic targets. They serve as a repository for data that can support research and aid in the design of clinical trials. Lack of patient data in standardized digital formats that is properly consented for use in research is a major hindrance to rare disease research and drug development. Most patient registries and natural history studies are funded by and initiated in the Western world, leaving out most patients on the fence as they remain unaware of the need to be visible in these research databases. There is an urgent need to create such research databases for patients with rare diseases in the rest of the world and LMICs. In the context of clinical trials and drug development, a large number of data points have to be collected from a small number of patients, often requiring wearable devices, sensors, and continuous streaming data to monitor endpoints such as seizures.

4.2 DEIA (Diversity, Equity, Inclusion, and Accessibility)

In rare disease research, ensuring diversity in clinical trials is crucial to understanding variabilities across different populations. Equity in access to trials and treatments can be hindered by geographical, economic, and sociocultural barriers.

4.3 The Human Genome Project and Its Impact on Rare Diseases

The Human Genome Project has revolutionized our understanding of genetic underpinnings in rare diseases, improving diagnostic accuracy and enabling targeted gene therapies. About 80% of rare diseases have a causal genetic mutation. This era of low-cost whole genome and exome sequencing has opened new avenues for high-throughput screening, diagnosis, and personalized medicine in rare disease treatment.

4.4 Treatment Modalities for Rare Diseases

Treatment options for rare diseases often involve gene/cell therapy, enzyme replacement therapies, and repurposed drugs. Innovative approaches such as combination modalities (e.g., a small molecule and biologic), CRISPR, and mRNA technologies are being explored to expand treatment possibilities.

4.5 Boon and Bane of Data and AI

While applying AI and ML in rare disease research, one critical thing to watch out for is bias in the data [14, 15]. With most of the data currently available from the western population, any AI and ML algorithms are likely going to suffer from sampling bias and are unlikely to represent the global population affected by rare diseases. The solution to this problem is not easy and needs to begin with (1) inclusion of global patients in patient registries and other research databases; (2) helping LMICs and the rest of the world to digitize their rare disease populations by facilitating the creation of high-quality patient registries and natural history studies serving as a means to minimize the data bias.

5 Rare Disease Clinical Trials Statistics

As of June 2024, there are over 7600 active clinical trials for rare diseases. Over 80% of these clinical trials are in the United States and European Union. For e.g., the number of active rare disease clinical trials in the most populous country in the world, India, is <1%. This has been a fundamental and persisting problem in the industry begging to be addressed. A majority of the rare disease clinical trials also happen to be focused on rare cancers. The FDA Asha program has created a mechanism for US-based cancer clinical trials to have a site in India to increase the representation of Indian patients proportional to their general population [16]. Unified and inclusive patient-centric clinical research platforms that can support decentralized, hybrid, or traditional brick-and-mortar clinical research protocols can be an important part of the solutions.

6 Chemistry, Manufacturing, and Controls (CMC) in Drug Development

This section will focus on the basics of CMC, with special emphasis on rare disease drug development.

6.1 What Is CMC

CMC is an acronym for Chemistry, Manufacturing, and Controls, which are crucial activities when developing new pharmaceutical products. CMC involves defining manufacturing practices and product specifications that must be met to ensure product safety and quality consistency between batches. CMC initiates after a lead compound is identified through drug discovery and continues through all remaining stages of the drug development life cycle to approval and covers post-approval changes as well. CMC is an integral part of the overall drug development process, whether in common therapeutic or rare disease (RD) areas. Typically, CMC activities begin following the nomination of a lead and a backup candidate compound (CAN) from the discovery function of an organization.

Sponsors are required to ensure that the quality of their finished products during all phases of drug development is maintained. This CMC information is required in Module 3 of the Clinical Trials Application (CTA), such as the Investigational Medicinal Product Dossier (IMPD) in Europe and the Food and Drug Administration's (FDA's) Investigational New Drug (IND) application in the US, as well as EU Marketing Authorization Applications (MAA) and FDA New Drug Application (NDA) or biologics license applications (BLAs). This is applicable to the clinical and dossier submission procedures for other countries as well. Following approval of drug or biologic, post-approval changes to CMC attributes also need to follow specific regulatory guidance to report changes and actions as applicable.

Regulatory authorities need to see detailed CMC standards to ensure consistency of identity, safety, quality, stability, purity, and potency between the product used for clinical trials and product batches produced for commercial purposes on an ongoing basis. This CMC chapter will focus on the FDA regulatory pathway only.

6.2 CMC in Discovery

The three components that contribute to the lead compound (CAN) nomination from CMC perspective are (1). Molecule Design Intent (MDI), (2). Manufacturability Attribute (MA), and (3). Potential Genotoxic Impurity (PGI) Risk Assessment. MDI includes a list of physico-chemical properties that are linked to potential Critical Quality Attributes (CQAs) of the lead compound. MA refers to the factors/parameters that govern the potential of facile manufacturing and eventual scale-up of the lead compound to provide: (1). toxicology study material in pre-clinical, (2). clinical trial material (CTM) in development phase, and (3). commercial scale for marketed product. PGI risk is important to assess for lead compounds as it can provide an estimate for potential formation of nitrosamines in drug substances and DPs.

6.3 CMC in Pre-clinical Development

Some key CMC pre-requisites during the non-clinical phase are noted below [17]:

- Enough drug substance (usually milligram-gram) batches must be manufactured for all non-clinical animal toxicity studies.
- For Good Laboratory Practice (GLP) studies, the drug substance (DS) must be qualified or follow Good Manufacturing Practice (GMP) for each batch produced.
- Dose and administration (route and frequency) of DS should be determined. Also, a prototype formulation can be explored that will be used in clinics.
- Assessment of the key physicochemical attributes (e.g., stability, purity, solubility) of the investigational compound.

DMPK is an integral part of IND-enabling studies and includes in vitro studies such as permeability, protein binding and partitioning, and stability assays [18].

6.4   CMC in IND Development Phase (1–3)

6.4.1   DS and DP CMC Data for Phase 1 IND

The amount of information to be included in IND pivots on the following factors: (1). Phase of the investigation, (2). Known or suspected risks, (3). Novelty of the drug, (4). Previous studies conducted, (5). Dosage form/route of administration, (6). Nature & extent of clinical study, and (7). Patient population. Figure 2 below indicates how CMC fits in the overall drug development life cycle from Discovery to commercialization.

Table 1 below lists the key CMC aspects for DS and DP in Phase 1 IND. As phase appropriate approach, details are not needed in most cases at this stage of clinical development[3]:

6.4.2   DS and DP CMC Data for IND Phases 2 and 3

Table 2 below summarizes the CMC requirements for IND Phases 2 and 3 [19] As noted, in contrast to Phase 1 (Table 1), the phase-specific CMC requirements are more complex and comprehensive as the development proceeds from Phase 1 to Phase 3. IND CMC data are expected to be presented in the CTD format of 3.2.S and 3.2.P sections for DS and DP, respectively.

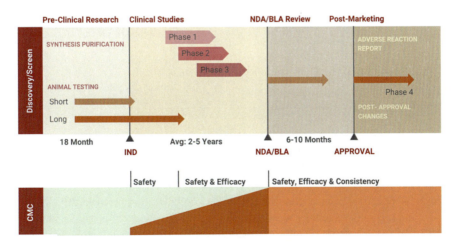

Fig. 2 CMC in Drug Development Phase (Guidance on CMC for Phase 1 and Phases 2/3 Investigational New Drug Applications Charles P. Hoiberg. DIA China, Beijing, China May 16–18, 2011. https://www.diaglobal.org/productfiles/25849/20110518/track4/11%20t4-4_chuck%20hoiberg.pdf)

Table 1 Summary of IND Phase I CMC Requirements for DS and DP

IND Phase 1 DS CMC Content	IND Phase 1 DP CMC Content
Reference to current edition of USP-NF, if applicable	Summary report containing the following items:
Authorized reference to a DMF, if applicable	Components and composition
Brief description, including physical, chemical, and biological properties	List of all components used in the manufacture of the investigational drug
Sufficient evidence to support chemical structure	Quality of inactive ingredients (e.g., USP/NF)
Manufacturer identified	Novel excipients additional CMC information may be needed
Method of preparation**	Brief summary of composition
Brief description of manufacturing process	Component ranges not needed
List of reagents, solvents, and catalysts	Manufacturer identified
Flow diagram –suggested	Method of manufacturing and packaging
** More information may be needed for well-characterized, biotechnology derived drugs and drugs extracted from human or animal sources	Brief description (sterilization process for sterile products, if utilized)
Specification: Proposed acceptance criteria supported by analytical data from clinical trial material	Flow diagrams –suggested
Brief description of analytical procedures	Specification: Proposed acceptance criteria supported by analytical data from clinical trial batch
Certificate of Analysis (COA) -suggested	Brief description of analytical procedures
Stability Brief description of stability study and analytical procedures used Preliminary stability data (tabular) may be submitted Detailed stability data not needed Stability protocol not needed	COA of the clinical batch -suggested DP Data
	Established specification or methods validation not needed
	Stability Brief description of stability study and analytical procedures used Preliminary tabular data -may be submitted Detailed stability data not needed Stability protocol not needed

6.5   Communication with FDA During Orphan Drug Development

The FDA encourages early communication and collaboration between drug developers and regulatory agencies to streamline the drug development process and facilitate regulatory approval. Few examples of engagement avenues with the FDA: pre-IND meeting, End-of-Phase 2 meeting, pre-NDA meeting, and as-needed communications [20].

Table 2 Summary of IND Phases 2 and 3 for DS and DP CMC content

IND Phases 2 and 3 DS CMC content	IND Phases 2 and 3 DP CMC content

Characterization and description
  More detailed description of the configuration and chemical structure for complex organic compounds
  Phase 3 complete description of the physical, chemical, and biological characteristics and supporting evidence to elucidate and characterize the structure
Manufacturer Phase 2 and 3 addition, deletion, or change of any manufacturer during Phase 1
  Also, include contract laboratories for quality control and stability testing
  Synthesis/method of manufacture and controls
  Starting materials (more information for Phase 3)
  Safety updates on reagents, solvents, auxiliary materials, and proposed changes identified during earlier phase(s) (more information for Phase 3)
  Synthetic and manufacturing process—general description (updated from a safety perspective, if changed) for Phase 2 and detailed description for Phase 3
  Flow diagram in-process controls reprocessing and pertinent controls
  Safety-related information for Phase 2 and description for Phase 3
Reference standard established in both phases
Specification Phase 2 and 3
  Complete description of the analytical procedure and supporting validation data ready for submission at drug Substance for Phases 2 and 3
  Any change in the tentative specification from earlier phase(s)
  List of the test method used test results, analytical data, and COA of clinical trial materials since original IND filing
  Specification Phase 3 impurities should be identified, qualified, and quantified, as appropriate suitable limits based on manufacturing experience should be established
	Components/composition/batch formula: Any change during earlier phase(s)
	  Established names and compendial status for components if any
	  Quantitative composition per unit dose
	  Batch formula
	  List components used and removed during the manufacturing of the DP for Phase 3
	  The formulations of certain DP delivered by devices (e.g., MDIs, DPIs, and nasal spray) should be similar to that intended for the marketed DP
	Specifications for components
	  Active: Any change during earlier phase(s)
	  Compendial inactive: Specify quality, if changed
	  Non-compendial Analytical procedures and acceptance criteria brief
	  Analytical procedures and acceptance criteria
	Brief description of manufacture and controls or an authorized reference to a DMF or NDA for Phase 2
	  Full description of the characterization, manufacture, control, analytical procedures, and acceptance criteria for Phase 3
	  Manufacturer any changes during earlier phase(s) including contractor
	  Method of manufacturing, packaging, and process controls
	  A brief general step-by-step description for the unit dose
	  Flow diagram information on specific equipment, packaging, and labeling process, in-process controls except for sterile products or atypical dosage forms not needed for Phase 2. Information on key equipment employed are needed for Phase 3
	  Reprocessing procedures and controls—safety-related information for Phase 2 and description for Phase 3.
	  Brief description of the packaging and labeling for clinical supplies for Phase 3
	Sterile products
	  Changes in the DP sterilization process
	  Other changes in the process to sterilize bulk drug substance or DP, components, packaging, and related items
	  Validation of the sterilization process is not needed
	Specification
	  Changes to specifications (tests and acceptance criteria)
	  Data updates on the degradation profile
	  Identification and qualification of degradation products for Phase 3

(continued)

Table 2 (continued)

IND Phases 2 and 3 DS CMC content	IND Phases 2 and 3 DP CMC content
Detailed list of tests performed General description of the USP analytical procedures Complete description of the non-USP analytical procedures with validation data container/closure brief information for Phase 2 and more detailed information for Phase 3 Stability: Phase 2 Stability-indicating method Stability protocol preliminary Stability data on a representative drug Substance for Phases 2 and 3 All stability data for the clinical material used in Phase 1 Phase 3 Detailed stability protocol Detailed stability data Stress studies—should be conducted	Summary table of the test results, analytical data, and COA for the lots used in clinical studies Container closure system Updates on information previously filed The container closure system of certain DPs delivered by devices (e.g., MDIs, DPIs) should be similar to that intended for the marketed product Name of the manufacturer and supplier DMF reference and authorization, if available Additional information may be recommended for atypical delivery systems (e.g., MDIs, disposable injection devices) Stability Stability protocol (detailed protocol for Phase 3) Preliminary stability data based on representative material for Phase 2 and detailed stability data for Phase 3 Phase 3 all available stability data for the clinical material used in earlier phase(s) Stress testing results for Phase 3 Container closure integrity tests for sterile products, where applicable discussion for Phase 3

6.6 CMC for RD Drug Development (RDDD)

It is expected that orphan drug CMC manufacturing development should move in parallel with clinical development. There are various CMC challenges encountered by the sponsors in the RD space. The key ones are highlighted below:

- Orphan drugs require unique supply chains (for direct access to patients), which lead to highly patient-centric personalized approach.
- Accelerated transition from clinical data to market (often including a compassionate-use program) usually maintains the process used for clinical supply unaltered through commercial launch. This dictates a very integrated complex approach to research, development, and manufacturing.
- The need for low-volume orphan drugs limits the number of batches manufactured and thus can make it challenging to build a significant data history of critical process parameters to optimize manufacturing yield.
- Cell/ gene therapies, antisense oligonucleotides, and therapeutic proteins are some of the complex technology platforms used by RD. However, the associated manufacturing technologies can be novel and highly specialized, with significant cost, time, and resource implication for the sponsor.
- Limited drug substance availability in the early stage makes establishing a reliable and consistent supply chain challenging, which can impact the CMC strategy. For example, the key global regulatory bodies require 3 distinct DS and 3

separate DP batches to perform confirmatory testing of standard nitrosamines and Nitrosamine Drug Substance Related Impurities (NDSRIs). However, such a number of batches may not be available at the time of dossier submission due to low clinical trial material (CTM) requirements. Often, CMC needs to partner with clinical supply chain to secure limited materials to run tasks that are required to meet some country-specific submission requirements.

- Keeping up with evolving regulatory requirements to file in global markets is a major challenge for small to mid-size RD companies due to the varied CMC HA requirements. Often, the sponsor may not be aware or have not anticipated the specific CMC requirements for a particular launch country due to lack of internal regulatory CMC global expertise. This can delay the submission process such as requirements for running specific stability programs for DS and DP, In-country Analytical Testing (ICAT), mandatory number of process validation batches, bridging studies etc.

- For the same RD company to work on multiple modalities (e.g., small molecules, biologics, and gene therapy) could pose a challenge for CMC, as it is difficult to have the same SMEs with equal bandwidth in all modalities (e.g., manufacturing, analytical, and formulation). This can put constraints on the CMC program as the company needs to then hire new SMEs, seek consultants, or build internal SME knowledge as part of employee development plans—all of which require time and budget.

- The DP must be developed in patient-centric dosage mode to suit a wide range of patient needs. The majority of RD drugs are meant to treat both adults and children. Medication for an RD indication in children has additional complexities throughout development, requiring specialist knowledge to effectively develop a formulation that is palatable and suitable for pediatric patients.

- The cost recovery of a full CMC development program of a new technological platform for a very RD is challenging. Developing an orphan drug is expensive due to the limited patient population globally and the need for specialized manufacturing processes. This can make it challenging to invest in the necessary CMC activities, such as analytical method development and validation, stability studies, and process validation. However, regulators are willing to work closely with manufacturers to tailor a CMC registration package that is 'fit for purpose' in a disease with high unmet need while maintaining product quality.

- Increasing competition in recent years in a few specialized, established rare-disease areas, such as lysosomal storage disorders, can also be a roadblock. Orphan drug designation establishes a period of 7-year exclusivity for companies that are first to market with a particular treatment modality for a specific indication. Other competing companies can break that exclusivity period if their new treatments are significantly different and offer a superior clinical benefit. So, a company needs a clear rationale to demonstrate that their new drug really will be better for patients for a new CMC development program with pre-existing treatment.

- As mentioned earlier, RD indications require low-volume production during development and commercialization, with fast-changing demands. However, many small RD companies do not have the capability to maintain their own manufacturing facility. Hence, finding a robust scale-up and manufacturing partner that can handle these demands is vital. This is also applicable for dossier submission assistance, as companies must seek the expertise of a local consulting agency (e.g., a US based RD company needing guidance from a local consulting firm in Japan to ensure right first time JNDA submission to PMDA). The same applies for individual consultants often utilized for CMC troubleshooting on complex modalities. These add up to the drug development CMC cost significantly.

- Orphan drugs often require complex manufacturing processes for biologics and gene therapy, which can increase the risk of variability and batch-to-batch inconsistencies. This can make establishing a robust and reproducible manufacturing process challenging, which is critical to ensuring consistent product quality.
- The larger amounts of drug requirements during Phases 2 and 3 may modify the manufacturing procedures and purification methods. Also, manufacturing responsibilities can transition after the initial nonclinical studies and/or clinical investigations (e.g., from a single investigator to an organization, from a small to large organization), which is common for RD. Any of these changes (even changes expected to be minor) can lead to unpredicted modification to drug characteristics (e.g., drug impurities, physical-chemical characteristics of proteins, etc.) as well. If the differences are significant, they may warrant additional nonclinical and clinical studies, which will add up time, cost, and resources.
- Changes in CMC critical quality attributes of the planned commercial drug after the clinical investigations might raise concerns that the safety and effectiveness findings do not apply to the newly manufactured drug. These concerns could drive additional studies (nonclinical, clinical, or both) to address the concern before marketing approval. Given the wide variety of RD drugs, some of which are complex, FDA advises sponsors to consult relevant guidance for industry.
- The type, number, and level of impurities in a drug used in clinical investigations and for commercial distribution should be comparable to the drug batches used in toxicology studies. Changes might raise concerns that the drug used in later clinical investigations has unknown toxicological characteristics. Additional toxicology studies may then be needed to evaluate the newly produced drug, delaying the clinical development program.
- Sponsors may need more guidance on the CMC strategy for orphan drugs, which can make it challenging to navigate the regulatory requirements. This can increase the risk of regulatory approval delays and impact the drug's commercialization timeline.
- Orphan drugs may require specialized storage and shipping requirements due to their inherently sensitive nature, impacting the drug's stability and shelf life. This can make establishing appropriate storage and shipping conditions challenging, which can impact the drug's overall quality.

FDA may accommodate certain flexibility on the type and extent of manufacturing information in submission and approval for certain CMC components (e.g., stability data updates, process validation strategies, inspection planning, manufacturing scale-up). The extent of flexibility is based on the regulator's confidence in the developed drug after considering factors such as (1). product critical quality attributes, (2). seriousness of the condition and medical need, (3). manufacturing processes, (4). the robustness of the quality system, and (5). the strength of the sponsor's risk evaluation and mitigation system (REMS).

6.7 CMC for Pediatric Drug Development

Developing age-appropriate formulations is necessary to maximize efficacy and design quality, promote safety, minimize risks, and increase patient adherence to treatments [21].

The CMC requirement for an ideal dosage form of pediatric medicines should be taken into consideration: (1) The amount of the DS is adjusted to the age needs of the child, and thus the intended dose volume and size are tailored to the target age group. (2) The acceptability of the dosage form. (3) The palatability of the DS, which may influence the choice of dosage form and its design. Ideally, the desired dosage form should be palatable by itself. However, adding excipients in the formulation is required for taste-masking purposes. (4) Minimum dosing frequency to guarantee the adherence to the dosing scheme both by caregivers and by older children. (5) The end-user needs such as water accessibility, which is important when a medicine needs to be dissolved, diluted, or dispersed prior to administration. (6) The geographical and cultural/socio-economic differences that may impact the preferred tastes and flavors.

6.8 CMC Innovation Platforms

Innovation platforms for research often involve the sharing of resources. Companies, federal agencies, and nonprofit patient groups are taking the initiative to build such new models for drug development for both common and RD. This section highlights such models [22–24].

6.8.1 FDA

FDA's Critical Path Initiative, CDRP program [25], ARC program [26], INTERACT program [27], START program [28], and several other programs [29] are available to sponsors for assistance in RD and CMC areas.

6.8.2   Repurposing Existing Drugs

Another innovation involves repurposing (also called drug repositioning, drug rescue, reprofiling, retasking, or therapeutic switching) old drugs for potential treatments of RD [24]. Without the need to repeat toxicological or pharmacokinetic assessments, a considerable portion of the costs of R&D pipeline can be saved. Furthermore, population safety, dosing, and adverse events are already known.

6.9   Role of CDMO in RDDD

Finding a single vendor who can execute development, CTM manufacture, and commercialization is the ideal scenario for orphan drugs. The CDMO should also have adequate bandwidth to reduce the cost of CMC, tackle regulatory queries, and accelerate CMC if expedited approval is granted. Several challenges include sponsor/CDMO interactions, including unexpected incidents, adherence to timelines, unexpected costs, and relationship management. A large, integrated CDMO can offer advantages to a sponsor, as it is likely to have access to an international network of state-of-the-art facilities, alongside a team of highly skilled, experienced technical experts.

6.10   CMC Due Diligence for In-licensed Compounds

A sponsor must do a comprehensive phase appropriate CMC due diligence prior to acquiring a molecule via paper-based exercise on information provided by the vendor or post-acquisition. This should be a part of sponsor's CMC Risk Evaluation and Mitigation Strategy (REMS) and involves identifying gaps in data or processes that could affect the commercial viability of DS and DP. This ensures that a given compound meets the technical, quality, regulatory compliance, and financial pragmatic elements essential for successful commercialization.

6.11   Artificial Intelligence (AI) in CMC

AI can assist CMC in the following areas [30]: (1) data accuracy and faster analysis, (2) improved analytical method development, (3) seamless integrated drug development workflow, (4) Time and cost-efficient production, (5) proactively minimize human errors in processes, (6) automate processes, (7) supply chain robustness to ensure meds reach patients timely, (8) forecast product quality, (9) robust quality control, (10) formulation optimization, (11) enhance informed and accurate decision-making in complex CMC scenarios, (12) reduce manufacturing waste,

(13) robust, resilient, and adaptive CMC strategy, (14) optimize energy consumption, (15) assist in reg-CMC dossier submission data package preparation, (16) safer drug with less side effects, and (17) higher yield of chemical synthesis, and (18) predict equipment failure.

While AI can assist in CMC processes, it must maintain the fundamental human factor in every aspect of its usage. Hence, a balanced approach between human expert involvement and AI automation may lead to the most productive and effective outcomes.

6.12   In-Silico CMC

In-silico CMC modeling is widely gaining popularity in the pharmaceutical industry in recent years. Various types and scales of modeling, including statistical DOE, principal component analysis (PCA)/partial least square (PLS), bioprocess, molecular modeling, mechanistic models based on first principles, and combination of models (where computational fluid dynamics (CFD), mechanistic chromatography modeling, molecular biophysics, and plant simulation) are explored and used in industry [30, 31].

7   Regulatory Pathways for Orphan Therapies

The US Food and Drug Administration (FDA) has several regulatory pathways to help orphan drugs get to patients faster, including expedited programs and grant funding:

- Fast Track: Expedites development and review, including rolling reviews
- Breakthrough Therapy: An expedited program
- Accelerated Approval: Uses surrogate endpoints as measures of clinical benefit
- Priority Review: Specifies an expedited review timeline of 6 months
- Orphan Products Clinical Trials Grant: A unique grant funding opportunity
- Humanitarian Use Device (HUD) program: An alternative pathway for medical devices that treat or diagnose conditions that affect 8000 or fewer people in the US

The FDA also facilitates patient-focused drug development (PFDD) meetings to gather patient experience data. The Center for Drug Evaluation and Research (CDER) reviews and approves new drug applications (NDAs) for orphan drugs, with review divisions organized by therapeutic areas.

To obtain orphan designation or a fast-track designation, clinical development plans must clearly outline and support the intent to treat a rare disease population.

However, orphan drug development can be more expensive and time-consuming than non-orphan drugs, and only about 17% of orphan drugs get regulatory approval and enter the market.

The US FDA has designated 7169 drugs as orphan and approved 1237 orphan drugs between 1983 and June 23, 2024. The FDA CBER has approved 37 cell/gene therapy products as of April 2024. These therapies have the potential for long-term relief from diseases such as blood cancers, hemophilia, sickle cell disease, thalassemia, and other severe conditions such as spinal muscular atrophy (SMA) and Duchenne muscular dystrophy.

8 Commercialization of Orphan Drugs

The commercial launch sequence for new drugs usually starts with an initial launch in the US following FDA review, followed by a launch in the EU, followed by EMA review, and then in Japan, following JPA review. There is a long tail before the drugs are launched in the rest of the countries including the more populous and low-and-middle-income countries (LMICs). Without global engagement for clinical R&D, it is unlikely to get commitment from biopharmaceutical sponsors to launch orphan products globally. It is inequitable, unfair, and unacceptable for orphan drugs to be developed entirely in the western world and leave 90% of the patients on the fence.

Pricing Orphan drugs and cell/gene therapies remain more expensive for 99% of the world's population to afford, calling for new models of R&D and commercialization. Value-based pricing, insurance reimbursement, and national health systems (governments) reimbursement are all possible when ultimately affordability and universal access are considered.

9 Future Directions

Orphan drug development has greatly accelerated in the US and EU due to regulatory frameworks such as the Orphan Drug Act of 1983 and the overall biopharmaceutical development ecosystem. Orphan drugs remain inaccessible to 90% of the world's population due to various barriers such as regulatory, geographic, supply chain, lack of patient data registries, economics, and lack of cross-border collaborations. Patient advocates in the US and EU have led the orphan drug revolution in the last 40 years since the Orphan Drug Act of 1983. It is now time to extend and engage patient advocates in the rest of the world globally to ensure the economies of scale can help accelerate the orphan drugs development process in a more equitable and universally accessible manner.

We need a new generation of empathetic, visionary leaders at large biopharmaceutical sponsors and patient-centric, innovative biotechnology companies that recognize the need to view each rare disease as a global public health issue to bring about meaningful change to achieve true diversity, equity, inclusion, and access.

Additional Reading by the Author

AI and the Lack of Diversity in Data: Implications and the Path Forward for Rare Disease Research [6].

AI's Role in Advancing Rare Disease Research [32].

References

1. Korth-Bradley JM. Regulatory framework for drug development in rare diseases. J Clin Pharmacol. 2022;62(Suppl 2):S15–s26.
2. Ramanan S, Dave K. Orphan drug development for rare diseases. Regulatory Affairs Professional Society; 2021.
3. US-FDA. Guidance Documents for Rare Disease Drug Development 2023 [Available from: https://www.fda.gov/drugs/guidances-drugs/guidance-documents-rare-disease-drug-development.
4. Chirmule N, Feng H, Cyril E, Ghalsasi VV, Choudhury MC. Orphan drug development: challenges, regulation, and success stories. J Biosci. 2024;49
5. Vashishta L, Bapat P, Bhattacharya Y, Chakraborty Choudhury M, Chirmule N, D'Costa S, et al. A survey of Rare Disease awareness among healthcare professionals and researchers in India. BioRxiv. 2023;April.
6. Rajhamsa RK. AI and the lack of diversity in data: implications and the path forward for rare disease research. MedTech Intelligence [Internet]. 2023. Available from: https://medtechintelligence.com/feature_article/ai-and-the-lack-of-diversity-in-data-implications-and-the-path-forward-for-rare-disease-research/.
7. Patterson AM, O'Boyle M, VanNoy GE, Dies KA. Emerging roles and opportunities for rare disease patient advocacy groups. Ther Adv Rare Dis. 2023;4:26330040231164425.
8. Chopra M, Modi ME, Dies KA, Chamberlin NL, Buttermore ED, Brewster SJ, et al. GENE TARGET: a framework for evaluating Mendelian neurodevelopmental disorders for gene therapy. Mol Ther Methods Clin Dev. 2022;27:32–46.
9. Stein S, Bogard E, Boice N, Fernandez V, Field T, Gilstrap A, et al. Principles for interactions with biopharmaceutical companies: the development of guidelines for patient advocacy organizations in the field of rare diseases. Orphanet J Rare Dis. 2018;13(1):18.
10. Panahi S, Rathi N, Hurley J, Sundrud J, Lucero M, Kamimura A. Patient adherence to health care provider recommendations and medication among free clinic patients. J Patient Exp. 2022;9:23743735221077523.
11. Rajasimha HK, Shirol PB, Ramamoorthy P, Hegde M, Barde S, Chandru V, et al. Organization for rare diseases India (ORDI) – addressing the challenges and opportunities for the Indian rare diseases' community. Genet Res (Camb). 2014;96:e009.
12. Khera HK, Venugopal N, Karur RT, Mishra R, Kartha RV, Rajasimha HK. Building cross-border collaborations to increase diversity and accelerate rare disease drug development – meeting report from the inaugural IndoUSrare Annual Conference 2021. Ther Adv Rare Dis. 2022;3:26330040221133124.
13. Krishnaraj P, Rajasimha HK. Cross-border rare disease advocacy: Preethi Krishnaraj interviews Harsha Rajasimha. Dis Model Mech. 2024;17(6)
14. He D, Wang R, Xu Z, Wang J, Song P, Wang H, et al. The use of artificial intelligence in the treatment of rare diseases: a scoping review. Intractable Rare Dis Res. 2024;13(1):12–22.
15. Sankar A, Ravi Kumar YS, Singh A, Roy R, Shukla R, Verma B. Next-generation therapeutics for rare genetic disorders. Mutagenesis. 2024;39(3):157–71.

16. US-FDA. FDA – Asha Program (Collaboration to increase cancer clinical trial access in India)2024. Available from: https://www.fda.gov/about-fda/oncology-center-excellence/project-asha.
17. Eupati OC. Key aspects of Chemistry, Manufacturing, Control (CMC) during non-clinical development 2000 [Available from: https://learning.eupati.eu/mod/book/view.php?id=308&chapterid=153.
18. WuXi-AppTec. What Is IND-Enabling Testing & What Does It Include? 2025 [Available from: https://labtesting.wuxiapptec.com/2022/08/01/what-is-ind-enabling-testing-what-does-it-include/.
19. Hoiberg CP. Guidance on CMC for Phase 1 and Phases 2/3 Investigational New Drug Applications, DIA meeting Proceedings. 2011 [Available from: https://www.diaglobal.org/productfiles/25849/20110518/track4/11%20t4-4_chuck%20hoiberg.pdf.
20. US-FDA. Formal meetings between the FDA and sponsors or applicants of PDUFA products guidance for industry 2025 [Available from: https://www.fda.gov/media/172311/download.
21. Domingues C, Jarak I, Veiga F, Dourado M, Figueiras A. Pediatric drug development: reviewing challenges and opportunities by tracking innovative therapies. Pharmaceutics. 2023;15(10)
22. Amsberry K, Lai Y. Navigating early cell and gene therapy: CMC perspectives, Cardinal Health 2025 [Available from: https://www.cardinalhealth.com/en/services/manufacturer/bio-pharmaceutical/drug-development-and-regulatory/resources-for-regulatory-consulting/cmc/cmc-perspectives.html.
23. Cauchon NS, Oghamian S, Hassanpour S, Abernathy M. Innovation in chemistry, manufacturing, and controls—a regulatory perspective from industry. J Pharm Sci. 2019;108(7):2207–37.
24. Field MJ, Boat TF. Development of new therapeutic drugs and biologics for rare diseases and orphan products: accelerating research and development rare diseases. https://www.ncbi.nlm.nih.gov/books/NBK56179/: National Academies Press (US); 2010.
25. US-FDA. Chemistry, Manufacturing, and Controls Development and Readiness Pilot (CDRP) Program 2025 [Available from: https://www.fda.gov/drugs/pharmaceutical-quality-resources/chemistry-manufacturing-and-controls-development-and-readiness-pilot-cdrp-program.
26. US-FDA. Accelerating Rare disease Cures (ARC) Program 2025 [Available from: https://www.fda.gov/about-fda/center-drug-evaluation-and-research-cder/accelerating-rare-disease-cures-arc-program.
27. US-FDA. OTP INTERACT Meetings 2025 [Available from: https://www.fda.gov/vaccines-blood-biologics/cellular-gene-therapy-products/otp-interact-meetings.
28. US-FDA. Getting START'ed: New FDA programs aim to advance drugs to treat rare diseases 2023 [Available from: https://www.hoganlovells.com/en/publications/getting-started-new-fda-programs-aim-to-advance-drugs-to-treat-rare-diseases.
29. Henderson L. DIA 2023: FDA Rare Disease Town Hall 2023 [Available from: https://www.appliedclinicaltrialsonline.com/view/dia-2023-fda-rare-disease-town-hall.
30. von Stosch M, Portela RMC, Varsakelis C. A roadmap to AI-driven in silico process development: bioprocessing 4.0 in practice. Curr Opin Chem Eng. 2021;33:100692.
31. Roush D, Asthagiri D, Babi DK, Benner S, Bilodeau C, Carta G, et al. Toward in silico CMC: an industrial collaborative approach to model-based process development. Biotechnol Bioeng. 2020;117(12):3986–4000.
32. Rajhamsa RK. AI's role in advancing rare disease research 2023. Available from: https://www.clinicalleader.com/doc/ai-tools-for-advancing-rare-disease-research-0001.

N = 1 Drug Development Pipeline for Rare Diseases

Yiwei She

Abstract Advancements in genetic sequencing technologies have brought about a new era in rare disease biology, facilitating the identification of an increasing number of such conditions. Scientists and patient advocates have begun tackling rare diseases with patient populations as low as one to two patients. Drug development for such diseases requires a personalized approach, often involving N = 1 clinical trials, which need to customize and alter study designs and statistical techniques commonly employed in standard population-based clinical trials. N = 1 trials focus on a single patient, posing challenges to extrapolating the results to a broader population. The limited patient size associated with rare diseases often results in reduced interest from pharmaceutical companies and academic researchers in pursuing such clinical trials. This often results in families of patients with rare diseases to take responsibility of designing and adapting the trial to one subject. In this chapter, we discuss drug development pipelines for rare diseases, with an emphasis on a case study involving a family engaged with developing a drug for their son with a rare genetic disorder.

Keywords Rare diseases · Personalized medicine · Genetic mutation · N = 1 trials · *TNPO2* gene · Anti-sense oligonucleotides (ASOs) · Neurodevelopmental delay

1 Introduction

N = 1 drug development, also known as personalized medicine or precision medicine, is a tailored approach to drug development and treatment that focuses on the individual. Such trials are relevant for developing drugs to treat rare and ultra-rare genetic diseases.

Y. She (✉)
TNPO2 Foundation, Sacramento, CA, USA
e-mail: yiwei@tnpo2.org

© The Author(s), under exclusive license to Springer Nature Switzerland AG 2025
N. Chirmule, V. V. Ghalsasi (eds.), *Approved: The Life Cycle of Drug Development*,
https://doi.org/10.1007/978-3-031-81787-8_13

As with trials for common diseases, N = 1 trials are aimed at developing a safe and effective treatment for patients. Therefore, such trials leverage study design and statistical techniques commonly used in standard population-based clinical trials, and both drug development pipelines consist of common steps [1].

The first step involves understanding enough of the disease biology to design a drug that would prevent or reverse disease progression. The next step is to identify a suitable modality or platform to address the root cause of the disease. Some platforms or modalities include monoclonal antibodies, cell therapies, and gene therapies. One or more modalities are chosen based on the disease biology and lead compounds that have the potential to treat the root cause of the disease are identified.

Once this is ready, it is subjected to preclinical trials in several biological systems, such as flies, mice, and patient cell lines. These studies are aimed at analyzing the pharmacology—pharmacokinetics and pharmacodynamics—of the drug. Additionally, they also assess the safety profile of the drug and establish the tolerated dosage. After finalizing the lead compound through preclinical testing, it must be formulated to ensure its stability and manufactured without compromising its purity.

Once preclinical trials are successfully completed, researchers can file for an investigational new drug (IND) application to regulatory agencies to obtain approval for human trials.

N = 1 trials differ from standard clinical trials in that the trial is conducted in a single patient rather than a population cohort. Such a trial can generate high-level data for an individual. However, since they are conducted in only one patient, N = 1 trials do not assess inter-individual heterogeneity. As such, data from one person may be difficult to generalize to other patients.

Since the patient population is too small to support any economic incentive, organizations from academic labs to small biotechnology companies to large pharmaceutical companies all generally hesitate to design drugs for these patients.

Despite this, families of patients with ultra-rare diseases take on the challenge by designing and adapting standard clinical trials for one subject. One such family is that of Yiwei She, whose son, Leo, was diagnosed with an ultra-rare disease a few months after his birth.

2 The Diagnostic Odyssey

2.1 First Signs

During a routine 33-week ultrasound appointment in 2021, Yiwei She noticed that the ultrasound technician spent a long time trying to measure the baby's head circumference. The technician left the room, probably to consult the obstetrician. When the obstetrician did not raise any alarms, the patient left to go home and take care of her first child, a happy and energetic one-year-old.

Baby Leo was born on October 4, 2021, with height, weight, and head circumference all in the less than third percentile. Two weeks later, at a regular checkup the pediatrician diagnosed Leo with microcephaly—his head circumference was "off-the-charts" too small. Moreover, it was growing at an alarmingly slow rate. This manifestation was a symptom with only a 10% chance of being idiopathic (where there is a family history of benign microcephaly). The other 90% of cases foretold moderate to severe intellectual disability for Leo's future.

The family followed the standard workup for this symptom: with test after test; labs, CT scans, MRIs, consultations with specialists (including brain surgeons), all returning little additional information. Leo's symptoms increasingly pointed toward a genetic root cause, in which case according to his pediatrician: "nothing could be done."

2.2 Getting to an Actionable Diagnosis

Leo spent the next few weeks getting tested for several other causes of microcephaly. At 11 weeks old, he suffered from his first full-body seizure, which resulted in him being prescribed the first-line anti-epileptic drug—Keppra (generic name Levetiracetam).

Two months later, on a review of his seizure presentation, doctors diagnosed him with infantile spasms. These are very often caused by genetic mutations, with over 100 different causative genes [2]. The family immediately enrolled Leo in a clinical trial to carry out triband sequencing for infants with epilepsy. Four months after he was born and almost 2 months after his first seizure, Leo was given a genetic diagnosis—he had a pathogenic mutation that caused guanine to be replaced with adenine (G > A) at c.466, in one copy of his *TNPO2* gene. At the protein level, this replacement resulted in converting the amino acid aspartate (D) to arginine (N) at the 156th position (D156N).

3 *TNPO2* Gene

The gene *TNPO2* codes for a protein called transportin-2. This protein is responsible for shuttling cargo proteins, between the cell's nucleus and cytoplasm (Fig. 1) [3]. Many of these proteins are involved in cells of neuronal development. Public datasets show this protein is highly expressed in the brain [4].

As of early 2022, there were fewer than 20 patients documented with pathogenic mutations in this gene. These patients usually present with global intellectual developmental disorder accompanied by low muscle tone and impaired speech. All of the studied pathogenic mutations in *TNPO2* gene cause a single amino acid residue variation in the protein. These missense mutations often cause gain-of-function

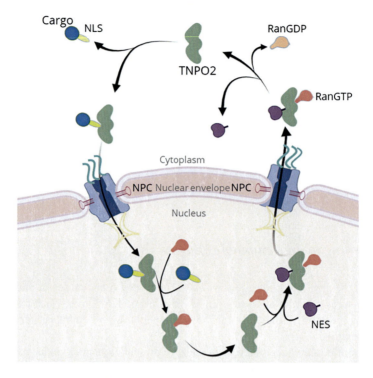

Fig. 1 Model of protein nuclear import and export via the transportin-2 (*TNPO2*). Imported cargoes containing nuclear localization signals (NLSs) form complexes with *TNPO2* in the cytoplasm, enter the nucleus through the nuclear pore complex (NPC), and are dissociated from the importins with the aid of RanGTP. Nuclear export of cargoes starts with the formation of trimeric complexes consisting of *TNPO2*, nuclear export signal (NES)-containing cargo, and RanGTP. The trimeric complex transits through the NPC and is dissembled in the cytoplasm upon the hydrolysis of RanGTP. (Adapted from [5])

effects, where the altered gene product causes toxicity. However, the specific molecular pathogenic mechanism is still unknown [4, 6].

Among these mutations, the variant D156N, that Leo carries, causes significant developmental toxicity [3]. Leo became the second-known person in the world to carry this variant with a severe prognosis, making him a patient suffering from an ultra-rare disease.

4   Rare Diseases

The Rare Diseases Act of 2002 defines rare diseases as those that affect fewer than 200,000 people in the United States [7]. There are over 10,000 known rare diseases that affect about one in 10 people, or 30 million people in the United States [8].

Although the cause of several rare diseases is unknown, many (including over 80% of pediatric rare diseases) are caused by genetic mutations, as in Leo's case.

The only other person in the world known to carry the same genetic mutation as Leo was a six-year-old boy from the Netherlands, also known as Proband 6. Leo's family contacted Proband 6's family and discovered that Leo's symptoms closely match those of this boy's. Preclinical research in *Drosophila* flies and other animal models suggested that the D156N mutation is the most toxic among all the variants studied [3].

The small patient population and paucity of basic research around the gene made it clear that no one else would do anything to help Leo. That's when Leo's family decided to go ahead and try to develop a drug for Leo themselves.

Three months after receiving Leo's diagnosis, the family incorporated the TNPO2 Foundation, a nonprofit to organize the growing research effort to help Leo and other children like him. For more information about the TNPO2 Foundation, see https://www.tnpo2.org.

5   The Next Step

Upon the advice of their friends and mentors, Leo's family contacted patients who had experience tackling the same problem they had encountered. They connected with parents of other children who were diagnosed with rare diseases, and who had undertaken the challenge to attempt a therapeutic rescue. "We learned from setbacks by saying, this is what not to do," said the family. "And we learnt from success by saying, this is what we emulate, and try to improve upon."

They drew inspiration and hope from Julia Vitarello, whose daughter Mila was the first person to receive an individualized drug for her rare neurodegenerative disease. Researchers designed and administered a drug for Mila at record speed, condensing a timeline of about a decade into a few months [9]. "This set a timeline. They went from diagnosis to treatment in the clinic in 10 months. Maybe we could even hope to improve on the 10 months," the family said. They hoped the trail opened by others could help Leo and those who came after.

Many parents of children with ultra-rare conditions helped Leo's family. The family of Michael Pirovolakis, who suffers from spastic paraplegia (SPG50), guided Leo's parents toward experts who could help them. They consulted several other families including Valeria's (KCNT1), who had successfully reached the clinic for her toxic gain-of-function mutation that prevented any cognitive or motor development. Lydia Seth (KCNQ2) suffered from a dominant negative mutation, which is a particularly severe type of gain of function mutation. Since Leo was also suspected to suffer from a gain of function mutation, the family learned valuable lessons from Valeria's and Lydia's families. "The open sharing of their journeys left a trail and the generosity of these patient families will always be appreciated," said Leo's mom.

5.1   Stronger Together, Learning from Others, and the Network

The open sharing of information on the internet and social media helped Leo's journey much as it helped those patients who came before. Leo's family joined a group on Facebook that was dedicated to patients suffering from *TNPO2* mutations. The small group consisted at first, of five to six families of children with *TNPO2* mutations and a couple of researchers. By the time the Leo's drug reached the clinic, there were over 15 patients and families.

There, Yiwei met A.M., the mother of the first child ever to be diagnosed with a *TNPO2* mutation. A.M.'s grit led her to achieve a precision diagnosis for her child. Along the way, A.M.'s efforts got the first study of pathogenic *TNPO2* mutations published in 2021. [3] Leo's family could connect with Proband 6 only due to this study and this publication. Connecting with parents of other children with *TNPO2* mutations helped Leo's family understand his disease and potential prognoses, it also helped them understand what would be involved in attempting a rescue by precision medicine.

5.2   Consulting the Experts

While talking to other parents helped her understand Leo's disease, Leo's family knew that they needed to look for experts in the field who could design and develop a drug for Leo. They reached out to experts and leading physician scientists. Friends and friends of friends helped them connect with academic scientists and biotech companies that specialized in designing and developing the types of drugs that could be personalized for Leo.

However, not everyone could help; a research lab that had previously designed an individualized treatment for a child with a rare genetic disease was skeptical about taking Leo's case because they were concerned about the disease biology and Leo's rescuability. But some other scientists did offer to help.

After the wet lab experiments were in progress, the family began reaching out to parents of other children with rare diseases to learn more about disease biology. They also started reading up and looking for relevant information sources.

5.3   Literature Review and Bioinformatics

Reaching out to other parents helped the family collect a lot of data about rare diseases that was not publicly available. "They were all willing to share data with us because we were Leo's family. Not a drug developer, not a research scientist, but parents," they said.

Aside from this personalized data, the family scoured through other resources that could help them understand Leo's condition and how to treat it. This included both non-peer-reviewed articles in blogs and peer-reviewed journal articles. The family went through large public -omics datasets that were made available to citizen scientists to analyze.

Although some papers were behind paywalls, most other information resources were open, so the family could not only access literature but also gain a better understanding of biology. Being from a mathematics background, the family had to start from the basics of biology.

The family compiled a short document detailing the biology of the disease. Here, they included all the data they had collected so far. They used bioinformatics tools from the Baker Lab and AlphaFold to model the mutated protein in silico. While studying more about the mutated protein, they found some useful information about nonsense mutations, the kind that cause premature termination of translation, leading to a partially or completely inactivated protein. Coarse "high throughput" studies from mice had shown that loss of the *TNPO2* gene does not result in a phenotype [10]. Additionally, data from six humans (from the UKBB database) with a frameshift mutation in their *TNPO2* gene indicated that they showed no disease phenotype [11]. This data, along with Leo's symptoms of seizures and microcephaly strongly indicated that he had a toxic gain of function mutation.

Reading further about treating gain of function mutations, the family found out more about how drugs for one disease could be repurposed for some other diseases. While doing a literature survey of drugs, they also found out more about the kind of drugs that would eventually help Leo—antisense oligonucleotides, or ASOs.

6   ASO in Rare Diseases

ASOs are short, synthetic, single-stranded DNA fragments that can bind to specific portions of the RNA to alter protein expression (Fig. 2). ASO-mediated therapies have a higher chance of success than drugs that target products later in the biological process. This is mainly because ASOs can target the source of pathogenesis—the mutation [12].

The team treating Leo suspected (with good evidence) that targeting the mutated copy of the *TNPO2* gene to reduce the production of mutant protein would improve Leo's symptoms. A common method to knock down the mutated gene product is by targeting the RNA transcript.

Double-stranded small interfering RNA (siRNA) can knock down the expression of target genes by mediating targeted mRNA degradation to reduce faulty and harmful protein levels [13]. Although the U.S. Food and Drug Administration (FDA) has approved some siRNA-based therapies, several barriers limit the clinical applications of siRNA. They often require a viral vector for targeted delivery which can cause an immune reaction in the host. siRNAs can sometimes induce off-target effects that can lead to dangerous mutations in essential genes. They also tend to display poor stability and pharmacokinetic behavior [14].

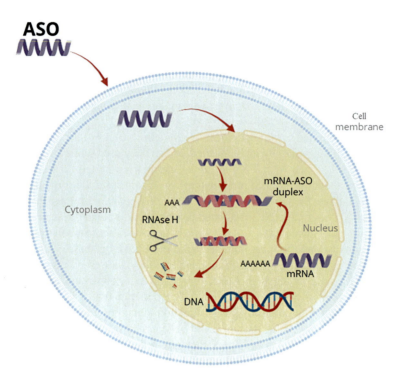

Fig. 2 The mechanism of action of an antisense oligonucleotide (ASO): The single-strand ASO enters the cell and the nucleus and binds to the complementary mRNA target. This attracts endogenous RNase H to this site, which cleaves the target mRNA, thereby reducing the expression of the target gene. (Adapted from [15])

By contrast, ASO drugs bind to plasma proteins preventing their loss via excretion and increasing their availability for tissue uptake [13]. Experts advised the family and the TNPO2 Foundation that designing and treating patients with an ASO would be technically less challenging, with fewer hurdles than other methods such as using siRNAs. They were also armed with the data from six people showing that haploinsufficiency of *TNPO2* protein does not cause a phenotype, indicating that a dramatic reduction in protein production (observed sometimes with ASO treatments) would not harm Leo. All of these factors prompted the family to concentrate their efforts toward developing and trialing an ASO.

7 Designing Leosen

The research team partnering with the TNPO2 Foundation for Leo's treatment relied on a new AI-based method to design the sequence and chemical modifications of the ASO drug. Using in silico tools, the team generated over 80 lead ASOs, with an effort to select one with minimal hepatic, nephrotic, neuro, immune, and genetic toxicity risk.

The next steps involved quantifying the downregulation that the candidate ASO could offer and to establish whether the drug showed a dose-dependent response. To analyze gene expression, the team used a digital gene expression-based approach. This method generates outputs of gene expression and enables analyses with high sensitivity [16].

In parallel with the ASO effort, the TNPO2 Foundation also attempted to develop disease models for translational and basic science experiments. Because so little was understood about the basic biology, any additional scientific understanding could lead to biomedical strategies to not only the ASO treatment but drug repurposing or other clinical management changes that could help Leo.

As with several drug development pipelines, the team used yeast cells, *Saccharomyces cerevisiae* as the first-line approach in drug discovery. They studied the endogenous expression of the protein KAP104, which is the yeast ortholog for human *TNPO2* [17]. They engineered the yeast to produce a mutated protein, like Leo's body produced. However, since the goal was to get as close to a human phenotype as possible, they also employed more complex model organisms.

They engineered the production of human wild type (WT) and mutated *TNPO2* in the nematode *Caenorhabditis elegans*. However, the mutated *TNPO2* gene C. elegnas showed more robust fitness than WT *TNPO2 C. elegans*, which led them to abandon this model organism.

Overexpression of the D156N protein in *Drosophila melanogaster* was previously shown to induce lethality. The *Drosophila* model is still in progress and utilizes conditional and humanized models. Unfortunately, there is no way to administer ASOs to this model reliably, so the Foundation focused on other models.

The next step was to test the ASO in a living mammalian system. For these experiments, they introduced the single nucleotide polymorphism (SNP) that Leo carried using CRISPR gene editing into the mouse genome. The resultant mice that carried a heterozygous mutation would not propagate naturally, rendering the line difficult to develop.

The time constraints that the research groups had to administer the drug to Leo meant that they could not fully rely on *Drosophila* or mice models, which can take months, to conduct the testing. This challenge compelled them to choose a model system that would most resemble the mutation in Leo's cells. The most promising, "Leo-like" model for testing the ASO came from Leo himself. Researchers drew blood and biopsied Leo's skin to isolate fibroblasts. They then reprogrammed these into induced pluripotent stem cells (iPSCs) which were further differentiated into neurons or used to establish brain organoids.

The miniature "Leo-brains-in-a-dish" not only gave the first disease model for answering pharmacology questions but also recapitulated the first clinical observation of microcephaly—brain organoids derived from Leo's stem cells were smaller in size compared to brain organoids derived from people with normal *TNPO2* protein expression. The phenotype was so severe that it was even clear to the (trained) naked eye.

By this time, the team had a lead drug candidate—dubbed leosen—that could selectively reduce the amount of mutant transcript, leaving over 80% of the wild type transcript in cell models. Treating the microcephalic brain organoids with this lead compound led to an increase in their size. And although the rescue was not dramatic, it was an indication that the candidate ASO could potentially help treat Leo's condition.

A few months later, the researchers developed one more clinical model to test the drug. They corrected the mutation in Leo's iPSCs using CRISPR gene editing. They then generated neurons from these edited cells that did not carry the mutation and compared these with neurons developed from stem cells carrying the mutation. Cellular phenotypes appeared but were difficult to quantify.

By this stage, the team had tested the ASO in WT mice to check for toxicity, and initiated GLP testing.

Once they had data on the safety and efficacy of the ASO drug, the team applied for a single-patient investigational new drug (IND) to the FDA. The IND package consisted of

 (i) Evidence supporting the mechanism of action of disease pathogenesis (gain of function mutation)
 (ii) Pharmacological demonstration of efficacy
(iii) Toxicology reports
(iv) Process of design and production of the ASO.
 (v) The clinical trial protocol

The approval took about three months and several rounds of discussions with the agency. This was a landmark event which took 17 months from receiving the diagnosis (Fig. 3).

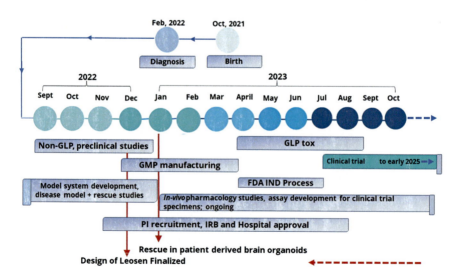

Fig. 3 Timeline to get Leosen to clinic since Leo's diagnosis

7.1   Designing a Trial, Not Just a Drug

Through the process of getting a safe and effective drug for Leo, the family has found unique problems in the drug development and clinical delivery pipeline for ultra-rare disease. With no documented safety or other pharmacology parameters, they had to develop and optimize everything from scratch.

According to the family, they were doing much more than designing just a drug—they were designing an entire trial. Although there was only one patient "enrolled" in the trial, he generated large amounts of data every time he had any signs of the disease. Quantifying the constellation of symptoms that Leo showed— from the seizures, to strabismus, the rash on his face and body, to the developmental regressions, to tongue thrusting, to changes in his demeanor, to hyper- and hypo- tonicity in his muscles—could provide them with clinical endpoints that provided scientific evidence to design a trial.

Documenting Leo's symptoms in detail frequently helped the family and the research team in this process. They could record when his symptoms were worsen- ing and inspect if Leo showed any signs of improvement when he got the drug treatment.

8   Administering Leosen

The final steps before Leo could receive the drug were for the human-grade product to undergo quality control checks for purity. The drug was ready to be administered to Leo within one and a half years of his diagnosis. Looking back, the family recog- nized that they could accomplish this in such a short time by taking a patient- centered approach.

Leo received his first dose 3 months before his second birthday. The doctors dispensed the drug intrathecally, into the space that holds the cerebrospinal fluid (CSF) in the spinal column.

The first dose was just a fraction of the full dosage that he was scheduled to receive eventually. The first 24 hours after the first treatment were critical—the team observed him closely for any adverse effects. He did not display any immediate adverse effects, and the team continued to monitor him through EEG, MRI, and blood and urine tests.

Leo's trial timeline consisted of a monthly dose escalation phase, with doses doubling each time (Fig. 4). He ultimately received his first full dose 3 months after the first fractional dose, just around his second birthday. Since all the markers of potential adverse effects in his blood, urine, and CSF remained normal, the doctors cleared him to receive more doses periodically. According to the treatment plan, Leo would receive quarterly maintenance doses followed by the first full dose.

Considering the dose escalation schedule, pharmacokinetic and pharmacody- namic properties of the drug, and poorly understood stability of the target protein,

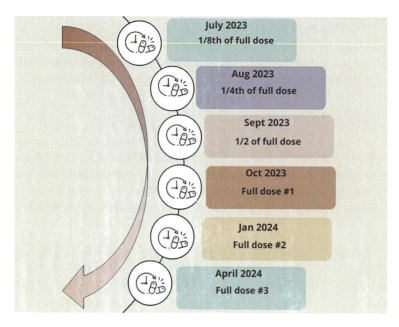

Fig. 4 Dose escalation schedule followed for Leo's treatment

the team expected Leo's symptoms to improve anywhere between 6 weeks to 6 months after the first dose. Added to this, the primary endpoint that the team hoped to achieve was the rescue of the neurodevelopmental delay, which is a difficult goal to quantify.

But Leo started showing significant signs of improvement a few days after his second escalation dose. The frequency of his seizures reduced, and his parents noticed that he was more present. He showed even more improvement after his third escalation dose and first full dose.

He crossed several developmental milestones. His physical movements had improved—he could hold his head up and sit up unassisted for a few minutes when somebody helped him get into that position. He started to make more complex sounds, including repeated consonant sounds. He could associate actions and objects with words (such as "kiss," "hug," and "raspberry") and was able to anticipate pleasure and respond (smile in anticipation) accurately.

Although these milestones excited the family, they are still unsure of Leo's long-term prognosis.

They also observed that the initial improvements that Leo showed were followed by slight decline—he showed some reversal of the initial developmental progress—by the time he was due for his next dose. This observation provided some hints about the stability and life of the drug in Leo's body.

8.1 Further Studies into the Drug Properties

Although brain organoids generated from Leo's cells offered insights about the efficacy of the ASO drug, they did not offer information about the pharmacokinetic properties of the drug. To study more about the pharmacology of the drug, measured through absorption, distribution, metabolism, and elimination (ADME) of the drug molecules, the research team employed WT rats.

According to published literature, each ASO's chemistry and sequence dictates its affinity to the different proteins which bring the ASO into and out of cells. [13] After leaving a cell, it gets eliminated from the CSF within 24 h, with any remaining active compounds sequestered in cells. Empirically in study animals, Leosen has a very short half-life, with approximately only $10^{(-2)}$ remaining after 30 days. This extrapolates to a very short half-life.

This data supported the clinical observations of Leo suffering from more seizures and developmental setbacks in the second half of his dosing schedule.

The short half-life combined with the clinical observation indicated that Leo would need more frequent dosing, which meant more paperwork and discussions with the FDA, another unanticipated development in the process.

8.2 Hurdles to Cross

Throughout the process of designing a drug and treating Leo, the family and the research teams had to overcome multiple challenges.

One of the first (and unforeseen) roadblock they encountered was the paperwork and the material transfer agreements that impeded Leo's fibroblasts from being transported between the researchers working on this project. The family learned of the tendency for institutions and companies to retreat into siloes firsthand, and the difficulty of fostering collaboration. Even when individual researchers supported collaboration, these siloing tendencies had seeped into the legal and operating processes of the institutions themselves; making overcoming these barriers unnecessarily time-consuming and expensive. One academic institution necessitated 30 staff hours to complete the paperwork for sharing a one patient cell line with one outside collaborator.

Other expected challenges were just as frustrating and time consuming. Though the FDA has issued clear guidelines for the trial to proceed, these still took the same shape as those required of commercial efforts. And although some requirements were simplified, the family had to execute complex chemistry, manufacturing, and control (CMC) assays as well as drafting the required regulatory documents. This seemed burdensome to the family for a compound that was intended for a single patient. The cost in dollars was significant, but more burdensome was the cost in time—time during which Leo continued to suffer developmental regressions and life-threatening seizures. The constraints imposed by a GLP study in particular implied that many weeks and months long steps could only be run sequentially. Since the N = 1 trials are something that the biomedical industry hasn't tackled frequently, there were many unique challenges that Leo's family faced.

In moments of desperation the family considered bypassing FDA jurisdiction if that meant Leo could receive the treatment that he needed sooner. This thought process is a common circumstance in the rare disease community. Hunter Davis was diagnosed with spinal muscular atrophy, a fatal rare disease, 8 weeks after he was born. His parents identified a clinic in Mexico that could treat him with an experimental ASO drug. Hunter received his first dose within 8 weeks of getting diagnosed [18]. The family found qualified individuals willing to aid in the development and administration in such an ex-FDA compound.

In hindsight, the family sincerely believes that regulatory agencies have a profound opportunity to help patients like Leo and Hunter, and their families. Hunter's mother, Khrystal Davis, learning from their experience, went on to become a passionate advocate, and frequently testifies on behalf of patients at the FDA; highlighting how such agencies can help other important stakeholders in the process of precision drug development for rare diseases. The successful demonstration of efficacy in disease models within months in some other N = 1 trials motivated Leo's family to further question the additional CMC requirements.

Consensus from families of children with rare diseases is that regulatory agencies could alter their policies for N of 1 trials such that the drugs are trialed and/or approved significantly sooner. Such agencies, in the family's opinion, can revise their guidelines such that documentation about the pharmacology of the drug and its efficacy should suffice, without the need for additional data on the drug's mechanism. Downstream approval policies and processes could be modified to appropriately address patient populations who suffer from devastating diseases, facing very different risk-benefit and profiles than traditional biopharma indication markets.

This was a major roadblock when the researchers acquired data about leosen's short half-life. Coupled with Leo's declining health in the second half of his dosing regimen, the family quickly realized Leo needed more frequent dosing than previously thought.

Thorough lab work and MRI scans had already confirmed that Leo did not suffer from any adverse effects of the drug. Yet, after submitting documentation supporting the need to amend the protocol for more frequent dosing, the approval took over two dosing cycles to be granted.

9 Lessons in Hindsight

Leo's family firmly believes that the patient and his or her family are the stakeholders most aligned with the mission of a successful treatment. This differs significantly from the motivations of other stakeholders; researchers, doctors, the pharmaceutical industry, or even regulatory agencies. The family members want what is best for their loved ones, and they will always prioritize making positive steps forward in the process. If and when mistakes happen along the way, they will sidestep the issues of responsibility and blame, in favor of making the best forward-looking decisions for their loved ones. Drug regulatory agencies like the FDA have an opportunity to recognize this and amend their guidelines to accelerate the process of drug development for rare or ultra-rare diseases.

Being outsiders in the field of drug development (former academic mathematicians), Leo's family could approach the problem with a fresh perspective. They realize that they were not tied down with having to follow a dogmatic or outdated process about the order in which experiments would happen. They could approach the problem using only first principles; asking and answering questions with only Leo's best interest in mind, unencumbered by process which came about for unrelated reasons; parallelize experiments, run experiments at risk, and try new technologies, i.e., AI in the discovery phase, to optimize their time.

Ideally, experiments would start in simpler organisms like the yeast *S. cerevisiae* and move toward more complex organisms like the fly *D. melanogaster*, mammalian systems like rodents, and eventually neurons or organoids derived from Leo's iPSCs. However, conducting experiments on different model systems in parallel helped them save a significant amount of time. They could reach their goal much faster than the time it would have taken if the experiments were conducted sequentially.

They are also cautiously optimistic about the immense potential that new developments in combining AI and with drug development. They believe that convergence of many different technologies and interdisciplinary science can be leveraged to benefit ultra-rare disease patients. Technologies like AlphaFold and RobettaFold and AI-driven drug development have the potential to accelerate timelines and lower costs by an order of magnitude or more.

Technologies can also be used to build automated tools to quantify neurodevelopmental delay or progress, something difficult to measure clinical trials with current tools.

Other than making the most of innovative technologies to accelerate drug development, they also believe that some administrative revisions can help make this process faster. Currently, only a commercial entity or an investigator can apply for an IND approval with the FDA. Leo would have benefited from the investigator applying for the IND approval since that would be faster, but it placed immense liability on one person and their institution. Much of the time gained by choosing the investigator-initiated pathway was lost to bureaucratic delays; the family needed to obtain the approval of the institution and hospital. When the specific investigator could no longer conduct the trial for unrelated reasons, the family had to start the entire process; legal, IRB, and so, on all over.

Revising guidelines so that a nonprofit organization or the patient's family can file a sponsor-investigator-initiated IND would not only help distribute the liability but could also potentially help speed things.

There is an unacknowledged tension between the FDA's desire for equity and the stringency of the IND application process. They require that the paperwork be filed by a single treating physician for every patient that could benefit from a particular N = 1 treatment. This effectively reduces the ability to commercialize (and thus scale) these efforts.

Overall, Leo's family and other patient advocacy groups observe that regulatory agencies may not have moved forward at the same pace that the science and biotechnology field has moved. The advancements in sequencing the human genome have enabled people like Leo to be diagnosed with diseases that very few other people suffer from. But the frameworks put in place (both written and unwritten) for drug development may not always account for such exceptional cases. Though each individual patient and disease is rare, the combined patient population makes up a significant fraction of patients, and consumes an oversized portion of healthcare resources.

However, Leo's and the families of other children with rare genetic diseases are hopeful that regulatory agencies will revise their guidelines such that they are more beneficial for all the various stakeholders in such situations. After all, while there is plenty of awareness of the risks of trialing an impure compound, the patients and family are often the only individuals who face a life of trialing nothing at all. Leo and his family join the patients and advocates who came before them in asking regulators that they not face these disease burdens alone.

9.1 Lessons on N = 1 Clinical Decision-Making

Leo's family applied basic probability to his circumstances. Once there was microcephaly, the probability of a severe prognosis rose to 80–90%. The remaining 10–20% consisted of "hereditary" microcephaly, which was not present in Leo's family. Seizures added more confidence in a poor prognosis. The decision tree from there forward became relatively easy to follow. Infantile spasms are not only known to have genetic etiologies but ultra-rare genetic etiology.

The rationale for tGoF came from the UKBB data. The drug development decision-making was driven by the desire to "load the probabilistic dice" as much as possible in Leo's favor. Knowing that most ASOs are toxic, the family chose to begin the funnel with a company whose founding principle was modeling and precomputing for safety, a way to strongly bias the first set of leads ASOs for safety. Every assay performed thereafter would order (and thus eliminate the poorest performing) leads according to some known toxicology assay. The family said that there will always be unknown risks, "but we attempted to mitigate all known risks to our best ability."

9.2   Lessons on N = 1 Clinical Trial Design

Going through this process has taught the family many lessons on clinical trial design. Where classical clinical trial designs have the option of increasing N "horizontally" to power their conclusions, this option is not available for N = 1 clinical trials simply due to lack of known patients. However, there is a way forward in increasing N "vertically," that is, increasing the data collected from an individual patient.

Current design for clinical trials takes measurements infrequently, on the order of 1–6 months in between clinical visits. Data collection at this low frequency means each patient adds very little statistical significance. For neurological disorders where seizures are present, families may keep a seizure diary as another source of data. Seizure logs, though a daily metric, are imperfect. It is very difficult to observe the patient continuously, not to mention that the seizures themselves can be sporadic and noisy.

Stakeholders can collaborate to develop technology for robust home data collection. For example, lab tests that can analyze hundreds of biomarkers in a drop of blood (Snyder lab/Stanford), or clinical grade EEG with dry-electrodes that can be put on as a hat. These tools allow robust n-of-few studies, possibly even improving on scientific rigor and becoming more widely used for higher-n clinical trials. AI can be used to analyze video and derive biomarkers for gross motor function, spasticity, or even clinical seizures.

Patients and their caregivers are in the driver's seat when it comes to data collection and should be empowered partners in the process. It takes many years of preclinical work for new interventions to reach patients in the clinic. During these trials, human-drug interactions should be measured thoroughly and frequently, to maximize the information that can be gleaned. While this is necessary in the n-of-few setting, the same tools and principles can be leveraged for extracting more and deeper data from traditional clinical trials with more participants. With higher resolution data collection, inter-individual differences in response in traditional clinical trials can also be analyzed and leveraged to guide clinical decision making such as prescribing and dosing for future patients. Clinical trials with responders of different levels can be analyzed to derive more precise and scientifically actionable hypotheses.

10   The Way Forward

The patient's family has established a non-profit called the TNPO2 Foundation that aims to bring lifesaving and life-changing precision medicine to patients with the highest unmet need. The foundation's goals are to advocate for rare diseases and work together with other rare disease-associated organizations to reduce the barriers to treatment. They aspire to make treatment more accessible by making drugs more affordable and reducing the bureaucratic challenges to getting a disease-modifying therapy. The foundation's work will begin with diagnosis, the first opportunity to improve the standard of care for ultra-rare disease patients, and follow through until they can be treated.

In the long run, Leo's family hopes to help others who face similarly challenging and devastating diagnoses with modern science, medicine, and technology.

References

1. Chirmule N, et al. Orphan drug development: challenges, regulation, and success stories. J Biosci. 2024;49:30.
2. https://www.ninds.nih.gov/health-information/disorders/infantile-spasms
3. Goodman LD, Cope H, Nil Z, et al. TNPO2 variants associate with human developmental delays, neurologic deficits, and dysmorphic features and alter TNPO2 activity in Drosophila. Am J Hum Genet. 2021;108(9):1669–91. https://doi.org/10.1016/j.ajhg.2021.06.019.
4. https://omim.org/entry/603002
5. Yang Y, Guo L, Chen L, Gong B, Jia D, Sun Q. Nuclear transport proteins: structure, function, and disease relevance. Signal Transduct Target Ther. 2023;8(1):1–29. https://doi.org/10.1038/s41392-023-01649-4.
6. https://www.informatics.jax.org/glossary/gain-of-function
7. https://www.congress.gov/107/plaws/publ280/PLAW-107publ280.pdf
8. https://rarediseases.info.nih.gov/about
9. Keshavan M. Saving Mila: How a tailor-made therapy, developed in a flash, may have halted a young girl's rare disease. STAT. Published October 22, 2018. https://www.statnews.com/2018/10/22/a-tailor-made-therapy-may-have-halted-a-rare-disease/
10. https://www.mousephenotype.org/data/genes/MGI:2384849
11. https://docs.google.com/spreadsheets/d/1s_Md8FGaAzl5qxsLSdw6LdEBYknQvu_4YbbEQKfAh-c/edit?gid=910020264#gid=910020264
12. Rinaldi C, Wood MJA. Antisense oligonucleotides: the next frontier for treatment of neurological disorders. Nat Rev Neurol. 2017;14(1):9–21. https://doi.org/10.1038/nrneurol.2017.148.
13. Hu B, Zhong L, Weng Y, et al. Therapeutic siRNA: state of the art. Signal Transduct Target Ther. 2020;5(1):1–25. https://doi.org/10.1038/s41392-020-0207-x.
14. Cring MR, Sheffield VC. Gene therapy and gene correction: targets, progress, and challenges for treating human diseases. Gene Ther. 2020;29:1–10. https://doi.org/10.1038/s41434-020-00197-8.
15. Gareri C, Polimeni A, Giordano S, Tammè L, Curcio A, Indolfi C. Antisense oligonucleotides and small interfering RNA for the treatment of dyslipidemias. J Clin Med. 2022;11(13):3884. https://doi.org/10.3390/jcm11133884.
16. Rodríguez-Esteban G, González-Sastre A, Rojo-Laguna JI, Saló E, Abril JF. Digital gene expression approach over multiple RNA-Seq data sets to detect neoblast transcriptional changes in Schmidtea mediterranea. BMC Genomics. 2015;16(1) https://doi.org/10.1186/s12864-015-1533-1.
17. Quan Y, Ji ZL, Wang X, Tartakoff AM, Tao T. Evolutionary and transcriptional analysis of Karyopherin β superfamily proteins. Mol Cell Proteomics. 2008;7(7):1254–69. https://doi.org/10.1074/mcp.m700511-mcp200.
18. https://www.congress.gov/117/meeting/house/112551/witnesses/HHRG-117-IF14-Wstate-DavisK-20210504.pdf

What Entrepreneurs Need to Know About Intellectual Property When Building a Strong Patent Portfolio

Joanna Brougher, Eliza Hong, and Gray Ah-Ton

Abstract Intellectual property (IP) is crucial for the growth and innovation of technology companies, serving to protect the core ideas behind commercial inventions and products. Securing IP ensures that these valuable assets remain with the company, making it an important factor for investors and partners assessing a company's value.

This chapter discusses the four main forms of IP—copyrights, trademarks, patents, and trade secrets—each providing unique protections. The focus then narrows to patents, which are essential for safeguarding the fundamental ideas of a business. Patents incentivize innovation by granting exclusive rights for 20 years. This exclusivity allows companies to recover their investments and drive the discovery of new treatments and therapies. Additionally, patents promote collaboration by requiring inventors to disclose their inventions to the US Patent and Trademark Office (USPTO), fostering a knowledge-sharing ecosystem that benefits the public. The chapter will also examine strategies for building strong patent portfolios that protect a product's market position, maximize patent longevity, and withstand post-grant challenges. Effective patent strategies balance broad and narrow claims to protect products while safeguarding against invalidation. Such strategies enhance IP protection, ensuring sustained innovation and competitive advantage in the market.

Keywords Patents · Copyrights · Trademarks · Innovation · Patent prosecution · Protection · Competitive advantage · Intellectual property · Patentability · Nonobviousness · Novelty · Ownership · Patent portfolio · Strategy · Brand recognition · Prior art · Trade secrets · Investments · Infringement · Enforcement

J. Brougher (✉) · E. Hong · G. Ah-Ton
Biopharma Law Group, PLLC, Fairport, NY, USA
e-mail: jbrougher@biopharmalaw.com

© The Author(s), under exclusive license to Springer Nature Switzerland AG 2025
N. Chirmule, V. V. Ghalsasi (eds.), *Approved: The Life Cycle of Drug Development*,
https://doi.org/10.1007/978-3-031-81787-8_14

1 Introduction

Intellectual property (IP) is fundamental to the success and growth of technology companies. At its core, IP protects ideas, the basis upon which commercial inventions and products are built. By securing IP, companies protect their assets, which is crucial for attracting investors and partners. These stakeholders often assess a company's IP portfolio to determine its value and potential for growth. Consequently, successful businesses prioritize IP as a way to safeguard their innovations and maintain competitive advantages.

This chapter will begin by exploring the four main forms of IP: copyrights, trademarks, patents, and trade secrets. Each type of IP provides different protections and serves various purposes:

1. Copyrights protect the expression of ideas, such as creative works including speeches, dances, and songs.
2. Trademarks protect the origin of ideas, enabling consumers to associate products with specific brands.
3. Trade secrets safeguard confidential business information like formulas, processes, and designs that provide a competitive edge.
4. Patents cover the ideas themselves, such as inventions, drugs, and devices.

The chapter will then focus on patents, which are critical in protecting the core ideas of a business. Patents serve dual purposes. First, they incentivize innovation. In industries like pharmaceuticals, patents grant exclusive rights for 20 years, encouraging companies to invest in research and development. This exclusivity ensures that companies can recoup their investments, which, in turn, drives the discovery of new treatments and therapies. Without patents, there would be little incentive for companies to undertake the costly and risky process of innovation, as competitors could easily replicate their advancements.

Patents further facilitate knowledge sharing. Patents require inventors to disclose their inventions to the US Patent and Trademark Office (USPTO). This transparency promotes collaboration within the scientific community by allowing researchers to build on each other's knowledge. By fostering a collaborative ecosystem, patents help expedite the development of new and improved technologies.

To obtain a patent in the United States, an invention must meet specific criteria outlined in 35 U.S.C. Sections 101, 102, 103, and 112. These requirements include:

1. Subject Matter: The invention must be of a type that is eligible for patent protection.
2. Utility: The invention must be useful.
3. Novelty: The invention must be new or novel.
4. Nonobviousness: The invention must not be an obvious improvement or modification of an existing invention.
5. Written Description and Enablement: The invention must be sufficiently described in the patent application to enable others to understand and replicate it.

The chapter will also examine strategies for building strong patent portfolios that protect a product's market position, maximize patent longevity, and withstand post-grant challenges. The strategy of a portfolio should align with the product's business goals and the company's financial resources. Key strategies include filing patents in multiple countries to protect international markets, filing patents to protect different features of a product over a product's lifecycle, and securing new patents for improvements to a product.

The chapter will conclude with best practices for obtaining patents. Although the process can be complex, costly, and time-consuming, securing patents is important for entrepreneurs to protect their ideas and inventions. Entrepreneurs should begin by conducting thorough search to ensure their invention is truly novel and unique over the existing prior art. Once confirmed, the patent application process can begin, which requires a detailed description of the invention, including its technical features, functionality, and potential applications. Collaborating with patent attorneys can help navigate this process and increase the chances of approval. It is essential to avoid errors in drafting, filing, or prosecuting patent applications, as mistakes can lead to delays, increased costs, or weakening of the patent's enforceability.

By understanding and leveraging the different forms of IP, particularly patents, technology companies can secure their innovations, attract investment, and maintain a competitive edge in the market.

Intellectual property is essential to the development and growth of most technology companies. At its very center, intellectual property is the protection of ideas. Since ideas are the foundation of commercial inventions and products, the outcome of business and financial opportunities often depends upon whether these ideas are protected. This is because, to some extent, protection of intellectual property guarantees that certain assets belong to the company. As such, investors and potential partners turn to a company's intellectual property portfolio to determine the value of a company and whether a partnership with the company should be pursued. Successful companies and entrepreneurs, therefore, place a high value on ownership rights granted to intellectual property.

Intellectual property comes primarily in four forms: copyrights, trademarks, patents, and trade secrets. Each of these covers a company's idea from a different angle. For instance, copyright protects the expression of an idea and includes creative works such as speeches, dances, and songs. Trademarks, on the other hand, protect the origins of ideas, helping consumers associate a product with its brand. The third form of intellectual property is trade secrets, which protect secrets such as formulae, practices, processes, designs, instruments, patterns, or compilations of information not known to the public. Patents go one step further to protect the idea itself, such as the drug, device, vaccine, composition, or whatever it is that the company is developing or selling.

This chapter will briefly cover copyright, trademark, and trade secret law, before delving into the most useful domain of intellectual property for entrepreneurs: patent law. The chapter will conclude with discussing best practices on how to obtain a patent.

2 Copyright

Copyright has been broadly conceived as incentivizing the production of creative works by enabling a creator to derive financial gain from their creations. Specifically, copyright gives authors "negative right," which prevents others from making use of the work without permission. An author of a copyrighted work has the *exclusive* right to produce copies or reproductions of the work, sell copies, import or export the work, create derivative works, perform or display the work publicly, sell or assign copyright rights to others, and transmit or display the work by radio or video. The author can also resort to legal action to stop others from doing any of the above. The author can furthermore trade these rights for value through contract.

At its core, copyright protects the expressions of ideas. This includes drawings, writings, or even speeches and architecture. The key word here is "copy." Independently coming to the same musical arrangement does not amount to copyright infringement. In practice, this either means having a red-handed admission from the infringer that they copied the author's work, or, more commonly, it means proving that an infringer had access to the author's work and that their work is similar to the author's. Courts have suggested that they will require a base level of similarity but that identical works may be viewed as obviously copied, even without evidence of access.

To prove infringement, not only the work must have been "copied" but also the copier must have taken too much. This is measured "intrinsically" and "extrinsically." Works are *intrinsically* similar if an ordinary person would look at the two and find them substantially similar. At trial, juries, sitting in the shoes of an ordinary person, will be the ones evaluating intrinsic similarity. Works are *extrinsically* similar if a plaintiff can point out a sufficient amount of copyrightable content that has been replicated in the allegedly infringing work. If what was copied from the work is not copyrightable (perhaps it does not count as "creative," "original," or "expressive"), then there is no extrinsic similarity. Both types of similarity must be present.

While these are powerful rights, they do not attach to all creations. As alluded to above, copyright law remains the domain of "creative" and "original" expression. "Creativity" requires merely that the choices made in creating the work were not "inevitable" or entirely "tradition[al]." For example, it is not "creative" to list names and telephone numbers in alphabetical order. The "originality" requirement prevents already-existing facts from being copyrighted—they are not original; they were merely discovered. Further, it prevents someone from copyrighting something that they themselves copied from another work.

The United States Copyright Act of 1976 helps us by listing the types of things that legislators had in mind as deserving copyright protection: literary works (books, periodicals, tapes, disks), musical works (the composition of a song, the lyrics, the melody), dramatic works (a play, an acting performance), pantomimes and choreographic works (a complex, choreographed dance number immortalized on video), pictorial, graphic, and sculptural works (illustrations, maps, plans, sketches, photographs), or motion pictures (TV, movies, soundtracks), sound recordings (the recording of a musical composition, the recording of a speech), and architectural works (architectural plans, the overall form and design of a building).

The rights received under the Copyright Act are automatic. The author owns the copyright to creative, original expression simply by expressing it in a tangible medium. For example, as soon as an author writes a story, that story is copyrighted. But if the author wants to be able to enforce his copyright rights, the work must be registered with the Copyright Office. Registration with the Copyright Office involves paying a fee and submitting an application along with a nonreturnable copy of the work. Unlike trademarks and trade secrets, copyrights have fixed lifespans. Generally, copyright protection lasts for the life of the author plus 50 or 70 years. In the United States, because the American government has toyed with the exact duration of that lifespan, the copyright protection depends on when the work was created.

- For works registered or published *before* 1923, all copyrights have expired, and the works are in the public domain. As such, the work belongs to the public, not to any individual owner. Anyone can copy, quote, or use these works as the basis for other works without permission.
- For works created *after* 1923 and *before* January 1, 1978, the duration of copyright depends on a variety of factors.
- For works created *on* or *after* January 1, 1978, the duration of the copyright lasts for the lifetime of the author plus 70 years.

There is one large exception in copyright protection. The "fair use" doctrine permits certain use of copyrighted works. For example, a copyrighted work can be used for personal consumption, education, creating parody, or for any other use that encourages the flow of ideas. Because the purpose of copyright is to encourage creativity of expression, the law does not want to limit uses that do not interfere with the author's rights. Instead, the fair use doctrine allows public access to the copyrighted work for the purpose of copying, performance, display, sale, and distribution that would otherwise be prohibited. The fair use defense exists to preserve free speech against copyright claims that mistake "copying" for "replacing." A quintessential fair use case would be one where the alleged infringer has created a parody of the original work, a news piece, or is criticizing the original work. These are all instances in which the allegedly infringing work is likely to be "substantially similar" to the original work but in service of a different creative purpose.

To determine if a defendant's work involves a sufficiently distinct use from the plaintiff's original work—and thus constitutes a "fair use"—courts consider (1) to what degree the purpose and character of the use differs, (2) the nature of the original work (for instance, it is more difficult to claim fair use when copying a highly expressive work), (3) the amount and substantiality of what was copied in relation to the original work as a whole, and (4) the effect of the use on the original work's market—potential or existing.

2.1   Case Study: Copyright and Artificial Intelligence

Recently with the advent of artificial intelligence (AI), the Copyright Office has been slowly paying more attention to copyright issues in AI. Generative AI, such as Open AI's text generator ChatGPT or image generator Midjourney, takes prompts (usually strings of words) and outputs texts, images, and now sometimes videos. Such AI models generate outputs based on highly complex patterns distilled from vast bodies of data, including copyrighted works of various authors and artists, often without permission. The generated output is not directly drawn from any one work but can be made to resemble existing works, such as a prompt asking for an output "in the style of" an artist.

In several lawsuits, the debate over copyright infringement and AI is ongoing. Artists and AI companies are currently fighting over whether the works produced by generative AI infringe on the copyrights of the works they "scraped." What we do know is that AI-generated works will not directly be registered for copyright. For example, if a prompt is entered that generates an image, the Copyright Office will not register the image itself. However, works containing AI-generated work that contains other elements of human originality can be copyrighted because of the human contribution. For example, a comic that contained AI-generated images could be copyrighted for the story and creative layout but not for the images themselves.[1]

To submit a work for registration that contains AI-generated text, images, or videos, the author must disclose the AI-generated portions in the application. The portions of the work submitted as "author created" must be portions not directly generated by AI.[2] Finally, the AI itself cannot be listed as the author or a coauthor. The Copyright Office does not recognize any nonhuman authors.

In summary, the Copyright Office and associated regulations are slowly developing standards for generative AI, while the extent of copyright protection applied to AI is currently being litigated. This area of technology is developing more quickly than the law, so novel interpretations of copyright will continue to emerge in relation to AI.

---

[1] United States Copyright Office. Correspondence regarding "Zarya of the Dawn" (Registration # VAu001480196). February 21, 2023.

[2] Federal Register. Copyright Registration: Guidance: Works Containing Material Generated by Artificial Intelligence. *Federal Register*. 2023 Mar 16;88(52):15884–15894. Available from: https://www.federalregister.gov/documents/2023/03/16/2023-05321/copyright-registration-guidance-works-containing-material-generated-by-artificial-intelligence

3 Trademark

While copyright law protects expression of ideas, trademark law protects the origin of those ideas. In that sense, trademark law governs the whole world of brand signaling and aims to protect (1) consumers from being confused about what they are buying and (2) companies from having opportunists reduce the brand recognition they built.

Famous trademarks like Coca-Cola®, Apple®, Google®®, Amazon®®, and others protect consumers by ensuring consistent quality and safety standards across their products. Through stringent quality control measures, these brands maintain their integrity and safeguard consumers against potential risks or counterfeit products. These brands also provide clear labeling and packaging to help consumers identify genuine products, thus fostering trust and confidence in their brand.

There are several subcategories of trademark law:

- *False Advertising:* Whenever a commercial advertisement contains a false claim, it may qualify as "false advertising." To prevail on a deceptive-advertising claim under the Lanham Act, a plaintiff must establish that (1) the defendant made a factually incorrect statement in a commercial advertisement; (2) the false statement actually deceived or had the tendency to deceive a substantial segment of its audience; and (3) the plaintiff has been or is likely to be injured as a result of the false statement. The claim must therefore be convincing enough to deceive a substantial segment of people, it must be injurious, and the deceit must not be insignificant. Even advertisements that merely *imply* a falsehood can be deceptive.
- *Likelihood of Confusion:* Branding materials might also cause a "likelihood of confusion" between two brands. A court's inquiry will focus on customer confusion—specifically a consumer's sensory experience of the competing products or other brand materials. In what context will the consumer view them? Do they look similar? Sound similar? Will the consumer think they mean similar things?

Branding components that are archetypal for that industry or mere descriptors of a product/service do not receive as much protection from the trademark law. Courts do not want to prevent all cafes from using the word "coffee" in their names or all wine producers from having grape iconography on their bottles. Some branding components are so widespread that they have become "generic," meaning that a whole industry now uses them to point to a general class of products. And some advertising efforts have been so successful that they turned a previously protected trademark into a "generic," unprotected one. This defeats the company's trademark; thus, the phenomenon is named "genericide." Examples include Chapstick, Kleenex, Aspirin, Xerox, and Popsicle. Each of these terms used to be protected trademarks but now is generic because of widespread use.

- *Likelihood of Dilution:* Brands that are "famous" have more legal tools at their disposal. Even if consumers are not "confused" about whether a product or service comes from another brand, their perception of a famous brand might be "blurred" or "tarnished." These are both actionable claims. A company can "blur" a famous brand by producing brand materials that approximate it and cloud its distinctiveness. "Tarnishment" looks more negative: dangerous or lower-tier products that reference a famous brand can "tarnish" its reputation.

Trademark disputes are often settled by timing and registration because whoever establishes their brand first gets "priority" and can sue the other for trademark infringement. To establish priority, a company will want to, first, use its branding materials in commerce—by advertising, shipping, and selling branded products—and, second, register its trademark with the United States Patent and Trade Office (USPTO).

Famous luxury companies such as Louis Vuitton and Gucci frequently encounter trademark dilution issues, which stem from the unauthorized use of its distinctive brand elements and designs. When counterfeiters produce counterfeit goods bearing the Louis Vuitton and Gucci logos and motifs, this dilutes the brand's distinctiveness, damages its reputation for luxury and exclusivity, and undermines consumer trust and confidence in the authenticity of genuine items. To combat this issue and safeguard its brand integrity, these luxury brands need to constantly be vigilant in monitoring their goods and employing various legal measures, such as filing lawsuits, against counterfeiters.

4   Trade Secret

The third form of intellectual property is trade secret. Trade secrets consist of information unknown to the public and give the owner an economic advantage over its competitors. The best trade secrets are the ones that cannot easily be reverse-engineered or those not susceptible to discovery using modern investigative or analytical techniques. Unlike patents and trademarks, trade secrets do not require the filing of any applications. However, they still require the owner to take reasonable steps to keep the information a secret. The owner can maintain its secrets by limiting access to the information and requiring those with access to the information to sign nondisclosure agreements and noncompete agreements.

Trade secrets are subject to both state and federal protection. At the federal level, trade secrets are governed by the Defend Trade Secrets Act (DTSA). At the state level, trade secrets are generally governed by the Uniform Trade Secrets Act (UTSA) (47 states have adopted the UTSA). Under the UTSA, information can only qualify as a "trade secret" if it is a formula, practice, process, design, instrument, pattern, or compilation of information. Under the DTSA, a broader range of information counts as a trade secret. It could be information of any form, regardless of "how [it is] stored, compiled, or memorialized physically, electronically, graphically, photographically, or in writing," and of any type: "financial, business, scientific, technical, economic, or engineering information." A database of DNA, computer programs, occupancy levels, and technical know-how can all be trade secrets under the DTSA.

Trade secrets can live on indefinitely as long as they are hidden with reasonable measures and not disclosed. But through any number of voluntary or inadvertent actions, they may be extinguished in an instant. This is why companies must take great care to prevent inadvertent disclosure. Because once a trade secret is "out" to the public, it will no longer be a trade secret.

One advantage to trade secrets is the greater ease with which one can obtain an injunction for misappropriation as compared to infringement of a patent. Under the patent law, a permanent injunction will only be issued after a stringent four-factor test is applied to consider (a) whether the plaintiff has suffered irreparable injury; (b) if remedies available are adequate to compensate for the injury; (c) whether a remedy in equity is warranted, considering the balance of hardships between the plaintiff and defendant; and (d) whether public interest would be disserved by a permanent injunction. Even if the plaintiff has satisfied all four factors of the test, the court may still refuse to issue a permanent injunction. However, under trade secret law, courts will only consider whether (a) the plaintiff has prevailed (or will likely prevail) on the merits of his misappropriation claims; (b) the plaintiff will suffer irreparable injury if the permanent injunction is not issued; and (c) the balance of the equities weighs in favor of the entry of the permanent injunction. Under the trade secret law's three-factor standard, therefore, courts may be more willing to issue a permanent injunction.

Wherever the patent law is unsettled, it is important to investigate the trade secret law as an alternative to patent protection. For example, the manufacturing processes of biologics may not be patentable, but they can remain trade secrets if reasonable efforts are made to keep them confidential. Moreover, because seeking patent protection can involve disclosure of the information to competitors, some companies may find it more effective to avoid the patent filing process entirely.

4.1   Case Study: KFC and Coca-Cola®

One famous trade secret is KFC's fried chicken recipe with 11 herbs and spices. The original handwritten recipe is kept in a 770-pound safe guarded by video cameras. Employees who know the recipe sign confidentiality agreements. KFC has even sued a couple claiming to have found a handwritten original recipe, claiming that it is inaccurate. When other claims of the secret recipe surface, KFC fervently denies its authenticity. KFC claims that the recipe is one of the best-protected trade secrets in the world, and the care they take in protecting it helps KFC maintain the exclusivity of their product.[3]

Another famous trade secret is the recipe of Coca-Cola®. This secret formula has been the same for more than a century and is known only to two executives of the

---

[3] "Is This the Top-Secret KFC Recipe?" The New York Times. 2016 Aug 26. Available from: https://www.nytimes.com/2016/08/26/dining/is-this-the-top-secret-kfc-recipe.html.   Accessed April 2, 2024.

company who are not allowed to travel together. The written formula is secured in a vault in the Coca-Cola® Museum in Atlanta, Georgia. By protecting their trade secret so cautiously and combining it with a distinctive trademark, Coca-Cola® has ensured their position in the carbonated beverage market for over 125 years.

5 Patents

Patents protect ideas that are often the most valuable core of a business. Patents serve dual roles. First, patents play an important role in incentivizing innovation within the pharmaceutical industry, particularly in drug development. These legal protections provide pharmaceutical companies with the exclusive rights to manufacture and sell their inventions for a term of 20 years. This exclusivity in turn encourages companies to invest resources in research and development, as they can recoup their investments through market exclusivity. Without patents, there would be little incentive for pharmaceutical companies to invest in the costly and risky process of drug discovery and development, as competitors could easily replicate their innovations. Thus, patents foster a competitive environment that drives innovation in the pharmaceutical sector, ultimately leading to the discovery of new treatments and therapies for various medical conditions.

Patents also serve to facilitate knowledge sharing and collaboration within the scientific community. In exchange for providing a term of 20 years of exclusivity, patents require that the inventors disclose their inventions to the USPTO. This act of filing for patent protection allows inventors to share their proprietary information with others while maintaining control over their ideas. This collaborative approach not only accelerates the pace of drug development but also allows researchers to build upon existing knowledge and leverage each other's expertise. In this way, patents promote innovation by fostering a collaborative ecosystem where ideas and insights can be exchanged freely, ultimately benefiting the public by expediting the development of new and improved technologies.

This section will outline the requirements for obtaining a patent in the United States. The requirements to obtaining a patent are similar across countries, but here the focus will be on the United States. In the United States, the subject matter must fulfill each of the following as directed in 35 U.S.C. Sections 101, 102, 103, and 112.

- Subject Matter: The invention must be of the subject matter allowed to be patented.
- Utility: The invention must be useful.
- Novelty: The invention must be new or novel.
- Nonobviousness: The invention has to be nonobvious, not simply a trivial modification of an existing invention.
- Written Description and Enablement: The invention must be sufficiently described in the patent application.

5.1 Patentable Subject Matter: Section 101

Many types of ideas can be patented, including processes (such as the method of making a drug), machines (complex physical articles with moving parts), manufacture (simpler physical articles with few or no moving parts), compositions of matter (chemical inventions, such as pharmaceutical drugs), and even improvements to any of these types of inventions. In the United States, morality is not considered when issuing patents.

Patents can cover a wide range of inventions, and little has been excluded historically. But the exceptions are notable. First, laws of nature, like gravity, are unpatentable. Conversely, a useful application *of* a law of nature is patentable, like using a naturally occurring chemical to treat a disease.

Second, natural physical phenomena, like naturally occurring microorganisms or chemicals, are unpatentable. However, genetically modified microorganisms are patentable. Moreover, a natural process with the addition of known elements is not patentable. For example, the process of metabolizing naturally occurring compounds in the body is not patentable. Human beings are not patentable, but certain gene sequences are patentable.

Third, abstract ideas are not patentable. "Abstract ideas" can include mathematical equations and their equivalent computer software form, and purely mental processes such as perpetual motion machines. However, if the idea is made, modified, or transformed by humans, then it can be patentable.

Over the recent years, the courts have taken issue with certain technologies, particularly those in the personalized medicine and diagnostic space. In 2012, for instance, the US Supreme Court invalidated claims directed to in vitro diagnostics in *Prometheus Laboratories, Inc. v. Mayo Collaborative Services, 566 U.S. ___ (2012)*. More specifically, the Court invalidated claims directed to the method of giving a drug to a patient, measuring metabolites of that drug, and with a known threshold for efficacy in mind, deciding whether to increase or decrease the dosage. In its ruling, the Court called the correlation between naturally produced metabolites and therapeutic efficacy and toxicity to be an unpatentable "natural law."

Following the *Mayo* decisions, the courts and the USPTO tried to clarify their positions on subject matter eligibility. In 2014, the Supreme Court created a two-step test, later known as the *"Alice/Mayo"* test, in *Alice Corp. v. CLS Bank Intl.*, 573 U.S. 208 (2014), which involved a software patent. Under the *Alice/Mayo* test, an invention would have to: (1) be directed to a law or product of nature and (2) include an additional, inventive step to be patent-eligible. The steps were intended to evaluate patent applications as a whole and accept claims covering an otherwise patent-ineligible invention, provided that a practical application had been added. Exactly what that additional inventive step is, however, remains unclear.

Despite how well-intentioned the *Alice/Mayo* test was, it nevertheless threatened, and continues to threaten, the patentability of many discoveries concerning personalized medicine based on the detection or correlation of naturally occurring phenomena, using new and useful methods.

5.1.1 Case Study: The Diagnostics Dilemma

The effect of the Supreme Court's decisions in these cases can be seen in the 2015 Federal Circuit decision in *Ariosa Diagnostics, Inc. v. Sequenom, Inc.*, 788 F.3d 1371 (Fed. Cir. 2015). In *Ariosa*, Sequenom is the exclusive licensee of US Patent No. 6,258,540, claiming methods for using cell-free, fetal DNA (cffDNA) circulating in maternal plasma (cell-free blood) to diagnose fetal abnormalities. Sequenom's invention allowed doctors to detect fetal abnormalities by testing DNA found in maternal blood rather than testing DNA found in fetal blood, which can only be accessed through invasive procedures such as amniocentesis and which subsequently increases the risk of miscarriage. Sequenom's discovery was, arguably, a drastic improvement in the current state of the art.

Sequenom's patent claim involved two steps: first, amplifying (by polymerase chain reaction), and second, detecting paternally inherited DNA from the mother's plasma sample. The technology for amplifying and detecting DNA was already well known and generally used to detect DNA.

Claim 1 of US Patent No. 6,258,540 is provided below for reference:

1. A method for detecting a paternally inherited nucleic acid of fetal origin performed on a maternal serum or plasma sample from a pregnant female, which method comprises:

   - amplifying a paternally inherited nucleic acid from the serum or plasma sample and
   - detecting the presence of a paternally inherited nucleic acid of fetal origin in the sample.

In evaluating the claim, the Federal Circuit considered the detection of fetal DNA a natural law and rendered the preparation and amplification of cell-free fetal DNA a conventional, noninventive activity. Accordingly, the Federal Circuit ruled that the method of prenatal diagnosis of fetal DNA was not patentable subject matter, and thus it is invalid.

This decision proved detrimental to Sequenom, as the company was acquired by LabCorp following the sudden drop in value as a result of the invalidation of its core patent. It is important to note that while Sequenom's patent was invalidated in the United States, it remained valid and enforceable in other jurisdictions including Australia, Germany, the United Kingdom, and other European countries, highlighting the differences in patent laws across jurisdictions.

5.2 Utility: Section 101

For an invention to be patentable, it must also have a purpose or use. Sometimes this requirement may seem obvious depending on the subject matter. For example, a stapler is useful because it can attach pieces of paper together quickly. But some inventions have no clear use. For example, a drug compound that only "produces the intended product" instead of treating a clearly articulated disease has no clear "use."

While the hurdle to satisfying utility is not steep, problems may arise when particular drug products have unknown uses or functions. This problem was at the center of the controversy in *Brenner v. Manson*, 383 U.S. 519 (1966), a case involving a process for making a steroid compound that did not disclose the use of that compound. In that case, the Supreme Court ultimately held that a process for making a steroid compound without disclosing the use of the compound failed to satisfy the utility requirement. The Supreme Court rejected arguments that the requirements for utility were satisfied "because it works—that is, produces the intended product," or "because the compound yielded belongs to a class of compounds now the subject of serious scientific investigation." The Supreme Court's decision in *Brenner* was, in part, motivated by trying to avoid allowing overly broad patents.

While most inventions can satisfy the utility requirement, it is important to describe the purpose of the invention when drafting the patent application to avoid potential issues down the road.

5.3 Novelty: Section 102

A third requirement for patentability is novelty. For an invention to satisfy the novelty requirement, it must be original and cannot resemble any "prior art." Prior art refers to any information available to the public that may be relevant to the invention's claims of novelty before the filing date of the claimed invention. These could be prior patents and existing products on the market. But prior art can also be anything that has been described, shown, or created before. It does not necessarily have to be patented, commercially available, or even exist physically. Even a prototype created for an academic study or knowledge used in traditional medicine will be "prior art." Trade secrets are not prior art, because they are not known.

Of course, each invention builds on prior ideas. The test is whether an invention can be anticipated by a person skilled in the subject matter if they had access to the prior art. If an invention contains an element not found in a prior art reference, then it is likely novel even if it contains some of the features from the prior art. For example, if a patent of a method contains steps A and B and a prior patent contains steps A, B, and C, then the method is likely not novel because both A and B were in the prior art. However, if the prior patent had only step A, then a patent with steps A and B is novel because B is a new element.

In the biopharmaceutical industry, new composition of matter claims will not normally present a novelty issue. However, it is more difficult for drugs like cannabis products to overcome the novelty hurdle. Because cannabis has only been recently legal in some jurisdictions, obtaining reliable prior art references is difficult to ascertain the patentability of cannabis claims. The first cannabis patent, US Patent No. 9,095,554, entitled "Breeding, Production and Use of Specialty Cannabis," issued to BioTech Institute, LLC., broadly claims strains with a specific cannabidiol content. This claim broadly could cover 50–70% of cannabis strains on the market. Patenting such products can lead to prior art challenges based on novelty in the future.

Additionally, prior art can expressly or inherently disclose parts of the claimed invention. This is called the problem of inherency. Express disclosure means that the prior art directly expresses the same idea. Inherent disclosure means that the prior art inherently possessed that property, characteristic, or function, even if not explicitly realized. For example, claims related to uses of a known composition are not patentable. If a prior art drug results in the production of a metabolite in the human body when administered, that metabolite is not novel. In this situation, the prior art inherently claims the metabolite because it is a result of the drug being administered, even if it is not explicitly stated. To be novel, the idea must be original or contain at least one truly original element.

In *Schering Corp. v. Geneva Pharmaceuticals, Inc.* 339 F.3d 1373 (Fed. Cir. 2003), for example, the Federal Circuit invalidated a patent based on the formation of a metabolite after consumption of the drug as anticipated based on prior art disclosure of administering the drug. The compound at issue was the antihistamine loratadine, known as Claritin®. The patent, which was assigned to Schering Corporation, covered a metabolite of loratadine called descarboethoxy-loratadine (DCL), its fluorine analog, and its salts. Schering Corporation also owned a patent covering a class of compounds, including loratadine itself, which was prior art to the metabolite patent. This prior art patent did not expressly disclose DCL and did not refer to metabolites of loratadine. The Federal Circuit, nevertheless, found inherent anticipation because the metabolites of loratadine DCL were "necessarily and inevitably" formed in every patient who consumed the drug, and, thus, the metabolite was inherently anticipated by the original patent.

5.4   Nonobviousness: Section 103

A further requirement of patentability is that the invention cannot be obvious in view of the prior art. In contrast to the novelty requirement of Section 103, which focuses on the absolute newness of an invention, nonobviousness considers whether the invention involves some sort of inventive step beyond what would be considered obvious to those skilled in the art. For example, one inventor added electronic sensors to adjustable pedals and the court asked if this combination was "obvious to try." If a person with ordinary skill in the field would think to implement an invention, then the invention may be obvious.

To assess obviousness, several factors (called the *Graham* factors after a Supreme Court decision) are considered together: the scope and content of the prior art, the level of ordinary skill in the art, difference between the claimed invention and prior art, and objective evidence.

Some secondary arguments can also be applied to demonstrate nonobviousness. If the invention is commercially successful, the inventor can argue that someone else would have invented it long ago had the concept been obvious. Another possible argument is that the invention addresses a long-felt but unsolved need, which explains its commercial success. The inventors can show that others were skeptical of their solution to the problem, and therefore the invention was nonobvious. Next, if others have attempted and failed at developing the invention, the inventor can point out shortcomings in the research of the time to bolster a nonobviousness argument.

Another strong argument to overcome obviousness is to show that the invention had unexpected results. For instance, the drug worked at a lower dosage than was expected, or it worked in a different indication than expected. Supporting this argument with experimental evidence is crucial to its success. Below are several examples of drugs that are better known for their unexpected results than what they were originally designed to do.

- Viagra® (sildenafil) was originally developed as a treatment for hypertension and angina; however, it was found to have an unexpected side effect—improved erectile function. This unexpected result led to the development of Viagra® as a treatment for erectile dysfunction.
- Botox® (botulinum toxin) was originally used to treat muscle spasms and disorders, but it was found to have a cosmetic benefit—reducing the appearance of wrinkles. This unexpected result played a significant role in expanding the use of Botox® for cosmetic purposes.
- Rogaine® (minoxidil) was originally developed as an oral medication to treat high blood pressure, but it was unexpectedly found to stimulate hair growth. This unexpected result led to the development of a topical formulation for treating hair loss.

5.5   Written Description and Enablement: Section 112

As discussed earlier in this chapter, patent law is based on a trade-off between disclosure and exclusivity. When applying for a patent, the inventor must disclose their invention to the public, and in return, they receive a period of exclusivity—20 years of patent term. Thus, the patent office requires that an inventor should "enable" the public to reproduce their invention by describing the invention in sufficient detail in the patent application. The description needs to allow a person skilled in the relevant field to reproduce it without "undue experimentation." If a skilled person cannot make the described invention work without further experimentation, then the inventor has not disclosed the process in sufficient detail and the claim is said to be indefinite.

The claims are also important in defining the scope of the invention. By knowing how narrow or broad the claims are, others will be able to know what related inventions would be outside of the patent's scope. The words of a claim confine what an inventor can exclude. Even if an invention appears close to a patented one, the patent may not cover it if the claims are not broad enough.

The written description also proves that the inventor actually created the invention themselves because they can describe it in enough detail. If the inventor only described one type of invention but attempts to claim a broad scope, problems may come up. Recently, courts have leaned toward narrowing the scope of patent claims.

One important case involving written description involves a blockbuster drug, Celebrex® (celecoxib). Celebrex ®, a COX-3 inhibitor and nonsteroidal anti-inflammatory drug (NSAID), was developed by G. D. Searle & Company, which was later acquired by Pfizer. The University of Rochester sued Searle claiming that Celebrex® infringed its patent, US Patent No. 6,048,850, claiming a method of inhibiting COX-2 in humans using a compound. The problem, however, was that the University's patent did not identify any COX-2 inhibitors. Since the University of Rochester provided no written description of a compound able to inhibit COX-2, the Federal Circuit held that the University had failed to satisfy the written description requirement, and it invalidated the patent.[4]

5.5.1 Case Study: Means-Plus-Function Claims as a Way of Overcoming 112 Hurdles

While traditional patent claims describe their inventions in terms of the structure, such as the specific parts of a machine, another way of describing an invention can be by function. This type of claim is authorized by Section 112(f) and commonly called means-plus-function, or MPF, claims. This approach provides greater flexibility in patent drafting, especially in fields with rapidly evolving technology or where specific structures may vary while achieving the same function. Therefore, means-plus-function claims can often provide more specificity and clarity than simply structural characterization.

Means-plus-function standards in Section 112(f) are applied when the claim uses language such as "means for" or "step for." If this type of language is not used, the standards for reviewing and interpreting means-plus-function claiming will not be applied, and the functional aspect will not be adequate.

Means-plus-function claiming also requires that the patent specification adequately disclose the corresponding structures or materials that perform the claimed function. If the specification fails to sufficiently describe such structures, the claim may be deemed indefinite.

In the biopharmaceutical industry, a means-plus-function claim could be of a combination of substances or a combination of steps in a process. For example, this could be a composition containing an antibody and a *means* for achieving some desirable outcome activity of the antibody and a pharmaceutically acceptable carrier, such as better binding. In such a claim, the corresponding structure of the

---

[4] University of Rochester v. G.D. Searle & Co., 358 F.3d 916 (Fed. Cir. 2004).

antibody must be identified along with its function in the patent specification. The patent specification, therefore, would need to show examples of binding data that correlates to specific antibody structures.

6 Building a Strong Patent Portfolio

To generate value from patent portfolios, a variety of techniques can be employed to develop a strong and effective patent portfolio. A strong patent portfolio will cover the features and embodiments of the product to protect its position in the market, prevent competitors from designing around the product, maximize the length of the patents, and insulate against potential post-grant challenges. Each patent portfolio will depend on the product, drawing on the product's strong points. Patent portfolios also come at a steep price, so the size and strategy of a portfolio should be fitted to the projected market value and to a business's finances. Being aware of the main strategies can help an inventor develop a robust patent portfolio from the outset.

In addition to protecting a company's position in the market, a strong patent portfolio must also withstand challenges to it. Recent changes in patent law have allowed patents to be invalidated more easily through both legislation and court rulings. If the patent is invalidated, the protections will no longer apply and the patentholder will lose market exclusivity.

One major change is that the America Invents Act of 2011 allows third parties to challenge issued patents through a process called inter partes review (IPR). IPR challenges are heard in front of the US Patent and Trademark Office's Patent Trial and Appeal Board (PTAB), which renders a decision in 12–18 months. The third parties can be other companies who challenge patents based on "prior art" invalidity either under Section 102, novelty, or Section 103, obviousness. The lower cost of IPR challenges (as opposed to suing in district court) encourages more challenges. In addition, the PTAB only upholds patents around one-third of the time, meaning that nearly two-thirds of patents challenged by IPR claims are invalidated.

6.1 Goals for Designing a Strong Patent Portfolio

When designing a strategy for a strong patent portfolio, it is important to keep in mind both the strengths and limitations of the company, as well as company's principal patent goals.[5] The right patent strategy will accurately represent company's priorities because each action taken to improve patent coverage will involve financial and other resources. Once a company has identified the products in need of

---

[5] Brougher J. Billion Dollar Patents. JTB Publishing LLC; November 2019.

patent protection, the focus should turn to drafting strong patent applications. A good patent will protect a company's product in the marketplace by retaining exclusive use and production of the invention for the company.

There are two main concerns when writing a patent. First, the patent must be broad enough to cover the features of the product so that competitors cannot sell a similar product. Second, it must be narrow enough so that it is not invalidated. Broader patents are more easily invalidated because they will not be supported by the patent specification and data. The ideal patent is narrow enough to specifically identify the invention but broad enough so that competitors cannot design a similar competing product.

Additionally, in each individual patent, the subject matter should be covered in a variety of scopes. Having multiple claims around different aspects of the product shields it from invalidation challenges. If one claim is invalidated, the other claims can still protect the product. For example, in the biopharmaceutical context, claiming an antibody by function can be strengthened by including narrower claims to the antibody sequence. Functional elements can be combined with structural elements in the same claim to create a stronger patent that is less likely to be completely invalidated.

Building a strong patent portfolio requires awareness of the limitations and benefits of certain strategies. A strong patent portfolio should only protect different features of a product in different patents. Strategies such as evergreening, temporally staggering claims, and continuation applications can each extend the life of the patent portfolio. Being aware of a company's goals and resources along with these common strategies will allow a company to build a strong patent portfolio that withstands invalidation claims from the outset. Below are several strategies for building a strong patent portfolio.

6.2   Go Global

Filing patents in multiple countries can also protect the patent portfolio. Each patent only protects the product in the issuing country. For example, a patent in the United States only protects the product from infringing activity in the United States. If a company wants to protect its invention in multiple countries, the company should consider which countries to file additional patents in.

To decide which countries to file patents in, several considerations arise. First, the company should file the patent application in a country where the company is physically located. This ensures that the company will have legal jurisdiction over its product in the country of origin.

Second, the company should file the patent application in countries where the product will be manufactured, developed, or commercialized. This may be in multiple countries. Filing patents in these countries can help prevent infringing activities anywhere the company's product is physically located.

Finally, patents should be filed in countries where there could be competition for the products. This should include countries where competitors may try to transfer or commercialize the product. Weighing the cost of the patents and the potential benefits should be a company- and product-specific strategic decision.

To expedite the process of filing patents in multiple countries, the Patent Cooperation Treaty allows the filing of one international patent application in a single language that becomes a placeholder application for later applications in other member countries. Thirty months from the initial priority date, national stage patent applications must be filed in every country that the patentholder wants to pursue patent protection in. Filing in multiple countries can be costly but could be worthwhile depending on the commercial product.

6.3   Evergreening: A Strategy to Extend the Patent Term

One strategy that companies use is called evergreening, which extends the life of the patent by filing multiple patents for different features of a product. Evergreening is controversial because it delays competition by extending the patent term around the same product. For example, in the pharmaceutical industry, extending the patent term delays generic drugs from entering in the market, which keeps drug prices high. This market exclusivity creates incentives for companies to develop drugs and is the reason why companies use evergreening.

For example, to develop a patent portfolio around a drug using evergreening, a company may initially patent the active ingredient. Several years later, the company may patent the optimum dosage of the drug. The dosage patent would be a new patent and will expire several years after the patent expires on the active ingredient. Several years later, the company may file patents to the method of using the drug or to a method of manufacturing it. In this way, the company may file many different patents around the same drug, all staggered temporally. Competitors will not be able to use the drug to treat patients until the second patent expires, thereby extending the life of the patent.

A company can also combine multiple products into one, as part of evergreening. For example, in the pharmaceutical industry, multiple drugs can be combined into one. The patient then only has to take one drug, leading to more compliance and thus improvement of the product. One example of such a combination drug is Pfizer's Caduet®, which is used to treat high cholesterol and blood pressure. Caduet® is a combination of a calcium channel blocker, Norvasc® (amlodipine), and a cholesterol-lowering agent, Lipitor® (atorvastatin). Norvasc® and Lipitor® had individual terms, expiring in 2007 and 2011, respectively, but in its combined form, Caduet® had a patent term extending into 2018.

Another related strategy is product hopping, sometimes called brand migration. When a company makes incremental changes or improvements to products, they can sometimes secure patents for the new versions, thus extending the patent term.

The company can then discontinue the previous version, leaving no reference products for which competitors can substitute.

One problem encountered with evergreening and product hopping is satisfying the nonobviousness prong of patentability. The new product cannot be such that a person of ordinary skill in the art may find the underlying invention obvious. Patents associated with the obviousness problem are most likely invalidated upon a challenge. However, a company may still choose to evergreen with a patent with the obviousness problem if that patent adds valuable time beyond the initial patent.

6.3.1 Case Study: Humira®

One very strong patent portfolio that has withstood many challenges is Abbvie's blockbuster drug Humira®. Humira® is not protected by just one patent. It is protected by more than 100 patents until the year 2034.

Humira®'s patent portfolio is large and diverse. Abbvie listed 22 patents directed to various diseases or methods of treatment, 14 directed to the drug's formulation, 24 encompassing the manufacturing practices surrounding the production of the drug, and 15 "other" patents. The Humira® patents protect various aspects of the product to make it less susceptible to competitors designing around the patent. Abbvie has protected the formulas, manufacturing processes, and improvements, even when these ideas surrounding the main patent are designed after the initial patent filing. By patenting the inventions around the main patent to cover the scope of ideas associated with the product, Abbvie has created a robust patent shield.

Humira®'s patent portfolio is not only large and diverse, but the patents are also filed over different times. This staggered patent strategy aims to extend the patent term. Because the drug has multiple aspects, Abbvie strategically filed for additional patents by adding additional inventive elements to the original filed application. Abbvie thus extended the life of the patent far longer than the 20-year term of the initial patent.

6.3.2 Case Study: India

In India, evergreening for incremental improvements in pharmaceutical drugs is not allowed. India enacted the Patents (Amendment) Act 2005 to its Patent Act 1970 (India), effectively disallowing evergreening drug patents for public health. Specifically, this law interprets "invention" and "inventive step" to only enhancements of the substance. This results in certain drug improvements to be unpatentable, such as dosage change, combination drugs, or new use of already known substances. Furthermore, Indian law defines substances to include a broad range.

With this amendment in place, patenting features of previously known products or those that are not truly novel face significant hurdles in India, even if those features have arguable benefits. India's laws prevent companies from filing patents to improvements or modifications to existing products. This means that companies cannot extend the patent-protected life of an existing drug even if the drug has some newly discovered features that provide additional benefits for consumers. Below are examples of products that have been denied patent protection in India:

- In 2008, the Indian Patent Office rejected a patent application for Viramune Suspension® (nevirapine hemihydrates), a syrup form of Viramune® (nevirapine), allowing children living with HIV who are unable to swallow tablets a way to take their medication. The Patent Office concluded the syrup form to be a "new form" of a "known substance."
- In 2009, the Indian Patent Office rejected patent applications on two ARVs, Viread® (tenofovir) and Prezista® (darunavir), despite arguments that the drugs— which consist of a previously known compound— demonstrated enhanced efficacy.
- In 2010, the Indian Patent Office overturned a patent for Roche's drug, Valcyte® (valganciclovir hydrochloride), as lacking an inventive step and not showing increased therapeutic efficacy as required under Section 3(d). Valcyte® is a modification of an existing drug, Cytovene® (gancyclvoir), used to treat common, opportunistic infections associated with HIV, called cytomegalovirus.
- In 2011, the Indian Patent Office rejected an application filed by Abbott Laboratories for Kaletra®/Aluvia® (lopinavir plus ritonavir), a heat-stable version of Abbott's earlier drug. The Patent Office concluded that the drug was not a new invention, even though it allowed the drug to be stable at higher temperatures.

If a company wants to evergreen a drug by patenting benefits around the drug, they will not be able to do so in India.

6.3.3   Case Study: The Battle over Glivec®

India's application of its laws has resulted in a backlash from companies trying to obtain patent protection in India. Several years ago, Novartis challenged India's laws on various grounds, including being noncompliant with the Trade-Related Aspects of Intellectual Property Rights "TRIPS" Agreement. Despite ultimately losing, Norvartis' case was significant in showing how far countries can go to protect their interests while remaining compliant with TRIPS.

The patent case associated with Novartis' anticancer drug called Glivec® or Gleevec® (imatinib mesylate) was controversial. This drug is used to treat patients suffering from chronic myeloid leukemia. Since the drug controls the cellular action by which cancer grows but does not act to cure cancer, the drug must be taken by patients for the rest of their lives. The price for the drug is between about US $25,000 and US $50,000 per patient per year; however, generic versions are available from Indian generic drug manufacturers, such as Ranbaxy, Cipla, Natco, and Hertero for about US $2100 per patient per year. Glivec® is an important drug for Novartis, generating more than $1.5 billion in global sales in 2018, down from about US $4.7 billion in 2015, before generic drugs entered the market.

Glivec®'s patent dispute involved the modification of an existing drug. The patent covering the existing drug was filed in the United States and other countries in 1993 but not in India. It was directed to imatinib as a "freebase" molecule and disclosed the salt as imatinib mesylate. In 1998, Novartis discovered a new application for a beta crystalline form of imatinib mesylate and filed a patent application for this new salt. This second application was filed in India. When the patent application came up for examination in 2005, a pre-grant opposition was filed by several organizations, including Natco Pharmaceuticals, Alternative Law Forum, and Lawyers Collective, on behalf of the Cancer Patients Aid Association. The opposition challenged the Glivec® application, based, in part, on Section 3(d) of the Patents Act, claiming the application concerned only a modification of an already existing drug and did not improve its efficacy.

Under India's patent laws, isomers are considered the same substance, unless they differ significantly in properties regarding efficacy. Accordingly, to overcome the patentability issue under Section 3(d), Novartis had to show that Glivec® differed significantly from the existing drug regarding its efficacy. To prove efficacy under India's strict standards, Novartis demonstrated that there was enhanced bioavailability of 30% in studies conducted on rats. However, Novartis was unable to demonstrate how the enhancement in efficacy was critical to the performance of the drug or the difference it made compared to known efficacy. The Patent Office remained unconvinced by Novartis' arguments and found Glivec® unpatentable under Section 3(d).

Novartis challenged the Patent Office's rejection of the Glivec® patent application with the Intellectual Property Appellate Board (IPAB). If Novartis were to win, it would have to show that the 30% increase in bioavailability was an enhanced efficacy and the beta crystalline form of the mesylate salt was not an obvious form of the free base form. After nearly 7 years of fighting over Glivec®, on April 1, 2013, the Supreme Court of India found that Glivec® did not show enhanced efficacy under Section 3(d), and thus did not meet India's standards for patentability. By rejecting Glivec's® patentability, India solidified its stance against allowing incremental improvements.

7   Obtaining Patent Protection

Securing patents is a crucial step for entrepreneurs looking to protect their ideas and inventions. The process of obtaining a patent can be complex, costly, and time-consuming. It therefore requires a strategic approach. This approach would first require an entrepreneur to understand what idea or invention the entrepreneur believes they have. This would require the entrepreneur to first conduct thorough research to ensure that their invention is truly novel and unique and not already in the prior art. Such research would involve searching through existing patents, scientific literature, and other relevant sources.

Once the entrepreneur has determined that their invention is novel, entrepreneurs can begin the process of applying for a patent. This process can take several years but starts with drafting and filing a patent application with the appropriate patent property office, such as the United States Patent and Trademark Office (USPTO) or the European Patent Office (EPO). A complete patent application would include a detailed description of the invention, including its technical features, functionality, and potential applications. Entrepreneurs may also need to include diagrams, drawings, or prototypes to illustrate their invention more effectively. It is important to ensure that the patent application is well-written and meets all the requirements set forth by the patent office to increase the chances of approval. Patent applications that fail to satisfy the patentability requirements discussed earlier in this chapter may have a harder time getting approved by the patent office. Working with patent attorneys or agents who specialize in patent law can help entrepreneurs better navigate the complexities of the patent process and maximize their chances of success.

Table 1 describes various steps of the patent process. By following these steps, entrepreneurs can navigate the patent process successfully and secure valuable protection for their ideas, thereby gaining a competitive edge in the market.

During the patent drafting and prosecution process, it is important to avoid certain mistakes as those mistakes could have serious implications on the outcome of patent applications. Errors in drafting, filing, or prosecuting patent applications can result in delays, increased costs, or even the rejection of the patent. Moreover, mistakes may weaken the patent's enforceability or even increase its susceptibility to challenges. Given the competitive nature of the patent landscape, it is essential to pay meticulous attention to detail, adhere to legal requirements imposed by the patent offices, and take proactive measures to prevent mistakes throughout the patent prosecution process. By avoiding mistakes and improving the quality of the patent application, inventors and companies can maximize the potential value of their patent assets and mitigate potential risks associated with patent challenges.

Table 2 describes several mistakes that inventors should avoid when drafting and prosecuting a patent application.

8 Conclusion

In summary, patents offer entrepreneurs vital legal safeguards, ensuring that their inventive concepts remain protected from unauthorized use or replication. By obtaining patents, entrepreneurs gain a distinct advantage in the market, empowering them to leverage their original ideas without concerns of infringement. Patents, furthermore, serve as catalysts for continued investment in research and development, as they provide avenues for recovering expenses and securing exclusive rights. The presence of patents enhances entrepreneurs' attractiveness to potential

Table 1 Description of the patent prosecution process

Conduct a preliminary patent search	The patent process should begin by conducting a thorough search of the prior art to ensure the invention is novel and not already patented by someone else. Online databases, such as those provided by international patent offices, as well as scientific literature databases can be used to conduct this search
Evaluate patentability	Next, it is important to evaluate whether the invention satisfies the criteria for patentability, which typically includes patentable subject matter, utility, novelty, and nonobviousness. A patent attorney or agent can help evaluate the patentability of the invention
Document the invention	It is important to keep detailed records of the invention, including written descriptions, diagrams, sketches, and any experimental data that demonstrates its functionality and uniqueness. Proper documentation can be crucial for preparing and later prosecuting a patent application
File a provisional patent application (*optional*)	While not required, a provisional patent application can be filed to establish an early filing date and secure a "patent pending" status. This application is not examined by the patent office but instead provides temporary protection while the non-provisional patent application is finalized
Prepare a non-provisional patent application	A non-provisional patent application should be prepared which includes a detailed description of the invention, along with any necessary drawings, diagrams, or prototypes. This application must also contain the claims that set forth the invention. This application must meet all the legal requirements set forth by the patent office and will be the application that gets examined by the patent office
Submit the patent application	No later than 1 year after the filing of a provisional application, the non-provisional patent application must be filed with the appropriate intellectual property office, such as the USPTO or the European Patent Office (EPO). There are certain government fees and paperwork that will be required with the submission of the patent application
Respond to office actions	After submitting the patent application, the patent office will send office actions requesting additional information or amendments to the claims. The inventor will be expected to address any concerns raised by the examiner and amend the claims as needed
Prosecute the patent application	The back-and-forth with the patent office will likely continue for several years. During this time, the inventor should be prepared to defend the novelty and inventiveness of the invention
Obtain patent approval	Once the patent office is satisfied with the application, a patent will be granted. The invention will be protected for a specified period, typically around 20 years from the filing date of your application
Maintain the patent	Once a patent is granted, there will still be payments required to maintain the patent in force throughout its term. Each country has different requirements for these maintenance fees. Some require yearly payments while other countries require payments every few years

Table 2 Mistakes that inventors should avoid when drafting and prosecuting a patent application

Avoid disclosing the invention publicly before filing	Public disclosure can jeopardize an inventor's ability to obtain a patent, as it could invalidate the invention's novelty. Prior to any public disclosure, a patent application (especially a provisional application) should be filed
Avoid delaying the filing of the patent application	In a first-to-file patent system, delaying the filing of a patent application could result in losing patent rights to another party who filed an application to a same or similar invention. It is therefore important to file the patent application as soon as possible to establish priority and secure rights to that invention
Avoid relying solely on verbal agreements	When collaborating with others on the invention, it is important not to rely solely on verbal agreements or informal arrangements regarding patent ownership and rights. Such rights should be clearly documented in an agreement in writing to avoid disputes later on
Avoid overlooking the importance of patent search	Patent searches are important in uncovering prior art that could impact an invention's patentability. By not uncovering such prior art, time and resources could be spent on an application that is unlikely to be granted
Avoid filing a poorly drafted patent application	Patent laws are specific in their requirements and thus a poorly drafted patent application can dramatically reduce the chances of obtaining patent protection. It is therefore important that a patent application is clear, concise, and accurately describes the invention's patentable characteristics
Avoid ignoring office actions or deadlines	It is very difficult to recover from a missed deadline. If the patent office sends office actions or requests for additional information from the inventor, it is important to respond timely to these requests to avoid abandonment of the patent application
Avoid assuming automatic approval	It is unlikely that a patent application will be granted immediately. It is therefore necessary to work diligently and patiently to address any rejections, objections, or other requests that the patent office may require

investors and collaborators, showcasing the value and potential of their intellectual property. Ultimately, patents are integral to fostering a climate of innovation, allowing entrepreneurs to safeguard their creations and thrive amidst competition in the business landscape.

The Power of Data Analytics: Manufacturing Quality and Compliance

Ravindra Khare

Abstract Decisions based on data yield superior results than those made on hunches. Manufacturing processes are no exception to this axiom. Statistical data analytics plays a major role in manufacturing. Process and product quality are majorly driven by the efficacy of associated data and scientific methods of analysis and inferencing. This text discusses scientific data methods used in process development, process qualification, measurement qualification, and screening of incoming material for desired quality. The topics here cover design of experiments. Efficient ways of experimentation lead to process insights and enable control over quality in an efficient way. Design and analysis of factorial experiments are discussed here. Process once developed needs to be qualified for its suitability to perform in a sustained manner. Methods of process capability and stability evaluation are discussed. An outline of how process management is implemented through statistical methods is given. Various tests and studies that go into robust and usefully accurate measurement of analytes are enumerated. Acceptance sampling is an important aspect of regulating the quality of incoming material with sampling. Acceptance sampling terminologies and their applications are discussed here.

Keywords Design of experiments · Process capability · Process control · Process qualification · Continued process monitoring · Method validation · Acceptance sampling · Statistical methods

R. Khare (✉)
Symphony Technologies Pvt Ltd., Pune, India
e-mail: ravi@symphonytech.com

1 Introduction

Statistical data analysis and making decisions based on data are vital to assure manufacturing process quality. We discuss here scientific data analytics methods that go into various stages of the manufacturing process.

2 Decisions Based on Data

The importance of basing decisions based on scientific premises can never be overstated. Scientific premise is brought home by banking upon data and scientific analysis of data. Statistical analytics is all pervasive in any scientific quality and manufacturing operation. The drug development and manufacturing process is no exception. In this text we discuss statistical methods that relate to development and manufacture.

3 Scope

The following areas related to development and manufacture are focused upon.

- Statistics for Manufacturing and Quality: The principles of process design, process qualification, and continued process verification.
- Measurement and method validation
- Acceptance sampling

In all the above we shall base our discussions on best practices in the industry, guidance by the U. S. Food and Drug Administration (USFDA) and related standards like the ANSI standards which are cited in the guidance as well as those used by the industry.

3.1 Process Development and Manufacturing

The USFDA guidance on process validation [1] under current good manufacturing practices (CGMP) talks of three stages: process design, process qualification, and continued process verification.

Toward process design we shall explore the statistical technique of design of experiments abbreviated as DOE. Scientific design of experiments is a major tool in the arsenal of drug development and manufacture. It is used in both product development and process development. To understand the practice of DOE better we shall take examples from process development where DOE is commonly used.

The second stage of process qualification is an evaluation of whether the process is capable of repeatable commercial manufacture. Here, aspects of process capability, process stability, and process robustness are brought to the fore. We shall discuss the concepts and statistics of process capability evaluation. The indices Pp and Ppk and Cp and Cpk and the applications of these will be discussed.

The third stage of continued process verification deals with continued monitoring of the process for stability of critical quality attributes and critical process parameters. We shall discuss those elements of statistical process control that deal with monitoring of process stability. The role of Shewhart control charts will be brought forth.

3.2 Method Validation

The purpose of analytical methods is to quantitatively or numerically evaluate analytes and biomarkers. Method validation assures that the intended method reliably measures the intended analyte. Validated methods are important to evaluate and assure the safety and efficacy of drugs, and biological products. We shall discuss the various validation tests and their purpose. USFDA Guidance to Method Validation [2] details this aspect.

3.3 Acceptance Sampling

Acceptance sampling deals with accepting or rejecting a lot based on the outcome of inspected samples. The ASQ/ANSI Z1.4 [3] and Z1.9 [4] provide tables for acceptance sampling.

4 Design of Experiments

Design of experiments is a data-oriented technique for efficient experimentation. Process development involves experimenting with multiple factors to arrive at optimum settings of critical process parameters (CPPs) to deliver the best output in terms of critical quality attributes (CQAs) on the product being manufactured. Apart from achieving CQAs it is also important to attain efficiencies in the process. Good process development will also aim at efficiencies like optimum processing time, economizing on resources expended, and maximizing the yield of the process.

4.1   Efficient Experimentation

The number of settable parameters in a process can be dauntingly large. To experiment with each one independently and evaluate the impact of each on the desirable outcome can become a lengthy and expensive process. The key to efficient experimentation with DOE is to be able to simultaneously change all the candidate process parameters that we believe will impact the outcome of the process. Efficient experimental patterns that the DOE technique gives us are called orthogonal arrays.

Let us take an example of a process where we want to maximize the yield. Factors that are believed to be impacting are pH, agitation rate, and temperature. We want to experiment with various settings of these factors to find the setting where the yield is maximized. We set up an experiment at two levels of each factor. The factors and their settings are shown in Fig. 1.

An efficient experimental pattern comprising eight runs is shown in Fig. 2.

The pattern shown above is a full-factorial experimental design for three factors and two levels. Eight experiments are run with settings determined by the experimental pattern, and the response (yield, in this example) for each run is recorded. The following characteristics of the experimental pattern can be observed.

1. Each factor setting pattern (column in the array) is distinct from other factor columns.
2. Each experimental run setting pattern (row) is distinct from other rows.

The array is thus orthogonal. The experiment gives us independent information about the impact of each factor on the response (outcome of the experiment).

It can also be seen from the experimental pattern that the experiment is balanced. Each factor at each level expresses itself an equal number of times with all combinations of other factors. Each factor is expressed four times at the '+' coded level and four times at the '−' coded level. Taking an example of pH, the factor pH expresses itself at the '+' coded level two times when agitation rate is coded '+' (runs 1 and 2). pH expresses itself at coded level '−' too two times when agitation rate is coded '+' (runs 5 and 6). Such balance holds good in the experimental pattern for any combination of factors. A balanced experiment delivers unbiased information about the impact of each factor on the response.

It can be seen that the experimental pattern that each factor is set to each level four times. If we were to experiment with each factor independently, we would need to run 24 experiments (three factors run four times at two levels each) to get the

Fig. 1 Factors and levels of experimental design

Factor	Levels	
	-	+
pH	7.0	7.4
Agitation Rate(RPM)	8	10
Temperature °C	25	32

Run	pH	Agitation Rate	Temperature	Response- Yield
1	+	+	+	
2	+	+	-	
3	+	-	+	
4	+	-	-	
5	-	+	+	
6	-	+	-	
7	-	-	+	
8	-	-	-	

Fig. 2 Three-factor two-level full-factorial experimental design

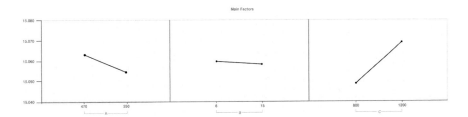

Fig. 3 Effect plots

same amount of information. It is possible to get the same amount of information in eight experiments due to the orthogonal nature of the experimental pattern. The experimental pattern can be run multiple times to get more data. These multiple repetitions are termed replicates. Replicates consume costs and time. Experimenters must trade off between benefits that accrue out of more replicates and the costs involved.

Responses in experiments are the CQAs that are required to be optimized in the process being developed. Responses are not only a function of multiple factors (CPPs) in the process but also interactions of a pair of factors. A full factorial experiment is able to resolve how the response gets impacted by individual factors as well as by the interaction of two factors. It is said that interactions of three factors or more do not exist in the real world. Hence evaluating an experiment for setting main factors and two-factor interactions is adequate.

Figure 3 shows a typical effect plot resulting from analysis of DOE data. An effect plot visually displays the magnitude and sensitivity of the relationship between each CPP (factor) and the CQA (response).

4.2   Fractional Factorial Experiments

The efficiency of experimentation can be further increased by packaging more factors in the experimental design than a full factorial experiment can address. This is done by confounding the setting additional factors typically with the interaction terms. Such experiments serve as screening experiments. These serve to quickly eliminate a large number of unimportant factors. Once unimportant factors are screened out, a full factorial experiment can be conducted to characterize the process with the main effects and interaction effects of each factor. Guidance on process validation recommends that detailed experimentation be done so as to evaluate the impact of all candidate factors and their interactions. Hence full factorial experiments are the recommended practice during process design.

4.3   Dealing with Noise

Factors in process development are CPPs. The experimenter needs to take cognizance of potential noise factors in the experiment. Noise factors are ones which contribute toward variation in the CQAs but are not of interest from the setting perspective. The strategy to deal with noise factors is one of the robust process designs where overall settings are done to minimize the variation transmission due to noise factors upon the CQAs.

4.4   Predictive Modeling

The model underlying the design of experiments is one of multiple linear regression. This is especially true when CQA targets are sought to be achieved as numerical values. When CQA targets like concentrations, content percentages are to be achieved, it is essential to determine the CPP-to-CQA relations for each CPP and CQA in terms of a transfer function. The transfer function is a linear prediction equation that has linear coefficients for each factor and each interaction to arrive at the value of the CQA.

In such cases in DOE, rather than using coded units of '+/−' for levels, uncoded units, which are the actual numerical settings of the factors are used. Categorical factors are assigned binary numerical values using a technique called dummy variables.

Analysis of DOE data leads to formulation of the prediction equation. In the process the model is validated using ANOVA, and the statistical significance of coefficients of factors is tested at a predetermined level of confidence. A detailed mathematical explanation of the process is beyond the scope of this text.

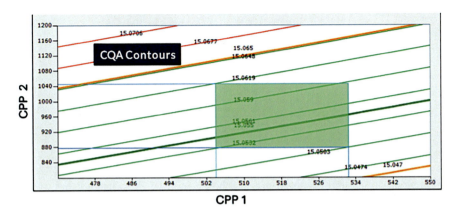

Fig. 4 Setting CPPs with contour plots

There are usually multiple CQAs asked from a process. Prediction equations that connect each CQA with CPPs are evaluated for multi-response optimization, to arrive at optimum settings of CPPs to deliver the best conformance to multiple CQAs.

Figure 4 shows a contour plot that plots contours or contoured surfaces of CQAs on a two- or three-axes plot. Each axis represents a factor. The figure illustrates how specifications on CPPs can be derived based on the tolerance limits on the CQA plotted.

Parameter settings will be in the normal operating range of the process which is narrower than the proven acceptable range which meets the quality criteria. Characterizing the process across the proven acceptable range will help in understanding the impact of deviations in process behavior.

5   Process Qualification

For process qualification, it is necessary to demonstrate that the process is capable of reproducible commercial manufacture. A good approach to the demonstration of such a capability is to evaluate the short-term and long-term process capability of the process under a commercial manufacturing environment.

5.1   Process Capability Evaluation

Statistical process capability evaluation however requires an adequately large number of batches, the number being typically 40 batches. Manufacturing such a large number of batches is usually not feasible prior to the commencement of commercial

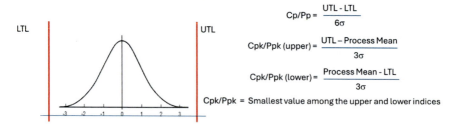

$$Cp/Pp = \frac{UTL - LTL}{6\sigma}$$

$$Cpk/Ppk \text{ (upper)} = \frac{UTL - \text{Process Mean}}{3\sigma}$$

$$Cpk/Ppk \text{ (lower)} = \frac{\text{Process Mean} - LTL}{3\sigma}$$

Cpk/Ppk = Smallest value among the upper and lower indices

Fig. 5 Process capability evaluation

production. Commonly observed practice in the industry is to manufacture three batches to demonstrate that the short-term variability is not significantly large and that commercial production can commence.

Process capability can however be evaluated as the batches progress, and there is adequate data to calculate. Thereafter, a running process capability evaluation can be done over a sliding window of the last specified number of batches. This can be done as a continued activity along with continued process verification for stability monitoring.

Process capability evaluation is done by comparing process variation and process centering with the tolerance limits. Evaluation of process capability indices Cp/Cpk and Pp/Ppk is done.

Figure 5 shows how process capability is evaluated by superimposing the process parameter or CQA histogram on the respective specification or tolerance limits.

Figure 6 visually shows various cases of process variation compared with tolerance limits. It can be seen that a process that is low variation and centered between tolerance limits is a capable one.

Another metric that is often talked about is parts per million. Parts per million is the area under the process distribution curve that extends beyond the tolerance limits. This represents the probability that the process would go out of tolerance. A low PPM value signals a good process.

The value of both Cp and Cpk (or Pp and Ppk) needs to be higher than 1. Commonly accepted industry practices say that the values should be higher than 1.33. A high value of process capability index Cp establishes that the process variation is narrower than what is allowable by specifications. The value of Cpk being close to that of Cp assures that the process is well-centered.

5.2   Two Sets of Process Capability Indices (Cp/Cpk and Pp/Ppk)

Cp/Cpk and Pp/Ppk differ from each other based on the way they are calculated.

If the process standard deviation (σ) is calculated by the root-mean-square method, the process capability indices evaluated are Pp and Ppk.

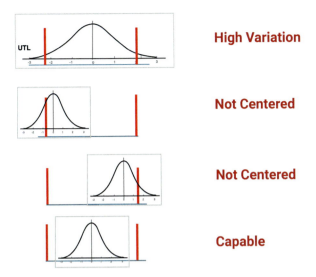

Fig. 6 Cases of process variation and centering

Process standard deviation can also be estimated from control charts that demonstrate the process to be stable. The process capability indices that are evaluated using this standard deviation are Cp and Cpk.

While the evaluation of the process capability indices Cp and Cpk is similar to the evaluation of indices Pp and Ppk the indices Cp and Cpk are used for different purposes than Pp and Ppk.

Pp and Ppk provide anecdotal information about the status of the process when data was collected. On the other hand, Cp and Cpk if coming from a stable process have a predictive value. The stability of the process is demonstrated by a Shewhart control chart being under control. If the process that is under control (meaning stable over time) continues to remain stable with the same control limits, it is an assurance that its Cp and Cpk will continue to be maintained, and it will remain capable in the long term too. Donald Wheeler brings out this point well in his literature.

Data that will be collected for continued process verification can also be used to evaluate running Process Capability in a moving window of time.

6 Continued Process Verification with Control Charts

The third stage of validation is continued process verification. It aims to demonstrate that the process remains in a stable state of control (validated state) throughout commercial manufacture. The guidance for continued process verification applies to CQAs as well as CPPs. Any instability or departure from the established state of control makes it necessary to requalify the process.

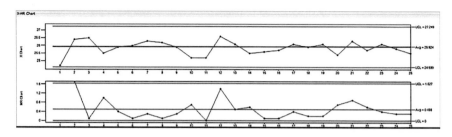

Fig. 7 Individual/moving-range control charts

A well-established tool for evaluating process stability over its progress over batches is the Shewhart control charts. Control charts were invented by Walter Shewhart in the 1920s to evaluate the stability of processes as they progress. The process being in control as signaled by a control chart tells us that the process is stable over a period. No significant change has occurred in the process.

When applied to pharma manufacturing process, the individual and moving-range charts work the best. It can be applied to numerically specified CPPs as well as CQAs. Every parameter monitored must have its own control chart.

6.1   Using a Control Chart to Monitor Process Stability

For every CPP or CQA monitored, the process should be observed for about 25 batches, and a baseline acceptable process established. After verifying that the Pp and Ppk indices of the processes are acceptable, Shewhart control limits are applied to the control charts (both individual and moving-range charts). All subsequent readings are plotted on the charts with carried-forward control limits. Charts within control demonstrate that no significant change has taken place in the process. When subsequently captured data points show an out-of-control signal with pre-established control limits, it is time to tweak and adjust the process.

Figure 7 shows individual and moving-range control charts with control limits.

Once control limits are established, process stability is monitored with preestablished control limits. Process capability indices Cp and Cpk are also monitored over a moving process window.

Figure 8 shows a schematic of the process of continued process verification.

Apart from out-of-control signals where data points go beyond control limits, rules that signal a drift in the control chart (Western Electric Rules or WECO rules) provide an early warning about drifts or changes in variation in the underlying process.

Process capability evaluation and control charting are valuable tools to comply with the guidance on process validation.

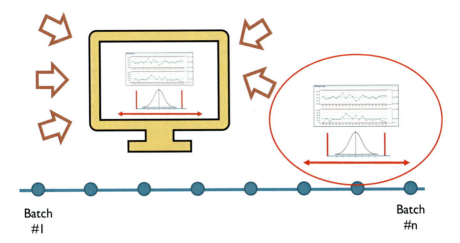

Fig. 8 Continued process verification over a moving process window

7   Method Validation

Method validation evaluates whether analytical methods are suitable for their intended analytical purposes. We enumerate here the various parameters of validation of methods and their importance. USFDA guidance on bioanalytical method validation describes these in detail.

- *Precision:* Precision is the degree of repeatability of the method when a homogenous analyte is repeatedly measured.
- *Accuracy:* Evaluates how close to the ground truth the analytical method measures.
- *Linearity:* Evaluates whether the analytical method measures in a linearly proportional way.
- *Specificity:* Evaluates whether the analytical method specifically measures the intended analyte and no other substance. It evaluates the ability to measure correctly in the presence of other substances.
- *Range:* Range is the interval within which the method measures with the desired accuracy, precision, and linearity.
- *Limit of Detection (LOD):* LOD is the lowest concentration of the analyte that can be detected by the method.
- *Limit of Quantitation (LOQ):* LOQ is the lowest concentration of the analyte that can be quantitated by the method within the required accuracy and precision.

Methods are measurement systems for analytes. A validated method would drive manufacturing processes and in turn deliver the intended process and product quality.

Detailed statistical treatment of method validation can be found in the USFDA guidance on bioanalytical method validation.

8 Acceptance Sampling

Manufacturing involves a lot of material that is bought out. It spans across active pharmaceutical ingredients (APIs), buffer solutions, solvents, culture media, and discrete materials like packaging and dispensing devices. The quality of incoming material needs to be verified before accepting it into the manufacturing process. Bulk material and reagents would come with a certificate of analysis. Sampling inspection may be necessary for the rest.

8.1 Sampling

Sampling is a method of inspecting a few samples in a lot and making decisions about accepting the lot based on the observations made on the sample.

Sampling is predominantly used for incoming lots of material received from suppliers.

Sampling is a trade-off between inspecting the entire lot and being 100% confident about the decision to accept or reject it.

Inspecting a sample from the lot and making a decision of lot acceptance is based on what you see in the sample. This entails a calculated risk of either wrongly rejecting a good lot or wrongly accepting a not-good lot. Inspecting the entire lot is not economically feasible due to constraints on time and inspection resources. Hence, acceptance sampling is practiced. This is done by designing the sample in a scientific manner, in such a way that the risk of incorrect acceptance as well as the risk of incorrect rejection is contained within a predetermined limit.

The concept of sampling risk is illustrated in the Fig. 9.

A good sampling plan strikes a judicious balance between both extremes.

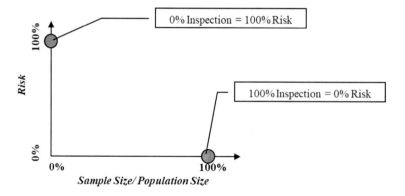

Fig. 9 Sampling and risks

8.2   Types of Sampling

Sampling and consequently sampling plans are classified as:

1. Attribute Sampling Plans: Sampling is done in order to classify product characteristics as acceptable and not acceptable.
2. Variable Sampling Plans: Sampling is done to measure a value of a parameter. Acceptance is based on whether the value lies within the limits of a specified limit of acceptance.

8.3   ANSI Z Standards

The most commonly recommended and used standards for acceptance sampling are the ANSI (American National Standards Institute) standards published by the American Society of Quality in 2003.

The ANSI/ASQ Z1.4 standard defines sampling procedures and tables for inspection by attributes. It provides tightened, normal, and reduced inspection plans for attribute inspection for percent nonconforming as well as nonconformities per 100 units.

The ANSI/ASQ Z1.9 defines sampling procedures and tables for inspection by variables for percent nonconforming. It provides tightened, normal, and reduced inspection plans for variable measurements.

8.4   Sampling Plans: Definitions of Terms

- *Lot:* A group of units or pieces submitted for inspection.
- *AQL:* This was defined as an acceptable quality level by earlier standards. This is replaced by the acceptance quality limit by ANSI Z 1.4. This is the fraction defective that is associated with acceptable quality.
- *LTPD:* Lot tolerance percent defective. This is the fraction defective that is associated with unacceptable quality. This is also referred to as RQL (rejection quality limit) or LQ (limiting quality).
- *N:* Number of units or pieces in a lot submitted for inspection.
- *n:* Number of units or pieces in a sample.
- *c:* Acceptance number. The maximum allowable number of defective units or pieces found in a sample for a lot to be accepted.
- *p:* Fraction defective of the process from which the lot was manufactured.
- *M:* Number of defects in the lot.
- *Pa:* Probability of acceptance of a lot at a specified fraction defective p.
- *α:* Producer's risk. The probability that a lot will be rejected at AQL. Also called a type I error.

- β: Consumer's risk. The probability that a lot will be accepted at LTPD.

8.5 Operating Characteristics Curves

Operating characteristic curves are a representation of risks associated with sampling plans.

On one end, AQL defines the fraction defective in the lot that defines an acceptable quality. A lot with AQL fraction defective should be rejected with an α probability, which means that the probability of acceptance of a lot with AQL fraction defective should be $(1 - \alpha)$. For example, if AQL is set to 2.5% and α risk set to 5%, a lot with 2.5% defective or 0.025 fraction defective should be accepted 95% (95% = 100%–5%) of the times the lot is inspected.

On the other end, LTPD defines the fraction defective in the lot that represents the quality that must be rejected most of the time. A lot with LTPD fraction defective could be accepted with a low β probability. For example, if LTPD is set to 7% and β risk set to 10%, a lot with 7% defective or 0.007 fraction defective should be accepted only 10% of the time the lot is inspected. This means that such a lot should be rejected 90% of the time it is sampled.

An OC curve passes through both the AQL and LTPD points and shows the probability of acceptance of the lot for intermediate values of fraction defective between AQL and LTPD.

The OC curve is a graph that has

- X-Axis: Percent nonconforming or nonconformities per 100 units.
- Y-Axis: Probability of acceptance of the lot in percentage.

The calculations of an OC curve are done based on sample size and lot size on the basis of underlying statistical distributions. A quality professional may look up values from the ANSI Z tables and configure sampling plans most suitable to the required quality norms.

Figure 10 shows a typical OC curve with explanations of the AQL and the LTPD points.

9 Modern Methods: A Note on Machine Learning

Machine learning has taken statistical methods to the advanced era of computers. Statistical methods have been practiced with dedicated statistical software developed and deployed for the purpose. These software have been programmed by expert statisticians and programmers to perform calculations and draw inferences from data.

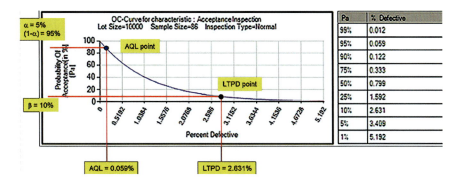

Fig. 10 Operating characteristics curve for a sampling plan

Matters have progressed now where computers are becoming autonomous and learning to function without having to be specifically programmed for the task. Computers will observe processes and phenomena and learn to perform complex tasks. This is in a nutshell, the world of machine learning.

In the scientific sector, machine learning is dominantly used in exploratory data analysis and development spheres. The use of machine learning for compliance is not yet common in times when this text is written.

9.1 Machine Learning Methods

Machine learning methods have evolved over the years. Major heads under which they can be viewed are:

1. *Supervised Learning*: Supervised learning is based on data that has labels. To train a machine for prediction, data that is used comprises predictors which are the factors causing the end results as well as the labels which are observed results. The machine observes data sets and figures out the underlying relationships between factors and outcomes. Methods of regression fall under this category. Predictive models that predict the classification of the outcome in multiple classes rather than predicting values are termed classification models. Linear regression, logistic regression, support vector machine (SVM), naïve Bayes classifiers, random forest, and K-nearest neighbor are some of the supervised learning methods that are practiced.

2. *Unsupervised Learning*: Unsupervised learning is not based on labels but aims at finding similarities in data sets and clustering the data sets by multiple similar dimensions. The methods explore similarity in various attributes in the data and cluster similar data points together. The aim is to find patterns in data that make data sets behave in a similar manner. K-Means, density-based clustering, and hierarchical clustering are some of the unsupervised learning methods that are practiced.

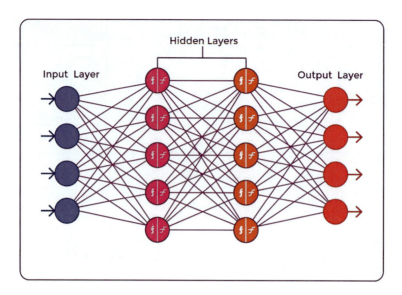

Fig. 11 Artificial neural network schematic

9.2 Deep Learning

Modern methods of machine learning are gravitating toward deep learning that uses artificial neural networks. Artificial neural networks are computer emulations of the human brain. Figure 11 shows a schematic.

A human brain is structured by neurons that are connected together through synapse networks. A computer equivalent is a network of connected perceptrons. A perceptron is a small computer program that performs a specific mathematical function.

Connected perceptrons are arranged in layers. Each layer gives the network some learning power. Mathematical functions in perceptrons, weights assigned to their connections and biases assigned to each layer build the math that helps the neural network to learn and build into it an ability to predict.

Neural networks have variants based on their architecture and functions. Convolutional neural networks (CNNs) are used for image recognition and processing and recurrent neural networks (RNNs) are useful where information is assimilated in steps and streams, with little information feeding into each layer of the network.

Advancements in neural networks have moved to generative AI where neural networks are able to creatively generate.

10 Conclusion

Statistical approaches described in this text aim to look at the goals of analysis and then discuss methods that satisfy these goals. The focus is on applications rather than a deep dive into statistical mechanisms and algorithms. These are described in detail in textbooks dedicated to statistics.

References

1. Health UDo, Services H. Process validation: general principles and practices. Washington, DC: US Department of Health and Human Services; 2011.
2. Fda U. Bioanalytical method validation guidance for industry. US Department of Health and Human Services Food and Drug Administration Center for Drug Evaluation and Research and Center for Veterinary Medicine; 2018.
3. Gore KL, Smith DL, editors. AOQL metric's robustness to manufacturing process instability. In: 2024 annual reliability and maintainability symposium (RAMS). IEEE; 2024. (ASQ/ANSI Z1.4:2003 (R2018) standard: Sampling Procedures and Tables for Inspection by Attributes).
4. Avramova T, Vasileva D, Peneva T, editors. An overview of the basic concepts and terms related to manufacturing process capability evaluation. In: AIP conference proceedings. AIP Publishing; 2024. (ASQ/ANSI Z1.9:2003 (R2018) standard: Sampling Procedures and Tables for Inspection by Variables for Percent Nonconforming).

Quality Risk Analysis: The Key to Failure Prevention

Ravindra Khare

Abstract Risk is defined as a potential for hazards and failures. Risk analysis done proactively helps mitigate the possibility of failures well before they occur. We explore here a scientific methodology of risk identification and proactive actions on causes that lead to failures. The focus of risk analysis and management is on prevention of occurrence of hazards rather than acting on failures after they occur. Structured approach of quality risk management (QRM) as recommended by USFDA guidelines is taken as a baseline for discussions. Our discussions address scope of risk management and scientific principles of the approach. We explore the structure of risk through all its elements and enumerate the process that leads to risk management and minimizing potential failures in processes. Failure mode and effects analysis (FMEA) is widely recognized as the underlying method of risk analysis and is practiced widely in the industry. We delve in detail into FMEA methods and their workflow. Risk analysis tools of fault tree analysis and criticality matrix evaluation are touched upon.

Keywords Risk · Potential failures · Failure effects · Failure causes · Preventive actions · Pharma risk · Quality risk management

1 Introduction

Quality risk management (QRM) deals with the prevention of hazards. Proactive prevention of causes that could lead to failures is an effective strategy to prevent failures from occurring. Risk management lays out a scientific methodology and a structured approach to attaining hazard-free processes and product realization.

R. Khare (✉)
Symphony Technologies Pvt Ltd., Pune, India
e-mail: ravi@symphonytech.com

531

2   Defining Risk

Risk is defined as a possibility of occurrence of an event leading to undesirable consequences.

USFDA Guidance ICHQ9 defines risk as the combination of the probability of occurrence of harm and the severity of such harm [1]. In the pharmaceutical business, there are multiple stakeholders who are subject to risk, the patients, drug manufacturers, medical practitioners, and the government to name a few. The focus of risk management is primarily on patient protection. The focus of risk is on product quality and availability in the hands of the patient. What is perceived as an undesirable event or harm can be thought of as arising out of a defective product, defective process that produces the product, or defects in the environment under which product manufacture, formulation, packaging, distribution, or dispensing is done.

It is known that defects occur due to failure of a process, a product, or the environment under which the product is manufactured, distributed, stored, and dispensed.

The essence of risk management therefore is to focus sharply on failures of all sorts.

3   The Potential Nature of Risk

Since the definition of risk is about the probability of harm, the nature of risk is potential, which means that risk management deals with failures that can potentially occur rather than the failures that have been encountered at some point in the past. In the pharmaceutical business, severity of harm (caused by one failure or another) will often be very serious and life-threatening. It is logical therefore that risk analysis be done by preempting what can potentially fail and lead to patient harm. Waiting for every type of failure to show up, then fixing it and preventing recurrence is not a recommended strategy of risk management. Failure prevention or defect prevention and consequently prevention of harm are the watchwords in risk management. Prevention is the key, and a proactive scientific approach is the methodology that needs to be adopted.

4   USFDA Guidance on Quality Risk Management

USFDA has defined quality risk analysis in their guidance ICH Q9 (R1) [1]. The version stood as updated in May 2023 at the time of writing this text [2] [3]. We shall refer to this guidance as *Guidance* in the text that follows.

5   Scope and Areas of Application Under the Guidance

The Guidance defines the scope of application of quality risk management tools to multiple aspects of pharmaceutical industry applications. The scope encompasses the following processes:

- Development
- Manufacture
- Distribution
- Inspection and submission review

Products covered by the guidance range are the following:

- Drug substances
- Medicinal drug products
- Biological and biotechnological products

The scope for medicinal drug products and biological products also extends to the raw materials, solvents, excipients, packaging, and labeling of such products.

6   Principles of Quality Risk Management

The Guidance underlines two basic principles of quality risk management.

1. *Quality risk evaluation should be based on scientific principles, and mitigation actions should ultimately aim at protection of the patient from harm.*

   Risk evaluation and management need to be essentially carried out by the process or the product owner (expert) rather than some risk expert. We shall understand further that to evaluate and prevent failures in product processes or systems, it is necessary to first understand the functions of what is under evaluation. Failure is a negation or derating of a function. Functions of products are best understood by those who design them. Similarly, functions of processes and systems are best perceived by those who design them or operate them.

   Risk management entails understanding the following:

   - Structure of risk.
   - Scientific tools and methods of risk evaluation.
   - Scientific tools and methods of risk mitigation leading to failure prevention.

     The role of a risk expert is to introduce and institutionalize scientific tools in the risk management process and teach scientific methods to product and process owners. Armed with such scientific understanding, technical personnel will work to deliver products and processes with fewer defects and of low intensity (severity). This is the pathway to protecting the patient from harm. A risk expert, however, should not perform risk analysis in isolation and without the involvement of process or product owners. Risk analysis works best when performed by a cross-functional team. Multiple facets of risk and its mitigation are evaluated and acted upon by members of the team. Interactions and discussions between team members are vital to bring about effective risk management.

2. *The amount of effort put into risk management should be driven by the level of the risk perceived.*

The second principle asks us to prioritize and focus on preventing failures that lead to high severity effects. To do so effectively, it is necessary to understand the structure of risk. We shall understand further that the structure of risk is multidimensional. An understanding of where in the structure of risk, the severity of potential harm, and the probability of occurrence of causes leading to such harm are perceived and acted upon.

7 Quality Risk Management Process

A systematic QRM process is a series of risk assessment, evaluation, control, and review actions. At every stage communication with other stakeholders is emphasized. Importance is given to the assessment of residual risk after each stage of mitigation and containment of risk. Residual risk in the system (product or process) under assessment needs to be communicated to all stakeholders to ensure visibility into the nature and quantum of risk that remains in the system after current mitigation actions have been taken.

Figure 1 is the flowchart of the risk management process, reproduced from the Guidance.

We shall read further in this text how the QRM process maps to the failure mode and effects analysis (FMEA) methodology. Each action block in the flowchart is mapped to steps in the systematic FMEA process.

The process comprises three action blocks.

- Risk assessment
- Risk control
- Risk review

At each action block there is a communication to stakeholders that is necessary.

8 Structure of Risk

As seen earlier, quality risk management can be done for products as well as processes that lead to product manufacture, storage, distribution, and dispensing (these process activities can be collectively termed as product realization). A structured representation of all elements of risk helps us effectively navigate through risk assessment, risk control, and review.

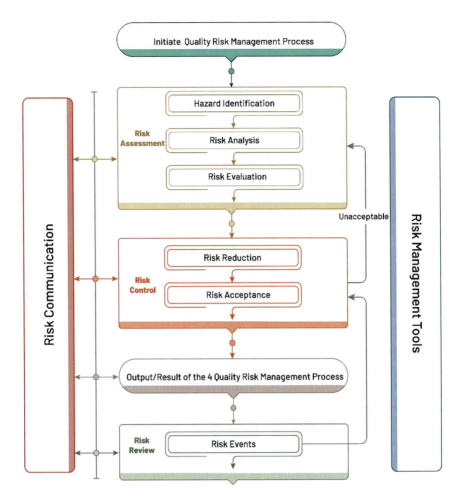

Fig. 1 Risk management. (Reproduced from the ICH Q9 (R1) guidance)

A continuous review process needs to be in place to monitor the following:

1. Whether actions taken to control risk have indeed resulted in reduction of risk to a level that is not significant to cause harm to the patient.
2. Whether the mitigation actions have resulted in occurrence or severity reduction of harm in a sustained manner. Continued surveillance becomes valuable in evaluating the sustenance of results.
3. Whether newer risks have come to light. Such risks need to be dealt with through the same structured approach.
4. Whether actions taken previously have led to unintended effects leading to new pathways of failure. Such failures will need another set of risk control actions.

Fig. 2 Multidimensional structure of risk

5. Whether prevention of the dominant failure has uncovered new types of failures that could not potentially prevail earlier.

9 Risk Assessment

The first step in risk assessment is to identify the hazard. Hazard is anything that may lead to undesirable impact upon products or processes and consequently upon patients.

Risk assessment stage in the QRM process guides the analyst to evaluate the following three questions in regard to harm.

1. What might potentially go wrong? What can potentially fail?—This is risk identification.
2. What is the probability that it will go wrong? What is the probability of failure occurring?
3. What is the severity of the consequences of such a failure?

A product or a process can fail to deliver its intended function in multiple ways. Each failure can be triggered by multiple causes. Each failure can lead to multiple consequences (effects). Risk is thus a multidimensional structure. Each failure can be represented as shown in the diagram below (Fig. 2).

Evaluation of risk needs to be done considering data-supported evidence about the three questions listed above.

9.1 Hazard Identification

Risk identification will be done considering what the team of analysts believes can potentially go wrong with the product or process under analysis. Here, the team's understanding of the pharmaceutical process becomes vital. Understanding functions of various elements of the process and its sub-processes and their causal relationships with the outcome is the starting point of hazard identification. This is called function analysis. Function analysis identifies the functions of the product and process. A function defines what a product or process should do. An effective hazard analysis never jumps to failures without first identifying functions. Functions identified should not only identify the primary intent of a product or a process. Some of the functions that go beyond the primary product or process intent would be the following:

- Primary intent of the product or process
- Patient safety and operator safety in plant operations
- Application ergonomics
- Packaging and shipping
- Quiet, noise-free operations of equipment and facilities
- Service life of the plant and peripheral equipment
- Customer usage profile for internal and external customers

The more detailed the function identification becomes, the richer, and more encompassing the hazard identification gets. Hazard identification is potential, and what can potentially go wrong rather than what has been seen to be going wrong is needed to be identified.

Negation of every function amounts to a potential failure also known as a hazard. A function can fail in multiple ways. To do a comprehensive hazard analysis, analysts need to identify all such potential hazards. Not all failures that may potentially occur can be identified at the initial round of hazard analysis. Risk analysis is a recursive process that is live through the operating life of products, processes, and infrastructure. New hazards will be identified as time progresses, risks analyzed and mitigated, making operations failure-free to a greater degree with every recursion.

9.2 Risk Analysis

Risk analysis links the risk with the likelihood of occurrence of causes leading to the hazard and the severity of harm. Here causal relationships of cause events leading to failure and failure further leading to harm are identified. This process is focused on identifying causes and effects of failure and linking them.

9.3 Risk Evaluation

Further to understanding the causal relationship of events leading to failure and the impact of such failure risk analysis focuses on evaluation of the outcome and prioritizing the actions that are needed to be taken to mitigate such risks. The outcome of evaluation can be a numerical priority number (like the risk priority number in FMEA) or a qualitative descriptor like "high," "medium," or "low." The outcome of risk evaluation serves as the basis on which priority of actions can be evaluated. More about prioritization is discussed further in this text while exploring the FMEA methodology.

10   Risk Control

Risk control works toward reducing the risk to a low level where the risk that remains in the system would not materially lead to harm or jeopardize patient safety. Risk control is the action part of risk management where actions are taken to reduce the risk that is identified in the previous stages.

Questions that drive the actions of risk reduction are:

- Is the risk material enough to deserve risk mitigation interventions?
- What actions can be taken to prevent the hazard from occurring?
- Are new hazards uncovered as a result of mitigation of dominant hazards?

10.1   Risk Reduction

Risk can be reduced only by actions that reduce the probability of occurrence of the causes that lead to failure. Such actions are actions of prevention of hazard. In case no immediate preventive intervention is possible, actions that help timely detection of causes that may lead to potential impending failure could be taken, and detection controls (alarms) could be instituted in the system. Such controls are, however, only interim measures. The focus of risk reduction needs to be on prevention of hazards rather than detecting them. Severity of harm is an inherent part of the nature of failure. So if the potential hazard does actually show up as a failure, the severity of its outcome is a given and cannot be reduced by mitigation actions. Actions should therefore focus on prevention.

10.2   Risk Acceptance

Risk being a probability of harm cannot be entirely eliminated. Reducing the probability of its occurrence to an acceptably low level and accepting the low residual risk as nonmaterial is therefore necessary. The specified acceptable level is decided upon by the analyst based on its level of impact and small probability of occurrence.

11   Residual Risk and Its Communication

Risk that is perceived at every stage in the risk management process as well as the residual risk that remains in the system needs to be communicated to all stakeholders. Industry, regulators, healthcare providers, and patients need to be made aware of the risks. They need to be updated about the risk mitigation status and residual risks in the system.

12   Risk Review

Risk management is an ongoing process. As actions are taken to mitigate risks and reduce failures in the system, new hazards are uncovered. A continuous review and systematic intervention make the system safer and less prone to failure. Risk management process thus is active through the lifecycle of the product or the process it addresses.

13   Failure Mode and Effects Analysis

The failure mode and effects analysis (FMEA) method, if practiced in the proper earnest, maps accurately to the workflow and intent of the ICH Q9 guidance. To practice FMEA correctly and derive risk mitigation and failure prevention results, it is essential to understand the structure of FMEA. It is also important to understand the meaning of every element of FMEA and the intent it serves. Figure 3 summarizes the structure of FMEA.

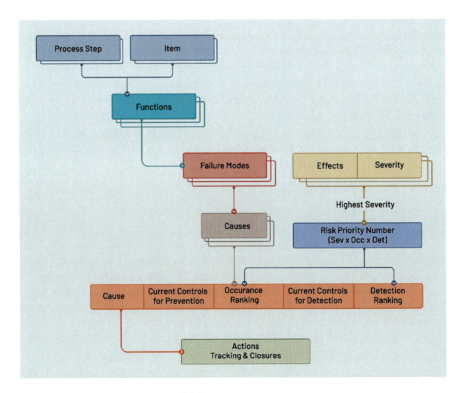

Fig. 3 Multidimensional structure of risk

13.1   Elements of FMEA

Each block represented in the diagram above is an element in the FEMA process. Let us look at each element and the purpose it serves.

13.2   Item or Process Step

Risk analysis can be done on products that are produced as a result of manufacturing operations as well as on processes. Under the context of this discussion, we focus FMEA done on processes. These can be manufacturing processes like growing cells in a bioreactor, supporting infrastructure processes like the HVAC system maintaining the required environment of auxiliary processes like shipping and packaging. Process step is a starting point of the analysis. What you perform FMEA upon is defined as the process step.

13.3   Function

Function is the useful work that the process should perform in order to deliver the output expected from the process.

Examples of functions are maintaining temperatures in an air conditioning system or delivery of metered fluids and drugs through an intravenous delivery mechanism. A function description is the answer to the question, "what should the product or process do to satisfy process intent?"

A lot many times the analysts stop at the primary function description. Functions may go way beyond only the primary one. Examples of a variety of functions are listed in the hazard identification section earlier.

13.4   Failure Mode

A failure mode is the negation of the function. It tells us how a function can potentially fail to work. Failure modes may describe various ways functions can fail. A function may not work at all, work only partially, or may work in a derated manner. All such failure modes need risk mitigation initiatives to prevent them from occurring.

Once failure modes are identified, the analysis further branches off into two branches. One branch is the effects branch and the other causes branch.

13.5   Failure Effects

Once a failure mode is identified, its effects will be analyzed in detail. A single failure will lead to multiple effects. Some effects are more severe in consequences, others less. The analyst must identify and log each one of the effects identified with the failure mode and assign a severity rating to each.

13.6   Severity Ratings

The severity rating is a number on the 1–10 scale that tells you about how serious that effect of failure is. Rating 1 denotes the least severe failure, while rating 10 denotes the most severe one. A failure mode will lead to multiple potential effects. Potentially a failure mode can lead to each one of the identified effects including the effect with the highest severity rating. Hence while deciding the priority for actions of the most serious effect, the highest severity rating among all is considered. Failure modes leading to fatality or serious disability of the patient are assigned the highest possible severity rating.

13.7   Causes and Current Controls

Causes are mechanisms or events that lead to failure. These are identified by the analyst. It is causes that actions are taken upon. Actions on causes should be aimed at preventing the cause from occurring, or even if it occurs existing controls of prevention should prevent it from leading to the failure mode. Current controls of prevention in the system are identified. These prevent the cause from occurring. Current controls of detection identify the current controls that help the analyst in detecting the causes. An early detection will help the causes to be identified and enable the analyst to take timely action to arrest the progression of failure.

13.8   Occurrence Rating

Occurrence rating (scale 1–10) is a rating that tells you how frequently a failure due to this cause can occur, despite the current control of prevention being in place. Modern thinking on this kind of analysis puts occurrence rating as the measure of effectiveness of the current control of prevention rather than an absolute number of occurrences of the cause.

13.9   Detection Rating

Detection rating (scale 1–10) tells you how easy or difficult it is to identify the cause or failure once it occurs, despite the current detection controls being in place. Similarly, detection rating is a measure of how effective the current controls of detection are.

13.10   Risk Priority Number

Risk priority number (RPN) is the multiplication of

- Highest severity rating associated with the failure mode
- Occurrence rating associated with each cause
- Detection rating associated with each cause

There has been a criticism about the RPN that it just becomes a number on a scale of 1–1000. It masks the relative importance of severity occurrence and detection ratings. Modern thinking and methods have replaced the risk priority number with a simpler action priority. Action priority is worked out on the basis of the same three criteria above. Action priority is expressed as one of the three categories: "high," "medium," or "low". It is easier to adapt and causes less confusion.

13.11   Actions

Improvement and risk mitigation actions are taken on causes. The author recommends that rather than prioritizing actions on a risk priority number, the following approach should be preferred.

- Prioritize actions based on severity.
- Aim toward putting in place controls of prevention. These will prevent failures from occurring.
- Detection controls can be used only in the interim where effective controls of prevention are not immediately found.

13.12   Multidimensional FMEA Structure

FMEA thus does not follow cause-effect relationship structures. FMEA is a treelike structure. A failure mode can lead to several effects. There is no way to prevent any of the effects from occurring once the failure mode shows up.

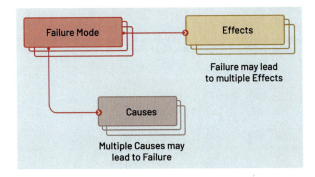

Fig. 4 Relationship between failure causes and failure effects

Also a failure mode can be caused by multiple causes. Either a single cause, a combination of causes, or a sequence of causes can lead to failure mode. For an encompassing risk analysis, all causes and their combinations need to be acted upon. Figure 4 shows the relationship between failure causes, failure modes, and failure effects.

14 Where Do You Find Failure Modes, Causes, and Effects?

Failure modes, causes, and effects are all varieties of failures. They are termed as modes, causes, and effects based on where they occur in the failure chain. Let's take an example of a failure chain from an orthogonal area to understand the terminology well. This thinking is described well in the *Verband der Automobilindustrie* guidance on failure mode and effects analysis.

14.1 Failure Chains

Figure 5 is a diagrammatic representation of the functioning and failures encountered in an ice cream freezer.

The element of focus upon which the risk-of-failure analysis is being performed is the freezer. Failure associated with the element of focus is the failure mode.

Failures in the elements and mechanisms that drive the freezer operations lead to the connected failures in the freezer. The connected failures in the elements and mechanisms thus become the failure causes. Before identifying the failures as causes, however, the causal relationships between the many failures in elements and mechanisms with the failures in the focus system need to be identified and verified.

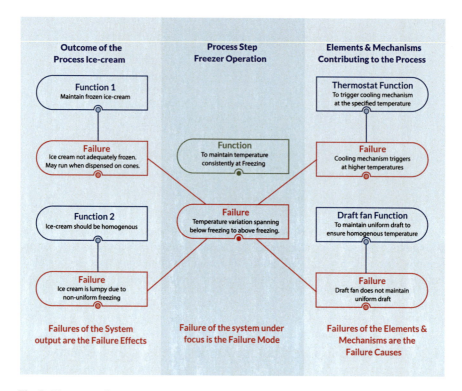

Fig. 5 Diagrammatic representation of failure chains

Failure in the focus element leads to multiple failures in the quality of the product or outcome of the system; in the example, the ice-cream becomes the failure effect. Again, the causal relationships between the failures of the focus element and the failures of the outcome need to be established before identifying failure effects in such a way.

15   Risk Analysis: An Ongoing Process

Risk analysis is an ongoing process. In a system that is seen to be running trouble-free, there is always a possibility of new hazards showing up. Preempting new potential hazards and instituting controls of prevention that will ensure that the hazard (potential failure) does not actually occur is the essence of risk management.

At every stage in the lifecycle of risk management, it is essential to communicate the residual risk in a system to all stakeholders and maintain visibility through the system.

16 Risk Evaluation Tools

There are several risk evaluation tools that support the mainstream risk analysis process. Some important ones are enumerated here.

16.1 *Fault Tree Analysis*

When risk analysis is done in complex systems, it is important to understand how failure in unit elements of the system leads to system failure. Here aspects of redundancy of mechanisms and critical elements are highlighted. A system redesign in terms of prevention of occurrence of high-impact failures will improve the risk potential of such systems. A sample fault tree depicting the risk in a firefighting system is shown in Fig. 6.

Calculations through the AND/OR gates give a good idea of probabilities of system failures based on individual failure event probabilities.

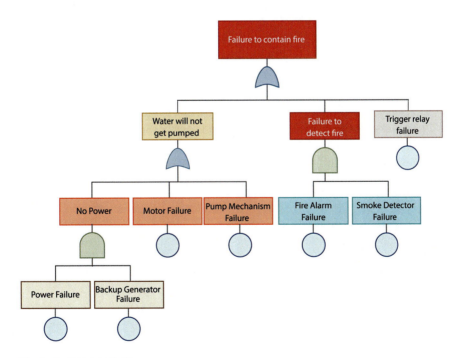

Fig. 6 Fault tree analysis

Fig. 7 Criticality matrix

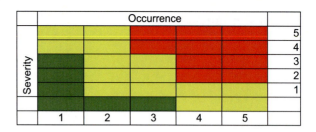

16.2   Criticality Matrix

Criticality matrix presents a summary of failure modes by classifying them in two ways. Each failure mode is classified on the basis of the severity of effects it leads to. Each one is also classified on the basis of the probability of occurrence of the causes leading to the failure mode. A failure mode is thus categorized in a box of the criticality matrix. The boxes are color coded to easily visualize the action priority. A sample criticality matrix is shown in Fig. 7.

17   Conclusions

Facets of risk analysis are discussed in the text above. The following bullet points summarize the essence of risk analysis:

- Risk analysis deals with potential hazards.
- Risk analysis is a structured scientific process that has a well-defined flow.
- Prevention of potential failures and failure avoidance is a preferred method of mitigating risks rather than building controls to detect impending failures.
- It is important to maintain visibility and transparency through the risk analysis process. All stakeholders in the system need to be updated about the risks that have been identified, analyzed, and mitigated. The residual risk in the system for every identified hazard should be communicated to stakeholders at every stage in the process.
- Risk analysis is a continuous process that is live through the lifecycle of the product or process that is under analysis. So long as the product is in active consumption and process in active state of running, risk analysis needs to go on. Newer hazards are uncovered and mitigated in the course of time and the incidence of failures is reduced with the progress of time.
- The FMEA process maps with the risk analysis guidance from regulators and works well toward the intent of risk mitigation.

References

1. Guideline IH. Quality risk management Q9 (R1). 2005.
2. Q9(R1). Quality risk management guidance for industry—U.S. Department of Health and Human Services. Food and Drug Administration; 2023.
3. O'Donnell K, Kartoglu U. QRM, knowledge management, and the importance of ICH Q9 (R1). Pharm Technol. 2024;48(7):20–31.

Regulatory Strategy

Narendra Chirmule

Abstract A well-executed regulatory strategy is essential for successful drug and biologics development and approval. Regulatory professionals, such as Global Regulatory Leads (GRL), navigate complex processes governed by the US Code of Federal Regulations (CFR) and relevant regulatory standards. Critical steps in this strategy include investigational new drug (IND) applications, biologics license applications (BLA), and new drug applications (NDA), where adherence to the U.S. Food and Drug Administration's (US-FDA) guidelines, such as Good Clinical Practice (GCP) and Current Good Manufacturing Practice (cGMP), is paramount. The FDA utilizes Prescription Drug User Fee Act (PDUFA) dates to streamline review timelines and hosts multiple meeting types (Type A, B, and C) for direct dialogue, often to resolve issues preemptively or ensure continuous development support.

Key stages also involve the Center for Biologics Evaluation and Research (CBER) and the Center for Drug Evaluation and Research (CDER), both pivotal in determining safety and efficacy standards. Regulatory flexibility and thorough preparation for Advisory Committee (AdCom) meetings, FDA audits, and inspections mitigate risks of Form-483 observations or Warning Letters. Effective strategies must prioritize chemistry, manufacturing, and controls (CMC) compliance and address Institutional Review Board (IRB) requirements, with clear action plans for managing all communication channels.

Keywords Job titles (GRL) · CFR · IND · PDUFA · BLA · AdCom · Regulatory Flexibility · IRB · GCP · CMC · CBER · CDER · Form-483 · Warning letters · Type 1, 2, 3, A, B, C. meetings · FDA audit

N. Chirmule (✉)
SymphonyTech Biologics, University of Pennsylvania, Philadelphia, PA, USA
e-mail: Narendra.Chirmule@symphonytech.com

© The Author(s), under exclusive license to Springer Nature Switzerland AG 2025
N. Chirmule, V. V. Ghalsasi (eds.), *Approved: The Life Cycle of Drug Development*,
https://doi.org/10.1007/978-3-031-81787-8_17

1   What Is a Regulatory Strategy?

Regulations, guidances, and laws of the land (Code for Regulation [CFR]) evolve and are developed to continuously ensure that drugs are safe and efficacious. Hence, the regulatory strategy constantly evolves over time. Since there are no drugs that have *no* side effects, the risks of potential adverse events of the drug have to be balanced with the efficacy and the unmet need for the disease indication. In a personal conversation with Peter Marks, M.D. (Head of CBER at US-FDA, for his brief bio see the section in the Introduction chapter), he mentioned a term "regulatory flexibility" to help figure out ways to address issues that arise during development. The degree of latitude of flexibility is dependent on the degree of unmet need for the disease. The regulatory agency is a partner in the drug development process, using a science and risk-based approach to ensure drug safety. The US-FDA website provides a wealth of information about all aspects of regulatory science [1].

The centerpiece of a regulatory strategy in a company is the culture of the organization. While all aspects of drug development have equal importance (much like all parts of our body), the cultures of organizations are highly influenced by the leaders of the organization, and the macro- and microenvironment. The culture of compliance, quality, and ethics are pillars to any organization. Pharmaceutical companies are unique in that they operate at the intersection of science, medicine, and commerce, with intense regulatory oversight and fierce competition of an ever-evolving market dynamics. Organizational cultures formed by legendary leaders such as Steve Jobs (Apple), Jack Smith (General Electric), Jeff Bezos (Amazon), Albert Bourla (Pfizer), Maurice Hilleman (Merck), and many more have been extensively studied. Hence one of the most important aspects of regulatory strategy is the people and their roles and responsibilities.

2   Who Does the Job and What Does the Job Entail
in Regulatory Departments?

To understand regulatory strategy, let's evaluate what the various types of jobs are required in regulatory departments in companies and also in the regulatory agencies (Table 1).

The entry-level position Typically, the educational requirements for entry-level positions in regulatory departments are Masters and PhD, in various fields of science and engineering, or physicians. Some knowledge, and better yet—experience, in any aspect of drug development is important. The required jobs at an entry-level position in the company first involve getting trained in all the regulations, the laws of the land, standard operating procedures (SOPs), and several hundred guidance documents in the various aspects of drug development. Most of these documents are available on agency websites as a search [2], such as documents for pharmacoge-

Table 1 Stages in the career in regulatory science

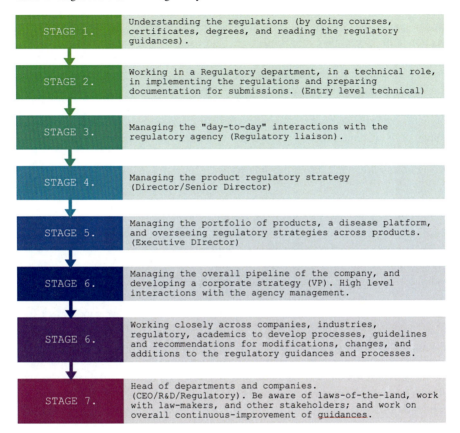

STAGE 1.	Understanding the regulations (by doing courses, certificates, degrees, and reading the regulatory guidances).
STAGE 2.	Working in a Regulatory department, in a technical role, in implementing the regulations and preparing documentation for submissions. (Entry level technical)
STAGE 3.	Managing the "day-to-day" interactions with the regulatory agency (Regulatory liaison).
STAGE 4.	Managing the product regulatory strategy (Director/Senior Director)
STAGE 5.	Managing the portfolio of products, a disease platform, and overseeing regulatory strategies across products. (Executive DIrector)
STAGE 6.	Managing the overall pipeline of the company, and developing a corporate strategy (VP). High level interactions with the agency management.
STAGE 6.	Working closely across companies, industries, regulatory, academics to develop processes, guidelines and recommendations for modifications, changes, and additions to the regulatory guidances and processes.
STAGE 7.	Head of departments and companies. (CEO/R&D/Regulatory). Be aware of laws-of-the-land, work with law-makers, and other stakeholders; and work on overall continuous-improvement of guidances.

nomics [3] and toxico-genomics [4]. The job can involve reading, commenting, and documentation of subject matter documents. Daily activities can include interacting with subject matter experts within the company on review and discussions on data, attending project meetings, and communicating any regulatory updates from the FDA, white papers, and guidances. Typically, early career regulatory professionals may support one or two drug projects in one therapeutic area. With gained experience, these individuals may explain the regulations, policies, and procedures to their counterparts in the various departments in the company and ensure compliance with the regulations. The entry-level position is predominantly "internally facing," i.e., majority of the interactions are with individuals and teams within the company.

Mid-career professionals Interpretation of regulatory guidances and regulations requires extensive training and, more importantly, experience. Hence, the understanding of regulations and implementation of regulated processes requires experiential learning. The next level of jobs in the industry can be the managers of regulatory personnel, who may be overseeing multiple products in various stages of development, and across several therapeutic areas (e.g., oncology, inflammation, metabolic diseases, etc.). These folks have extensive knowledge of the regulations and policies of the government, and they have excellent verbal and written communication skills and ability to train the staff and to deal with ambiguous situations. These level managers are liaisons and interact regularly with their equivalent counterpart personnel at the regulatory agency. They are involved in preparation for submissions and respond to information required by the agency. Usually, the key point of contact for day-to-day communications with the agency for the product. These mid-career roles are both internally and externally focused, with an emphasis on the goings-on within the company. The external interactions involve communications with the project manager at the regulatory agency, attending conferences, and comparing notes with peers across the industry.

Senior management The managers of managers, i.e., mid-management, are accountable for developing and implementing the regulatory framework of the drug development process for a drug. Usually, with the title of "global regulatory lead" (GRL) for the program, these individuals hold responsibility for coordination of all aspects of the drug development of a particular drug, such as mechanisms of disease pathogenesis and drug action, pharmacology, toxicology, product quality, and clinical trials. The GRL works closely with the product development and project management teams to plan the overall activities required for the five modules of the BLA, timelines, anticipate the risks, and plan risk mitigation strategies. The role oversees regulatory publications teams which have regulatory writers, database administrators, and scientific experts in the different functions, who are responsible for the final publishing of the BLA and ready for electronic submission to the FDA. These middle management professionals have equal involvement in internal goings-on in the company and also an eye on external happenings on regulatory activities in the agency, competitor companies, and academia. These individuals are representatives in professional organizations, such as the American Association of Pharmaceutical Sciences, BIO, and World Vaccine Congress, and also contribute to white papers and regulatory guidances.

Regulatory leadership roles The highest management level of regulatory professionals in companies (with titles such as Senior Vice President, Global Head) has several decades of experience in regulatory submissions, approvals, and especially failures and challenges. Their role in the company is predominantly external facing. They participate in meetings across the industry, and senior leadership of the regulatory agencies to evaluate the development of new regulations, modifying existing ones and addressing challenges in the regulatory landscape in the land.

In summary, the roles of personnel in a career contribute to understanding the regulatory strategy in a company. It involves much more than project activities and timing planning exercises. It involves multiple activities including (i) continuous interactions with the agency for advice on various steps; (ii) regular scientific and operational review of the progress by advisory committees; (iii) identifying and mitigating risks in the program, process, and product; (iv) effective team communications; and (v) training of staff and many more. Bringing all of these activities together, by developing goals, targets, and accountabilities, can comprise a regulatory strategy. And central to the regulatory process is the art of working with ambiguity, and constant change.

2.1 IND Submission

The journey of developing a drug is a multifaceted process that begins with rigorous preclinical testing and culminates in regulatory approval for marketing. At the heart of this process lies the investigational new drug (IND) application, a critical milestone that allows sponsors to initiate clinical trials in humans under the oversight of regulatory authorities such as the U.S. Food and Drug Administration (FDA). The IND application serves as a request for regulatory approval to conduct clinical trials of the investigational products in humans. It is a comprehensive submission to regulatory authorities that provides detailed information on the drug substance, drug product, manufacturing processes, preclinical pharmacology and toxicology studies, and proposal protocol for clinical trials. The IND application is designed to demonstrate the safety, efficacy, and quality of the investigational product while ensuring the protection of human subjects participating in clinical trials.

The preparation of an IND application is a meticulous process that requires careful planning, coordination, and documentation. Sponsors must compile all relevant data and information to support the safety and efficacy of their investigational product. The investigator's brochure (IB), a comprehensive document summarizing the preclinical and clinical data, plays a crucial role in supporting the IND application. Preclinical data include pharmacology and toxicology studies conducted in laboratory animals. These studies provide critical information on the safety profile, pharmacokinetics, and pharmacodynamics of the investigational product. These studies establish the range of doses of efficacy, and maximal tolerated doses for safety. These studies are conducted in accordance with regulatory guidelines and standards to ensure the validity and reliability of the data submitted in the IND application. The chemistry, manufacturing, and controls (CMC) section of the IND application provides detailed information on the composition, manufacturing process, and quality control of the investigational product. This includes data on the drug substance, drug product formulation, manufacturing facilities, and stability testing. The product must be manufactured under GMP regulatory standards to demonstrate process consistency and reproducibility.

The clinical protocols are detailed plans that outline the objectives, study design, patient population, treatment regimens, endpoints, and statistical analysis plan for each proposed clinical trial. The protocols and processes of conducting trials require adherence to Good Clinical Practices (GCP) guidelines. These protocols are submitted as part of the IND application and are reviewed by regulatory authorities to ensure safety and scientific validity while minimizing risks to human subjects participating in the trials. Before initiating any clinical trials, approval from an Institutional Review Board (IRB) or Ethics Committee must be obtained for each study site. The IRB review process evaluates the ethical, scientific, and regulatory aspects of the proposed research to ensure the protection of human subjects.

Once the IND application is submitted to the U.S. Food and Drug Administration (FDA), the agency conducts a comprehensive review to assess the safety, efficacy, and quality of the investigational product. The FDA evaluates the preclinical and clinical data, as well as the CMC information, to determine whether the proposed clinical trials can proceed. The FDA review process may involve multiple rounds of communication and clarification between the sponsor and the agency to address any questions or concerns.

Once the IND is approved, the sponsor has ongoing responsibilities to ensure the safety and integrity of the research, which includes monitoring and reporting adverse events, submitting protocol amendments, and providing updates to regulatory authorities as required. The approval of the IND is a landmark event in the life cycle of the drug development process. The Table 2 below shows the list of activities involved in preparing for the IND submission and the potential costs involved in each of the activities.

These are estimated times and costs, which will give the reader an approximate expense planning template. The monetary values will be different depending on several aspects, including geographies (United States, EU, Japan, etc.). The time taken depends on the clinical indications, number of subjects being treated, large population studies versus rare diseases, etc. *Note: The timing in years for each step is extremely conservative, and for a mid-sized company with limited past experiences. The goal of efficient drug development is to reduce the timelines from discovery to approval from at least two decades to 5 years. This is a formidable task in operations but has been done in COVID-19 vaccines and some rare diseases.

3   The PDUFA Date

The PDUFA Act stands for the Prescription Drug User Fee Act [5]. It is a United States law passed in 1992 and subsequently reauthorized several times, with the latest reauthorization being in 2017. The primary purpose of the PDUFA Act is to allow the U.S. Food and Drug Administration (FDA) to collect fees from drug manufacturers in order to fund the review process for new drug applications. These fees help to expedite the drug approval process by providing additional resources to the FDA. In exchange for these fees, the FDA agrees to certain performance goals, such as specific timelines for reviewing and acting upon drug applications. The PDUFA Act has been credited with improving the efficiency and timeliness of the drug approval process in the United States.

Table 2 A list of activities and milestones for IND submission

Development of the drug	Timeline (years)	Cost ($M)
Biology (pathogenesis of disease, and mechanism of action of drug) (basic research)	3	1–2
Use of AI/ML		
Product		
Gene therapy (AAV)		
Cell therapy (stem cells, engineered cells)		
mRNA		
Biologic/small molecule		
ASO (exon-skipping)		
Process development	2	2–4
Non-GLP drug substance		
Non-GLP drug product		
Analytical development, critical attributes testing, and stability testing		
Pharmacology studies (efficacy)	2	1–2
In vitro binding and functional assays		
In vitro cell-based assays and organoid models		
Animal model 1		
Animal model 2		
Analytical development (potency assays, lot-release assays, characterization assays)		
Risk assessment and mitigation		
Manufacturing of GMP DS/DP manufacturing & packaging	2	3–5
Toxicology studies (safety)	1	1–2
POC toxicology: cell-based/in vitro/rodent		
Nonhuman primate study 1		
Nonhuman primate study 2		
IND application (MOA, process development, pharmacology, toxicology, and phase 1 protocol)	5–8	1
Phase I clinical studies (safety)	2	1–3
Analytical (PK/PD, toxicology, immunogenicity assays)		
Regulatory reporting		
Clinical studies (Confirmatory studies, phase II and III) using validated clinical endpoints	5	10–30
Regulatory reporting		
BLA filing		20–50

3.1 The BLA Submission: The War Room

It is a special area in a company, generally a set of conference rooms that are dedicated to the team which is preparing for submission of the million-page biological licensing application (BLA) filing to the regulatory agency. The initiation of the BLA writing and collating process starts when the positive results of the phase III clinical trials have been achieved. Public announcements have been made (within 24–48 h of unblinding of pivotal clinical trials—lest the results have impact on insider trading in the stocks of the company), patients are eagerly anticipating the approval and access of the drug, and the individuals and teams who have contributed to these successes for the company are expecting rewards and recognition. At this stage, management approves the commit-to-file stage of the drug development stage (i.e., approves and releases the funds required for the BLA filing). The teams are assembled, and there is an "auspicious" start, after the leadership (generally heads of R&D, manufacturing, quality, etc.) addresses the team and gives a motivational pep talk to initiate and complete the BLA filing process (which usually may take 6 months of extremely tight timelines). The project management team maps out a very detailed plan of all the activities, timelines ("beat the timelines"), quality checks, and quality assurance audits. Highly experienced project managers effectively juggle many balls to help the team reach its goal.

The core team consists of global leaders for overall clinical development (GDL), generally an M.D., PhD., product manufacturing (GPL), marketing (GML), pharmacology, toxicology, biomarker, and early development. These teams are supported by project management, regulatory affairs, statistical teams, and medical writing teams. The quality control and quality assurance teams ensure the review and appropriateness of the documents that are to be submitted. Each team has subteams, e.g., (i) the pharmacology teams, which are responsible for writing the summary-of-pharmacology section of the BLA and collect information and reports from bioanalytical teams, biomarker teams, pharmacogenomics teams, quantitative pharmacology teams; (ii) the manufacturing teams, which are responsible for the chemistry-manufacturing-controls (CMC) section of the BLA and obtain reports from process development teams (upstream, downstream, formulation), analytical teams, and supply chain teams. Hundreds of individuals are involved in the BLA filing process. There are at least three levels of reviews: the immediate manager, their manager, and the department head. There are many iterations of comments and responses. All reviews are approved and signed off (which means, taken accountability). A digital system, such as an electronic notebook, becomes necessary for the volume of work that needs to be accomplished. The energy and enthusiasm of the team is palpable. The people resources team ensures that each member of the team has all the social, psychological, and personnel support required for the filing. Tensions rise when timelines come close and the risk of missing them increases. Escalation of "issues" to the heads of departments and above occurs more frequently with time. Effective prioritization of projects and workloads and utilization of allocated resources, decision-making, and verbal and written communication are critical skills required. No one (or department) wants to be on the "critical path." My personal experiences on the "critical path" are final clinical study reports waiting for results follow-up safety data from the last few patients in the trial, manufacturing teams, collecting the last data sets of product stability studies (the longer the stability data, the better claims for distribution and marketing after approval). Finally, it all comes together. The regulatory publishing teams collate the five modules of the BLA. The FDA, by law (the PDUFA Act), takes 18 months to review the BLA, performs a detailed (re) analysis of the data sets, and sends questions, known as "information requested (IR)," for which responses need to be provided within 24–48 h. There can be as many as a thousand IRs during the course of the review. These requirements of timelines are very stringent. The FDA also has an equivalent team structure, with clinical teams, manufacturing teams (which review the CMC section), pharmacology teams, and toxicology teams. There is a project manager, and project management team, which is the single point of contact for communication with the company regulatory representatives. FDA management is also heavily involved in the review processes. There is a lot of coordination and teamwork with regard to the IRs.

It is a complicated, multifunctional process, which cannot be easily explained in a book, or lecture, it has to be experienced!

How does one prepare to write a BLA and respond to questions from the FDA? What training is required during academic training processes? All the steps in the regulatory interactive process are generally learned "on the job." Each employee experiences these BLA submissions a couple of times during their career. Each filing teaches what to do better, and more importantly, what not to do. It's an endless cycle of learning. The team learns. Management learns. The FDA also learns.

3.2 The FDA Advisory Committee (AdCom)

One of the central processes of the evaluation of the BLA submission by the US-FDA for novel drugs is the advisory committees (AdCom). The presentation to the AdCom is accessible to the general public at the US-FDA offices in Silver Springs, MD, and broadcast live on YouTube. The committee is comprised of independent experts of various fields in various aspects of drug development. The composition of the committee includes clinicians, researchers, statisticians, industry representatives, and also patient representatives, who are voting and nonvoting members. The goal of the committee is to conduct a thorough, extensive, and balanced analysis of the data and data analysis presented in the BLA filing on safety and efficacy of the drug.

The meeting begins with the FDA staff presenting the rules and regulations of the process of enquiry during the meeting. The briefing documents, prepared by the company, which summarize the major aspects of the BLA, i.e., the preclinical, clinical, and manufacturing data to establish efficacy and safety, are made available to the public on the FDA website, a few days before the meeting. The FDA also invites anyone wishing to make comments about the drug, disease, or any relevant topic, to present at this public meeting (albeit with prior appointment). Transparency of the review process is paramount in the entire process.

The first presentation is on the disease indication by an expert. Each presentation is followed by a question and answer session. The next presentation is typically done by the company which has submitted the BLA. This presentation details the entire drug development process, from the mechanism of the drug to product characteristics, pharmacology, toxicology, and clinical trials. This 2-h presentation is one of the most important activities by the company team for each drug. The presenters have been trained for this presentation for several months. These training sessions involve rigorous practice sessions of answering hundreds of questions, and PowerPoint slides. Typically, very senior members of the team (Vice Presidents of departments) present in AdComs. The skills required include crystal clear verbal communication skills, a calm demeanor, charisma, knowledge of the entire drug development process for the drug, and the ability to delegate the responses to the questions to appropriate experts. The support team sits in the "dug-out" in order to answer questions that follow the presentation. [Note: one can watch many recorded sessions of live AdComs on YouTube to get an idea of the gravitas of this important presentation].

The next presentation is done by the FDA review committee. This is the FDA team that has thoroughly reviewed the BLA filing and prepared a summary in the FDA version of the briefing document. This presentation and document is a version of the data and interpretation of the BLA filing by the FDA. Many, if not most of the time, not all aspects of the interpretation of the data by the company and FDA are the same. Hence it is up to the AdCom to ask appropriate questions to get to the "truth" and decide, by voting, if the evidence provided in the BLA supports the drug is safe and efficacious.

Before the vote by the AdCom, the general public has an opportunity to provide comments during the meeting. Each presenter is given 3 min, and there can be more than 20 public comments, usually for and against recommendation for approval.

Finally, it is time to vote. The committee votes on the question "Is there reasonable assurance that the DRUG is safe for use in patients who meet the criteria specific in the proposed indication?" and "Do the benefits of the DRUG outweigh the risk for use in the patients who meet the criteria specified in the proposed indication?" The AdCom votes and provides the FDA with the justification for their vote. It is a nail-biting, usually 8-h (9 am to 5 pm) meeting, and the fate of the decade's years of development, and millions of dollars spent, comes down to this vote.

3.3 An FDA Audit

Experiencing an FDA audit can be a rigorous and sometimes stressful process for pharmaceutical companies, as it involves thorough scrutiny of their manufacturing processes, quality control systems, and compliance with regulatory requirements. Let's walk through the stages of an FDA audit, from receiving the audit plan to addressing findings and responses.

Typically, a pharmaceutical company will receive advance notice from the FDA regarding an upcoming audit. This notice includes the audit plan, which outlines the scope, objectives, and schedule of the audit. The audit plan may specify which facilities will be inspected, the areas of focus, and any particular documents or records that the FDA inspectors will review. The audit can either be planned, e.g., after completion of the scientific review of the BLA, or annual audit, or a for-cause audit.

On the day of the audit, FDA inspectors arrive at the company's facilities to conduct the inspection. The inspection team may consist of subject matter experts in areas such as manufacturing, quality control, regulatory affairs, and compliance. The audit starts with an opening meeting with the management team and auditors. The "most responsible person" of the manufacturing site is defined; it is usually the CEO, if not COO, or head of manufacturing. After the introductions of all the key personnel, the audit plan is reviewed, and the audit is initiated. There is a dedicated room or area designated for the auditors to conduct the review. Inspectors conduct interviews with personnel, review documentation, observe manufacturing processes, and assess the overall compliance with regulatory requirements. Throughout the audit, inspectors may request for information, directly or indirectly related to the BLA submission.

An example of an audit situation could involve the following. The auditor requests for the deviation log for the fermenter. Upon review of the log, the auditor notices that the fermenter temperature control was out of calibration. The auditor may request the preventive maintenance SOP for the fermenter temperature control process. At this time, the regulator may request the qualifications and competence of the author of the SOP and the training records of the operator. Such detailed audits enable the auditor to examine the overall quality management processes involved in the manufacturing of the product.

After completion of the review process, the auditor can give the following types of responses, in the document From FDA 483. These observations may range from minor procedural issues to more significant deviations from regulatory requirements. If the FDA determines that the deficiencies identified during the audit are significant and pose potential risks to public health, they may issue a warning letter. A warning letter outlines specific violations of regulations and requests prompt corrective action from the company. It may also indicate potential regulatory enforcement actions if the issues are not resolved. A complete response letter (CRL) is issued by the FDA to inform a pharmaceutical company that the agency will not approve their application in its current form. It outlines specific deficiencies or issues that need to be addressed before the application can be approved. Another type of letter issued by the FDA is refusal to file (RTF) when incomplete or inadequate information is provided for review. It indicates that the application will not be accepted for review until certain deficiencies are addressed. The other types of responses can be categorized into critical, major, and minor observations. The company is typically given a specified timeframe to respond to the observations outlined in the Form FDA 483. Responses may include corrective actions taken to address the deficiencies, plans for implementing corrective measures, and documentation supporting compliance efforts. In some cases, the FDA may conduct follow-up inspections to verify that corrective actions have been implemented effectively and that compliance has been achieved.

After all the responses to the FDA questions have been sufficiently provided, the FDA provides an approval letter to notify a pharmaceutical company that their application has been approved. It grants permission to market and distribute the drug. Residual deficiencies that may still exist after the review process are required to be addressed in post-marketing requirements (PMR). These are studies or clinical trials required by the FDA as a condition of approval for a drug. They are conducted after the drug has been approved and marketed to gather additional information about its safety, efficacy, or optimal use.

During an FDA audit, inspectors may ask a wide range of questions related to various aspects of drug development, manufacturing, and quality control. While the specific questions asked can vary based on the nature of the audit and the specific circumstances of the pharmaceutical company being inspected, here are some examples of questions that may be asked during an FDA audit:

1. Can you provide an overview of your company's quality management system (QMS) and its implementation?
2. How do you ensure that your manufacturing processes comply with current good manufacturing practices (cGMP)?
3. Can you describe your procedures for handling deviations, investigations, and corrective and preventive actions (CAPA)?
4. What controls are in place to ensure the identity, strength, quality, and purity of your drug products?
5. How do you monitor and control environmental conditions in your manufacturing facilities?
6. Can you provide documentation for the validation of your manufacturing processes and analytical methods?
7. How do you ensure that raw materials and components used in your drug products meet specifications?
8. What measures do you take to prevent cross-contamination during manufacturing?
9. Can you describe your procedures for cleaning and sanitation of equipment and facilities?
10. How do you ensure the integrity and security of electronic records and signatures?
11. Can you provide documentation for the training of personnel involved in manufacturing, testing, and quality assurance?
12. What measures do you take to ensure the safety and efficacy of drug products during storage and distribution?
13. How do you handle complaints and adverse events associated with your drug products?
14. Can you provide documentation for the qualification and monitoring of suppliers and contract manufacturers?
15. What measures do you take to ensure the safety of personnel working in your manufacturing facilities?

16. How do you conduct risk assessments for your manufacturing processes and drug products?
17. Can you describe your procedures for conducting stability studies to support expiration dating of drug products?
18. How do you ensure that changes to manufacturing processes or facilities are properly evaluated and controlled?
19. What measures do you take to prevent counterfeiting and ensure the authenticity of your drug products?
20. Can you provide documentation for the annual product quality review (PQR) of your drug products?

These questions cover a broad spectrum of topics relevant to pharmaceutical manufacturing and regulatory compliance and are typical of those asked during FDA audits to assess a company's adherence to regulatory requirements and standards.

3.4 Regulatory Communications

Communication with the regulatory agency is a critical aspect of the drug development process. These communications involve formal processes and milestones that guide the interaction between pharmaceutical, biotechnology, and the FDA. The formal processes and milestones involved in regulatory communications with the FDA, exploring the stages from pre-submission meetings to post-market surveillance.

Pre-submission Meetings Before submitting a formal application to the FDA, companies often engage in pre-submission meetings to discuss their product development plans, regulatory strategies, and any potential issues or questions. These meetings serve as an opportunity for companies to seek clarification on regulatory requirements, obtain feedback on their development plans, and address concerns early in the process. The FDA encourages sponsors to request pre-submission meetings to ensure that their submissions are complete and well-prepared. These meetings can help streamline the regulatory review process and mitigate potential delays or deficiencies in the application. Key topics discussed during pre-submission meetings may include study design, endpoints, patient populations, manufacturing processes, and labeling.

Investigational New Drug (IND) Application In addition to the topics discussed in the section above, the IND application is the first detailed written communication to the FDA which details the intent of the drug for treatment of the clinical indication of the disease. Upon submission of the IND application, the FDA conducts a thorough review to assess the safety and feasibility of the proposed clinical trial. The agency evaluates the study protocol, the qualifications of the investigators, and the adequacy of the manufacturing facilities. If the FDA finds the IND application acceptable, the sponsor can proceed with the clinical trial.

From the time of the IND to the approval of the BLA, there are several formal interactions between the sponsor and the FDA. The table below lists the major types of meetings.

Meeting type	Purpose
Type 1	Clarification on FDA regulations, guidance documents, and general advice on development plans
Type 2	Discussion of IND submission, including study design, preclinical data, and plans for clinical trials
Type 3	Review of NDA submission, including clinical data, manufacturing, labeling, and post-marketing plans
Type A	Clarification on regulatory requirements, labeling, and general advice on development plans
Type B	Discussion of premarket submissions, including 510(k), PMA, IDE, De Novo, or HDE applications
Type C	Meetings not covered by types A or B, including device modifications, quality system issues, etc.

Each of these meetings is strategically planned. The questions, along with the background data, information, and interpretation are provided to the FDA. The responses by the FDA can be written responses or in-person meetings. The preparation for each of these meetings requires careful review processes within the company. The first drafts are written by the subject matter experts, edited by experienced regulatory writers, and reviewed by management. The larger the company, the more layers of review.

Biologics License Application (BLA) Once clinical trials are completed, sponsors submit the BLA to the FDA for review. These applications contain comprehensive data on the safety and efficacy of the investigational product, including results from preclinical and clinical studies, manufacturing information, and proposed labeling. The FDA conducts a rigorous review of the BLA to evaluate the benefits and risks of the product, followed by the AdCom, and site audit (described above). The review process involves multiple disciplines within the FDA, including medical reviewers, pharmacologists, statisticians, and manufacturing experts. The agency assesses the quality of the data, the robustness of the study results, and the adequacy of the risk management plan and sends questions termed as "information requested (IR)-and the number" (e.g., IR-121). The turnaround time for responses to the IR is usually days. So every time the regulatory liaison receives the IR from the agency, the question is sent to the appropriate subject matter expert to respond. During a BLA review process, there can be a few thousand questions, over a period of 12–18 months. A formidable, full-time stressful job.

Advisory Committee Meetings As described in the above section, in some cases, the FDA may convene an advisory committee meeting to solicit expert advice on a particular product or regulatory issue. Advisory committees are composed of external experts in relevant fields, such as clinical medicine, pharmacology, and biostatistics. These committees review data presented by the sponsor and provide recommendations to the FDA regarding the approvability of the product.

Post-approval Requirements Following approval of a drug or biologics product, sponsors are required to fulfill certain post-approval requirements to ensure ongoing compliance with regulatory standards. These requirements may include post-marketing surveillance studies, risk evaluation and mitigation strategies (REMS), and periodic safety reporting. Post-marketing surveillance studies are conducted to monitor the long-term safety and effectiveness of the product in real-world clinical practice. These studies may involve large-scale observational research, registry-based studies, or post-approval clinical trials. The FDA uses the data from post-marketing surveillance studies to assess the risk-benefit profile of the product and make regulatory decisions as needed.

Risk evaluation and mitigation strategies (REMS) are comprehensive risk management plans designed to ensure that the benefits of a drug or biologics product outweigh its risks. REMS may include elements such as restricted distribution programs, patient education materials, and healthcare provider training. The FDA may require sponsors to implement REMS for products with significant safety concerns or potential for misuse.

Periodic safety reporting involves the submission of regular safety updates to the FDA to monitor adverse events, emerging safety signals, and other relevant safety information. Sponsors are required to submit periodic safety reports at specified intervals throughout the product lifecycle, as outlined in the approved risk management plan.

Thus, effective regulatory communications with the regulatory agency which involves formal processes can guide the sponsor during the life cycle of drug development. The resources to understand the communication processes include information on the agency websites (as guidances) or training programs by professional agencies such as RAPS.

4 Case Study 1. Amgen Versus Regeneron in the Race for Anti-PCSK9 Antibody Development

Development of biotherapeutics for treatment of cardiovascular diseases has been undertaken to transform the management of this disease. Among these, the inhibition of proprotein convertase subtilisin/kexin type 9 (PCSK9) to reduce cholesterol has emerged as an important pathway in managing hypercholesterolemia. Amgen and Regeneron, two biopharmaceutical giants, have been at the forefront of this battle, racing to bring their respective anti-PCSK9 antibodies to market. As a case study for regulatory strategy, this "story" describes the intricacies of this competition, exploring the scientific innovations, regulatory hurdles, and market dynamics that are involved.

Understanding PCSK9 and Its Role in Cardiovascular Health PCSK9 is a protein primarily produced in the liver, playing a crucial role in regulating plasma low-density lipoprotein cholesterol (LDL-C) levels by modulating the degradation of LDL receptors. PCSK9 chaperones LDL-C and binds to LDL receptors leading to the degradation of cholesterol, downregulation of LDL receptors, and PCSK9. In the absence of PCSK9, LDL-C binds to LDL-R and is internalized into endosomes in hepatocytes; the LDL receptor is recirculated back to the cell surface, enabling higher level of clearance of cholesterol. Monoclonal antibodies targeting PCSK9 have been shown to reduce circulating LDL-C levels. Both Amgen and Regeneron were developing anti-PCSK9 antibodies for treating patients with cardiovascular disease.

In 2008, the enthusiasm for development of the anti-PCSK9 antibody as an approach for the treatment was validated by the discovery of a rare clinical indication, with individuals with a mutation in PCSK9. These subjects demonstrate extremely low levels of cholesterol and reduced cardiovascular incidence. The Amgen team had initiated preclinical studies and demonstrated unprecedented 60–80% reduction in LDL-C in nonhuman primate studies. Senior management in the company prioritized the development of this molecule to fast-track. Each department was instructed to determine the operational steps that could enhance the speed of development. The process development team implemented processes with reduced timelines by more than 6 months, enabling the availability of product for clinical trials. Phase I studies were initiated in record times, and the dramatic results of LDL-C reduction observed in preclinical studies were faithfully reproduced in humans. Predictive pharmacology and adaptive clinical trial approaches were developed to support the dose determination studies in parallel with phase II studies. The energy in the team was palpable.

Regeneron, a much smaller company at that time, was slowly developing their own anti-PCSK9 antibody through preclinical studies and phase I. With the potential for this molecule to have an impact on cardiovascular disease, Sanofi, a pharmaceutical giant, joined Regeneron, enabling substantial resources to the program. The Regeneron/Sanofi team was able to recruit participants in the phase I and II trials to match Amgen. Both companies presented their findings at the same international meeting. The race for the phase III trial was on. Marc Stephen Sabatine, Harvard Medical School, was the lead principal investigator for the Amgen trials, and Professor Jennifer Robeson of the University of Iowa was for Regeneron/Sanofi. Both companies recruited subjects from worldwide clinical sites. Investigative research showed sites in South Africa were very proficient in recruiting subjects. Both companies bee-lined to that country. While these double-blinded studies were progressing, both teams were preparing for regulatory submissions, launch planning, and high-profile presentations and publications.

Amgen completed its phase III trial in August 2015 and submitted their BLA in November 2014, and Regeneron/Sanofi submitted it in January 2015. Both clinical trials were published in the prestigious *New England Journal of Medicine*, on 16 April 2015.

A very interesting thing happened. Regeneron/Sanofi bought a voucher from BioMarin for $67 million, requiring the FDA to enhance the review time for their BLA. A voucher is given by the FDA to any company that gets approval for a drug for a rare disease indication. BioMarin had done that. The submission of the voucher enabled the FDA to approve the Regeneron/Sanofi BLA in July 2015. Amgen's anti-PCSK9 antibody was approved in August 2015. Why does a month matter? First-to-market and launch is one of the most important steps in marketing strategy (out of the scope of this book). Amgens' drug is called Repatha (evolculumab), and Regeneron/Sanofi's is Praluent (alirocumab).

5   Case Study 2. Amgen Versus Novartis in the Development of Molecules that Block the IL-17 Cytokine Pathway in the Treatment of Various Autoimmune Diseases

Interleukin-17 (IL-17A) is a soluble, pro-inflammatory cytokine that is critical in the pathogenesis of psoriasis. Its receptor is found on the surface of keratinocytes and inflammatory cells like neutrophils, as well as other immune cells. Blocking of IL-17 cytokine and IL-17 receptor interaction had been hypothesized in thc pathogenesis of autoimmune diseases such as psoriasis, psoriatic arthritis, and ankylosing spondylitis. Only the clinical trial would prove it. There were several complexities, such as there are many isoforms of IL17, A, B, C, D, E, F, and multiple forms of the receptors to which each of the IL-17 s bind. To which cytokine, or which receptor should the drug be developed? This was a very difficult decision that needed to be made by doing large sets of early pharmacology experiment. If interested, there is a vast literature on this topic.

Harnessing the potential of IL-17 inhibition, Amgen and Novartis embarked on a journey to develop monoclonal antibodies (mAbs) targeting IL-17 (brodalimumab) and IL-17 receptor (secukinumab). The regulatory strategy of both companies involved preclinical development requirements and clinical design using standard clinical endpoints. *Phase I trials* focused on establishing safety profiles, determining optimal dosing regimens, and exploring early signs of efficacy. *Phase II trials* expanded patient cohorts to further assess efficacy and safety signals observed in phase I. Finally, *phase III trials* were pivotal in demonstrating the clinical efficacy of IL-17-targeted mAbs and supporting marketing authorization applications.

Unexpectedly, after the unblinding of the phase III study, a few more subjects treated with brodalimumab (the Amgen drug) exhibited suicidal ideation, compared to the subjects in the control arm [6]. This safety side effect was not predicted through the mechanism of action of the drug. Amgen made a decision not to file the BLA for approval. This side effect was not observed in the secukinumab trial, and Novartis obtained approval. The drug Cosentyx has become the standard of care for treatment of psoriasis [7].

Amgen licensed brodalimumab to Leo Pharma. A detailed reanalysis of the data was conducted, and the BLA demonstrated that although cases of suicidal ideation and behavior were reported, no causal association between treatment and increased risk of suicidal ideation and behavior has been established. Its launch was supported by post-marketing pharmacovigilance activities to capture and follow up on any reports of safety events.

6   Case Study. Antibody Drug Therapies for Treatment of Alzheimer's Disease

An estimated 35 million people worldwide have Alzheimer's disease. There was no drug approved for treatment of Alzheimer's disease for almost 20 years. Suddenly in 2021–2024, there has been a spate of monoclonal antibody drug therapy approvals, after significant regulatory challenges and controversies. Regulatory agencies require robust evidence from multiple phases of clinical trials before granting approval. Phase III trials, which are pivotal in this process, demand extensive data on safety, efficacy, and patient outcomes. This case study describes a controversial regulatory approval and the subsequent processes and drugs that overcame the challenges, at least in part.

Case study of development and approval of aducanumab In order to expedite clinical development, this drug, which targets the amyloid beta protein, disrupts amyloid aggregation, and facilitates clearance of β-amyloid plaques, was granted accelerated approval status. This regulatory pathway allows drugs to be approved based on surrogate endpoints reasonably likely to predict clinical benefit. The risk is that long-term studies may not fully capture the clinical benefits of the treatment. This was evident in the controversy surrounding the approval of aducanumab which faced criticism despite its accelerated approval based on amyloid plaque reduction.

The clinical trial design involved phase I safety assessment studies, a dose-determining phase II study, and two pivotal phase III studies. The evaluation of safety adverse events such as amyloid-related imaging abnormalities (ARIA), including edema (ARIA-E) and microhemorrhages (ARIA-H). The patients selected for these studies were restricted to mild dementia. Biomarker analysis included amyloid PET imaging and cerebrospinal fluid (CSF). The phase III clinical endpoints included cognitive and functional measures focused on patients with mild cognitive impairment (MCI) or mild dementia. The primary outcome included change in Clinical Dementia Rating (CDR) Sum of Boxes score, an integrated metric of cognitive and functional status. Secondary outcomes included reduction of β-amyloid on PET imaging. Two phase III clinical trials (ENGAGE and EMERGE) with identical designs were done in about 2200 patients [8].

ENGAGE showed no significant changes in clinical outcomes, while EMERGE showed significant improvement in clinical outcomes versus placebo. Both trials showed significant reduction in β-amyloid levels. The conflicting clinical efficacy results led the manufacturer to stop the trials. But after further post-hoc analyses, the manufacturer determined that those who received higher doses in both studies had a significant reduction (−23%) in CDR Sum of Boxes score versus placebo recipients, leading to the manufacturer applying for FDA approval. The FDA reviewed the BLA submission and as per protocol had an advisory committee (AdCom) to provide feedback and guidance on the analysis and interpretation of the data submitted (the AdCom can be watched in this YouTube).

In November 2020, the FDA advisory committee voted against approval based on the trial data presented by Biogen. Ten of the eleven members of the independent expert panel voted "no" to the question "Does the data presented provide sufficient evidence to demonstrate a positive risk-benefit ratio of efficacy versus safety?" Despite this vote, in 2021 the FDA approved aducanumab for treatment of all stages of Alzheimer's disease [9]. The approval was based on its ability to reduce amyloid plaques in the brain, a surrogate endpoint. The decision was controversial due to mixed results from clinical trials regarding its impact on cognitive decline. One of the two phase III trials failed to meet the predefined acceptance criteria. Further, aducanumab was tested in patients with early and mild Alzheimer's, but the approval included advanced forms of the disease. Several advisors in the FDA advisory committee publicly resigned [10] in protest. The approval highlighted the tension between the urgent need for new Alzheimer's treatments, strongly advocated by the patient support community, and the rigorous standards required for demonstrating clinical benefit. The FDA required Biogen, the drug's manufacturer, to conduct a post-approval clinical trial to verify its clinical benefit, reflecting the ongoing uncertainty and scrutiny surrounding the drug.

Subsequently, the approval was revised to target only those patients with mild dementia, similar to the phase III clinical trial participants. The advisory committee provided detailed guidance on the prescription and use of the drug [8, 11]. Then, in 2022, and 2023, two other monoclonal antibodies, *lecanemab (Biogen/Eisai), [approval date July 2023],* and *donanemab (Eli Lily), [approval date March 2024],* which showed a statistically significant reduction in the amyloid plaques and slowed cognitive decline in patients with early Alzheimer's, were approved by the FDA. Since the approval of these drugs, aducanumab has now been discontinued from the market; the company states "to reprioritize its resources in Alzheimer's disease."

The Grand Finale   "Approved". The letter which is sent by the regulatory agency and addressed usually to the regulatory liaison, and head of the drug development process, is one of the biggest accomplishments of a career scientist. The ecstasy of approval is unparalleled. Hundreds, sometimes thousands of folks contribute to the approval process. Yet, no single person knows it all. Not even on the side of the regulatory agency. We hope that the book provides you, the reader, with some background information on each of the steps and empowers you to do your job better.

References

1. US-FDA. Regulatory information; 2018.
2. FDA. Search Page Link for guidance documents; 2024.
3. FDA. Clinical pharmacogenomics: premarket evaluation in early-phase clinical studies and recommendations for labeling; 2024.
4. F.G. Document. S2(R1) genotoxicity testing and data interpretation for pharmaceuticals intended for human use; 2012.
5. US-FDA. Prescription drug user fee act (PDUFA); 2017.
6. Lebwohl MG, Papp KA, Marangell LB, Koo J, Blauvelt A, Gooderham M, Wu JJ, Rastogi S, Harris S, Pillai R, Israel RJ. Psychiatric adverse events during treatment with brodalumab: analysis of psoriasis clinical trials. J Am Acad Dermatol. 2018;78:81–89.e5.
7. Ratner M. IL-17-targeting biologics aim to become standard of care in psoriasis. Nat Biotechnol. 2015;33:3–4.
8. Rabinovici GD. Controversy and progress in Alzheimer's disease - FDA approval of aducanumab. N Engl J Med. 2021;385:771–4.
9. Mullard A. Landmark Alzheimer's drug approval confounds research community. Nature. 2021;594:309–10.
10. Belluck P, Robbins R. Three F.D.A. advisers resign over agency's approval of Alzheimer's drug. New York: New York Times; 2021.
11. Alexander GC, Knopman DS, Emerson SS, Ovbiagele B, Kryscio RJ, Perlmutter JS, Kesselheim AS. Revisiting FDA approval of aducanumab. N Engl J Med. 2021;385:769–71.

Epilogue

The new era in drug development is being driven by groundbreaking advancements in data collection (and generation) processes across omics technologies. Analysis, interpretation, and application of these data using artificial intelligence and machine learning have already begun to decrease timeline and influence regulatoryprocesses (Fig. 1).

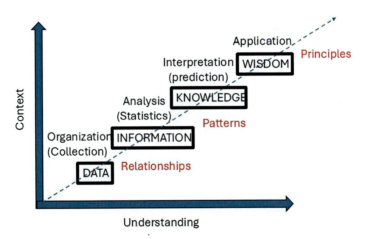

Fig. 1 Progression of the steps in drug development. The process starts with data collection. The advancement of technologies has enabled large data omics data sets. These data sets require systematic organization, to be analyzed to discover relationships between structure-function, input parameters, and outcomes. These analyses enable the development of algorithms which can be used to interpret the analyzed data to develop predictive results, rather than wait for observed data. The final step in the process involves application of these analyzed data sets and predictive patterns to develop safe and effective drugs that are precise for disease conditions, and indications, and ultimately provide value for patients

N. Chirmule, V. V. Ghalsasi (eds.), *Approved: The Life Cycle of Drug Development*, https://doi.org/10.1007/978-3-031-81787-8

For graduate students and early-career biotech professionals, understanding these changes in the drug developmentprocesses is critical to navigating the future of this field. The integration of these technologies is not just enhancing the efficiency of drug development; it is fundamentally reshaping how we think about the discovery, development, and distribution of therapeutic interventions. The aim of this book was to provide a glimpse into all aspects of drug development, written by highly experienced authors.

AI and ML are no longer just theoretical tools; they are now integral components of the drug development pipeline. Data collection through digital platforms is becoming critical. Understanding and learning the methods to analyze large datasets is now a must-have skill (e.g., R, Python, and related methods). By applying sophisticated algorithms to vast datasets, AI and ML can identify novel drug targets, predict drug-receptor interactions, and optimize molecular structures with unprecedented speed and accuracy. These technologies are also being used to model complex biological systems, allowing for more accurate predictions of drug behavior in vivo. As a result, the traditional trial-and-error approach to drug discovery is being supplanted by data-driven methodologies that significantly reduce time and cost. Genomics and other omics technologies, such as transcriptomics, proteomics, and metabolomics, are at the heart of the precision medicine revolution. These technologies allow for a comprehensive analysis of biological systems, providing insights into the molecular mechanisms underlying health and disease. By integrating these omics data with AI and ML, researchers can develop highly personalized therapeutic strategies tailored to the unique genetic makeup of individual patients.

The first five chapters in the textbook provide detailed information in drug discovery, in the areas of vaccines, biologics, cell, and gene therapy. The following chapters address the clinical development of drugs in industry—process development, pharmacology, biomarkers, toxicology, and clinical trials. These chapters detail the process steps involved in these areas and discuss the science, the people involved, and processes. The purpose of the content of these chapters is to provide the readers with an overview of the kind of workflow and thought processes of folks in these departments.

Precision medicine is the ultimate goal of drug development. While the potential of these technologies is immense, there are significant challenges that must be addressed to ensure that the benefits of innovation are accessible to all. The cost of drug development remains high, and without careful consideration, the fruits of these advancements could exacerbate existing disparities in healthcare. It is essential that regulatory frameworks evolve to support the development of affordable therapies while ensuring their safety and efficacy. The chapters on rare diseases and n-of-1 studies shed light on the development of drug for very small groups of patients. Lessons from these studies will have direct implications for developing precision medicines in the future.

The global nature of modern drug development necessitates international collaboration. Regulatory harmonization, data sharing, and collaborative research efforts are key to bringing new therapies to market efficiently and equitably. The future of drug development lies at the intersection of AI, ML, genomics, and global regulatory strategies. For those entering the field, the challenge is not just to innovate but to do so in a way that balances scientific progress with the ethical imperative of global access and affordability. By embracing a multidisciplinary approach and fostering international collaboration, the next generation of biotech professionals can ensure that the transformative potential of these technologies is realized for the benefit of all.

As you move forward in your careers, remember that the decisions made today will shape the future of medicine. It is a time of extraordinary opportunity, but also of significant responsibility. By staying at the forefront of scientific innovation and advocating for equitable access to healthcare, you can contribute to a future where the latest advancements in drug development are available to all, regardless of geographic or economic barriers.

Reflections on Contemporary Issues in Data Integrity

Perihan Elif Ekmekci

Francis P. Crawley

Data integrity is the core concept that embraces accuracy, consistency, reliability, quality, security, resilience, and legal compliance throughout the entire life cycle of data. Data integrity has an inherent value, together with its instrumental value in terms of providing grounds for data-driven decisions in various domains of life. The challenges in establishing and maintaining data integrity become pronounced in the digital age, where the need and utility of data are enhanced not only in terms of human use but also for machine learning purposes. This challenge is even magnified by the introduction of synthetic data and its broad use cases in the realm of generative artificial intelligence (AI) development and deployment.

The main issues in terms of data integrity can be listed as data quality assurance, data security, regulatory compliance, and data governance. Data quality assurance is a comprehensive strategy that starts from the initial data collection phase and extends throughout the entire data life cycle to maintain accuracy, consistency, continuity, and completeness. Data quality assurance serves as a building block for the credibility, validity, and reliability of data-informed decisions.

Data security is another challenge that gains particular significance as digitalization spreads rapidly across various domains, drawing attention to cyber-security and data confidentiality. AI technologies enable new ways of generating, storing, and transferring data that urge us to redefine and restructure data security elements to address threats issued by them.

P. E. Ekmekci
Department of History of Medicine and Ethics, TOBB ETÜ, Ankara, Turkey

© The Editor(s) (if applicable) and The Author(s), under exclusive license to Springer Nature Switzerland AG 2025
N. Chirmule, V. V. Ghalsasi (eds.), *Approved: The Life Cycle of Drug Development*, https://doi.org/10.1007/978-3-031-81787-8

Effective data governance and regulatory compliance are the cornerstones for data integrity that set the legal terms and administrative competence necessary for maintaining data integrity. However, these two concepts become vogue and derive contextual changes too due to the vast improvements in AI. Digitalization, and access to big data, combined together with disruptive AI technology in diverse domains with unforeseen aims that have not been thought of in the initial phases of the data life cycle, requires new approaches to data governance. Another aspect that adds to the current need for and actual act of change in data governance is the increased capacity to derive data from unorthodox sources, such as real-time data flow from wearable devices or data flow from social media accounts, which can lead to profiling. Considering the technological advances that facilitate constant, uninterrupted data exchange, sharing, and storage, data integrity becomes a more challenging issue not only in terms of proper and effective governance but also in terms of the cultural and moral change induced by technological developments.

Data integrity may encompass some variations depending on the domain from which the data is derived or used. Biological data is one of these domains that demands diligent care and elaboration. This demand emerges from the concept of uniqueness. Uniqueness is easy to understand when the biological data sources are actual human beings. Even though they are still alive or have lost their lives, data derived from human subjects should be treated with particular attention to respect human dignity and privacy as well as respecting the autonomy of the human beings in terms of their voluntary consent for the collection, use, reuse, or transfer of their data. This aspect has been shadowed by the argument that any data derived from human beings can be used, or sold for diverse purposes including commercial ones, if the data is fully anonymized, that is, it cannot be traced back to the individual provided the data in the first place. Although this approach was considered quite handy for overcoming ethical issues such as informed consent and individual privacy as well as legal compliance for the use of sensitive data, recent developments in technology and science reveal the fact that full anonymization of data is not possible for a considerable type of biological data. Moreover, even if full anonymization is possible in a set of biological data, it does not provide always allow for the use or reuse of this data for any purpose.

Another discussion point emerges from the need for the identification of data sources to meet the aims of the data used in some domains. This need is evident in research that requires source profiling. In this respect, anonymization zeroes the value of data, hence avoided together with collateral incidences that may becloud the identifiability of the data. In addition, some measures may be implemented to enhance the profiling capability of data, which may introduce threats to data integrity that complicate the data integrity and compliance with the ethical issues mentioned above.

This short discussion on biological data has some additional connotations on data integrity if we portray biological data in bioinformatics in a broader sense to embrace DNA sequence data, population data, and even ecological data. Considering the data access, collection, storage, sharing, and transfer, securing and maintaining data integrity becomes a complex process that needs to be updated regularly due to developments in technology and the unwanted consequences that threaten data integrity.

The relevance of the respect for the uniqueness of data sources and the discussions that accompany this concept, which are listed shortly, is that data integrity cannot be ensured or sustained without fulfilling the moral duties to data sources. Some of these duties are well-defined because of the tragic examples in recent history, such as the immortal HeLa cells or instrumentalization of several vulnerable human beings in infamous cases like the Willow Brook State Hospital. Although these tragic experiences created a moral awareness of data integrity in biomedical research, it is not the case in terms of other types of sensitive data that are being collected, stored, used/reused, or commercialized in other domains such as studies in climate change, environment, agriculture, or social science research including, but not limited to, sociology, psychology, or anthropology. It is beyond discussion that a huge amount of sensitive data is being processed within these domains that require particular attention to data privacy and confidentiality and urges reflection on how new emerging threats induced by AI technology and digitalization may be addressed.

Bias is considered one of the major threats to data integrity. The existing data on several domains may have a potential bias due to data collection facilities concentrated in certain regions, institutions, or communities. Apart from these reasons, bias can be induced by the user profile, as data provision is very much influenced by the availability of the data collection tools to potential data sources. Difficulties in accessing technology impact data bias by leaving disadvantaged regions, communities, or domains out of data. Another element that impacts bias in data is technology literacy, which is considerably lower in senior-aged populations compared to young generations, or in low-resource countries compared to high-income countries. Moreover, technology literacy may differ in closed communities that set a boundary for data flow and deepen data bias. As data collection means moving from conventional ways to real-time contemporary methods, access, familiarity, and frequency of technology use become a major factor that determines data flow, which works against data integrity by increasing data bias. Currently, synthetic data has become one of the major sources to train AI systems. The synthetic data mimics "real data" and copies it to have a broader data set so that AI systems can be trained more effectively. It is not difficult to see how data bias would be amplified and can become a bigger problem of data integrity if the original "real data" is biased beyond acceptable limits. To overcome this potential risk, validity tests to determine and flag imbalances in synthetic data should be developed and a skeptical attitude should be maintained to detect other unforeseen risks of data integrity induced by synthetic data.

Since AI technology is improving very fast to offer novel opportunities in terms of data utility, effective landscaping of technological developments and continuous reflection on maintaining data integrity are needed. Therefore, it is more appropriate to view data integrity not as a checklist or predetermined set of rules but as an active process that is constantly adjusted according to a moving target that technological development presents.

Organization Structures

Amitava Saha

Organizational structure is the foundation of every company as it depicts how employees are placed, decisions are made, careers can progress, and overall accountability is shared. There are many types of organizational structures as they need to match the type of business, operations, and future growth strategies while keeping in mind the state of evolution of the company.

Organizational structures in their absolute state are industry agnostic, but the nature of business and key focus areas does make a strong impact on how the organizational layers are built. Drug development within the pharma industry is a classic example of confluence of multiple factors as it operates at the intersection of science, business, medicine (lifesaving in most cases), and commerce with an intense regulatory oversight. To make matters more complex, the organizational structure in this case should be flexible and adaptive yet have clear lines of authority and responsibility and provide employees with room for innovation and career development.

Organizational structures can be centralized or decentralized, vertical or flat, hierarchical or circular depending on the needs of the business. Each of these has its merits and shortcomings and the decision should be based on the business strategy.

The following types of organizational structures are commonly used in different industries:

1. Functional/role-based structure
2. Matrix structure which could be product/process/client/geography based

I have worked in multiple industries like textiles, IT (Information Technology), ITES (Information Technology Enabled Services), banking, and pharma and have seen both structures in different forms at various phases of each organization. In pharma, most companies start with the functional structure as it provides maximum clarity on roles, responsibilities, decision-making, career development, and depth of knowledge.

N. Chirmule, V. V. Ghalsasi (eds.), *Approved: The Life Cycle of Drug Development*, https://doi.org/10.1007/978-3-031-81787-8

A case in point is Biocon, which also started its R&D journey using this model. The organizational structure followed a product development process flow. The talent pool was distributed in laboratories that mimicked smaller subsets of the entire process development journey. There were several such individual units like clone development (molecular biology), upstream (fermentation process development), downstream (product purification process), formulation, and analytical sciences (developing analytical tools and methods for other groups). Each lab was headed by a lab head with deep expertise in the lab's core function. This structure ensured that Biocon developed deep expertise in each core area.

The major shortcoming of this model is cross-functional collaboration with other teams in the business which limits the growth of the employee across various streams as they tend to work in silos. Over the next two decades, with the growth in the number of molecules, clients, geographies, and processes the model became more complex and meeting timelines became more challenging. It was imperative to develop a model that focused on different clients and projects in a customized fashion. The company needed collaboration to ensure that the responsibility and accountability for drug development to delivery was shared by cross-functional teams from different functions. The HODs of sales, operations, and quality along with myself and the CEO decided to break the model and create a matrix structure which addressed the above.

Functional Role-Based Organogram

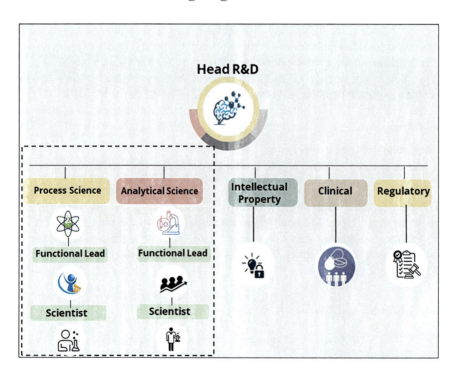

As a first step, the functional model was dissolved, and project teams were created with members from each core function of R&D. Teams were aligned on external customer projects and in-house projects, and differentiation was also made in terms of Biosimilar or Novel molecules. The talent pool was distributed based on each project's progression, project lifecycle, and of course work intensity at that point in time, e.g., early development or late development phase. The number of people in each pool was determined by earlier project experience, and it was optimized depending on the stage and complexity with a flexibility to allocate more resources from other teams at predetermined milestones. This matrix structure created more roles (as each group had their own leaders (PG lead), Program Manager (PM), and Project Management Office (PMO) to name a few) for career development compared to the earlier functional model. This helped in succession planning for both middle and senior management roles and enabled role rotation to create more versatile leaders in the future.

As part of the new work style, a stage-gate approach was taken for every project and progress was reviewed every fortnight to ensure there was no lag. This brought in sanity to overall project management which was earlier chaotic depending on the timeline and complexity. Keeping in mind the business and R&D strategy, the next step was to introduce the element of geography to make the delivery model sharper and more client-friendly.

The flip side of a matrix structure is multiple reporting, issues in decision-making, and of course, increasing the overall headcount as roles are duplicated in different project groups. However, this model helped in addressing attrition-related issues and gave multiple growth opportunities to the talent pool. This model has been tested and replicated in most pharma companies across the globe, and Biocon ensured that we learned from both successes and failures of other companies and our own internal teams.

Matrix Structure in BRL Organogram

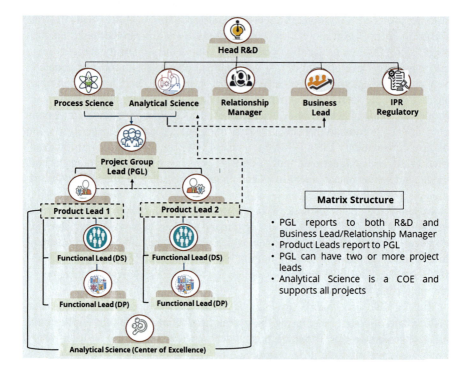

Acknowledgements Figures were made by Riya Bhattacharya.

An Academic Entrepreneur's Journey

Samir Mitragotri

Entrepreneurship is all about building a business enterprise ground up, all the way to taking the product to a point of realizing its impact. It involves extraction of economic value from an idea and building a sustainable enterprise. While entrepreneurship, regardless of the specific nature of the product, requires common skills, entrepreneurship in the biopharmaceutical field poses certain unique challenges. As several chapters in this book have described, drug development is a long process and it brings several risks which require continuous efforts of entrepreneurs over a long period of time.

In drug development, unlike some other sectors, the contribution of fundamental research arising in academia is often more pronounced. Historically, many transformative therapeutic ideas were first demonstrated in academic research well before a product even seemed feasible. These discoveries laid the foundation for entrepreneurs to start their journey, which often involved the very scientists who invented the founding idea, to advance these transformative ideas to the market.

The journey of academic discoveries from conception to patient acceptance, however, is quite arduous. It begins with a spark—an idea that ignites the mind of researchers, often in unexpected moments and places. But thereon, the journey of this idea to marketplace is a rollercoaster ride of excitement, despair, persistence, and resilience.

Having traveled this road many times for biotech startups in my career, 20 general steps of that journey are summarized here. Individual examples may deviate from the generalized steps listed here, but they capture the essence of most endeavors. One can appreciate the required perseverance and high-level execution from the team in completing this journey. Looking at each step, which must be successfully completed in series, one can also appreciate the challenges in reaching the finish

S. Mitragotri
School of Engineering & Applied Sciences and Wyss Institute, Harvard University,
Boston, MA, USA

line. If each step is executed with a 90% probability of success, the overall probability of success is about 10%. One can appreciate the risk when you consider that if the probability of success for each step drops to 80%, then the probability of overall success drops to 1%. One has to admire the audacity and enthusiasm of entrepreneurs who take on the journey regardless of these low odds.

1. **Birth of an Idea**

Academic research is inspired largely by the pursuit of knowledge. While the ultimate goal of improving patient lives may always underline that pursuit, the immediate goal is rarely to make a product or earn a profit. Breakthroughs are often so disconnected from the market gold standards, that making a product based on such breakthroughs may even seem inconceivable. The birth of a transformative idea rarely takes place within the confines of a lab or office. It may emerge during a late-night brainstorming session, a casual conversation, or even while sipping coffee at a local café. The researcher suddenly sees a possibility—a solution to an unmet medical need.

2. **Student Enthusiasm and Collaboration**

A typical academic research group is led by a principal investigator, who provides intellectual inspiration and resources to the research group. The research group membership spans a wide range of experiences and expertise—from undergraduate students and graduate students to postdoctoral fellows, all of whom collectively provide the necessary intellectual and logistical power to move research forward. When the idea strikes, it gets discussed with the research group members to explore a path to realize it. The idea gains momentum as discussions unfold. Various versions of the idea get discussed to sharpen the path. The youthful energy of the research group fuels the fire. The team dissects the concept, challenges assumptions, and envisions its impact. Their enthusiasm is contagious, and without it, the idea simply cannot move forward.

3. **Grant Applications and Rejections**

Academic research is enabled by grants. These grants are largely disbursed by government agencies through various programs. Often, private industries contribute to academic research endeavors. When the principal investigator decides to pursue the newly conceived idea, the first step is to secure a grant to advance the idea to a point of demonstrating its impact. Reality sets in when grant applications are submitted. Rejections follow—sometimes due to lack of perceived innovation, other times because feasibility seems unclear. It is so easy to shoot down ideas that do not have sufficient evidence to counteract skepticism. The researchers feel the impact of such skepticism and rejection. But setbacks only strengthen the resolve.

4. **Working with Limited Resources**

In academic settings, money is most often a scarce resource. Grants, when they arrive, often come with constraints. Grants are also rarely awarded after the first submission, often requiring revisions and resubmissions which add delays. During this time, research group members are also in transition. A given student is rarely available for prolonged times within the group. As research is a part of student's broader academic journey, they are in flux arising from graduation and career pursuits. Transitions can often delay the progression of ideas due to knowledge gaps and training times. Despite these odds, research continues. Undeterred, the students roll up their sleeves. With shoestring budgets, they experiment, fail, and iterate. Doubts creep in, but they persist.

5. Navigating Failure

In the journey of scientific discoveries, where hypotheses meet experimentation and face the reality of being disproven, the idea stumbles. It is akin to a climber losing footing on a treacherous slope, questioning whether the summit is worth the struggle. Experiments yield disappointing results. The team questions whether the idea stands on a solid ground. Doubts gnaw at the team's determination. In these moments, resilience emerges. Students recalibrate their plans, reframe failure as feedback, and forge ahead.

6. Signs of Success

Among the frequent occurrences of failures, a breakthrough arrives to provide a glimpse of validation of the idea. This breakthrough transforms the uncertainty into a possibility. The idea begins to show promise. The team rallies, fueled by the first sign of success. They dig deeper, enthused by the promise demonstrated by early experiments. The results need vetting for reproducibility and validity. The findings need corroboration with additional experiments and trials. Experiments expand to enable a more thorough assessment of the hypothesis. The once-doubtful outlook now appears hopeful and excited about what the success might look like.

7. Sharpening the Idea

Having reproducible results alone is not enough. The findings need to be spun into a story so that it can be narrated to stick in people's minds. Sometimes, more experiments are needed just so that the message is effectively conveyed. During these efforts, idea sharpens. It evolves, adapts, and gains clarity. Students gather more data, refine hypotheses, and inch closer to a demonstration of impact.

8. Manuscript Submission and Review

Research articles are the yardstick for assessing academic research. As the manuscript gets written, the newly accomplished results get spun into a story. Most suitable journals are identified, and the manuscript is submitted. The academic publication process has a very important quality check—peer review. Reviewers often offer mixed feedback. More experiments are conducted per reviewers' suggestion, and the manuscript is revised. This process can sometimes go through several iterations. A successful publication is often an end-point of fundamental academic research. However, when academic research has an element of innovation and product potential, publication is just the beginning.

9. Patents and Business Conversations

When research has the potential to turn into a business idea, it requires legal protection. Intellectual property rights provide this protection at multiple levels, offensive and defensive. The academic institution files for a patent. The filed patent application goes through its arduous journey of examination and grant. Since patent protection is an ongoing process over an extended tie period, business conversations start in parallel. These conversations take on many flavors ranging from those led by the inventors themselves to those led by the outside entrepreneurs to the university business development offices. With the potential to receive patent protection, business leaders convene, drawn by the idea's potential. Conversations shift focus from research to business.

10. Licensing and Fundraising

As the place of origin for research and innovation and the owner of the resultant intellectual property, the host academic institution is a key player in the commercialization process. Discussions ensue among the founding team, potential investors, and the institution. A business entity is established. Its initial cap table is decided. The business lead of the startup negotiates the terms and conditions of the patent license from the institution to the company. Capitalization of the company gets formalized. Unless the funding climate is in its boom phase, raising funds is always a challenge. But that is not the only obstacle. The team has to deal with the skepticism about the business model and market need. Just a few months ago, the discussions were focused around experiments and models. But the topics now evolve to include regulatory strategy and business development. The innovation is still considered "crazy" and "risky." Innovators try to defend by saying "all cool products were crazy and risky at one time!"

11. Forcing Ahead with Limited Resources

Most startups work on a shoestring budget. Regardless of the size of the capital raise, the funds can support the runway only for a short time period. The goal of the business is to reach the next value inflection point using available funding. The environment is often scarce of resources, but the motivation of its founders is relentless. The discussions are mostly about the milestones and derisking of the idea. For a typical biopharmaceutical product, the primary derisking event is clinical demonstration. Discussions about the path to the clinic take priority. Progress is incremental, but momentum builds.

12. Preclinical Development

In therapeutic development, no matter how exciting it may sound in preclinical stage, the future value depends on clinical success. Preclinical development paves the way for a successful clinical demonstration. Preclinical development requires a considerable amount of funds and expertise depending on the nature of the product. Conversations with regulatory agencies are initiated. Successful preclinical development ends with the approval from the regulatory agency to initiate clinical trials. At this stage, the technology is no longer a dream. It is on its way to clinical impact.

13. Clinical Demonstration

Once the regulatory approval is in place, patients are recruited. Early clinical trials aim to demonstrate safety and ideally, an early indication of efficacy. This sometimes takes months to years to fully implement. The entire process of clinical development including registration, regulatory approval, patient recruitment, patient follow-up, and finally data analysis could take a significant amount of intellectual and financial resources. But finally, clear clinical success emerges. The technology is now labeled "innovative" and "disruptive." Investors take notice.

14. Business Development Takes Center Stage

At this stage, the discussion about the logistics of taking the drug to patients takes center stage. Commercialization discussions intensify. Partnerships materialize with pharmaceutical giants to propel the idea into tangible marketable product. Strategic discussions emerge as to whether the business has aspirations and competency to launch the product vs. partnering with companies with existing sales and marketing strengths.

15. Advanced Clinical Trials

Despite the success in early clinical trials, later-stage trials are needed to position the product in the landscape of clinical comparators. Funds must be raised to support advanced clinical trials. While fundraising is never easy, raising funds at this stage of the technology is often more streamlined compared to that at the earlier stages. The technology is nearly de-risked. Hope appears to be on its way to becoming reality.

16. Society's Acceptance

If the company elects to go all the way to the launch, the product reaches the hands of patients. The real test of the product begins. The word spreads. The marketing and sales personnel of the company now become the driving force behind the idea. Healthcare practitioners prescribe the product to patients and benefits are evident. Public health analysts look at the high-level data and evaluate health benefits. Society embraces the product. A health problem that was once difficult to address is now addressable. The idea that was once labeled crazy and risky now appears indispensable.

17. Acquisition or Full Commercialization

With market success, pharmaceutical giants take notice, if not already. The company is acquired or it proceeds to full commercial launch. The idea has transcended academia—it is now a force in healthcare. The idea, which was in lab notebooks some years ago, is now a practical solution to an important problem.

18. The Product

The product stands tall, out in the market, speaking for itself. It is no longer an idea. It is a reality that is sparked by the idea and realized by the grit. The once-fragile concept now impacts patient lives, enabled by those who dared to dream it into existence.

19. Utility and Opinions

After prolonged use, reviews and opinions start surfacing. Users—patients, doctors, insurers—endorse its utility. The product transforms lives. The transformation is not sudden. It is the gradual acceptance demonstrated by the nod from practitioners, the reimbursement approvals, and the patient testimonials.

20. From Crazy to Obvious

In the complex process of drug development, there exists a defining moment—a revelation when society collectively nods and declares, "Of course." This moment arrives after years of relentless pursuit. The once-crazy idea—the audacious vision that raised eyebrows—now stands as an indispensable solution.

The entrepreneurial journey, akin to a marathon, culminates here. The pursuit of innovation, the grit of the team, and the constant battle in the face of the unknowns materialize into a tangible product. The idea is no longer speculative; it is integrated into the fabric of healthcare. This destination in fact drives the journey of most entrepreneurs, many of whom could have chosen a path of less risk and a comfortable profession. Yet, they chose the arduous path of entrepreneurship with no guarantees other than the opportunity to pursue their dream. And so, the journey—from mind to market—is complete.

In this journey, the entrepreneurial spirit thrives. The once-crazy becomes the obvious—an evolution enabled by science, resilience, and unwavering belief.

Acknowledgments

I appreciate and acknowledge insights from numerous individuals who have inspired this article. Over my career, I had the fortune of working with numerous brilliant scientists, tenacious entrepreneurs, effective executives, enthusiastic team members, and savvy investors. This article is inspired by my interactions with them and the experiences that I gained with them. This article is based on my previous LinkedIn post summarizing 20 steps of the entrepreneurial journey. Dr. Vihang Ghalsasi composed the contents of this article from that post with editorial assistance from Copilot. I express my deep gratitude to him for his help.

Health Economics, Pricing, and Reimbursement

Sumati Rao

Demonstrating product safety, efficacy, and quality to regulatory agencies is no longer sufficient. Market access and drug reimbursement by demonstrating the value of the product are equally important to health technology agencies (HTA), payers, and government agencies.

Healthcare systems and reimbursement vary across the globe. In Europe, HTA require drug manufacturers to submit the cost-effectiveness of the product vs. the competition. In the United States, the healthcare system is based on Medicare/Medicaid and commercial payer reimbursement. The Center for Medicare and Medicaid Services (CMS) has guidelines on drug reimbursement while commercial payers review direct costs (drug price, health economics). Decision-making in the healthcare ecosystem in the United States has become complex with multiple stakeholders influencing patient access to treatment. A variety of factors influence treatment decisions, the main factors being clinical differentiation efficacy, safety, and cost; operational considerations (administrative burden, supply chain, and patient support); practice economics (value-based payment, GPO discounts), and decision resources (guidelines, HTAs, and cost analysis). The Inflation Reduction Act (IRA) in the United States is specific to Medicare to negotiate prices on a limited set of drugs by putting a cap on annual out-of-pocket spending at $2000 in 2025. Drug price is only one part of the equation in the healthcare ecosystem. In some countries, such as India, a significant proportion of the healthcare costs are paid, "out of pocket."

Pricing and Treatment Access That Varies Across the Globe

In Europe, drug pricing follows a government-supported system, while in the United States the drug price is negotiated with healthcare insurers. For example, Germany has a structured system that combines elements of market-driven and regulated

N. Chirmule, V. V. Ghalsasi (eds.), *Approved: The Life Cycle of Drug Development*, https://doi.org/10.1007/978-3-031-81787-8

pricing. New drugs are priced for the first year on the market. After this period the price is negotiated based on the added benefit over existing treatment as determined by the federal joint committee, an independent health technology assessment body. In South America, for example, Brazil has a mix of public and private healthcare systems influencing drug pricing dynamics. The government exerts considerable influence over drug prices through the National Agency for Sanitary Surveillance and Pharmaceutical Market Regulatory Champers. They set the maximum price for the drug, considering factors such as production cost and reference pricing in other countries. The United States has a complicated pricing and reimbursement system. Price negotiations happen at different steps of the drug supply chain. The drug manufacturer in the United States negotiates the price with a drug distributor with rebates. Pharmacy and the drug distributor continue to negotiate the price along with pharmacy benefit managers (PBMs).

Key Health Economics and Pricing Components

Formulating a pricing strategy requires developing a value story which is a clear articulation of the product's value with the evidence-based rationale for pricing. This includes health economics strategy and questions such as why the treatment is needed (clinical, economic, and humanistic disease burden), which treatment to use (comparative effectiveness; benefit-risk, cost utilization, and cost-effectiveness), and how best to use (treatment optimization, sequencing of treatments, benefit of early treatment use).

Pricing and Health Economics Model There are two potential approaches to defining the price. (1) Price to product which involves setting prices based on the stand-alone value of the product. This is defined by the patient, system, or other evidence generation to demonstrate the value of the product with health economics by quantitative analysis such as cost-effectiveness and quality-adjusted life years (QALY) and qualitative (e.g., current external landscape) metrics. (2) Price to market which involves setting the price based on existing or upcoming competitors. This process is defined by the performance of the drug relative to comparative therapy, considering how the product is differentiated vs. what is currently available and using evidence (e.g., non-inferiority, superiority) to support the proposed price. This analysis also requires that sufficient evidence is developed to support claims relative to peer products. Regardless of the model selected, a holistic view of possible consequences of the resulting proposed price is required (e.g., pricing in relation to payer willingness to pay, public perception, implications for other geographies, implications for other portfolio products, etc.)

Aspired Funding Flow Funding for payment includes defining principles around inpatient vs. outpatient, a critical factor for some of the new modalities such as cell and gene therapy in the United States where in the inpatient setting, insurers cover a capitated amount for the drug and associated treatment cost. However, these steps have generally not been profitable for most hospitals, reducing the CAR-T prescription. Meanwhile, in the outpatient setting (which covers 10–15% of all CAR-T), the reimbursement for the drug is based on the benchmark price plus a fixed percentage. This process is expected to increase patient volume as patient outcomes and provider economics improve. Health economics evidence required are cost-effectiveness, patient affordability, and budget impact models specific to payers in the European Union and the United States.

Lifecycle and Portfolio Implications In many therapeutic areas such as oncology and immunology (e.g., several PD1-targeted drugs), it becomes critical to provide guidelines for indication sequencing and approach to handle combination products. For example, many companies use an additive approach when it comes to products from different manufacturers and a slight reduction if the brand is from the same manufacturer (e.g., Opdivo and Yervoy are separately priced at $325K, but when given as a combination drug, the drug is priced at $293K, a 10% reduction in price). In terms of health economics data, payers and health technology agencies request economic models specific to sequences of treatment.

Innovative Contracting Another component to define upfront is the type of innovative pricing models and tradeable (e.g., paired diagnostic). With the rapid growth of high-cost specialty drugs, payers continue to face pricing pressure from multiple fronts (e.g., budget constraints, legislative threats, and public perception). When appropriately used, innovative contracts can effectively extend access. There are four types of innovative agreement: (1) *finance-based* which is either traditional discounts (e.g., over volume or time), capitation, or special financial terms such as flat rate or indication-based pricing which are more sophisticated; (2) *value- or outcome-based* where reimbursements are based on health outcomes and are effective in high-cost drugs where there is more uncertainty or variability around its efficacy but is complex to negotiate; (3) *service-based* where there is a value service associated with the product such as service to the physician or the patient and can be an additional source of income to the company but requires higher compliance risk evaluation; (4) *bundle-based* where the manufacturers offer a basket of different products from their own portfolio and sometimes from other companies which provide higher value to prescribers and payers.

Cross Geography Access Finally, companies may also want to define an approach to launch sequencing as some countries deal with reference pricing, where countries peg the prices based on other countries (predominantly in Europe where for example Italy and Spain use an international reference system, setting drug prices based on the average prices in several reference countries) and hence becomes critical to select countries to enter and decide when to enter. In addition, having a clear strategy for pricing in low- and middle-income countries could be important to ensure access. Most pharma companies have international differential pricing (IDP) principles in place. Typically, they are anchored to GDP/capital often with purchasing power parity adjustment. Economic theory suggests that IDP based on GDP/capital also called Ramsey pricing optimizes the balance between price and access globally. Another option would be a subscription model where there is a fixed payment to service a population independent of the exact volume of drug used. Vertex offers its current portfolio (and any future drugs) for cystic fibrosis at a fixed price per country. Companies also define intra-country differential pricing to expand access to public sector and out-of-pocket paying patients.

Acknowledgment These views in the article are my own and not of GlaxoSmithKline (GSK)

Recommendation of Books to Read

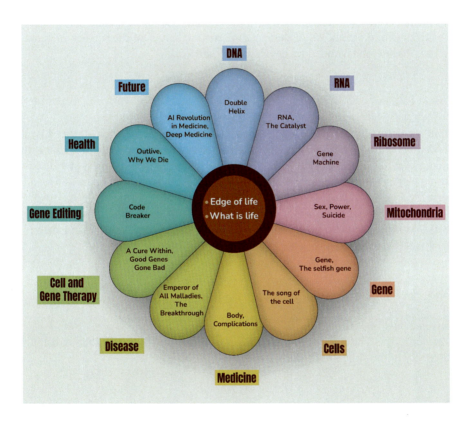

N. Chirmule, V. V. Ghalsasi (eds.), *Approved: The Life Cycle of Drug
Development*, https://doi.org/10.1007/978-3-031-81787-8

Book Recommendations

Before you start reading books, it may be important to consider "how to read a book." I was recommended this book (by Mortimer Adler, in 1937) by my friend Ajit Datar. It has transformed my ability to read books. I have written a short cryptic blog about this process. In summary, in order to read a book—properly and thoroughly—read it three times. The first time, listen to the author; the second time, discuss with the author; and the third time, argue and debate with the author. Take notes and mark important sections of the book. You will understand and remember the book much better (ref: "how to read a book": https://chirmule.wordpress.com/2021/11/06/how-to-read-a-book-book-review/).

The books below are a short list of topics that can be useful as reference materials for the various chapters in the book. The figure above of the analogy of the flower, and its petals, is that the books are interrelated. Many of these books discuss the topics within others. All of them are fascinating stories of the science of being human, disease, medicine, drug development, and health.

Origin of Life

What Is Life? With Mind and Matter and Autobiographical Sketches, Ervin Schoedinger, 1927—One of the great science classics of the twentieth century. It was written for the layman but proved to be one of the spurs to the birth of molecular biology and the subsequent discovery of DNA. A review of the book—75 years later. https://doi.org/10.1016/j.cels.2021.05.013

Life on the Edge: The Coming of Age of Quantum Biology, James Al-Khalili and John-Joe McFadden, 2016—It alters our understanding of our world's fundamental dynamics through the use of quantum mechanics.

DNA

Double Helix, A Personal Account of the Discovery of the Structure of DNA, James Watson, 1969—The classic personal account of Watson and Crick's groundbreaking discovery of the structure of DNA.

RNA

RNA, The Epicenter of Genetic Information, by John Mattick, Paulo Amaral, 2022—Described how science, scientific thought, and landmark discoveries revealed the central role of RNA in molecular biology and evolution.

The Catalyst: RNA and the Quest to Unlock Life's Deepest Secrets, Thomas R. Cech, 2024—Exploring the most transformative breakthroughs in biology since the discovery of the double helix, a Nobel Prize–winning scientist unveils the RNA age.

Ribosome

Gene Machine: The Race to Decipher the Secrets of the Ribosome, Venki Ramakrishnan, 2018—"Engaging and witty" (*Forbes*) memoir tells the riveting story of his race to discover the inner workings of biology's most important molecule.

Mitochondria

Sex, Power, Suicide: Mitochondria and the Meaning of Life, by Nick Lane, 2018— Cells that carry out the essential task of producing energy for the cell.

*Transformer: The Deep Chemistry of Life and Death, by Nick Lane, 2023—*A scientific renaissance that is hiding in plain sight—how the same simple chemistry gives rise to life and causes our demise.

Gene

*Gene, The Intimate History, Siddhartha Mukherjee, 2017—*A fascinating history of the gene and a magisterial account of how human minds have laboriously, ingeniously picked apart.

*The Selfish Gene, Richard Dawkins, 2016—*A classic exposition of evolutionary thought. Professor Dawkins articulates a gene's eye view of evolution.

*The Seven Daughters of Eve: The Science That Reveals Our Genetic Ancestry, by Bryan Sykes, 2002—*How we are descended from seven prehistoric women.

*Compatibility Gene: How Our Bodies Fight Disease, Attract Others and Define Ourselves, by Daniel M. Dovis. 2014—*The inside story of the Y chromosome's fatal flaw.

*Adams Curse, A Future Without Men, by Bryan Sykes, 2005—*The inside story of the Y chromosome's fatal flaw.

Cells

*The Song of the Cell, An Exploration of Medicine and the New Human, Siddhartha Mukherjee, 2023—*Blends cutting-edge research, impeccable scholarship, intrepid reporting, and gorgeous prose into an encyclopedic study.

Medicine

Body: A Guide for Occupants, by Bill Bryson, 2021—He guides us through the human body—how it functions, its remarkable ability to heal itself, and (unfortunately) the ways it can fail.

Complications: A Surgeon's Notes on an Imperfect Science, by Atul Gawande, 2003—Explores the power and the limits of medicine, offering an unflinching view from the scalpel's edge.

Being Mortal: Medicine and What Matters in the End, Atul Gawande, 2007—When it comes to the inescapable realities of aging and death, what medicine can do often runs counter to what it should.

Better, A Surgeon's Notes on Performance, by Atul Gawande, 2008—Explores how doctors strive to close the gap between best intentions and best performance in the face of obstacles that sometimes seem insurmountable.

The Checklist Manifesto: How to Get Things Right, by Atul Gawande, 2011—Shows what the simple idea of the checklist reveals about the complexity of our lives and how we can deal with it

Disease

Emperor of All Maladies, A Biography of Cancer, by Siddhartha Mukherjee, 2011—From its first documented appearances thousands of years ago through the epic battles in the twentieth century to cure, control, and conquer it to a radical new understanding of its essence.

The Breakthrough: Immunotherapy and the Race to Cure Cancer, by Charles Graebe, 2018—The astonishing scientific discovery of the code to unleashing the human immune system to fight in this "captivating and heartbreaking" book.

Cell and Gene Therapy

A Cure Within: Scientists Unleashing the Immune System to Kill Cancer, by Neil Canavan, 2017—Is the story of the immuno-oncology pioneers. It is a story of failure, resurrection, and success.

CAR T: A New Cure for Cancer, Autoimmune and Inherited Disease, by William Haseltine and Amara Thomas, 2023—Describes how CAR-T cell therapy works, the potential and possibilities of integrating mRNA technology, current uses to treat cancer, and ongoing efforts that adapt this system to treat other illnesses, including heart disease, HIV/AIDS, and rheumatoid arthritis.

Good Genes Gone Bad, A Short History of Vaccines and Biologics: Failures, Successes, Controversies, by Narendra Chirmule, 2021—Highlights seven such colossal failures in drug development, all of which culminated in the development of novel drugs, weaving together various analogies through the stories and thus allowing the reader to understand complex biological phenomena.

Gene Editing

The Code Breaker: Jennifer Doudna,Gene Editing, and the Future of the Human Race, by Walter Issacson 2021—Account of how Nobel Prize winner Jennifer Doudna and her colleagues launched a revolution that will allow us to cure diseases, fend off viruses, and have healthier babies.

Health

Outlive, The Science and Art of Longevity, by Peter Atria, 2023—A groundbreaking manifesto on living better and longer that challenges the conventional medical thinking on aging and reveals a new approach to preventing chronic disease and extending long-term health, from a visionary physician and leading longevity expert.

Why We Die: The New Science of Aging and the Quest for Immortality, Venki Ramakrishnan, 2023—A groundbreaking exploration of the science of longevity and mortality.

Vaccinated, From Cowpox to mRNA, the Remarkable Story of Vaccines, by Paul Offit, 2022—A fascinating story of modern medicine and pays tribute to one of the greatest lifesaving breakthroughs—vaccinations—and the medical hero responsible for developing nine of the big 14 vaccines which have saved billions of lives worldwide.

Lifespan: Why We Age—And Why We Don't Have To, by David A. Sinclair, 2019—Is aging a disease? Is aging treatable? Do sirtuins extend the lifespan in yeast, invertebrates, and vertebrates? https://www.ncbi.nlm.nih.gov/pmc/articles/PMC9669175/pdf/nihms-1841406.pdf

Future

The AIRevolution in Medicine: GPT-4 and Beyond, by Peter Lee, Carey Goldberg, Isaac Kohane, 2023—AI is about to transform medicine. Here's what you need to know right now.

Deep Medicine: How Artificial IntelligenceCan Make Healthcare Human Again, by Eric J. Topol, 2019—Reveals how AI will empower physicians and revolutionize patient care.

How Medicines Work and When It Doesn't: Learning Who to Trust to Get and Stay Healthy, Peter Wilson, 2023—Blending personal anecdotes with hard science, an accomplished physician, researcher, and science communicator gives you the tools to avoid medical misinformation and take control of your health.

Soft Skills

In order to work in any work environment where interactions with people are central to "getting the job done," soft skills are important. These skills are typically not taught in the science education curriculum. Below is a short list of books that are useful to understand oneself. Many podcasts and YouTube videos of interviews of the authors of these books also provide valuable insights into self-awareness and self-development.

How to Win Friends and Influence People, by Dale Carnegie, 1936—Where self-development all started.

The 7 Habits of Highly Effective People, by Stephen Covey, 1989—The classic book which started the process of self-help.

Outliers, The Stories of Success, by Malcolm Gladwell, 2011—Journey through the world of "outliers"—the best and the brightest, the most famous, and the most successful. He asks the question: what makes high achievers different?

The Power of Habit: Why We Do What We Do in Life and Business, by Charles Duhigg, 2012—Takes us to the thrilling edge of scientific discoveries that explain why habits exist and how they can be changed.

Grit, The Power of Passion and Perseverance, by Angela Duckworth, 2018—Hypothesis about what really drives success: not genius, but a unique combination of passion and long-term perseverance.

Thinking Fast and Slow, by Daniel Kahneman, 2018—Takes us on a ground-breaking tour of the mind and explains the two systems that drive the way we think. System 1 is fast, intuitive, and emotional; system 2 is slower, more deliberative, and more logical.

Atomic Habits: An Easy & Proven Way to Build Good Habits & Break Bad Ones, by James Clear, 2018—Practical strategies that will teach you exactly how to form good habits, break bad ones, and master the tiny behaviors that lead to remarkable results.

Loonshots: How to Nurture the Crazy Ideas That Win Wars, Cure Diseases, and Transform Industries, by Safi Bahcall, 2019—Reveals a surprising new way of thinking about the mysteries of group behavior that challenges everything we thought we knew about nurturing radical breakthroughs.

Nudge: Improving Decisions About Health, Wealth, and Happiness, by Richard H. Thale and Cass R. Sunstein 2021—It has taught us how to use thoughtful "choice architecture"—a concept the authors invented—to help us make better decisions for ourselves, our families, and our society.

Think Again: The Power of Knowing What You Don't Know, by Adam Grant, 2023—Examines the critical art of rethinking: learning to question your opinions and open other people's minds, which can position you for excellence at work and wisdom in life.

Supercommunicators: How to Unlock the Secret Language of Connection, by Charles Dugigg, 2024—A fascinating exploration of what makes conversations work and how we can all learn to be supercommunicators at work and in life.

Links to Blogs on Soft Skills

Topic	Short description	Link
Questions	A process to ask questions methodically	https://chirmule.wordpress.com/2022/02/02/a-process-on-how-to-ask-questions/
Decisions	A process to make decisions	https://chirmule.wordpress.com/2019/05/30/a-process-to-make-decisions/
Writing	Writing accomplishments Handwritten letters	https://chirmule.wordpress.com/2022/08/03/i-havent-done-anything-writing-accomplishments-as-a-ritual/ https://chirmule.wordpress.com/2020/12/30/writing-hand-written-letter-it-brings-happiness-at-both-ends/
Reading	Reading systematically	https://chirmule.wordpress.com/2021/11/06/how-to-read-a-book-book-review/
Speaking	The art of speaking clearly	
Art	Value of art as a hobby	https://chirmule.wordpress.com/2020/06/12/why-should-you-paint/
Music	Value of music as a hobby	https://chirmule.wordpress.com/2022/02/15/what-is-your-purpose-of-life/
A perfect CV	Steps in developing an introspective CV	https://chirmule.wordpress.com/2019/11/05/a-perfect-cv-its-the-journey/
Planning a career	A process to systematically plan a career	https://chirmule.wordpress.com/2020/10/20/a-self-mentoring-process/
Courage and fear	Understanding the physiological and neurological aspects	https://chirmule.wordpress.com/2022/05/10/courage-and-fear/
Purpose of life	Developing a process	https://chirmule.wordpress.com/2022/02/15/what-is-your-purpose-of-life/
Getting things done	Becoming more efficient	https://chirmule.wordpress.com/2020/10/16/getting-things-done-repeating-the-training/

A Self-Mentoring Process. Making Your "Perfect CV"

Narendra Chirmule

We all have made our curriculum vitae (CV), biodata, and resumes and used these terms interchangeably. Wikipedia describes a CV to be more detailed than a resume. Biodata is a detailed document that provides a comprehensive summary of a candidate's life accomplishments, along with career information. Resumes are more of an overview of a candidate's career. There is no template for a perfect CV; if there were one, it would be downloadable from the internet, and everyone would get their perfect job. Since there is no template, I have developed a template which gets very close to a perfect CV. This article articulates that process.

The Random Feedback-Process Saga

Many folks have shared their CV with me, as their mentor, to obtain feedback on the content and format. I felt obligated to give feedback. So, if the format had educational background on the top, I would recommend writing the personal statement; if it had a personal statement, I would recommend writing a bulleted list of major accomplishments; if it has accomplishments, I would say, "please write your educational background first" (Full circle). It was a random process of providing feedback.

After giving this random advice for several years (and feeling good about it myself at first), I started thinking "Why am I giving advice on the CV of someone else? It would not be *their* CV anymore. It would be my interpretation of their CV." Then I stopped giving advice on CV. In fact, I did not even correct the spelling mistakes.

Would you hire someone if they had a spelling mistake in their CV?

An exaggerated story (for impact): Why would someone purposefully make a spelling mistake in their CV? There is a mistake because they cannot see it. Their minds are on larger issues in life like solving poverty, making healthcare affordable, transforming children's education, etc. They cannot see small things like spelling mistakes. In this story, the folks who have no spelling mistakes in their CV are very focused below and do not see the "sky" for solving big issues in life. Now, whom would you hire? [forget this story, it was just to make you think].

Planning a career is stressful and can result in lots of psychological complications.

Planning a career involves a complex array of internal thoughts and diverse and specific advice from various mentors, friends, family, and strangers. Our brain collects all the "data" and analyzes it through conscious and unconscious mechanisms. The major parts of the brain include cerebellum, parietal, temporal, prefrontal cortex, thalamus, hypothalamus, hippocampus, amygdala, and striatum, all of which have been defined to have specific functions to "analyze" and "interpret" the information and execute hormone-induced actions, evoking emotional responses. These emotions such as fear, anxiety, frustration, and panic can lead to systems of depression, stress, and anger.

In order to develop your perfect CV, it is important to use a systematic approach, from collecting the data, analyzing it to get information, interpreting the information to get knowledge, and applying the knowledge to achieve wisdom, about yourself. It is an introspective process.

Collecting the Data

In order to collect the data, the process involves a systematic approach to answering the following seven questions.

Who am I?

I suggest defining yourself in four (or more) things that define you as a person. For examples:

- *Immunologist, with 30 years of experience in drug development, with a deep interest in Hindustani classical music.*
- *Mechanical Engineer, with diverse experiences ranging from designing cars, plastics, to software, and guitar-playing singer of international music.*
- *Arts Management Professional, with experience in building theaters and museums, with interests in traveling to exotic places.*

The goal is to define yourself so that the description becomes "uniquely you."

What have I learned?

I suggest you make an Excel spreadsheet of everything that anyone has ever taught you. List all your degrees, including high school, (even elementary school, if you think you learned something unique there). *In this case, if your mother has taught you cooking, write cooking, if your friend has taught you how to fish, fishing, etc. You get the idea.*

What am I good at?

List ALL the activities you can do. Don't worry about how good you are at them or how well you can do them (yet). But write down each and every activity, even the trivial ones, e.g.:

- *Writing haiku poems*
- *Performing molecular biology techniques*
- *Programming in python*
- *Watercolor painting*

Write at least seven things. Go ahead and write all the 100 things that you may think of. Why not? After all the exercise is to "tell yourself what you are good at." So, don't be shy.

What am I most proud of?

It should be easy to list your accomplishments, by now. Again, write down all of them including the ones in which you have "self-doubt," or consider not important, e.g.:

- *I got 100/100 in a surprise test in Mathematics in 5th grade. I remember this distinctly since my mother was a math teacher, and I was generally bad at math. This accomplishment is a very distinct memory of my "accomplishment."*
- *I got second place in the spelling bee competition in my high school.*
- *I climbed Mount Everest at the age of 20.*

What debilitating challenges have I had?

Everyone has challenges, some small, some big. These can range from being *bullied in school*, *losing a loved one*, and *getting seriously sick*. Write all your challenges from the beginning of (your) time. Again, don't worry about the magnitude (yet).

What advice has stuck with me?

We all receive advice from friends and family. Write down all the advice that you have got, that has stayed with you.

- *Focus on what you want, rather than what you should be*
- *Write weekly summaries*
- *Don't worry about small things*

What have been my inflection points?

Before determining these points, we will now "analyze our data."

Analyzing the Data to Create Information

[*Satisfaction quotient (SQ).* How satisfied do you feel with yourself? Satisfaction is multifactorial. It can be imagined in many ways, such as happiness, responsibilities, and money. A combination of all these factors can yield a subjective numerical value, ranging from 1 to 10. 1 being the lowest, and 10 being the highest.]

Making a scorecard

The next step in the introspective exercise is to rate each of the events listed in the data with a SQ number. *E.g. How satisfied do you feel about yourself at this time? (1–10). How was your high school experience? (1–10). How good are you at a skill? How do you feel about your accomplishments? How did you feel about your challenge? How good was the advice you got from your mentor?*

The next step is to make an X-Y graph. Y-axis is the SQ (1–10), and X-axis with each of the events in your "data" listed chronologically. The more detailed your data listing are complete, the more it will provide with the information of your SQ over time. The curve will not be a simple straight line; there will be lots of curves, going up and down. This graph is your emotional/satisfaction journey (Fig. 1). Examples of such graphs are shown.

– Next, to deconvolute the inflection points, examine the list of all the causes where the shape of your career/satisfaction curve has changed. Make a list of all the inflection points. These are your own thoughts, and actions that you have taken. To understand the reasons for failures/successes, e.g.:
– *Courage to confront my bully*
– *Strength to make a decision to move to boarding school*
– *Lack of focus (distraction), which resulted in low performance in college*

THIS DOCUMENT IS YOUR PERFECT (introspective) CV. Of course, there is no such thing as a perfect CV. It is an attempt to download all the thoughts you have in your short-term and long-term memory in your brain onto a piece of paper. When you SEE it, the information goes from your eyes to the occipital part of your brain. If you READ your CV aloud to yourself, that information goes to the auditory part of your brain, and if you WRITE by hand, the same information goes to your hypothalamus (touch sensor and the executor of actions).

Now you have collected all the data, analyzed it for information, and interpreted it for knowledge about yourself, it is time to use your WISDOM to make decisions of what you want to do.

For that, there is another introspective exercise, a self-mentoring process.

[Download the template of the Excel Spreadsheet here:

Excel Spreadsheet template for an Introspective CV

1	Title
2	Education
3	Skills
4	Challenges
5	Accomplishments
6	Mentor Advice
7	Inflection Point(s)

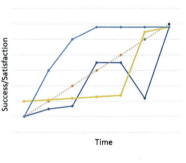

Planning Next Steps

The next step in the introspective process involves listing the careers you are considering.

[Download the template of the PowerPoint slides here:

PowerPoint Slide deck TEMPLATE for listing positive and negative points for each career]

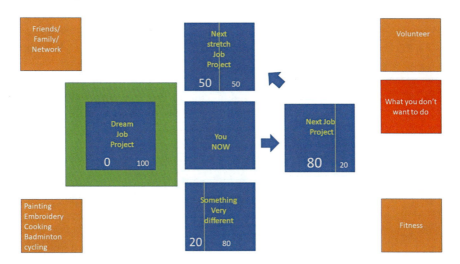

80/20: In this slide called "80/20" (plan A) write down the first thing you want to do. Be specific. Give a title to the position you want. Get a job description from Google. Next, write at least seven reasons of "why they should hire you" and then seven reasons "why they might NOT hire you."

50/50: For the next idea for your career (plan B) think in these terms. "What if I do not get the job in plan A, even after trying actively for more than 6 months? Then what will I do?" Here again, write down the title, job description, seven reasons why they should hire you, and seven reasons they will not hire you for THIS position.

20/80: The way to think about this option (plan C) is to answer the following question: "What would I do if I don't get the job in plan A, and do not get any connections for plan B, for more than a year? What then would I do?" There is always something "out of the box" that we want to do but never get around thinking about it seriously. In this process, write down the title and job description for this position, the seven reasons why you would do it, and seven challenges you might face. The process is to WRITE it. DOWNLOAD it from the unconscious parts of your brain.

0/100: The final plan (dream job). The way to think about this option is to consider, hypothetically, what you would do if you had unlimited money and unlimited access to resources. Here again, be specific. If there is more than one idea, make a separate slide for each one. Clarify your thoughts by writing the seven reasons why you like the plan and seven challenges. Write the obvious challenges, e.g., *not enough money*, *what will people say*, etc.

Once you have completed these four "slides" and clarified the pros and cons for each, you should complete the four activities below:

- *Fitness* (mental and physical). What CAN you do to maintain fitness (write this list on the left slide), and what ARE you doing (write these on the right side of the slide).
- *Hobbies*. Here again, write what you CAN and ARE doing.
- *Networking*. What are you doing to expand your network social media. Connecting with alumni. Write down the plan. Make Venn diagrams of people you know. See whom in your network you have not connected to in a while. Do not network only for a job. Build relationships.
- *Volunteer*. Do you volunteer? Have you volunteered? You always feel very happy. In fact, folks you volunteer for may not feel happy. But you will always do. The purpose of life is the pursuit of happiness. So, why would you NOT volunteer more often? Make a list of all the potential places you can volunteer. Even small amounts of volunteering are helpful (for the mind).
- *Do-not-want-to-do*. Sometimes, there are things that you will never do. If you have such a list, write it down.

Once you complete this exercise, **UNIVERSE WILL CONSPIRE TO MAKE YOUR PLAN HAPPEN**. This has happened so many times that I am a strong believer in this introspective process. So, go for it. Figure out the next way to document it. Use PowerPoint slides and Excel spreadsheets, and write by hand. Anything works. Just do it.

Author Biographies: The Experiences of the Authors Who Wrote the Chapters

Introduction

Narendra Chirmule is a founder-director at SymphonyTech Biologics, a data analytics company focused on engineering solutions for biology. As former Head of R&D at Biocon (Bangalore) and having held leadership positions at Amgen (Thousand Oaks, CA) and Merck Vaccines (West Point, PA), he has contributed to the clinical development of vaccines, biologics, and cell-and-gene therapies. During his academic career, he has worked on the development of a leprosy vaccine, the pathogenesis of AIDS, and gene therapy for several rare diseases. Dr. Chirmule is on the NIH advisory committee for HIV vaccines. He is a TEDx speaker and recently published a book, *Good Genes Gone Bad* by Penguin Press, both of which describe lessons learned from colossal failures in drug development. linkedin.com/in/narendra-chirmule-4968475

Biologics

Taruna Arora is currently Vice President, Biotherapeutics at Bristol Myers Squibb. Previously, Taruna held leadership roles in analytical, translational, and biotherapeutic discovery functions at Coherus Biosciences, Amgen, and NGM Bio. She has extensive experience in biotherapeutics discovery, protein engineering, development, and approval for a diverse range of biologic modalities. Under her leadership, teams delivered on multiple biologics IND candidates, which included bispecifics, T cell engagers, ADCs, agonistic antibodies, and enhanced cell-killing modalities. Notable contributions include approved products Blincyto and Aimovig. Taruna has been an active contributor to global regulatory filings and FDA meetings. Taruna earned her PhD in Immunology from the Mayo Clinic. She is also an

N. Chirmule, V. V. Ghalsasi (eds.), *Approved: The Life Cycle of Drug Development*, https://doi.org/10.1007/978-3-031-81787-8

inventor on issued patents and patent applications. Her interests lie in disruptive thinking, the accelerated advancement of therapeutics for patients' benefit, and organizational strategies that connect science, people, functions, and cultures. linkedin.com/in/arorataruna

Sanjay Khare founded ImmunGene in 2008, where he has been serving as President and CEO. Previously, Dr. Khare spent almost nine years at Amgen Inc with increasing responsibility culminating as Scientific Director in the Department of Inflammation/Immunology. While at Amgen, Dr. Khare co-discovered key immune co-stimulatory molecules such as B7RP1 and B cell activating factor (e.g., BAFF). Dr. Khare led several drug discovery programs related to immune cell costimulation and bifunctional molecules to optimize therapeutics. Dr. Khare is coinventor of several patents while working in Amgen and as an advisor to startup biotech companies. Several of these inventions have resulted in more than US$1B in potential deals from large pharmaceutical companies. Dr. Khare coauthored more than 50 scientific papers and abstracts and edited a book on the TNF superfamily. Dr. Khare has successfully capitalized ImmunGene from Angel investors, VC fundings, nonprofit organization (LLS), and the NIH-SBIR. Khare received his PhD in Immunology at the All India Institute of Medical Sciences, New Delhi, India, and did his postdoctoral fellowship in the Immunology Department at the Mayo Clinic and Medical School. He is a recipient of the ASHI scholar award in 1997. linkedin.com/in/sanjay-k-0241222

Vaccines

Amitabh Gaur is an accomplished biomedical scientist and leader in biotechnology, with expertise spanning assay development, vaccine innovation, immune regulation, autoimmune therapeutics, cancer immunotherapy, and stem cell research. He currently serves as a consultant and science advisor to biotechnology companies, driving advancements in drug development across pre-clinical and clinical stages. His pioneering work in developing assays to track immune responses to biological therapeutics, viral vectors, and transgenes has been instrumental in the success of biosimilars and the development of innovative cell and gene therapies. In industry, Dr. Gaur has held pivotal roles at Becton Dickinson (BD), a Fortune 500 company. As Business and Scientific Leader at BD Biosciences, San Diego, he spearheaded the development of cutting-edge assay solutions for biomedical applications across research, diagnostics, drug evaluation, and clinical trials. Under his leadership, the business achieved double-digit year-over-year growth, expanded its assay portfolio into critical areas such as leukemia and lymphoma diagnosis, vaccine monitoring, cell therapy, and biological therapeutics. He was a member of BD's Business Leadership Team that managed the Antibody and Assay Business, overseeing a team of over three hundred employees and successfully expanding the company's global reach in markets across Europe, Japan, Australia, Latin America, and Canada.

Prior to joining BD, Dr. Gaur led a pioneering research program at Neurocrine Biosciences, where he developed therapeutics for autoimmune disorders such as Multiple Sclerosis and Type I Diabetes, culminating in the selection of drug candidates that advanced to clinical evaluation. His earlier work on advancements in vaccine immunology was during his tenure at the National Institute of Immunology (NII) in New Delhi. Dr. Gaur's academic background includes a Ph.D. in Biochemistry and Immunology from the All-India Institute of Medical Sciences (AIIMS), New Delhi. Following his tenure at NII, he completed a post-doctoral fellowship at Stanford University Medical School's Department of Medicine, Division of Immunology. Dr. Gaur also served as an adjunct professor at the George Washington University School of Medicine and is a member of several professional organizations, including the American Association of Immunologists (AAI). An inventor with numerous patents to his name, Dr. Gaur has published many articles in top ranking scientific journals. linkedin.com/in/amitabhgaur

Priyal Bagwe is a driven researcher specializing in vaccine formulation. She is currently pursuing her PhD at Mercer University's College of Pharmacy in the Vaccine Nanotechnology Laboratory, Center for Drug Delivery and Research. Priyal possesses a wealth of experience across various biotechnological interventions, with expertise in vaccine formulation for infectious diseases and cell-based therapies for cancers. She has authored numerous scientific manuscripts focusing on innovative vaccine delivery strategies for combating infectious diseases. From conducting research to collaborating with industry experts, Priyal is wholly committed to improving global health and patient well-being through her scientific pursuits. linkedin.com/in/priyal-bagwe

Sharon Vijayanand is a Graduate Research Assistant in the Department of Pharmaceutical Sciences at Mercer University College of Pharmacy. She currently works in the Vaccine Nanotechnology Laboratory where she develops and tests inactivated polymer-based vaccines for COVID-19 and influenza and novel needle-free vaccine delivery strategies. Her work has resulted in multiple scientific publications, research awards, and presentations at scientific meetings. Sharon is motivated to continue her contributions to scientific research to improve overall patient health and awareness. linkedin.com/in/sharonvijayanand

Cell Therapy

Rahul Purwar is a pioneer in cancer immunotherapy and skin immunology research. He has completed his PhD from Hannover Medical School, Hannover, Germany, and postdoctoral studies from Harvard Medical School, Boston, USA. He joined IIT-Bombay as an Assistant Professor in 2014. He has published several papers in reputable journals and has received many awards. Cancer immunotherapy is an emerging and **transformative** approach for the treatment of cancer patients.

Immunotherapy is one of the best treatment options for patients with metastatic cancer; however, the benefits of available immunotherapy are limited because current immunotherapy focuses on life extension by only a few months. Therefore, there is an urgent need to explore novel strategies for the development of robust cancer therapy. In the context of this we have developed a robust CAR-T cell therapy platform for CD19+ malignancies. To mitigate the toxicity, we designed a novel humanized anti-CD19 CAR-T cells, that can efficiently eliminate tumor cells in a preclinical model. The novel humanized anti-CD19 CAR developed by our team has a favorable balance of efficacy to toxicity providing a rationale to test this construct in a phase I trial. Our research team is also working in the field of skin immunology and examining the role of effector T cells and its related cytokines in skin inflammatory disorders (vitiligo, atopic dermatitis, and psoriasis) and T cell lymphoma with skin involvement. linkedin.com/in/rahul-purwar-3187608a

Atharva Karulkar is the Head of Scientific Affairs and Co-founder of ImmunoACT. He obtained his doctoral degree in cancer immunotherapy from IIT Bombay under the mentorship of Prof. Rahul Purwar. During his tenure at IIT Bombay, Dr. Karulkar designed and developed India's first anti-CD19 CAR (NexCAR19), which subsequently underwent integrative process development and manufacturing under cGMP at ImmunoACT. Devoting six years of his budding career to the advancement of CAR technology, he has presented his groundbreaking research at both national and international conferences. His contributions have significantly impacted the landscape of cancer therapy in India, offering hope to countless patients. linkedin.com/in/atharva-karulkar-106771146

Karan Gera is the global business development manager at ImmunoACT, and a former researcher in the field of inherited metabolic diseases. Gera is trained in molecular, cellular, and developmental biology at Iowa State University (Master's) and Manipal Institute of Technology (Bachelor's). linkedin.com/in/karangera100

Aalia N. Khan is currently pursuing her PhD under the guidance of Dr. Rahul Purwar at IIT Bombay. Her early research endeavors include significant contributions to India's first indigenous anti-CD19 CAR-T cell therapy platform. Aalia's primary research focus revolves around the establishment of a novel CAR-T cell therapy platform for difficult-to-treat cancer. Honored with a distinguished Prime Minister's Research Fellowship for her innovative work, she is committed to leverage her cutting-edge research for the betterment of mankind. linkedin.com/in/aalia-khan-400813173

Sweety Asija is a final-year PhD scholar at the IIT Bombay specializing in the innovative field of designing, manufacturing, and producing CAR-T cells. With a passion for cutting-edge biomedical research, her work focuses on harnessing the power of immunotherapy to develop advanced treatments for cancer, particularly glioblastoma. Her ambition and determination drive her to explore novel methodologies and technologies in CAR-T cell therapy, aiming to enhance its efficacy and accessibility for needy patients. Her pursuit of scientific excellence and unwavering dedication to making a positive impact in the biotechnology field will help to revolutionize the landscape of cellular immunotherapy.

Ankit Banik is a fourth-year doctoral researcher pursuing his PhD from IIT Bombay. He has been working in the domain of cell and gene therapy for the last 3.5 years and contributed substantially to the field. He has prior experience and exposure to interdisciplinary sciences and has worked at Nanotech-Biology interface during his early research period. He is a CSIR-SRF, currently involved in research and development of affordable and superior CAR-T and CAR-NK cell therapy in India. linkedin.com/in/ankit-banik-784bb6182

Pratima Cherukuri is a highly accomplished leader with over 18 years of expertise in the development, manufacturing, and testing of viral vectors, including lentiviral, retroviral, and adenoviral vectors for gene and cell therapy. As Chief Scientific Officer at Genezen, she has been instrumental in transforming the company into a leading Contract Development and Manufacturing Organization (CDMO). Pratima is renowned for her ability to combine scientific and regulatory knowledge with strategic leadership, enhancing the company's competitive position within the biotechnology sector. In her role at Genezen, Pratima directs the company's scientific strategy, ensuring development priorities align with client objectives while maintaining compliance with cGMP standards. Her leadership drives continuous improvements in process development, viral vector production, and analytical capabilities. Prior to her current role, Pratima held key positions at Covance Laboratories and Indiana University's Vector Production Facility, where she pioneered innovative processes for viral vector production. She has authored 12 peer-reviewed articles, numerous white papers, and book chapters and has presented her research at major international conferences, contributing significantly to advancements in gene and cell therapy and biomanufacturing. linkedin.com/in/pratima-cherukuri-84ba8a67

Gene Therapy

Susan D'Costa is a molecular virologist with over 25 years of experience in virology. Over the past ten years she has been actively involved in viral vector analytics, process development, manufacturing, commercial strategy, and building successful teams. She is currently the Chief Technical and Commercial Officer at Genezen, a leading viral vector CDMO. Prior to Genezen, Susan was CTO at Alcyone Therapeutics, a biotechnology company pioneering next-generation CNS precision gene-based therapeutics for complex neurological conditions. At Alcyone, she was responsible for viral vector CMC, device development operations, and partnering on new technologies for both gene therapy and precision delivery. Dr. D'Costa has also held leadership roles of increasing responsibility at Thermo Fisher Scientific, Viral Vector Services and its predecessor companies—Brammer Bio and Florida

Biologix, working with different viral vectors, liaising with diverse biotech clients and building teams with scientific and operational excellence. Susan holds a PhD in biology, specializing in molecular virology, from Texas Tech University, an MS in biochemistry from Mumbai University (Grant Medical College), and a BS in microbiology/biochemistry also from Mumbai University (St. Xavier's College). linkedin.com/in/susan-d-costa-8877063

Hilda Petrs Silva is an associate professor at the Federal University of Rio de Janeiro (UFRJ), Brazil, with over 20 years of experience in research working with AAV viral vectors for gene therapy approaches. My experience with AAV vectors came from William Hauswirth Lab and the Vector Core in the Ophthalmology Department at the University of Florida. I have extensive experience in leading groundbreaking research in AAV vector-based gene therapy for retinal diseases, with a commitment to translating scientific discoveries into clinical applications. My research expertise centers on refining gene therapy approaches to treat retinal diseases, with specific emphasis on AAV modification enhancing transduction efficiency. I'm responsible for an AAV vector core at UFRJ. I'm an inventor on issued patent applications. I have collaborated with international teams to enhance the scope and impact of research, resulting in significant advancements and cross-border scientific integration. Due to my knowledge of gene therapy development, AAV production, and application, I have been acting as a consultant for the Brazilian regulatory agency, ANVISA, providing specialized expertise in the regulation of gene therapy products and clinical trials and guiding the development process to meet stringent regulatory requirements. Hilda holds a PhD in Biophysics, specializing in AAV vectors for retina gene therapy, from the Federal University of Rio de Janeiro in collaboration with the University of Florida, and a BS in Biomedical Science also from the Federal University of Rio de Janeiro. linkedin.com/in/hilda-petrs-silva-214558108

Mariana Santana Dias is a postdoctoral fellow in the Laboratory of Gene Therapy and Viral Vectors at the Federal University of Rio de Janeiro, Brazil. Mariana has experience in the manufacture, manipulation, and use of gene therapy viral vectors derived from adeno-associated virus (rAAV) for gene therapy, and investigation of gene therapy strategies in the context of neurodegeneration. Together with her supervisor Dr. Hilda Petrs Silva, they established bench-scale rAAV production in her current lab. Still in her undergraduate studies, she worked at the Gene Therapy Center of the University of North Carolina, where she first learned about AAV biology and vector manufacturing. Working on gene therapy applications with rAAVs in the retina, he received her master's and doctorate degree from the Carlos Chagas Filho Institute of Biophysics (IBCCF) at UFRJ, with a PhD in collaboration with the University Medicine Göttingen, in Germany. Afterward, she went back to the University of Göttingen for a postdoc, evaluating the role of a therapeutic rAAV on axonal transport and degeneration of the retinal ganglion cells in a glaucoma model. linkedin.com/in/mariana-santana-dias-8aa437201

Shashwati Basak is the VP and Head of Cell and Gene Therapy Unit at Intas Pharmaceuticals (Biopharma Division), Ahmedabad, India. She is leading the development of several programs in both gene and cell therapies, spanning from early-stage research to clinical stage. Prior to this, she served as the Head of Quality and Regulatory Operations at another leading Indian CGT company called Immuneel Therapeutics Ltd, focused on developing CAR-T cell therapies.

Dr. Basak has a PhD in Molecular and Cell Biology from the Indian Institute of Science, Bangalore. She did two postdoctoral fellowships from the Salk Institute for Biological Sciences and Stanford University, studying gene expression and regulation and cancer signaling pathways. She has 20+ years of scientific leadership experience in translational research, clinical biomarkers, analytical assays, technology platforms, quality, and compliance. She has worked in several biotech/biopharma companies including Biocon Bristol Myers-Squibb R&D Center, Aurigene Discovery Technologies, and Immuneel Therapeutics Ltd and held positions of increasing responsibilities, in varied roles. linkedin.com/in/shasho001

Natasha Rivas has more than 20 years of experience in the pharmaceutical and medical device industry, including 15 years working in the CDMO space. Natasha was previously the VP of Quality at SCA Pharmaceuticals and Director of Quality Assurance and Quality Control at Vetter Pharma's clinical manufacturing site in Chicago, IL. Prior to this, she held roles of increasing responsibility in quality control, quality assurance, and regulatory affairs at Abbott Laboratories and Hospira. She leads the design, implementation, and iteration of the quality systems and processes that meet customers' evolving needs and ensures delivery of safe and effective products at Genezen. linkedin.com/in/natasha-rivas-6262aa19

Jennifer Thiaville is currently the Associate Director of Business Development Operations for Genezen, a leading viral vector CDMO. Jennifer has over eight years of experience in the viral vector CDMO industry, with experience in quality control and commercial operations. Prior to Genezen, she held roles of increasing responsibility at Thermo Fisher Scientific, Viral Vector Services and its predecessor company Brammer Bio, first in quality control overseeing all contract laboratory testing for viral vector products, before joining the commercial operations team providing technical leadership for proposal generation for new programs and in support of ongoing viral vector manufacturing programs. Jennifer holds a PhD from the University of Florida in Biomedical Sciences, specializing in microbiology and microbiology and has over 12 years of research experience in microbiology and molecular genetics. https://www.linkedin.com/in/jenny-thiaville-22265683

Srinivas Rengarajan is a trained molecular biologist with a Master of Science degree from the University of Southern California, Los Angeles. He worked at Intrexon corporation, San Diego where he helped develop and test cancer immunotherapies using bispecific antibodies and CAR-T cells generated using nonviral gene delivery. He then had the opportunity to join Poseida therapeutics as one of the earliest members of the company and played an integral role in developing the company's autologous and allogeneic CAR-T programs against multiple cancers starting from preclinical research and in vivo testing through clinical trials. He led the teams that established lab operations for Poseida across multiple R&D facilities and GMP manufacturing plants. Additionally, he was also responsible for the manufacture of critical nucleic acid raw materials used in the production of clinical grade CAR-T therapies for up to phase 2 clinical trials. He has recently moved back to India with the objective of enabling commercialization of effective and low-cost cell and gene therapies in India. linkedin.com/in/srinivas-rengarajan-3b087151

Process Development

Ankur Bhatnagar is the Head of Process Sciences at Biocon Biologics Limited's Research and Development facility located in Bangalore, India. With a track record of 23 years in the industry, he possesses extensive expertise in diverse areas, including CMC (chemistry, manufacturing, and controls), process development, technology transfer, scale-up, regulatory filings, and the establishment of teams, laboratories, and pilot plants. He was an integral part of the team involved in getting global approvals for biosimilars like Trastuzumab, PEG-GCSF, and insulin-glargine. He holds a master's degree in Biochemical Engineering and Biotechnology from the Indian Institute of Technology (IIT) Delhi and an Executive MBA program at the Indian Institute of Management (IIM) Bangalore. Ankur has been recognized internally in various instances for his contributions in setting up a robust platform for the development of Biosimilars and as a people manager. Ankur won the Great Manager award conducted by the *Economic Times* and People Business in 2017, in the "Overall Category" across industries. linkedin.com/in/ankur-bhatnagar-5994645; (interview) https://youtu.be/UGq2rdQn-p4?si=xOG_CjmXsUadA8Rx

Dr. Nagaraja G is the General Manager, is Head of Cell Line Development group at Biocon Biologics Limited. He completed his master's degree in Biochemistry from the University of Mysore. He later obtained his PhD degree in Biotechnology from Kuvempu University. He has worked in the Indian Institute of Science and Jawaharlal Nehru Centre for Advance Scientific Research in Bangalore for 5 years and played a role in establishing molecular parasitology laboratory. During the tenure of 23 years at Biocon, he has contributed to establish therapeutic recombinant protein expression platforms in bacteria, yeasts, filamentous fungi, and mammalian systems. Also, it played a key role in the discovery and development of novel bispecific fusion antibody molecules targeting immuno-oncology space. These innovative efforts resulted in about 10 patents and 15 publications to his credit. linkedin.com/in/nagaraj-govindappa-a1a54740

Karthik Ramani graduated with a dual major in Pharmacy (Bachelor's, Hons) and Biological Sciences (Master's, Hons) from BITS, Pilani, India, and received his PhD in Pharmaceutical Sciences from the University at Buffalo, State University of

New York, USA. Immediately after his doctoral work, Karthik relocated to India and joined Biocon group in 2005. Subsequently, he established the Biologics formulation group at Biocon Biologics in 2006. He was also one of the founding members of the insulin devices team at Biocon Biologics. Karthik and his team over the years have played a pivotal role in the development and successful approval of the Biocon Biologics biosimilars (insulins and monoclonal antibodies) in EU, Japan, and the United States as well as domestic and emerging markets. This also includes the first-in-the-world approval of Trastuzamab biosimilar in India and first-in-US approvals for Trastuzamab and Pegfilgrastim and the world's first interchangeable insulin Glargine approved by the USFDA. He and his team were involved in drug product development activities focusing on formulation and associated technology transfer to Biocon's drug product facilities, supporting the manufacturing of clinical and commercial supplies and developing product specifications. Currently, Karthik is a portfolio and product developer leading a team of about 100 scientists focusing on both development (process and analytics) and product life cycle management of insulin analogues and monoclonal antibodies involving process improvements and supporting capacity expansion for both drug substance and drug product. He also serves as a member of the R&D leadership team at Biocon Biologics Limited. Karthik has served as an insulin expert panel member of United States Pharmacopeia (USP) from 2015 to 2020. He is also a board member (Member at Large) of the India chapter of the Parenteral Drug Association (PDA) since January 2024, in the panel of the PDA India Chapter, Board of Education, and has contributed as a tutor for their training programs. Karthik also serves as a reviewer of scientific manuscripts for the *Journal of Pharmaceutical Sciences*, USA, since 2021 linkedin.com/in/karthik-ramani-8526336 (Interview): https://youtu.be/XsV8EN0ce3A?si=6j7v Fa4MZAZB5Efd

Dr. Partha Hazra is the Head of Process Sciences and Innovation Technology in R&D, Biocon Biologics, India. He has 30+ year of experience in the field of drug manufacturing process development in lab and technology transfer. He has accomplished leadership responsibilities and successfully contributed to the CMC section, from lab development to final approval along with product life cycle management of biosimilar antidiabetic and anticancer drugs. Dr. Partha has versed experience in interaction with national and international regulatory agencies, like EMA, FDA, PMDA, and other emerging market regulatory bodies for EMA Scientific Advice, Pre-IND requests, Ph-1 IMPD, Ph-3 IMPD, etc., and is involved in BLA dossier submission. Dr. Partha completed PhD in Biochemistry from the Indian Institute of Chemical Biology, Jadavpur, and got postdoctoral experiences from the University of California, San Diego. He has authored around 20 scientific articles in peer-reviewed international journals and >15 granted patents (USA, EU). He has also contributed as an invited speaker at multiple international conferences. www.linkedin.com/in/partha-hazra-76984028/

Dr. Harish V. Pai is a General Manager/Project Group Lead, Analytical Sciences at R&D in Biocon Biologics. With a background in biochemistry and neurosciences and a wealth of experience in both research and industry, Dr. Pai brings a unique blend of end-to-end characterization of biosimilars. Drawing on his scientific expertise and business acumen, he has contributed, through functional in vitro characterization on a diverse portfolio of projects spanning multiple therapeutic areas, from diabetes and oncology to immunology at Biocon Biologics. Dr. Pai has been contributing with his team through the complexities of the drug developmentprocess, from clone to product development to preclinical research to regulatory approval and life cycle management. He was an integral part of the team involved in getting global approvals for biosimilars like Trastuzumab, PEG-GCSF, and insulin-glargine. Dr. Pai has authored around 22 scientific articles in peer-reviewed journals, including the biosimilars on which Biocon Biologics is working. He believes in the principle of making complex topics simple to understand and likes to engage in scientific discussions. https://www.linkedin.com/in/harish-v-pai-8870362

Dr. Navratna Vajpai is a Project Group Leader (Senior Director) at Biocon Biologics, a biopharmaceutical company focused on the development of affordable biosimilars. In different roles in R&D at Biocon Biologics (Bangalore), and Biological E Ltd (Hyderabad), he has contributed toward the development of biosimilars (insulins and mAbs) and complex generics. Earlier in the Discovery group at AstraZeneca, Alderley Park, UK, he contributed toward early-stage drug-discovery initiatives for oncology targets. His alma maters include IIT Madras, Biozentrum Basel, Switzerland, and The Salk Institute, California. His core expertise is the development of analytical tools for the assessment of biologics drugs. During his academic career, he has worked on the development of NMR-based analytical tools for structural elucidation of pharmaceutically relevant protein targets. Dr. Vajpai has authored around 20 scientific articles in peer-reviewed journals and is an invited speaker at multiple workshops and biologics forums across India. He is a co-guide for a few PhD and master's students in joint collaboration with IIT Bombay and BITS Pilani, Hyderabad. linkedin.com/in/navratna-vajpai-8846b118

Manufacturing

Dhananjay Patankar is an independent biopharmaceutical professional and advisor to several biopharmaceutical companies and research institutes. He previously worked as Vice President of Biologics CDMO business at Syngene and Chief Operating Officer at Intas Biopharmaceuticals and managed teams involved in the development and manufacturing of novel biologics and biosimilars. He has been intimately involved in the growth of the Indian biopharmaceutical industry since its early days. Over his career, he led teams that developed India's first biosimilar therapeutic product (EPO), India's first biosimilar approved for marketing in Europe (Filgrastim), India's first EU-GMP certified biologics manufacturing facility (Intas), and India's first commercial contract manufacturing of a novel biologic for the US market (at Syngene, for a US client). He has served as a Biologics Expert Committee member at the US Pharmacopeia from 2011 till 2020. By education he is a Chemical Engineer with bachelor's degree from IIT Mumbai and PhD from the University of Utah in the United States. linkedin.com/in/dhananjay-patankar-82a82a9

Clinical Trials

Prajak J Barde is a qualified physician, founder, and director of Med Indite Communications Pvt Ltd (MIC), and alumnus of Seth GSMC and KEM Hospital, Mumbai, and Government Medical College, Nagpur. Dr. Barde has 17+ years of rich and diverse clinical research and development experience in biotech and pharmaceutical companies and served in companies Rhizen Pharma, Glenmark Pharmaceuticals, Serum Institute of India (SIIL), and Lupin Limited in a senior position. Expert in clinical translation and clinical development of New Chemical Entities (NCE) and New Biological Entities (NBE), Dr. Barde has a direct interaction with regulatory authorities such as the US FDA, MHRA, TGA, Australia, and DCGI at different stages of clinical development. Dr. Barde was involved in the development of NCE compounds for multiple targets (e.g., PI3K, CRAC, PARP) in the hematology-oncology and solid tumors space, and novel vaccines (e.g., meningococcal A and rabies) which were developed in collaboration with WHO and PATH and was a key member of a team who successfully executed clinical development of H1N1 influenza pandemic vaccines, both nasal and intramuscular. Dr. Barde has more than 35 international publications to his credit. Currently, he is supporting various biotech and pharmaceutical companies in their development and strategy as a consulting Medical Director. https://www.linkedin.com/in/dr-prajak-barde-md-3a66668/

Mohini Barde is the founder and director of Med Indite Communications Pvt Ltd (MIC), has completed her MD from DY Patil Medical College, Pune, and MBBS from the Government Medical College Nagpur. Professionally trained, she has 10+ years of rich and diverse clinical research and development experience. Dexterous in clinical translation, her work mainly involves conceptualizing clinical study designs, planning, and exploring innovative approaches/ PK/PD models for proof-of-concept studies of new chemical entities (NCE). Experienced in clinical development, she has interacted with the Indian regulatory authority DCGI/Subject Expert Committee (SEC) and various research funding agencies (DBT/BIRAC) at different stages of drug development. She was involved in the development of multiple NCE compounds in various therapeutic areas. During the COVID-19 pandemic, she developed multiple protocols for SARS-CoV-2 and oversaw their execution. Currently, she is supporting various biotech and pharmaceutical companies for clinical development and strategy for global clinical development and pre-IND/scientific advice meetings; filing INDs in India and the United States; regulatory activities; medical monitoring; and development of the protocol/concept sheets and other essential documents. She is committed to the development of advanced cost-effective solutions to address unmet medical needs. linkedin.com/in/dr-mohini-barde-md-14572735

Pharmacology

Vibha Jawa is an Executive Director for Biotherapeutics Bioanalysis organization at Bristol Myers Squibb. Vibha is responsible for leading biotherapeutic and cell/gene therapy bioanalytical (BA) functions supporting DMPK and immunogenicity and providing strategic and scientific oversight for BMS developmental portfolio. Vibha was at Merck for 4 years where she led the Predictive and Clinical Immunogenicity group and at Amgen for 14 years supporting discovery to development for biotherapeutics. Vibha has 20+ years of experience in diverse fields of biologics, vaccine development, and gene therapy with successful support of 20 + IND, BLA, and MAA filings. Vibha is a recognized leader in bioanalysis and immunogenicity with 75+ peer-reviewed publications and serves as a reviewer and editor for *The AAPS Journal* and *Journal of Pharmaceutical Sciences*. She is an active member of multiple scientific societies and consortiums (IQ, SC space Consortium, and EIP). Within AAPS, she is Track Chair of Land O Lakes Bioanalysis Meeting, Chair of the Cell and Gene Therapy Bioanalysis and Biomarker working group, Steering Committee member of the Therapeutic Product Immunogenicity Community, and past chair of ImmunogenicityRisk Assessment and Mitigation Community and leads the IQ Consortium for Cell/Viral/Gene therapies. She was recognized as the AAPS Fellow 2022. Vibha is the President of STEAMpark a non-profit promoting STEM-based learning in underserved communities. She also likes to mentor high school students on STEM-related projects and early career scientists in her free time. linkedin.com/in/vibha-jawa-phd-faaps-6844555

Glareh Azadi is an Associate Director at the Pharmacokinetics and Drug Metabolism Department at Bristol Myers Squibb (BMS). She holds a PhD in Biomedical Engineering from Brown University, RI. Glareh's primary focus is on the discovery and development of biologics and quantitative translation through the application of modeling and simulation techniques. Prior to joining BMS, Glareh worked on PK/PD and translation aspects of biologics in Merck and AbbVie. Glareh has a keen interest in the application of mathematical fit-for-purpose models toward optimization and development of novel biotherapeutics. linkedin.com/in/glareh-azadi-78222b11

Afsana Trini is a Senior Scientist in Clinical Pharmacology, Pharmacometrics, and Bioanalysis (CPPB) at Bristol-Myers Squibb (BMS). She received her PhD in Pharmaceutics and Drug Design from Saint Joseph University in 2019. She also has an MPharm degree in Clinical Pharmacology from the University of Dhaka. At BMS, she is the Bioanalytical lead for Biotherapeutics from preclinical to clinical stages, supporting DMPK and immunogenicity functions. She has been the bioanalytical lead for multi-domain therapeutics and ADC in oncology and immunology space. She also prepares appropriate documentation and/or interacts with health authorities on the bioanalytical portion of the regulatory filings. Prior to joining BMS, she worked as a Senior Scientist at PPD Inc. and served as a Principal Investigator in multiple bioanalytical GLP and non-GLP studies. For her PhD, she worked in the field of metabolism and adiposity in rodent models. She is an active member of AAPS and has taken up the role of a co-lead in the Early career bioanalytical scientist (ECBS) community at AAPS in 2024. She volunteered as a STEM teacher to elementary and middle school students from underserved communities in 2022 and wishes to return as a teacher in 2024. linkedin.com/in/atrini

Laxmikant Vashishta is a Principal Scientist—Bioanalytics within the Clinical and Medical Affairs Department at Alvotech Biosciences India Private Limited, a subsidiary of Alvotech Iceland. He is the bioanalytical lead with 15 years of work experience for the biosimilar product development responsible for Pharmacokinetics (PK), Immunogenicity (Anti-drug Antibodies (ADA) and Neutralizing Antibodies (Nab)) and Biomarker Immunoassay development, validation, and regulated bioanalysis of samples from clinical trials. Prior to joining Alvotech, Laxmikant worked as Associate Scientific Manager at Biocon Biologics Limited contributing an integral part of successful bioanalytical studies for biosimilar Trastuzumab, biosimilar Pegfilgrastim, biosimilar Insulin Glargine, and Insulin Aspart. He has a master's degree in Applied Genetics from Bangalore University and has four international peer-reviewed publications and one granted international patent. linkedin.com/in/laxmikantvashishta

Kaushik Datta (KD), PhD, DABT is a Senior Scientific Director, Bristol Myers Squibb (BMS), is responsible for the strategic oversight and execution of nonclinical safety activities and health authority interactions to support BMS portfolio. He has been working in the pharmaceutical industry for approximately 24 years in the field of toxicology. Through the years, KD has been involved in several industry leadership and working group activities for scientific/regulatory enrichment and issue resolution and supported numerous drug submissions and post-marketing commitments. KD received his PhD in Toxicology at the University of South Florida, College of Public Health, and completed a postdoctoral fellowship in Biochemical Toxicology at the University of Texas at Austin. https://www.linkedin.com/in/kaushik-datta/

Biomarkers

Pradip Nair is currently employed at Syngene International, serving as the Head of Discovery Science for Bicara Therapeutics, a start-up based in Boston, United States. Bicara's focus is on developing novel multi-functional molecules in the immune oncology space. He collaborates closely with teams across discovery, development, translational research, and CDMO at Bicara for various projects. Pradip has played a significant role in the development of Bicara's lead asset, BCA101(Ficerafusp alfa) which has successfully completed Phase 1 clinical trials in head and neck cancers under a US IND and is now in Phase 2/3 registrational clinical trials. Pradip has over 20 years of experience in the Pharma industry associated with the development of Novel and Biosimilar biologics. BIOMAB™ (Nimotuzumab), an anti-EGFR monoclonal antibody approved for head and neck cancers, and the anti-CD6 monoclonal antibody ALZUMAB™ (Itolizumab) are novel molecules that he was actively involved with and led at Biocon respectively. He has 20 international, peer-reviewed publications and 10 granted international and national patents. Pradip has served as a reviewer for many international journals including Cancer Research, PLOS one, International journal of cancer, among others. Pradip pursued his academic journey by obtaining a Master of Science in Zoology with a specialization in molecular biology and biochemistry from Banaras Hindu University, Varanasi, India. He went on to complete his Ph.D. at the Regional Cancer Centre, Trivandrum, India, under a Council for Scientific and Industrial Research fellowship. His post-doctoral research was completed under a Department of Science and Technology Fast Track Young Scientist Award fellowship from the National Centre for Biological Sciences, Bangalore. Since 2005, Pradip has been an integral part of the Biocon group of companies. linkedin.com/in/pradip-nair-75b60b121

Bindhu OS holds a postgraduation in Medical Biochemistry from Manipal Academy of Higher Education and a PhD from the Regional Cancer Centre Trivandrum. She is currently serving as the Deputy Director at the Centre for Researcher Training and Administration, JAIN (Deemed to be University), Bangalore, and has accumulated 22 years of teaching and research experience. Her PhD research focused on the expression of molecular markers such as matrix metalloproteinases and their regulators (TIMPs and NFkB) to comprehend the neoplastic transformation and progression biology of oral cancer.

Presently, her research interests lie in the pharmacological and biotechnological prospects of plant latex proteases, wound regeneration biology, and biofluid-based metabolomics in oral squamous cell carcinoma (SCC). Prof. Bindhu has played a pivotal role in mentoring 9 PhDs and 4 MPhils, contributing to 27 international publications, 3 book chapters, and securing 1 process patent. As an educator, she has been actively involved in designing innovative courses and is dedicated to exploring teaching-learning pedagogies and assessment strategies. She has been instrumental in establishing several memoranda of understanding (MoUs) with diverse institutions, including St. John's Academy of Health Sciences, in Stem; NMR Center–Indian Institute of Science; IBMEC Group, Campinas, SP, Brazil; and International Stem Cell Services Limited. Prof. Bindhu also dedicates her expertise as a reviewer for reputable journals and holds memberships in several professional bodies, including Life Membership with the Indian Science Congress and the Indian Academy of Neurosciences and an Executive Membership in the Annual General Body Meeting of IAN Bangalore chapter. Notably, "Train the Brain" is a collaborative outcome of her association with NIMHANS, Bangalore, providing a platform for young minds across India to engage with 20 eminent neuroscientists from diverse research organizations. She is committed to continuous self-improvement, keeping pace with dynamic developments in the academic field. linkedin.com/in/dr-bindhu-o-s-104088144

Sayeeda Mussavira completed her PhD in Biochemistry from JAIN (Deemed to be University) in Bengaluru, India. She has successfully passed the Karnataka State Eligibility Test for lecturer/assistant professorship and was awarded the Jain University Junior Research Fellowship to support her doctoral studies. Dr. Mussavira's research interests revolve around the exploration of biofluids for disease markers using a metabolomics approach. She possesses proficient knowledge in 1D-H1-NMR spectroscopy-based biofluid metabolic profiling, including spectral processing and statistical analyses to identify differential metabolites. Furthermore, she has a foundational understanding of ethical practices associated with designing case–control studies and recruiting volunteers for research purposes.

In addition to her research skills, Dr. Mussavira also possesses excellent communication abilities, including public speaking and active listening, as well as a keen enthusiasm for acquiring new skills. She is adept at working collaboratively in team environments. Her professional experience includes proofreading research paper manuscripts to ensure adherence to journal-specific style sheets and clarity as a copyeditor at Macmillan Publishers India Ltd, Bengaluru. During her PhD, she worked closely with postgraduate students in their final-semester projects and also taught postgraduate students at the Oxford College of Science and Administrative Management College in Bengaluru. Currently, she is serving as a faculty member in the Biochemistry department at Maharani Cluster University, Bengaluru. Dr. Mussavira has received training in handling laboratory instruments and bioinformatics from Azyme Biosciences Pvt Ltd and the Centre for Systems Biology and Molecular Medicine at Yenepoya Research Centre (YRC), YENEPOYA (Deemed to be University). Her research contributions have been published in various national and international journals such as the *Research Journal of Pharmaceutical, Biological and Chemical Sciences*, *Research Journal of Life Sciences*, *Bioinformatics, Pharmaceutical and Chemical Sciences*, *Turkish Journal of Medical Sciences*, *Indian Journal of Physiology andPharmacology*, and *Biochemia Medica*. Furthermore, she has presented her research findings at numerous national and international conferences throughout her PhD journey. linkedin.com/in/dr-sayeeda-mussavira-0b21bb53

Toxicology

Padma Kumar Narayanan has extensive experience in the field of preclinical safety, toxicology, and pathology spanning over two decades. Starting as a Postdoctoral Research Fellow at Los Alamos National Laboratory in 1995, Narayanan progressed through roles at GlaxoSmithKline, Amgen, Ionis Pharmaceuticals, and The Janssen Pharmaceutical Companies of Johnson & Johnson. Currently serving as the Vice President and Head of Preclinical Safety at Wave Life Sciences, Narayanan has held various leadership positions, including Executive Director of Toxicology and Senior Director of NonClinical Safety. Padma Kumar Narayanan holds a Doctor of Veterinary Medicine (DVM) degree from Kerala Agricultural University, a Master of Science (MS) in Small Animal Surgery from Tamil Nadu Veterinary and Animal Sciences University, and a PhD. Immunophysiology/Immunotoxicology from Purdue University. linkedin. com/in/padma-kumar-narayanan-105a7711

Rare Diseases

Harsha Rajahansa is the founder member of the Organization for Rare Diseases in India and on the scientific expert panel for Rare Genomics Institute. He is academically affiliated as a faculty in the School of Systems Biology at George Mason University, Fairfax, VA. He is also the founder CEO of a Bioinformatics and genomics services company Jeeva Informatics Solutions based in the state of Maryland, USA. Since Nov 2013, Harsha is doing business development for Strand Life Sciences, implementing their end-to-end clinical genomics solutions (www.strandls. com) in the US hospitals. Prior to founding Jeeva, Harsha was a Sr. Director of Bioinformatics and Translational Research at Dovel Technologies, a Virginia-based health IT company, where he covered academic and US Federal Government contracts and advised the FDA on the STARLIMS implementation program at their office of regulatory affairs. Harsha has over a decade of experience working on various interdisciplinary projects involving genomics BigData as a consultant for clients including the National Cancer Institute, the National Eye Institute, Georgetown University, and Genome International Corporation. His research work has focused on the genomics and systems biology of diseases including cancer, infectious diseases, neuromuscular diseases, and retinal degenerative diseases. Harsha completed his MS in Computer Science and PhD in Genetics, Bioinformatics, and Computational Biology at Virginia Tech, where he developed and applied reusable simulation models of mitochondrial DNA heteroplasmy dynamics to various diseases. linkedin.com/in/harsharajasimha

Mohua Chakraborty Choudhury works on health policy research at DST-Center for Policy Research, Indian Institute of Science, focusing on Rare Disease Policy

Research. She has specific interest in understanding the challenges faced by the rare disease community in India and identifying the policy and public health interventions to improve their quality of life. She is a member of the WHO Global Collaborative Network for Rare Diseases. She is currently an MPH candidate at Johns Hopkins University as a JN Tata Scholar. linkedin.com/in/mohua-chakraborty-choudhury

Partha Sarathi Mukherjee is currently the Head of CMC Analytical Development at Amicus Therapeutics, an organization dedicated to rare disease patients. During his 27 years of career, he has worked in R&D divisions of major global biopharmaceutical organizations such as Pfizer, AstraZeneca, GlaxoSmithKline, Novartis, Bristol Myers Squibb, Merck, and Endo to cover branded, generics, and consumer health arenas. He held analytical and CMC leadership roles to take candidate molecules from discovery through development and regulatory approval to postcommercial phases for therapeutic areas including cardiovascular, CNS, GI, pain and inflammation, oncology, respiratory, antibacterials, oral, nutrition, smoking Cessation, etc. He led or contributed to 15 NDA, 3 BLA, 4 ANDA, 21 commercial global products, and also a total of 110 early- and late-stage development projects. The projects his global teams undertook included small molecule, biologics, hemisynthetic, radiopharmaceuticals, combination products, and diagnostics and spanned across various pharmaceutical dosage forms. During his career, he rolled out continuous improvement, innovation, operational excellence, quality by design, automation, and several centers of excellence. He has 13 peer-reviewed publications and 42 presentations at national and international scientific conferences. He also contributed to 22 INDs for startup and mid-size biotech and pharma companies in the United States, EU, and Asia-Pacific regions and 4–6 investment decisions/year by Venture Caps, as a CMC Consultant. He obtained his PhD in Pharmaceutical Analysis from the Medical College of Virginia, Virginia Commonwealth University where his thesis included a novel laser fluorescence-based ultrasensitive analysis of anticancer agents in biological matrices. Dr. Mukherjee has chaired numerous scientific R&D symposia and workshops during his career and has obtained numerous R&D Awards in Innovation and Leadership Excellence. He is an active member of USP Expert Committee and BioPhorum, and a Scientific Reviewer of WHO. linkedin.com/in/partha-s-mukherjee-7a009b1b

Data Analytics and Risk Assessment

Ravindra Khare (Ravi) is a Co-founder and CEO of Symphony Technologies Pvt Ltd based in India. He has worked extensively in statistics and data analytics. His focused areas of work have been quality engineering, design, and manufacturing in the pharma, engineering, and automotive sectors. He is a mechanical engineer with a master's degree in Data Science and Engineering from the Birla Institute of Technology and Science, Pilani, India. linkedin.com/in/ravi-khare-a5432714

Intellectual Property

Joanna Brougher is a patent attorney who focuses her practice on all aspects of services related to patents in the areas of biotechnology, pharmaceuticals, and medical devices, including patentability opinions, patent drafting, domestic and foreign patent prosecution, development and management of patent portfolios, and general client counseling during all phases of a product's lifecycle, from concept to commercialization. Joanna's patent experience covers a variety of complex and innovative inventions involving small molecule drugs, biologics, cell-based technologies, compositions, drug formulations and drug delivery systems, immunotherapeutics, medical devices, diagnostic tests, and immunology, particularly vaccines and antibodies. Joanna also has experience counseling clients on the Hatch-Waxman Act and is monitoring developments involving biosimilars under the Biologics Price Competition and Innovation Act. Joanna is also an Adjunct Professor at Cornell Law School and at the University of Pennsylvania School of Medicine. Previously, she was an Adjunct Lecturer at the Harvard School of Public Health. She regularly speaks or lectures on intellectual property-related topics and is a frequent author of articles related to patent law and healthcare. Joanna has published two books: *Intellectual Propertyand Health Technologies: Balancing Innovationand the Public's Health* and *Billion Dollar Patents: Strategies for Finding Opportunities, Generating Value, and Protecting Your Inventions*. linkedin.com/in/joannabrougher (interview) https://youtu.be/Zx-Y6KqZCNU?si=xsustNSTIm-ctELm

Eliza Hong is a third-year student at Cornell Law School. During law school, they served as a Note Editor for the *Journal of Law and Public Policy* and led access to justice technology chatbot and AI projects. They will be working in tax and transactional law after graduation. linkedin.com/in/eliza-hong

Gray Ah-Ton is a third-year student at Cornell Law School. She was the secretary of the Intellectual Property and Technology Students' Association (IPTSA) and is an editor at *Cornell's Law Review*. During her Communications minor at McGill University, she wrote and learned about media history. She has carried her interest in digital media—particularly, in video games and AI—into law school, and she will be continuing to work with the Intellectual Property practice at Debevoise & Plimpton after she graduates. linkedin.com/in/gray-ah-ton-785b99140

N-of-1

Yiwei She is the Founder and CEO of the TNPO2 Foundation. Yiwei trained as a mathematician and worked as an Assistant Professor at Columbia University. She dabbled in AI at a Bay Area startup during a prior "AI-hype-wave." She was enjoying the choice of being a stay-at-home parent when her second child Leo was diagnosed with a de novo ultra-rare neurodevelopmental genetic disease. Faced with no other options and unwilling to give up hope on a 4-month-old baby, Yiwei took on the project of attempting a pharmaceutical rescue of quality of life for her child. As parent, caregiver, advocate, and ad hoc drug developer, Yiwei brought together a team of basic and translational scientists from academia and industry to put a precision ASO into the clinic before Leo's second birthday. linkedin.com/in/yiwei-she

Entrepreneurship

Samir Mitragotri is a Professor of Bioengineering at Harvard University; faculty member at Wyss Institute of Biologically Inspired Engineering; researcher in drug delivery systems including transdermal patches, oral delivery systems, and nanotechnology-based targeted delivery systems; inventor and entrepreneur; and elected member of the national academies of engineering, medicine, and inventors. linkedin.com/in/samir-mitragotri-011971

Organizational Structure

Amitava Saha (Amit) is an HR professional with more than 25 years of experience in IT, ITES, banking, and pharma industries. He was President and Group Head HR for Biocon group of companies from 2013 to 2023. He is a member of the Governing Council of Krea University and Advisory team for the Global Executive MBA in Pharmaceutical Management program of Ahmedabad University. He is currently on a sabbatical, taking up occasional short-term HR consulting assignments. Amit is an Electrical and Electronics Engineer from Delhi College of Engineering and has an MBA from the Indian Institute of Management (IIM), Calcutta. He started his career in marketing with earthmoving equipment followed by textiles and finally Indian classic music before moving to HR domain with Infosys. Amit is passionate about building and managing large global organizations with a focus on talent development, DEI, and AI-enabled tools for HR. linkedin.com/in/amitava-saha-19427b2

Data Integrity

Francis P. Crawley is a philosopher with a career in bioethics and expertise in research policy, regulation, ethics, integrity, and methodology. His work focuses on data and artificial intelligence (AI), as well as virtual twins' ethics and law, with particular attention to the life sciences, including clinical trials, genomics/omics, and new technologies. He has extensive expertise in EU, US, international, and country-specific ethics and law, particularly regarding patient and community interests in health-related research. With strong experience working closely with patients, communities, researchers, and policymakers across disciplines, domains, and geographic regions, he has played a key role in establishing consortia. He has contributed to the development of patient registries, biobanks, and data repositories, as well as drafting data management and protection plans.

Crawley has a strong background in designing and reviewing health-related research methodologies. His effective communication and leadership skills, combined with diplomacy, enable him to influence changes in bioethics and law. He has also contributed significantly to research, guidance, and ethics development for global diseases affecting resource-poor settings and orphan diseases.

He has wide-ranging experience working with organizations such as UNAIDS, WHO, UNESCO, the European Commission, the Council of Europe, and various local organizations and industries. His contributions include developing health-related research projects, fostering collaborative engagements, conducting regulatory and policy outreach, and leading education and training initiatives in Europe, Africa, Asia, the Americas, and Eastern Europe & Central Asia.

Crawley has held leading roles in the development of international and national guidelines, capacity-building efforts, empowerment initiatives, and educational programs for health research. His expertise includes Good Clinical Practice (GCP), ethics review systems, genetics and biobanks, and data privacy and management. He is also a General Data Protection Regulation (GDPR) Data Protection Officer (DPO). His work focuses on the ethical, legal, and social implications (ELSI) of data, AI, digital twins, and organoids.

Since January 2020, Crawley has been highly active in research, ethics, data sharing, and policy discussions related to COVID-19. His contributions have centered on vaccines, repurposed medicines, and genetic modeling. He has organized more than 40 global webinars and coordinated research across high-income countries (HICs) and low- and middle-income countries (LMICs). Recently, he has been engaged in the policy, ethics, and regulation of data and AI in times of crisis.

He founded and coordinates the Ukraine Clinical Research Support Initiative (UCRSI) in collaboration with leading Ukrainian, European, and international organizations. Crawley is at the forefront of the European and global movement to reform and advance research assessment in universities and other research institutions. His work places a strong emphasis on research ethics and integrity, particularly in developing policies for data and AI in research assessment processes.

Perihan Elif Ekmekci M.D., PhD is the head of the History of Medicine and Ethics department at TOBB ETU School of Medicine. She was a research fellow at Imperial College, London, UK, in 2006. She has been a Fogarty Fellow at Harvard University and had her Fogarty/NIH Program Master's Certification in Research Ethics in 2014. She has been a fellow of WIRB International IRB Western Institutional Review Board Research Ethics Training Program, Seattle Washington (USA) since 2016. She served as the head of the EU relations department of the

Ministry of Health Turkey (2007–2016) and developed several projects in alliance with the EU. She was the Turkish representative for the European Center for Disease Control Advisory Board and served in this position between the years 2011 and 2016. Currently, she is the Chair of the International Unit in Bioethics/WMA Cooperation Center and Deputy Dean of TOBB ETU School of Medicine. She is chairing the Institutional Review Board of TOBB ETU, and she is a member of the open science committee of TOBB ETU, Co-chair of CODATA International Data Policy Committee, Co-chair of CoARA ERIP and the EOSC Future/RDA Artificial Intelligence and Data Visitation Working Group, and an associate member of EUREC. She is affiliated with the World Association for Medical Law and the International Forum of Teachers of the International Unit in Bioethics. She has several publications in distinguished journals on ethics and the history of medicine. Professor Ekmekci is the coauthor of the book titled *Artificial Intelligenceand Bioethics* published by Springer in 2020. She is teaching undergraduate and postgraduate courses on the history of medicine and ethics. https://www.linkedin.com/in/perihan-elif-ekmekci-42a03220/

Health Economics

Sumati Rao, PhD is an experienced health economics research executive with over 22 years of global pharmaceutical industry experience, leading transformation of health economics organization, delivering business initiatives, and executing research strategies. Sumati has successfully launched critical products such as Opdivo/Yervoy in Immuno-oncology and Humira in Immunology by leading teams of high-performing research professionals. For the past 22 years, Sumati has held research positions at various pharmaceutical companies including GSK, BMS, Abbvie, and Wyeth (now Pfizer) with global and US responsibilities for all aspects of health economics and outcomes research. She has led major transformations by conceptualizing and designing the Health Economics and Outcomes Research (HEOR) organization, formulating strategies for successful product launches in multiple therapeutic areas (Immunology, Immuno-oncology, and Clinical Oncology). Sumati believes in innovative research by designing and executing unique scientific methods and thinking outside the box on research initiatives. She is currently focused on leading health economics scientists and developing research strategies in oncology in the United States at GSK. https://www.linkedin.com/in/sumati-rao-ph-d-96008b3/

Graphic Design

Riya Bhattacharya is a biotechnologist who has been working actively in the field of antimicrobial resistance, novel drug design and development, essential oil-based novel formulations, and related allied areas of health sciences. Dr. Riya leads the biotechnology research vertical under the School of Technology at Woxsen University, Hyderabad, India. She also holds the position of Co-chair at the Centre of Excellence for Health Technology and Biodiversity at the same university. She has 18 patents to her credit and close to 40 international publications. Her PhD work was on the mechanism underlying the synergism between natural compounds and conventional drugs so that it will be possible to produce safe drug combinations and lessen the negative impact that antimicrobial resistance has on the health system. She is also the recipient of the young researcher award and other academic excellence awards from different organizations.. https://www.linkedin.com/in/dr-riya-bhattacharya-580a5420a/

Sibiraj Murugesan is a biotechnology graduate from Shoolini University of Biotechnology and Management Sciences, India, explores the intersection of art and science by blending his passion for scientific illustration with multidisciplinary research. As a freelancer, he collaborates with researchers from India, the USA, Nigeria, and Uzbekistan, transforming complex scientific concepts into compelling visual narratives. His enthusiasm for sustainability is reflected in his bioremediation research on novel bacteria-biochar composites made from Areca catechu husk, designed to remove aqueous lead. During his undergraduate studies, he coauthored three SCOPUS-indexed book chapters, highlighting his zeal for addressing critical environmental challenges. Beyond academics, he is committed to inspiring change in his community. Awarded the KECTIL Fellowship for youth in developing countries, he authored *The Tale of Hope*, a picture book aimed at raising awareness about muscular dystrophy among children in India. He is committed to making a difference in environmental and social causes through his active volunteering efforts with Healing NSS, Himalayas, YouWeCan, and the Indian Association for Muscular Dystrophy. linkedin.com/in/sibiraj-murugesan-570684201

Medical Writing

Sneha Khedkar is an Indian science journalist focused on life sciences and health. She has a bachelor's degree in Microbiology and Biochemistry and a master's degree in Biochemistry. Thereafter, she was a research fellow at the Institute for Stem Cell Science and Regenerative Medicine in Bangalore where she studied stem cells in the skin. Realizing that she likes writing about science more than doing science, she switched lanes to science journalism. Her articles have been published in *Scientific American*, *New Scientist*, and *Knowable Magazine*, among others. https://www.snehakhedkar.com/ (website) https://www.linkedin.com/in/snehakhedkar/

Pies of Life (YouTube Channel) of Interviews of Biotechnology Professionals

Ever wondered what makes the greats, great? What makes the successful, successful? What makes the brilliant, brilliant? Over the pandemic years, there were a lot of meet-ups with the celebrities of the Pharma industry and science. This YouTube channel has more than 50 videos of interviews with these pharma leaders, including Lt. Gen. Madhuri Kanitkar, Kiran Mazumdar Shaw, and Vijay Chandru. www.youtube.com/@PiesofLife

Index